AF362045

Mining Safety and Sustainability I

Mining Safety and Sustainability I

Editors

Longjun Dong
Yanlin Zhao
Wenxue Chen

MDPI • Basel • Beijing • Wuhan • Barcelona • Belgrade • Manchester • Tokyo • Cluj • Tianjin

Editors

Longjun Dong
Central South University
China

Yanlin Zhao
Hunan University of Science
and Technology
China

Wenxue Chen
University of Sherbrooke
Canada

Editorial Office
MDPI
St. Alban-Anlage 66
4052 Basel, Switzerland

This is a reprint of articles from the Topic published online in the open access journals *Sustainability* (ISSN 2071-1050), *Energies* (ISSN 1996-1073), *Minerals* (ISSN 2075-163X), *Sensors* (ISSN 1424-8220), and *Safety* (ISSN 2313-576X) (available at: https://www.mdpi.com/topics/Mining_Safety).

For citation purposes, cite each article independently as indicated on the article page online and as indicated below:

LastName, A.A.; LastName, B.B.; LastName, C.C. Article Title. *Journal Name* **Year**, *Volume Number*, Page Range.

ISBN 978-3-0365-4687-2 (Hbk)
ISBN 978-3-0365-4688-9 (PDF)

Contents

About the Editors

Longjun Dong

Longjun Dong, Ph.D, is a Professor and the Director of the Department of Safety Science and Engineering, Central South University, China. His research interests include rock mechanics, engineering seismicity, and applied geophysics in mining engineering. He obtained his PhD, MSc, and BSc in 2013, 2009, and 2007, respectively. He was appointed as Professor, Associate Professor, and Lecturer at Central South University in 2017, 2015, and 2013, respectively. He was invited as a research assistant at the Australian Center for Geomechanics (ACG) from 2012 to 2013. He presided over 15 projects of the National Key Research and Development Program, International Cooperation and Exchange of the National Natural Science Foundation of China (NSFC), etc. He is the recipient of the NSFC for Excellent Young Scholars (2018), the Young Elite Scientist sponsored by the China Association for Science and Technology (CAST) (2016), the leading talent of Science and Technology Innovation of Hunan Province (2021), and the Distinguished Young Scholars Fund of Hunan Province (2018). He has 138 publications in peer-reviewed journals, with 4159 citations and 37 h-index in Google Scholar, and has 45 Chinese patents. He is an individual ISRM Member and an IEEE Senior Member. He is a Topical Editor-in-Chief of the Arabian Journal of Geosciences (Springer nature); an Associate Editor of Journal of the Acoustical Society of America Express Letter, and Shock and Vibration; an Editorial Board Member of Safety Science (Elsevier), Soils and Foundations (Elsevier, the official journal of the Japanese Geotechnical Society), Scientific Reports (Springer nature), and International Journal of Distributed Sensor Networks; an Editorial Advisory Board Member of Archives of Mining Sciences, and a guest editor of Induced Seismicity in Scientific Reports. He was selected into the list of "2020 Highly Cited Chinese Researchers (Elsevier)" and the "Top 2% of Scientists in the World (Stanford University)".

Yanlin Zhao

Yanlin Zhao, professor and doctoral supervisor, is the Deputy Dean of the School of Resources Environment and Safety Engineering, Hunan University of Science and Technology. He mainly studies rock mechanics and rock ground control, multi-field coupling theory and the engineering response of deep mining rock mass. He presided over and completed two general programs of National Natural Science Foundation of China, more than 20 scientific research projects, such as provincial and ministerial level and enterprise cooperation, and participated in the basic theoretical research on the mechanism and prevention of coal mine water inrush (973 Program) and the key program of the National Natural Science Foundation of China. He is a visiting scholar at the University of Arizona, a young backbone teacher in ordinary colleges and universities in Hunan Province, an outstanding expert in Xiangtan City, an outstanding doctoral dissertation winner in Hunan Province, a Xiangjiang scholar at Hunan University of Science and Technology, a director of the Hunan Rock Mechanics and Engineering Society, a national technology forecasting expert of the Ministry of Science and Technology, an excellent master thesis advisor in Hunan Province, and a guest editor-in-chief of the international SCI journals "Frontiers in Earth Science" and "Geofluids". He was selected in the 2020 Elsevier Highly Cited Chinese Scholars and Top 2% Global Top Scientists in the field of "Mining Engineering". He has published more than 100 academic papers, 3 academic monographs, of which more than 80 papers were indexed by SCI and EI, 6 papers in international Top journals, 9 papers in the world's top 1% ESI highly cited papers, 4 hot papers, and 1 paper in "Frontrunner 5000" top articles in outstanding S&T journals of China (F5000), and 1 paper won

the 2019 Excellent Paper of the Chinese Society of Rock Mechanics and Engineering—Chen Zongji
Award.

Wenxue Chen

Wenxue Chen, Adjunct Professor of Sherbrook University, is Industrial Research Chair of
Canadian Natural Science and Engineering Research Fund (NSERC) "Innovative Composite Fiber
Materials Concrete Engineering Application Project" and the Principal Scientist of SFTec. His main
research fields include safety technology and engineering, underground engineering, structural
engineering and engineering application of new composite materials, etc. He is a committee member
of the Infrastructure Products Committee of the American Precast Association (NPCA-TIPC), a
committee member of the American Composites Manufacturing Industry Council (ACMA), and an
associate editor of the International Journal of Safety Science (IJSS). He has published 65 academic
papers, of which more than 20 were indexed by SCI. Additionally, he has had 25 authorized patents
and won a number of provincial and ministerial awards.

Preface to "Mining Safety and Sustainability I"

The mining industry provides energy and raw material guarantees for global economic development and social progress. Especially in recent years, with the increasing improvement of infrastructure facilities and people's living standards, the demand for mineral resources has gradually increased. However, with the increasing depth of mining, the difficulty of mining also necessitates higher requirements for mining equipment and safety. Detecting the mineral exploration environment, improving the safety of mining operations, developing intelligent tunneling equipment, and ensuring the coordination of the human–machine–environment relationship in the mining production system have become necessary conditions for promoting the transformation of mining methods and achieving the goal of "double carbon". It is urgent to explore safe, efficient, and sustainable mining methods of mineral resources. In the past, scholars around the world have carried out a lot of research work on safe and sustainable mining, but because mining operations are resource-consuming and involve productive production behaviors, applying the principles of safety and sustainability to mining itself is challenging.

It is of great significance to reduce the disaster risk of mining accidents, enhance the safety of mining operations, and improve the efficiency and sustainability of the development of mineral resources. Therefore, it is necessary to extract useful knowledge from the process of mineral resource development and resource integration by means of experimental techniques, simulation methods, data mining, theoretical innovation, technological development, and other methods. It will further provide theoretical support and technical support for guiding the normative, green, safe, and sustainable development of the mining industry. This Reprint aims to present new research and recent advances in the safety and sustainability of mining.

Last but not least, thanks to all the authors for the excellent and meaningful contributions to this topic. Additionally, more experts and scholars are welcome to present your new ideas in safety mining, sustainable mining, mineral resource management, technology of intelligent mining, research and development of intelligent mining equipment, sustainable development, new mining methods, etc.

Longjun Dong , Yanlin Zhao, and Wenxue Chen
Editors

sustainability

MDPI

Editorial

Mining Safety and Sustainability—An Overview

Longjun Dong [1],*, Yanlin Zhao [2] and Wenxue Chen [3]

[1] School of Resources and Safety Engineering, Central South University, Changsha 410083, China

[2] Hunan Provincial Key Laboratory of Safe Mining Techniques of Coal Mines, Work Safety Key Laboratory on Prevention and Control of Gas and Roof Disasters for Southern Coal Mines, Hunan University of Science and Technology, Xiangtan 411201, China; yanlin_8@163.com

[3] Department of Civil & Building Engineering, University of Sherbrooke, Sherbrooke, QC J1K 2R1, Canada; wenxue.chen@usherbrooke.ca

* Correspondence: lj.dong@csu.edu.cn

Citation: Dong, L.; Zhao, Y.; Chen, W. Mining Safety and Sustainability—An Overview. *Sustainability* **2022**, *14*, 6570. https://doi.org/10.3390/su14116570

Received: 13 May 2022
Accepted: 23 May 2022
Published: 27 May 2022

Publisher's Note: MDPI stays neutral with regard to jurisdictional claims in published maps and institutional affiliations.

1. Introduction

The mining industry provides energy and raw material for global economic development and social progress. Especially in recent years, with the increasing improvement in infrastructure facilities and people's living standards, the demand for mineral resources has shown gradual growth; however, the ensuing issues of mining safety and sustainable development are causing increasingly widespread concern and worries. For a more long-term development, a large amount of scientific research has been invested on these issues in the context of long-term development. By making full use of model building [1–4], experimental studies [5–8], field practice [9–12], theoretical innovation [13–16], data analysis [17–19], and technology development [20–23], the possibilities of safety and sustainability in the development of mineral resources have been explored. Useful knowledge is also obtained by reviewing existing studies and integrating resources [24–27], aiming to identify the future direction of the mining industry. This Special Issue aims to focus on the most recent theoretical, experimental, and technological advances in mining safety and sustainability. A brief summary of the articles published in this Special Issue and related recent works are presented in this editorial.

2. Guarantee for Mining Production Safety

Tailings dam failure is a great threat to life and property, and the diagnosis of the health of tailings dams is a complex nonlinear problem. Dong et al. [28] proposed a comprehensive, quantitative method for the diagnosis of tailings dam health based on dynamic weights and constructed a diagnosis index system for tailings dams with slope stability, deformation stability and, seepage stability as project layers. The proposed method was successfully applied to an actual engineering project. This study provides a new method for evaluating the safety of tailings dams.

Ma et al. [29] conducted a model experimental study of the surface settlement characteristics caused by coal seam mining using a special three-dimensional experimental setup. The surface settlement characteristics during mining were also studied in combination with field measurements. The results showed that the subsidence caused by mining disturbances below the coal seam was 79. These findings fully reflect that the three-dimensional test device provides a new experimental research tool that can be used to further study the surface subsidence characteristics and control caused by coal mining.

Combined with the movement principle of rock and soil layers in the respective study area and considering the influence of slope stability and additional mining slip on mining subsidence, Zhao et al. [30] proposed a probabilistic integral model-based surface subsidence prediction method for-coal seam mining in loess donga and verified its feasibility by field cases. A new, effective, and valuable tool is provided for the prediction of damage caused by underground coal seam mining.

Considering the difficulty of effectively identifying signals with low signal-to-noise ratios (SNRs) using microseismic monitoring, Fan et al. [31] proposed a wavelet scattering decomposition (WSD) transform and a support vector machine (SVM) algorithm. The artificial intelligence recognition model developed based on SVM and WSD not only provides a fast method with high classification accuracy, but is also suitable for online feature extraction of microseismic monitoring signals to achieve improved efficiency and accuracy of microseismic signal processing used to monitor rock instability and seismicity.

To better understand the mining characteristics during mining of shallow buried thick coal seams (SBTCS) under thick aeolian sand (TAS), Liu et al. [32] explored the ground damage characteristics and fracture development during mining under special geological conditions of TAS through theoretical derivation, numerical simulation, and field monitoring. The results revealed the essence of the development and the distribution of surface cracks caused by mining SBTCS, and depth-to-thickness ratio (DTR) was shown to be 13.43.

Considering that chemical corrosion and axial compression affect rocks' internal microstructure and mineral composition, which in turn affects their physical and mechanical properties, Xue et al. [33] used a combined dynamic and static load test apparatus to conduct cyclic impact tests on white sandstone immersed in chemical solution and studied the dynamic strength characteristics of white sandstone under the coupling effect of axial load and chemical corrosion. The results of the study provide a theoretical basis for safe and effective construction management of blasting projects under complex geological conditions.

To study the fracture patterns during rock fracture, Li et al. [34] investigated the acoustic emission characteristics and crack types of red sandstone during fracture by Brazilian indirect tensile tests (BITT), direct shear tests (DST), and uniaxial compression tests (UCT). They also discussed a relatively objective dividing line for tensile and shear crack classification and applied the dividing line to the analysis of the fracture source evolution and the damage precursor. The results of the study will provide a theoretical basis for rock stability judgment and prediction during mining.

Liu et al. [35] conducted a series of conventional triaxial unloading tests to analyze the mechanical properties, strain energy evolution characteristics, and failure modes of saturated rock masses, and their findings are of great significance for strength calculation, safety assessment, and disaster prevention and control.

3. Achievement of Sustainable Development

Considering the serious ecological pollution problems caused by acid mine drainage (AMD), Wu et al. [36] investigated the phytoremediation techniques and mechanisms of AMD through hydroponic experiments with six wetland plants. The results showed that the dominant plants for treating AMD were Juncus effusus, Iris wilsonii, and Phragmites australis; some of the pollutants in AMD were absorbed by plants and rest were removed by hydrolysis and sedimentation processes. These findings provide a theoretical reference for phytoremediation techniques for AMD.

Reinforced TSFs are beneficial for saving land resources, reducing environmental damage caused by mineral extraction, and achieving sustainable production in the mineral extraction process. Ding et al. [37] investigated the effects of freeze–thaw cycles on the mechanical properties and microstructural changes of cementitious material-reinforced tailings by performing unconfined compressive strength (UCS) tests, scanning electron microscope imagery, X-ray diffraction tests, and thermogravimetric tests. The results demonstrated that freeze–thaw cycles eventually reduce the UCS of all tested samples, and the higher the number of freeze–thaw cycles, the greater the damage to the surface morphology and matrix of the tailings.

Wang et al. [38] worked on the integrated management of coalbed methane and hydrogen sulfide at the working face in the coal seam distribution of abandoned oil wells in coal-mine resource areas. The study was conducted through parameter testing, gas composition analysis, source distribution site investigation, and determination of the influence range of gas and hydrogen sulfide in the coal seam within the influence area of the abandoned wells. The results of this work provide a theoretical basis for further understanding of gas- and hydrogen-sulfide-enrichment patterns at the mining face and the design of treatment measures within the influence of abandoned oil wells.

To address the low productivity, inconsistent management, administrative organization, high raw material waste, and negative social and environmental impacts faced by the Mexican marble industry, Alarcón-Ruíz et al. [39] systematically reviewed strategies and solutions used to address these problems between 2014 and 2021. They collected these surveys as well as industry experiences to propose a triple-helix intervention approach. The results of the study provide guidance for the sustainable development of the marble industry.

4. Optimization Design of Technology and Equipment

Yi et al. [40] investigated the effects of time, track shoe number, and grounding pressure, as well as other influencing factors, on the traction force of deep-sea crawler miner through a direct shear-creep experiment and the direct shear rheological constitutive model. They proved its effectiveness through the traction force experiment of a single-track shoe. The research results provide a scientific basis for the design and optimization of the deep-sea tracked miner.

Under the background that the cemented paste backfill (CPB) technology has been applied to solve the problems of stope instability and surface subsidence for so many years, Chen et al. [41] worked on the factors affecting the strength of CPB. They considered the coupled effects of curing conditions, which have received little attention, and used uniaxial compressive strength (UCS) as an important evaluation index of CPB. They successively performed mathematical modeling and laboratory verification of concrete strength. The findings suggest that the relationship between the UCS of CPB and curing stress develops the function of quadratic polynomial to develop with one variable, while the UCS of the CPB indicates a power function as the curing temperature increases. The conclusions obtained in this study have important implications for the safe design of CPB.

Rivera-Lavado et al. [42] proposed the use of RF split-ring resonators (SRRs) as downhole passive sensors for real-time crude oil monitoring through the estimation of the dielectric constant. The use of a low-cost SRR passive sensor for the real-time permittivity characterization of hydrocarbon fluids will contribute to solving the problem of performing difficult monitoring under harsh conditions such as high temperature and pressure.

5. Trends in Intelligent Mines

Considering the influence of the process parameters of fully mechanized caving on the recovery rate and gangue content of top coal, Liang et al. [43] used numerical simulation and a BP neural network to achieve the optimization of top-coal caving parameters for the actual situation of a working face. They demonstrated the effectiveness of this method using an in-lab similarity simulation test of the particle material. The findings of this paper effectively improve the decision-making efficiency of fully mechanized caving-process parameters and provide a basis for achieving intelligent, fully mechanized cave mining.

To solve the airflow reconstruction problem, Liu et al. [44] proposed a new algorithm of an independent cut set depending on the underlying graph structure and evaluated its effectiveness in practical applications. The results indicated that fewer than 30% of tunnels needed to have wind speed sensors set up to reconstruct the well-posed airflow of all the tunnels (>200 in some mines). The findings of this work provide some theoretical support for the implementation of intelligent ventilation.

Wang et al. [45] combined artificial intelligence techniques to analyze and model experimental data from circulating pipelines, using random forest machine learning algorithms to predict the pressure loss of slurry transport. The results of the study showed an accuracy of 0.9747, which met the design accuracy requirement. This finding will help to realize the optimal arrangement of deep-well-filling slurry-delivery pipelines.

To remedy the deficiencies of the previous studies, Wu et al. [46] proposed a neural network model consisting of one deep neural network (DNN) and four long short-term memory (LSTM) networks based on single-sensor and multi-sensor prediction results. They solved the amplitude-concentrated, expanded region-identification problem. The high-precision model for the automatic identification of amplitude-concentration-expansion zone provides the basis for the automatic identification of borehole depth.

In order to explore the explosion mechanism of coal and the factors that cause coal explosions, Khan et al. [47] used explosivity tests at different particle sizes and dust concentrations to construct a random forest algorithm, which was used to model the relationship between inputs (coal dust particle size, coal concentration, and gross calorific value (GCV)). To further understand the impact of each feature causing explosibility, the random forest AI model was further analyzed for sensitivity analysis using SHAP (Shapley Additive exPlanations). This work provides a reference for control factors to prevent coal dust explosions and improve safety conditions.

We sincerely thank all the above-mentioned authors for the excellent and meaningful contributions to this topic. Additionally, we hope that more relevant research will be conducted in the future to handle the issues about safety and sustainability in the mining industry. It will be helpful for providing further theoretical support and technical support in order to guide the normative, green, safe, and sustainable development of the mining industry.

Funding: This research received no external funding.

Conflicts of Interest: The authors declare no conflict of interest.

References

1. Sun, Y.; Li, G.; Zhang, J.; Xu, J. Failure Mechanisms of Rheological Coal Roadway. *Sustainability* **2020**, *12*, 2885. [CrossRef]
2. Dong, L.; Luo, Q. Stress Heterogeneity and Slip Weakening of Faults under Various Stress and Slip. *Geofluids* **2020**, *2020*, 8860026. [CrossRef]
3. Pretorius, J.G.; Mathews, M.J.; Mare, P.; Kleingeld, M.; van Rensburg, J. Implementing a DIKW model on a deep mine cooling system. *Int. J. Min. Sci. Technol.* **2019**, *29*, 319–326. [CrossRef]
4. Jiskani, I.M.; Cai, Q.; Zhou, W.; Lu, X. Assessment of risks impeding sustainable mining in Pakistan using fuzzy synthetic evaluation. *Resour. Policy* **2020**, *69*, 101820. [CrossRef]
5. Dong, L.; Zhang, Y.; Ma, J. Micro-Crack Mechanism in the Fracture Evolution of Saturated Granite and Enlightenment to the Precursors of Instability. *Sensors* **2020**, *20*, 4595. [CrossRef]
6. Esmaeili, J.; Aslani, H.; Onuaguluchi, O. Reuse Potentials of Copper Mine Tailings in Mortar and Concrete Composites. *J. Mater. Civ. Eng.* **2020**, *32*, 5. [CrossRef]
7. Dong, L.; Tong, X.; Ma, J. Quantitative Investigation of Tomographic Effects in Abnormal Regions of Complex Structures. *Engineering* **2021**, *7*, 1011–1022. [CrossRef]
8. Hefni, M.; Ahmed, H.A.M.; Omar, E.S.; Ali, M.A. The Potential Re-Use of Saudi Mine Tailings in Mine Backfill: A Path towards Sustainable Mining in Saudi Arabia. *Sustainability* **2021**, *13*, 6204. [CrossRef]
9. Ma, J.; Dong, L.; Zhao, G.; Li, X. Discrimination of seismic sources in an underground mine using full waveform inversion. *Int. J. Rock Mech. Min. Sci.* **2018**, *106*, 213–222. [CrossRef]
10. Teplicka, K.; Straka, M. Sustainability of Extraction of Raw Material by a Combination of Mobile and Stationary Mining Machines and Optimization of Machine Life Cycle. *Sustainability* **2020**, *12*, 10454. [CrossRef]
11. Dong, L.; Sun, D.; Li, X.; Ma, J.; Zhang, L.; Tong, X. Interval non-probabilistic reliability of surrounding jointed rockmass considering microseismic loads in mining tunnels. *Tunn. Undergr. Space Technol.* **2018**, *81*, 326–335. [CrossRef]
12. Funez Guerra, C.; Reyes-Bozo, L.; Vyhmeister, E.; Caparrós, M.J.; Salazar, J.L.; Godoy-Faúndez, A.; Clemente-Jul, C.; Verastegui-Rayo, D. Viability analysis of underground mining machinery using green hydrogen as a fuel. *Int. J. Hydrogen Energy* **2020**, *45*, 5112–5121. [CrossRef]
13. Hu, Q.; Dong, L. Acoustic Emission Source Location and Experimental Verification for Two-Dimensional Irregular Complex Structure. *IEEE Sens. J.* **2020**, *20*, 2679–2691. [CrossRef]
14. Johansen, K.; Erskine, P.D.; McCabe, M.F. Using Unmanned Aerial Vehicles to assess the rehabilitation performance of open cut coal mines. *J. Clean. Prod.* **2019**, *209*, 819–833. [CrossRef]
15. Dong, L.; Hu, Q.; Tong, X.; Liu, Y. Velocity-Free MS/AE Source Location Method for Three-Dimensional Hole-Containing Structures. *Engineering* **2020**, *6*, 827–834. [CrossRef]
16. Wang, D.; Shen, Y.; Zhao, Y.; He, W.; Liu, X.; Qian, X.; Lv, T. Integrated assessment and obstacle factor diagnosis of China's scientific coal production capacity based on the PSR sustainability framework. *Resour. Policy* **2020**, *68*, 101794. [CrossRef]

17. Wang, W.; Zhang, C. Evaluation of relative technological innovation capability: Model and case study for China's coal mine. *Resour. Policy* **2018**, *58*, 144–149. [CrossRef]
18. Dong, L.; Wesseloo, J.; Potvin, Y.; Li, X. Discrimination of Mine Seismic Events and Blasts Using the Fisher Classifier, Naive Bayesian Classifier and Logistic Regression. *Rock Mech. Rock Eng.* **2016**, *49*, 183–211. [CrossRef]
19. Cheng, X.; Drozdova, J.; Danek, T.; Huang, Q.; Qi, W.; Yang, S.; Zou, L.; Xiang, Y.; Zhao, X. Pollution Assessment of Trace Elements in Agricultural Soils around Copper Mining Area. *Sustainability* **2018**, *10*, 4533. [CrossRef]
20. Dong, L.; Sun, D.; Han, G.; Li, X.; Hu, Q.; Shu, L. Velocity-Free Localization of Autonomous Driverless Vehicles in Underground Intelligent Mines. *IEEE Trans. Veh. Technol.* **2020**, *69*, 9292–9303. [CrossRef]
21. Gallwey, J.; Robiati, C.; Coggan, J.; Vogt, D.; Eyre, M. A Sentinel-2 based multispectral convolutional neural network for detecting artisanal small-scale mining in Ghana: Applying deep learning to shallow mining. *Remote Sens. Environ.* **2020**, *248*, 111970. [CrossRef]
22. Dong, L.; Shu, W.; Sun, D.; Li, X.; Zhang, L. Pre-Alarm System Based on Real-Time Monitoring and Numerical Simulation Using Internet of Things and Cloud Computing for Tailings Dam in Mines. *IEEE Access* **2017**, *5*, 21080–21089. [CrossRef]
23. Brescia-Norambuena, L.; Pickel, D.; Gonzalez, M.; Tighe, S.L.; Azua, G. Accelerated construction as a new approach for underground-mining pavement: Productivity, cost and environmental study through stochastic modeling. *J. Clean. Prod.* **2020**, *251*, 119605. [CrossRef]
24. Dong, L.; Deng, S.; Wang, F. Some developments and new insights for environmental sustainability and disaster control of tailings dam. *J. Clean. Prod.* **2020**, *269*, 1562–1578. [CrossRef]
25. Hou, J.; Wang, Z.; Liu, P. Current states of coalbed methane and its sustainability perspectives in China. *Int. J. Energy Res.* **2018**, *42*, 3454–3476.
26. Dong, L.; Tong, X.; Li, X.; Zhou, J.; Wang, S.; Liu, B. Some developments and new insights of environmental problems and deep mining strategy for cleaner production in mines. *J. Clean. Prod.* **2019**, *210*, 1562–1578. [CrossRef]
27. Zhou, J.; Li, X.; Mitri, H.S. Evaluation method of rockburst: State-of-the-art literature review. *Tunn. Undergr. Space Technol.* **2018**, *81*, 632–659. [CrossRef]
28. Dong, K.; Mi, Z.; Yang, D. Comprehensive Diagnosis Method of the Health of Tailings Dams Based on Dynamic Weight and Quantitative Index. *Sustainability* **2022**, *14*, 3068. [CrossRef]
29. Ma, X.; Fu, Z.; Li, Y.; Zhang, P.; Zhao, Y.; Ma, G. Study on Surface Subsidence Characteristics Based on Three-Dimensional Test Device for Simulating Rock Strata and Surface Movement. *Energies* **2022**, *15*, 1927. [CrossRef]
30. Zhao, B.; Guo, Y.; Mao, X.; Zhai, D.; Zhu, D.; Huo, Y.; Sun, Z.; Wang, J. Prediction Method for Surface Subsidence of Coal Seam Mining in Loess Donga Based on the Probability Integration Model. *Energies* **2022**, *15*, 2282. [CrossRef]
31. Fan, X.; Cheng, J.; Wang, Y.; Li, S.; Yan, B.; Zhang, Q. Automatic Events Recognition in Low SNR Microseismic Signals of Coal Mine Based on Wavelet Scattering Transform and SVM. *Energies* **2022**, *15*, 2326. [CrossRef]
32. Liu, G.; Zou, Y.; Zhang, W.; Chen, J. Characteristics of Overburden and Ground Failure in Mining of Shallow Buried Thick Coal Seams under Thick Aeolian Sand. *Sustainability* **2022**, *14*, 4028. [CrossRef]
33. Xue, J.; Zhao, Z.; Dong, L.; Jin, J.; Zhang, Y.; Tan, L.; Cai, R.; Zhang, Y. Effect of Chemical Corrosion and Axial Compression on the Dynamic Strength Degradation Characteristics of White Sandstone under Cyclic Impact. *Minerals* **2022**, *12*, 429. [CrossRef]
34. Li, J.; Lian, S.; Huang, Y.; Wang, C. Study on Crack Classification Criterion and Failure Evaluation Index of Red Sandstone Based on Acoustic Emission Parameter Analysis. *Sustainability* **2022**, *14*, 5143. [CrossRef]
35. Liu, Z.; Yi, W. Experimental Study on the Mechanical Characteristics of Saturated Granite under Conventional Triaxial Loading and Unloading Tests. *Sustainability* **2022**, *14*, 5445. [CrossRef]
36. Wu, A.; Zhang, Y.; Zhao, X.; Li, J.; Zhang, G.; Shi, H.; Guo, L.; Xu, S. Experimental Study on the Hydroponics of Wetland Plants for the Treatment of Acid Mine Drainage. *Sustainability* **2022**, *14*, 2148. [CrossRef]
37. Ding, P.; Hou, Y.; Han, D.; Zhang, X.; Cao, S.; Li, C. Effect of Freeze-Thaw Cycles on Mechanical and Microstructural Properties of Tailings Reinforced with Cement-Based Material. *Minerals* **2022**, *12*, 413. [CrossRef]
38. Wang, X.; Ma, H.; Qi, X.; Gao, K.; Li, S. Study on the Distribution Law of Coal Seam Gas and Hydrogen Sulfide Affected by Abandoned Oil Wells. *Energies* **2022**, *15*, 3373. [CrossRef]
39. Alarcón-Ruíz, T.; Santiago-Jiménez, M.E.; Ruvalcaba-Sánchez, L.G.; Sánchez-Galván, F.; García-Santamaría, L.E.; Fernández-Lambert, G. A Triple-Helix Intervention Approach to Direct the Marble Industry towards Sustainable Business in Mexico. *Sustainability* **2022**, *14*, 5576. [CrossRef]
40. Yi, W.; Xu, F. Semi-Empirical Time-Dependent Parameter of Shear Strength for Traction Force between Deep-Sea Sediment and Tracked Miner. *Sensors* **2022**, *22*, 1119. [CrossRef]
41. Chen, S.; Wang, W.; Yan, R.; Wu, A.; Wang, Y.; Yilmaz, E. A Joint Experiment and Discussion for Strength Characteristics of Cemented Paste Backfill Considering Curing Conditions. *Minerals* **2022**, *12*, 211. [CrossRef]
42. Rivera-Lavado, A.; García-Lampérez, A.; Jara-Galán, M.-E.; Gallo-Valverde, E.; Sanz, P.; Segovia-Vargas, D. Low-Cost Electromagnetic Split-Ring Resonator Sensor System for the Petroleum Industry. *Sensors* **2022**, *22*, 3345. [CrossRef] [PubMed]
43. Liang, M.; Hu, C.; Yu, R.; Wang, L.; Zhao, B.; Xu, Z. Optimization of the Process Parameters of Fully Mechanized Top-Coal Caving in Thick-Seam Coal Using BP Neural Networks. *Sustainability* **2022**, *14*, 1340. [CrossRef]
44. Liu, Y.; Liu, Z.; Gao, K.; Huang, Y.; Zhu, C. Efficient Graphical Algorithm of Sensor Distribution and Air Volume Reconstruction for a Smart Mine Ventilation Network. *Sensors* **2022**, *22*, 2096. [CrossRef]

45. Wang, Z.; Kou, Y.; Wang, Z.; Wu, Z.; Guo, J. Random Forest Slurry Pressure Loss Model Based on Loop Experiment. *Minerals* **2022**, *12*, 447. [CrossRef]
46. Wu, Z.; Zhang, W.-L.; Li, C. Automatic Implementation Algorithm of Pressure Relief Drilling Depth Based on an Innovative Monitoring-While-Drilling Method. *Sensors* **2022**, *22*, 3234. [CrossRef]
47. Khan, A.U.; Salman, S.; Muhammad, K.; Habib, M. Modelling Coal Dust Explosibility of Khyber Pakhtunkhwa Coal Using Random Forest Algorithm. *Energies* **2022**, *15*, 3169. [CrossRef]

sustainability

Article

Optimization of the Process Parameters of Fully Mechanized Top-Coal Caving in Thick-Seam Coal Using BP Neural Networks

Minfu Liang [1,2], Chengjun Hu [3,4,*], Rui Yu [1,5], Lixin Wang [4], Baofu Zhao [4] and Ziyue Xu [1,*]

[1] School of Mines, China University of Mining and Technology, Xuzhou 221116, China; liangmf2014@cumt.edu.cn (M.L.); LB19020006@cumt.edu.cn (R.Y.)
[2] School of Economics and Management, China University of Mining and Technology, Xuzhou 221116, China
[3] School of Energy and Mining Engineering, China University of Mining and Technology (Beijing), Beijing 100083, China
[4] China Coal Tianjin Underground Engineering Intelligence Research Institute, Tianjin 561000, China; wanglixin@chinacoal.com (L.W.); zhaobaofu@chinacoal.com (B.Z.)
[5] Wangjialing Coal Mine, China Coal Huajin Group Co., Ltd., Yuncheng 043000, China
* Correspondence: huchengjun@chinacoal.com (C.H.); TS20020060A31@cumt.edu.cn (Z.X.); Tel.: +86-185-2623-3835 (C.H.); +86-195-5225-1610 (Z.X.)

Abstract: The method of fully mechanized top-coal caving mining has become the main method of mining thick-seam coal. The process parameters of fully mechanized caving will affect the recovery rate and gangue content of top coal. Through numerical simulation software, the top-coal recovery rate and gangue content, under different fully mechanized caving process parameters, were simulated, and the influence law of different fully mechanized caving process parameters on top-coal recovery rate and gangue content was obtained. A decision model for top-coal caving process parameters was established with a BP neural network, and the optimal top-coal caving parameters were obtained for the actual situation of a working face. On this basis, a in-lab similarity simulation test of the particle material was carried out. The results show that the top-coal recovery rate and gangue content were 86.56% and 3.45%, respectively, and the coal caving effect was good. A BP neural network was used to study the decisions optimizing fully mechanized caving process parameters, which effectively improved the decision-making efficiency thereabout and provided a basis for realizing intelligent, fully mechanized caving mining.

Keywords: top-coal caving mining; process parameters; decision model; BP neural network; similarity simulation test

Citation: Liang, M.; Hu, C.; Yu, R.; Wang, L.; Zhao, B.; Xu, Z. Optimization of the Process Parameters of Fully Mechanized Top-Coal Caving in Thick-Seam Coal Using BP Neural Networks. *Sustainability* **2022**, *14*, 1340. https://doi.org/10.3390/su14031340

Academic Editors: Longjun Dong, Yanlin Zhao and Wenxue Chen

Received: 31 December 2021
Accepted: 20 January 2022
Published: 25 January 2022

Publisher's Note: MDPI stays neutral with regard to jurisdictional claims in published maps and institutional affiliations.

1. Introduction

The 'World Energy Statistics Review', released in 2020, shows that although global coal consumption has decreased, coal still accounts for about 27% of primary energy, which is still the main source of energy [1]. Especially for China, with its characteristic 'rich coal, lack of oil and less gas', the status of coal is unshakable. According to statistics, thick-seam coal accounts for 44% of the proven workable coal reserves in China, and nearly half of the coal consumption in China is provided by thick-seam coal mining [2,3].

Although the loads of hydraulic supports should be monitored in the process of fully mechanized top-coal caving mining to ensure continued safe production [4], this method has become the main strategy for thick-seam coal mining because of its low energy consumption, high output, strong geological adaptability and economic benefits [5–7]. It has gradually become the main method of thick-seam coal mining in China, Vietnam, Australia, Turkey and other countries [8–14]. Improving the recovery rates for top coal and reducing the gangue content of top coal are two key points in the study of fully mechanized caving mining techniques. Zhang NB et al. studied the influence of the arch formed by top coal and gangue on top-coal recovery rates and put forward the method of eliminating these arches to improve recovery rates [15]. Yasitli, NE and Unver B used the FLAC 3D software to simulate the top-coal caving process and granular, fine-sand materials in simulations thereof, and proposed that presplitting blasting technology could improve top-coal recovery rates [9,16]. Ghosh AK et al. believed that compressive strength, advance abutment pressure and top-coal seam thickness are important factors affecting top-coal recovery rates and proposed using the combination of blasting and vibration to destroy coal arches and thereby improve recovery rates [17], while Klishin VI and Klishin SV explored the relationship between the support opening sequence and the subsequent top-coal recovery rate [18,19].

However, there are few quantitative studies on the process parameters of fully mechanized caving mining. In recent years, thanks to rapid developments in science and technology, advanced artificial intelligence and machine learning algorithms have been increasingly applied to coal production [20–22]. Fan YJ et al. used a BP neural network to establish a safety evaluation model for coal mines and put forth an effective safety evaluation method for them [23]. Meng XZ et al. proposed an early warning method for coal mine safety, based on a BP neural network, that could effectively extract the characteristics of a coal mine's fault state and issue early warnings thereabout for coal mine safety [24]. The application of artificial neural networks provides a new scientific method for conducting research in the field of coal mining.

Therefore, this paper took the No. 12309 working face of the Wangjialing mine in Yuncheng City, Shanxi Province, China as its engineering background, adopting the research methods of numerical simulation, similarity simulation and BP neural networks to establish an optimization decision model of the process parameters for fully mechanized caving mining of thick-seam coal. So as to realize optimized decisions concerning the parameters for the fully mechanized caving mining technique in thick-seam coal mining, we obtained the optimal process parameters for such mining. Finally, we used them to improve mining and caving efficiency and the efficacy of coal caving.

2. Engineering Background

At the No. 12309 working face of the Wangjialing coal mine, which has adopted the fully mechanized, low-caving method of coal mining, its advancing length and width are 1320 m and 260 m, respectively. The buried depth of the main coal seam is about 400 m, its average thickness is 6.1 m, its dip angle is 2° and the hardness coefficient of its top coal is 1.8 (f < 2). There, the interlayer thickness is 0.2 m, the mining height is 3.1 m and the top-coal caving height is 3 m. Thus, the ratio of mining height to top-coal caving height is 1.03:1. The coal caving step is 0.865 m for one cutting with one caving. In the normal operation cycle, after each coal cutting, the tail beam is recovered and the coal opening is opened for coal caving operations. When the immediate top rock is discharged from the coal caving opening, the opening is closed, to stop coal caving. The first pressure step is 35 m, and the natural caving method is used to control the roof of the goaf. The properties of the roof and floor rock in the working face of the coal seam are shown in Figure 1, and the layout of the working face is shown in Figure 2.

The top coal falls from the working face in a relatively timely manner, generally, just after the top beam of the support. At the Wangjialing mine, the size of the lumpiness of top coal, mostly, is approximately 40 cm × 30 cm × 30 cm. Occasionally there are larger pieces, but they can all be released smoothly. Field observations of the top-coal caving process are shown in Figure 3.

Roof and floor	Histogram	Rock name	Thickness
Main roof		Fine-Sandstone	4.20 m
Immediate roof		sandstone	3.54 m
2# coal seam		Coal	5.7~6.3 6.1 m
Immediate floor		Fine-Sandstone	1.57 m
		Carbon mudstone	0.55 m
Old floor		Fine-Sandstone	4.70 m

Figure 1. Comprehensive drilling histogram.

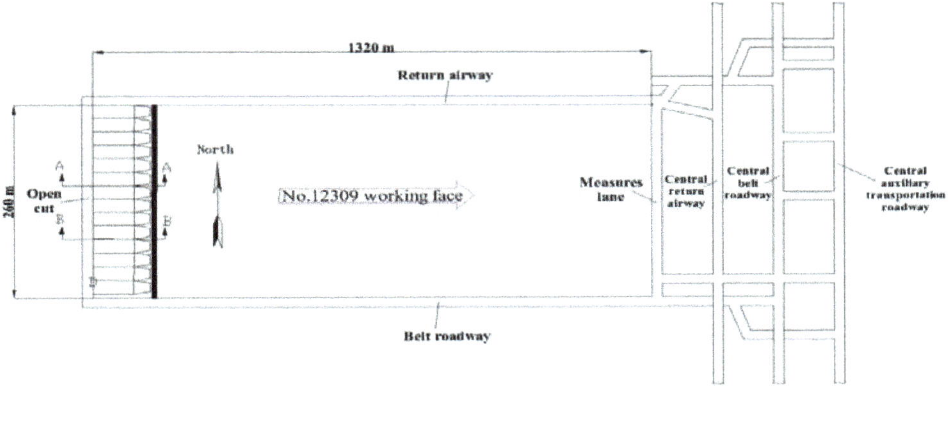

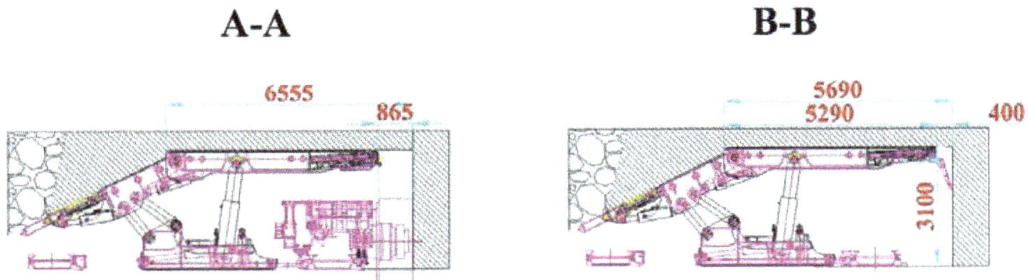

Figure 2. Layout of the roadway to and the fully mechanized top-coal caving face. (**A-A**) Maximum top control distance, (**B-B**) minimum top control distance.

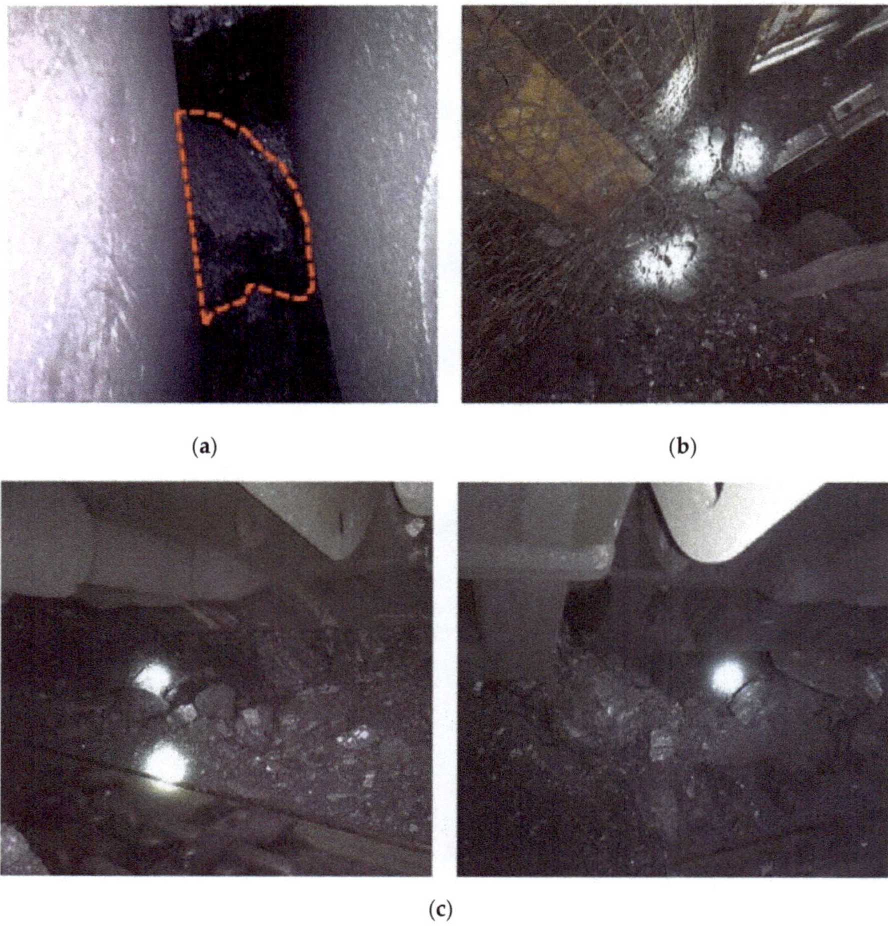

Figure 3. Top-coal caving condition of working face. (**a**) Broken top coal between supports; (**b**) coal caving at the top of the working face; (**c**) top-coal caving behind the conveyor at the front and rear of coal caving.

3. Methodology

3.1. Numerical Simulation Design

The two core indicators of the fully mechanized caving process are top-coal recovery rates and gangue content. Reasonable fully mechanized mining process parameters can effectively improve the top-coal recovery rates and reduce the gangue content. In order to study the influence of different fully mechanized caving process parameters on top-coal recovery rates and gangue content, the numerical simulation software PFC was used to conduct numerical simulation experiment schemes of different fully mechanized caving process parameters, with varying coal caving methods and procedures. The numerical simulation experiments of different fully mechanized top-coal caving process parameters were carried out by using the orthogonal experimental design method. By combining the principle of probability and statistics with computer technology, not only can the number of tests and calculation workload can be reduced, but the distribution of various influencing factors of fully mechanized top-coal caving mining is also more uniform within the test range, so as to achieve ideal results.

This numerical simulation study mainly focused on the top-coal releasing law without considering its crushing process. The top coal and immediate roof were assumed to be in a loose state. Therefore, both top coal and immediate roof adopted a linear constitutive model for 'Ball–Ball' and 'Ball–Facet' connections in numerical simulations. The elastic model was adopted for the bottom coal and main roof, and the linear contact model was adopted for each part and its internal contact model. The buried depth of the main coal seam was 400 m and the height of the numerical model was 18.84 m. Therefore, in the process of initial balance simulation, the weight of the 381.16 m-thick rock stratum should be applied to the upper surface of the model, and the average unit weight of the rock stratum was 25 KN/m^3, such that a boundary stress of $\sigma_z = 9.529$ MP was applied to the upper boundary of the model. Rigid walls slightly larger than the model were set up at the front, rear, left, right and bottom of the model, and its velocity was fixed at 0 m/s to server as a displacement boundary. The contact parameters are shown in Table 1, and the unit parameters are shown in Table 2.

Table 1. Contact parameters.

Contact Type	Constitutive Model	Firc	dp_Nratio	dp_Sratio	kn	ks
Immediate roof		0.5	0.3	0.3	4×10^8	4×10^8
Top coal	Liner	0.4	0.3	0.3	3×10^8	3×10^8
Ball–Ball		0.4	0.3	0.3	3×10^8	3×10^8
Ball–Facets		0.3	0.3	0.3	5×10^8	5×10^8

Table 2. Unit parameters.

Rock Stratum	Unit Type	Elastic Modulus/GPa	Bulk Density/kg·m^{-3}	Poisson's Ratio	Local Damping
Main roof	Zone	15.0	2660	0.34	/
Immediate roof	Ball	13.6	2660	/	0.7
Top coal	Ball	2.3	1400	/	0.7
Bottom coal	Zone	2.3	1400	0.26	/

The numerical simulation schemes of different fully mechanized caving process parameters with varying coal caving methods have considered various factors, such as coal seam thickness, mining and caving ratio, number of coal caving rounds, coal caving sequence, number of coal caving openings and top-coal particle size. Each factor was divided into three levels, as shown in Table 3. According to the orthogonal test method, a total of 18 models were established, and the numerical simulation schemes are shown in Table 4. The numbers 1, 2 and 3 in Table 4 refer to the corresponding factor level in Table 3.

The factors considered in the numerical simulation schemes of different fully mechanized top-coal caving process parameters with varying coal caving procedures include coal seam thickness, mining and caving ratio, coal caving procedure and top-coal particle size. Each factor was divided into three levels, as shown in Table 5. According to the orthogonal test method, a total of 9 models were established, and the numerical simulation schemes are shown in Table 6. The numbers 1, 2 and 3 in Table 6 correspond to the corresponding factor level in Table 5.

Table 3. Factor levels for the coal caving method.

Level	Coal Seam Thickness (m)	Caving Ratio	Number of Coal Caving Rounds	Coal Caving Sequence	Number of Coal Discharge Openings at the Same Time	Top-Coal Particle Size (m)
1	6	1:1	Single round	Sequential	Single opening	0.15–0.3
2	8	1:1.5	Two rounds	Group interval	Two openings	0.25–0.4
3	10	1:2	Three rounds	Interval return	Three openings	0.35–0.5

Table 4. Orthogonal simulation schemes for coal caving method.

Scheme No	Coal Seam Thickness	Caving Ratio	Number of Coal Caving Rounds	Coal Caving Sequence	Number of Coal Discharge Openings	Top-Coal Particle Size
1	1	1	1	1	1	1
2	1	2	2	2	2	2
3	1	3	3	3	3	3
4	2	1	1	2	2	3
5	2	2	2	3	3	1
6	2	3	3	1	1	2
7	3	1	2	1	3	2
8	3	2	3	2	1	3
9	3	3	1	3	2	1
10	1	1	3	3	2	2
11	1	2	1	1	3	3
12	1	3	2	2	1	1
13	2	1	2	3	1	3
14	2	2	3	1	2	1
15	2	3	1	2	3	2
16	3	1	3	2	3	1
17	3	2	1	3	1	2
18	3	3	2	1	2	3

Table 5. Factor levels for the coal caving procedure.

Level	Coal Seam Thickness (m)	Caving Ratio	Coal Caving Procedure	Top-Coal Particle Size (m)
1	6	1:1	One cutting with one caving	0.15–0.3
2	8	1:1.5	Two cutting with one caving	0.25–0.4
3	10	1:2	Three cutting with one caving	0.35–0.5

Table 6. Orthogonal simulation schemes for coal caving procedure.

Scheme No	Coal Seam Thickness	Caving Ratio	Coal Caving Procedure	Top-Coal Particle Size
1	1	1	1	1
2	1	2	2	2
3	1	3	3	3
4	2	1	2	3
5	2	2	3	1
6	2	3	1	2
7	3	1	3	2
8	3	2	1	3
9	3	3	2	1

3.2. Numerical Simulation Calculation

3.2.1. BP Neural Network

The artificial neural network is a nonlinear and adaptive information processing system composed of a large number of standardized neurons, which is capable of simulating a biological neural network, and it has been deeply studied and widely used all over the world [25]. At the same time, due to the different connections of artificial neurons, a variety of artificial neural network models have been developed. Among them, the BP neural network is the most widely used model in artificial neural networks, which is a multi-layer feedforward neural network trained according to the algorithm of error back propagation [26]. The BP neural network is composed of an input layer, hidden layer and output layer and based on Sigmod function for operation and application, and has a strong nonlinear mapping ability and flexible network results [27]. Therefore, we can establish a nonlinear evaluation model by using this technology to better solve the randomness of weight definition, and ensure the accuracy and scientificity of evaluation results and activities [28,29].

The neural network program was developed using Matlab platform; the program flow is as follows:

- Loading initialized training sample parameters, input array P and output array A, and randomly selecting 80% of the samples as the training group (P1, A1) and 20% of the samples as the test group (P2, A2).
- Defining a series of neuron parameters such as the number of neural network layers, the number of neurons per layer, the maximum number of trainings (net.trainParam.epochs) and the training target error (net.trainParam.goal).
- Using the feedforwardnet function in Matlab to establish the neuron model, and using the train function to train the input and output array of the samples, such that a corresponding network is obtained.
- Using the Sim function to calculate the error between the input data P2 of the verification group and the output A2 of the verification group in the network. According to the error condition, returning to the second step to adjust the neural network parameters and continuing to train until the error requirements are met.
- Taking the target parameters into the Sim function, calculating the output of the neural network net, the predicted results of the target parameters are obtained.

3.2.2. Cross-Validation

Cross-validation can obtain as much effective information as possible from limited learning data, so as to obtain more appropriate two-layer weights. Additionally, this method learns samples from multiple directions, which can effectively avoid falling into local minima. Dong L et al. [30] established a microseismic event and blasting event identification model based on a convolutional neural network by using cross-validation. The collected microseismic and blasting event waveforms were composed of a training set, test set and verification set, respectively. Compared with other machine learning methods, this method has high identification accuracy. In addition, Dong L et al. [31] proposed the LM-CAG-CDR method and recommended 16 combined methods to evaluate the level of clean and safe production of phosphate rock, which improved the development level of the clean and safe mining of phosphate rock.

In order to verify the effectiveness of the BP neural network, the parameter design and simulation results of 18 orthogonal experimental models described above were used as samples to train the BP neural network and obtain a neural network model (14 models as training set and 4 models as test set). The objective laws hidden beneath the orthogonal test samples can be discovered, the coal caving rates and gangue content of all mining conditions and fully mechanized top-coal caving process combinations can be predicted without numerical simulation (a total of 729 combinations were used as the verification set), and the effectiveness of the neural network program can be verified with reference to the analysis of the numerical simulation results.

3.2.3. Optimized Decision

Based on the function and characteristics of the BP neural network, an optimized decision-making model for top-coal caving mining process parameters based on the BP neural network was established. The model could make decisions on top-coal caving mining process parameters according to the actual mining conditions of the coal mine, and obtain the optimal mining process parameters.

According to the actual situation of the mine, the input vector P was the natural factors affecting the mining and recovery rates of top coal, mainly including the average thickness of coal seam X1 (m), firmness coefficient of top coal X2, development degree of interlayer joint fracture X3, buried depth X4 (m), lithology and thickness of coal seam roof X5, and dip angle of coal seam X6 ($°$). The output process parameters of top-coal caving mainly included coal caving sequence Y1, mining to caving ratio Y2, and coal caving step Y3 (m). In order to make the decisions on top-coal caving process parameters more universal, the conceptual description in input and output was numerically processed based on the geological and process parameters of working faces in multiple coal mines. The processing results are shown in Table 7.

Table 7. Learning samples.

Sample Parameter	Concept Description	Neural Network Assignment Range
Interlayer and joint fracture development degree	Interlayer thickness > 0.5 m, Joint fissure less developed	0–0.33
	Interlayer thickness 0.2–0.5 m Joint fracture development general	0.34–0.66
	Interlayer thickness < 0.2 m Joint fracture development	0.67–1.0
Roof lithology and thickness	Pressure step < 25 m Immediate roof thickness > 10 m	0.75–1.0
	Pressure step 25–50 m Immediate roof thickness 5–10 m	0.5–0.75
	Pressure step 25–50 m Immediate roof thickness < 5 m	0.25–0.49
	pressure step > 50 m Immediate roof thickness < 3 m	0–0.24
Coal caving sequence	Multi-round sequential coal caving	0.76–1.0
	Interval caving coal among multi-cutting	0.51–0.75
	Single-round sequential coal caving	0.26–0.50
	Single round interval coal caving	0–0.25

A 6-layer neural network was established, with 10 neurons, 6 input parameters and 3 output parameters in each layer, as shown in Figure 4. The three output parameters were the optimized process parameters.

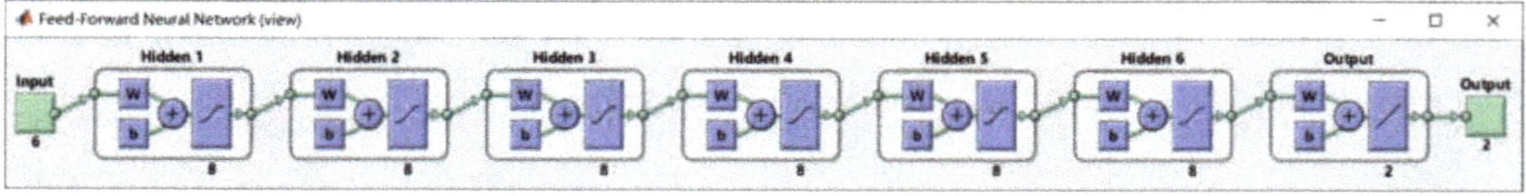

Figure 4. Parameters prediction neural network diagram.

3.3. Similarity Simulation Test

According to the geological conditions of the No. 12309 working face of the Wangjialing coal mine and the determined optimal fully mechanized top-coal caving process parameters, the similarity simulation test system for top-coal caving was designed and tested in the laboratory. The model frame was 2000 mm in length, 200 mm in width and 2500 mm in height, and the simulated material was composed of sand, lime and Bali stone. The height of the laid simulation material was 130.5 cm, and the geometric similarity ratio of the simulation experiment was C = 30/318 = 1: 10.6. According to the field observation, the top coal is easy to release, and there are few cases of large coal blocking. Therefore, the top coal, immediate roof and main roof in the experiment were laid into loose bodies, the coal seam was simulated by black particles, and the immediate roof was simulated by white particles. The similar material simulation test bench is shown in Figure 5, and the particle arrangement position is shown in Figure 6. The top coal was divided into upper, middle and lower layers by using marker particles. A 10 MPa uniformly distributed load was applied on the top layer of the model to simulate the load on the actual rock stratum (calculated according to the buried depth of 400 m). The opening and closing of the coal discharge opening were simulated by pulling out and pushing in the separator plate interposed between the supports, and the coal discharging was started and stopped under the action of the load.

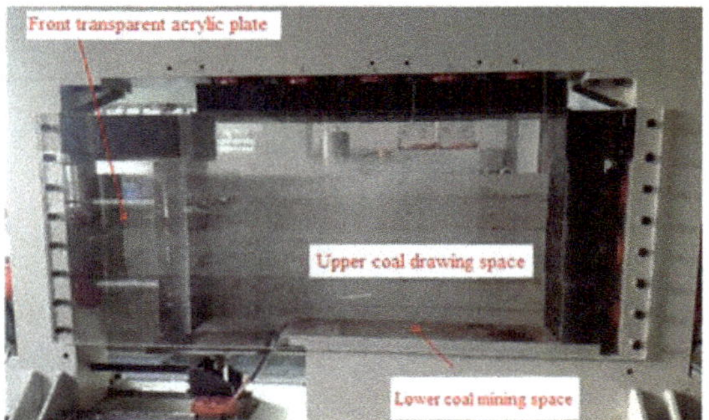

Figure 5. Similar material simulation test bench.

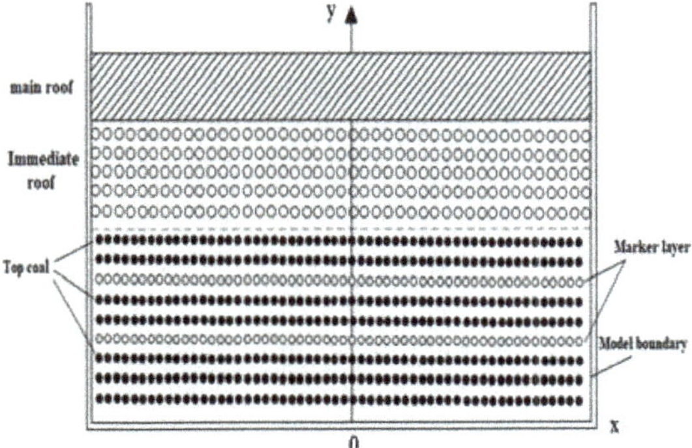

Figure 6. Diagram of particle position arrangement.

4. Results and Discussion

4.1. Numerical Simulation Experiment

4.1.1. Coal Caving Mode

The coal caving method mainly included three methods: sequential coal caving, grouping interval and interval return coal caving. The description with respect to Figures 7–9 is as follows: the coal–gangue boundary refers to the boundary between the top coal and the immediate roof; that is, the green particles represent the immediate roof and the blue particles represent the top coal. Residual coal refers to the part of coal left after top-coal caving.

(1)   Sequential coal caving

In order to study the flow characteristics of top coal under different simulation schemes, the representative scheme of the six numerical simulation schemes was selected for analysis, namely Scheme 1 in Table 4. The simulation process is shown in Figure 7.

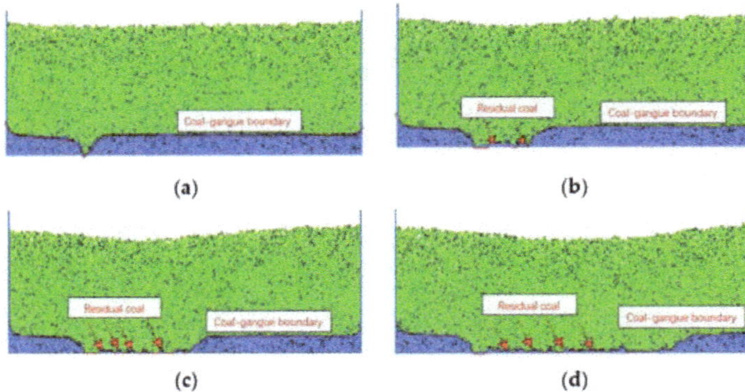

Figure 7. Simulation scheme 1 top-coal flow process diagram. (**a**) The 1# support coal caving end; (**b**) 7# support coal caving end; (**c**) 10# support coal caving end; (**d**) all supports finished.

The geological conditions and process parameters for simulation scheme 1 are as follows: the thickness of the coal seam is 6 m, the mining to caving ratio is 1:1 (mining height is 3 m, caving height is 3 m), single-round sequential coal caving, the number of caving openings is 1, and the particle size of top coal is 0.15~0.3 m. After the upper coal above 1# support is released, there is an obvious funnel-shaped caving space above the support (Figure 7a). When the support top coal was released in sequence, the boundary of the coal gangue dropped gently. After discharging all of the coal, the top-coal recovery rate and gangue content were 90.72% and 2.72%, respectively. Therefore, a single round of sequential caving could achieve a better caving effect for a short top-coal caving height.

(2)   Group interval coal caving

Six representative numerical simulation schemes with coal caving sequence 2 were selected for analysis, namely Scheme 8 in Table 4. The simulation process is shown in Figure 8.

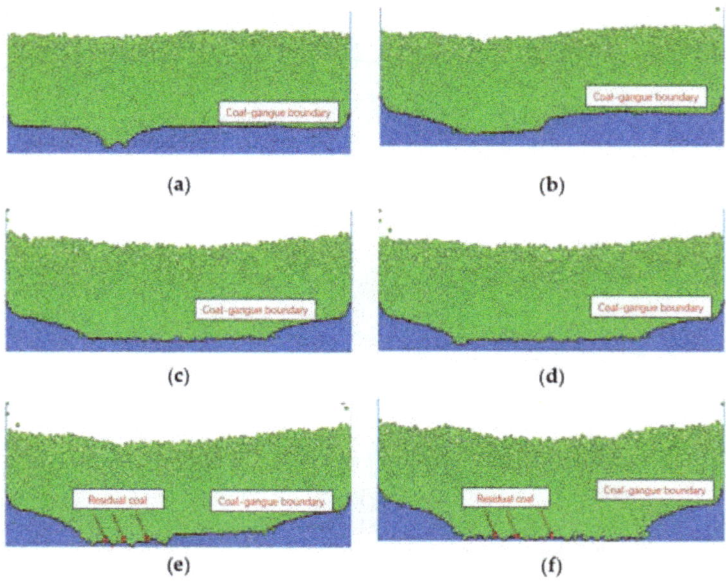

Figure 8. Simulation scheme 8 top-coal flow process diagram. (**a**) The 1# support coal caving end; (**b**) 10# support coal caving end; (**c**) end of the first round of coal caving; (**d**) 1# support end of second coal caving; (**e**) 10# support end of second coal caving; (**f**) end of the second round of coal caving.

Geological conditions and process parameters of simulation program 8: coal seam thickness 10 m, mining to caving ratio 1:1.5 (mining height 4 m, caving height 6 m), multi-round interval caving, the number of caving openings is 1, and the particle size of top coal is 0.35~0.5 m. Due to the use of multiple rounds of coal caving, the boundary between coal and gangue in the first round of coal caving decreased evenly (Figure 8a–c), preventing gangue from mixing into adjacent coal caving openings in advance. When the remaining top coal above the support continued to cave out, the flow characteristics of top coal were similar to those of simulation scheme 1 (caving height 3 m) due to the thin thickness of the remaining top coal. After all the supports were placed, there was less top coal missing in the goaf (Figure 8d–f). The top-coal recovery rate and gangue content were 91.88% and 4.05%, respectively. The coal caving effect was good. Thus, when the top-coal caving height is large (6 m), multiple rounds of coal caving should be adopted to ensure the uniform descent of the coal-gangue boundary and to prevent the gangue above the coal caving support from entering the adjacent coal caving opening too early.

(3)   Interval return coal caving

Six representative numerical simulation schemes with coal caving sequence 3 were selected for analysis, namely scheme 13 in Table 4. The simulation process is shown in Figure 9.

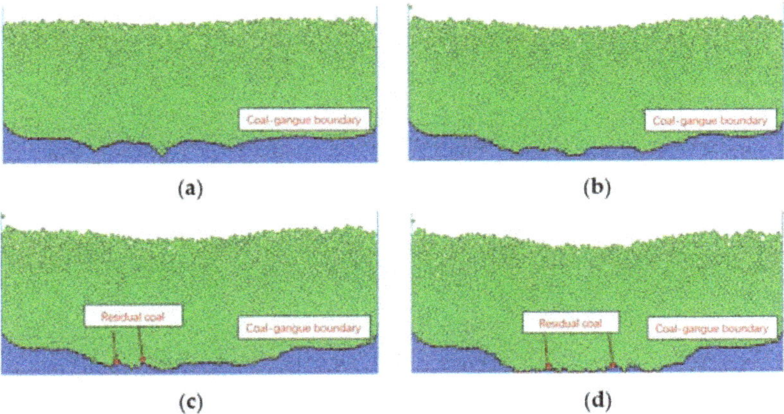

Figure 9. Simulation scheme 8 top-coal flow process diagram. (**a**) The 2# support end of first coal caving; (**b**) end of the first round of coal caving; (**c**) 2# support end of second coal caving; (**d**) end of the second round of coal caving.

The geological conditions and process parameters of simulation scheme 13 are as follows: coal seam thickness is 8 m, mining to caving ratio is 1:1 (mining height is 4 m, caving height is 4 m), the number of caving openings is 1, and the top-coal particle size is 0.35~0.5 m. Due to using group interval caving, the boundary line of coal and gangue descends unevenly after the first round of caving (Figure 9a,b), and the remaining top coal thickness is obviously different. There is more top coal left in the goaf after the second round of coal caving with all supports (Figure 9c,d). The top-coal caving rate and gangue content were 85.77% and 2.72%, respectively. In addition, compared with scheme 5 with the same coal thickness, the caving ratio (1:1) of scheme 13 was greater than that of scheme 5 (1:1.5), and the top-coal recovery rate of scheme 13 was better than that of scheme 5. Consequently, choosing a smaller caving ratio is conducive to top-coal caving under the condition of the same coal thickness.

The top-coal recovery rate and gangue content of different simulation schemes with varying coal caving methods were counted, as shown in Figure 10. Through the analysis of the 18 coal caving schemes, the top-coal recovery rate ranges from 73.43% to 95.41%, and the gangue content ranges from 1.09 to 10.21%. The difference for top-coal recovery rate and gangue content is obvious when adopting different coal caving sequences and different process parameters. Therefore, in order to obtain the ideal top-coal caving effect, the top-coal caving sequence and process under specific geological production conditions need to be analyzed. In addition, if schemes 7 and 11 are ignored, there is a positive correlation between the top-coal recovery rate and the gangue content. That is, with the increase in gangue content, the top coal release rate will also increase accordingly. On the other hand, as the gangue content decreases, the top-coal recovery rate will also decrease. Hence, properly increasing the gangue content could improve the top-coal recovery rate, and the critical point for gangue content needs to be determined according to the specific coal caving process parameters and actual production situation.

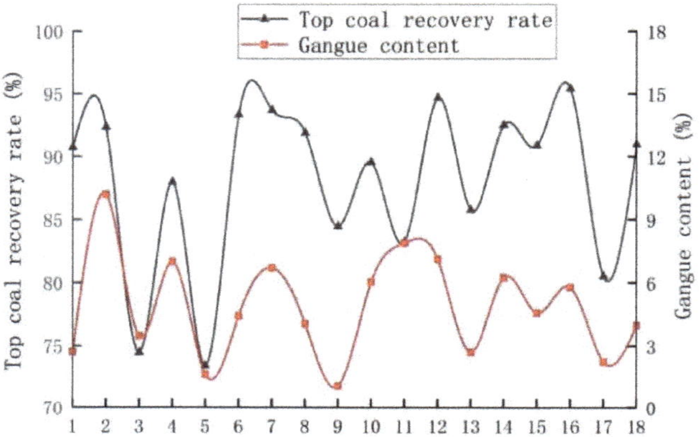

Figure 10. Top-coal recovery rate and gangue content curves.

4.1.2. Coal Caving Procedure

According to Tables 3 and 4, different simulation schemes with varying coal caving procedures were simulated. The top-coal recovery rate and gangue content of different simulation schemes were counted, as shown in Figure 11. The top-coal recovery rate ranged from 71.52% to 92.59% and gangue content ranged from 1.79% to 8.08%, respectively. There are great differences in top-coal recovery rate and gangue content when adopting different coal caving procedures and different process parameters. Consequently, in view of specific geological production conditions, in order to obtain an ideal coal caving effect, the coal caving step and coal caving procedure should be considered in detail. In addition, if scheme 4 is neglected, there is also a positive correlation between the top-coal recovery rate and the gangue content. That is, with the increase in gangue content, the top-coal recovery rate will also increase accordingly. On the other hand, as the gangue content decreases, the top-coal recovery rate will also decrease. When the coal caving step was taken as a single variable, it was found that with the increase in coal caving step, the top-coal recovery rate decreased and the gangue content increased; as the coal caving step decreased, the top-coal recovery rate increased and the gangue content decreased. Therefore, selecting a small caving step and appropriately increasing the gangue content could improve the top-coal recovery rate, and the critical point for gangue content needs to be determined according to specific caving process parameters and actual production conditions.

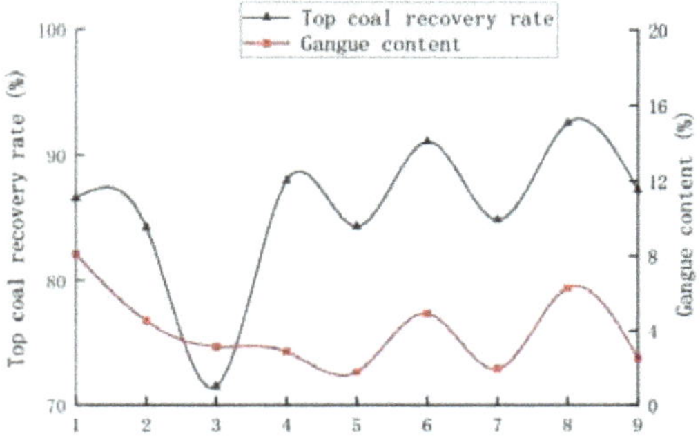

Figure 11. Top-coal recovery rate and gangue content curves.

4.2. Cross-Validation Results

All 729 possible model parameter combinations were input into the trained neural network model, and the optimized coal caving process parameters under different coal caving modes were obtained after ranking according to the comprehensive evaluation indexes, as shown in Table 8. The effectiveness of the BP neural network was verified.

Table 8. Optimized top-coal caving process parameters under different top-coal caving modes.

Parameters	Coal Seam Thickness	Caving Ratio	Number of Coal Caving Rounds	Coal Caving Order	Number of Coal Caving Openings	Top-Coal Particle Size	Top-Coal Recovery Rate	Gangue Content
Input parameters	1	1	0.5	0.5	0.5	0.25	92.48%	2.16%
Actual parameters	10 m	1:1	Three rounds	Three ports	Interval return coal caving	0.15–0.3 m	/	/

4.3. Optimized Process Parameters

According to the occurrence conditions of the coal seam in the Wangjialing coal mine, the main parameters of the Wangjialing coal mine were brought into the decision-making model (Figure 4) to obtain the decision-making process parameters thereof, as shown in Table 9. The optimized technological parameters for fully mechanized mining in the Wangjialing coal mine were single-round sequential coal caving, mining and caving ratio 1.09:1, and coal caving step distance 0.78 m.

Table 9. Determination of fully mechanized caving process parameters.

Coal Seam Occurrence Conditions (X)		Optimized Process Parameter (Y)	
Coal seam thickness (X1)	6.1 m	Coal caving sequence Single-round sequential	
Top-coal firmness coefficient (X2)	1.8		
Interlayer and joint fissure (X3)	0.4	Caving ratio	1.09
Depth of embedment (X4)	400 m		
Roof lithology and thickness (X5)	0.4	Coal caving step	0.78 m
Top-coal dip angle (X6)	2°		

4.4. Coal Caving Effectiveness

Figure 12 shows the experimental process of single-round sequential coal caving. Additionally, Table 10 shows the top-coal recovery rate and gangue content for each coal caving opening in single-round sequential coal caving.

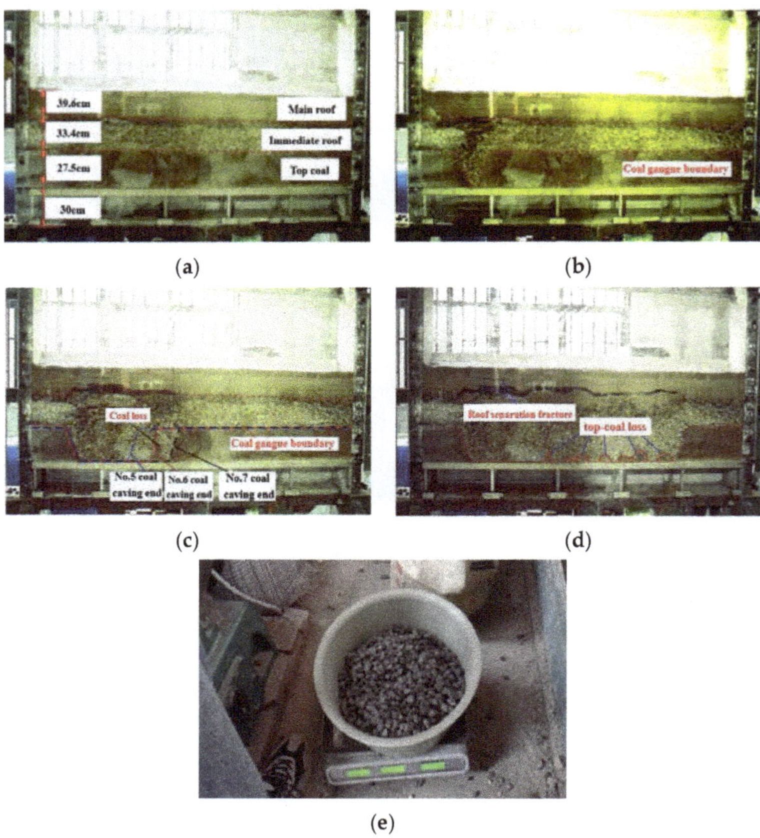

Figure 12. Single-round sequential coal caving process. (**a**) Initial state; (**b**) No. 3 coal caving end; (**c**) No. 5~7 coal caving end; (**d**) end of coal caving; (**e**) weighing the released coal.

Table 10. Single-round sequential coal caving results.

Coal Caving Opening Number	3	4	5	6	7	8	9	10	11	12	13	Sum
Coal quantity (kg)	2.82	2.82	2.82	2.82	2.82	2.82	2.82	2.82	2.82	2.82	2.82	31.02
Coal output (kg)	2.83	2.45	2.43	2.44	2.37	2.38	2.32	2.45	2.32	2.44	2.42	26.85
Gangue output (kg)	0.11	0.2	0.21	0.05	0.12	0.1	0.03	0.02	0.03	0.08	0.12	1.07
Top-coal recovery rate (%)	100.35	86.88	86.17	86.52	84.04	84.40	82.27	86.88	82.27	86.52	85.82	86.56
Gangue content (%)	3.90	7.09	7.45	1.77	4.98	4.26	1.06	0.71	1.06	2.84	4.26	3.45

Analysis of Figure 12 and Table 9 shows that the similarity simulation results of single-round sequential caving in the No. 12309 working face of the Wangjialing coal mine are consistent with the results of numerical simulations. When the top-coal caving height was small, the single-round sequential coal caving could achieve a better coal caving effect. When the first support (3# support) carried out coal caving, there was an obvious funnel-shaped coal caving space above the support (Figure 12b). When the top coal of the caving support was caved out sequentially, the coal gangue boundary descended gently; only a small part of top coal was left after caving all of the coal (Figure 12d). The top-coal recovery rate was 86.56% and the gangue content was 3.45%, with good caving effect. Thus, compared with the analysis and decision making of process parameters through industrial experiments, which takes a lot of time and consumes a certain amount of manpower and material resources, using the BP neural network to optimize the decision-making process of fully mechanized caving process parameters can effectively improve the decision-making efficiency and provide a basis for the realization of intelligent, fully mechanized caving mining.

5. Conclusions

In this study, the effects of different fully mechanized top-coal caving process parameters with different caving methods and different caving procedures on top-coal recovery rates and gangue content were studied. According to the occurrence conditions and actual production situation of the Wangjialing coal mine, the decision-making model for fully mechanized top-coal caving mining process parameters was established by using the BP neural network, and the optimized fully mechanized top-coal caving process parameters of Wangjialing coal mine were obtained. The in-lab similarity simulation experiment was carried out to verify the coal caving effect of the optimized fully mechanized top-coal caving process parameters. The following conclusions were drawn from the whole process:

(1) For different coal caving process parameters, the top-coal recovery rates and gangue content are obviously differen, the top-coal recovery rate could be improved by appropriately increasing the gangue content, and the critical point for the gangue content should be determined according to the specific coal caving process parameters and the actual production situation.

(2) In top-coal caving mining, the selection of a small caving step distance was conducive to top-coal caving. When the top-coal caving height was small, a better coal caving effect could be achieved by single-round sequential coal caving. When the top-coal caving height was large (6 m), using multiple rounds of coal caving was conducive to ensuring that the boundary between coal and gangue dropped evenly, and preventing the gangue above the coal caving support from entering the adjacent coal caving opening prematurely.

(3) Through the in-lab similar simulation experiment, it was indicated that the BP neural network can be used to study the optimized decision making of mining process parameters and can obtain good results, improving the benefit of process parameter decision making, and provide the basis for realizing the intelligent mining of fully mechanized top-coal caving.

(4) There are many factors affecting the top-coal recovery rate and gangue content in addition to the fully mechanized caving process parameters studied in this paper. The calculation of a relaxed ellipsoid can also be considered in future research.

Author Contributions: M.L., C.H., R.Y. and L.W. designed the study and experiment; C.H., B.Z. and Z.X. collected the data; C.H., R.Y., L.W. and Z.X. conducted the data analysis; R.Y., L.W. and B.Z. provided the statistical methods; M.L., C.H., R.Y. and Z.X. drafted the paper; M.L., C.H., R.Y. and Z.X. edited the paper. All authors have read and agreed to the published version of the manuscript.

Funding: This research was funded by the National Natural Science Foundation of China (Nos. 52004273 and 51874276), the Natural Science Foundation of Jiangsu Province (No. BK20200639), the China Postdoctoral Science Foundation (No. 2019M661992) and the Fundamental Research Funds for the Central Universities (No. 2020QN38).

Institutional Review Board Statement: Not applicable.

Informed Consent Statement: Not applicable.

Data Availability Statement: All data and code used or analyzed in this study are available from the corresponding author on reasonable request.

Acknowledgments: The authors thank the editor and three anonymous reviewers of the paper. Their constructive suggestions and comments have considerably improved the quality of the paper.

Conflicts of Interest: The authors declare no conflict of interest.

References

1. Leonard, M.D.; Michaelides, E.E.; Michaelides, D.N. Substitution of coal power plants with renewable energy sources—Shift of the power demand and energy storage. *Energy Convers. Manag.* **2018**, *164*, 27–35. [CrossRef]
2. Wang, J.; Yu, B.; Kang, H.; Wang, G.; Mao, D.; Liang, Y.; Jiang, P. Key technologies and equipment for a fully mechanized top-coal caving operation with a large mining height at ultra-thick coal seams. *Int. J. Coal Sci. Technol.* **2015**, *2*, 97–161. [CrossRef]
3. Bhattacharya, M.; Rafiq, S.; Bhattacharya, S. The role of technology on the dynamics of coal consumption–economic growth: New evidence from China. *Appl. Energy* **2015**, *154*, 686–695. [CrossRef]
4. Herezy, Ł.; Janik, D.; Skrzypkowski, K. Powered Roof Support—Rock Strata Interactions on the Example of an Automated Coal Plough System. *Studia Geotech. Mech.* **2018**, *40*, 46–55. [CrossRef]
5. Wang, Z.; Zhang, G.; Zhao, L. Recognition of rock–coal interface in top coal caving through tail beam vibrations by using stacked sparse autoencoders. *J. Vibroeng.* **2016**, *18*, 4261–4275. [CrossRef]
6. Wu, J.; Qin, Y.; Zhai, M. Mining safety of longwall top-coal caving in China. In Proceedings of the 8th U.S. Mine Ventilation Symposium, Rolla, MO, USA, 11–17 June 1999.
7. Wang, J. Development and prospect on fully mechanized mining in Chinese coal mines. *J. Coal Sci. Eng.* **2014**, *1*, 253–260. [CrossRef]
8. Le, T.D.; Mitra, R.; Oh, J.; Hebblewhite, B. A review of cavability evaluation in longwall top coal caving. *Int. J. Min. Sci. Technol.* **2017**, *27*, 907–915. [CrossRef]
9. Yasitli, N.E.; Unver, B. 3D numerical modeling of longwall mining with top-coal caving. *Int. J. Rock Mech. Min. Sci.* **2005**, *42*, 219–235. [CrossRef]
10. Alehossein, H.; Poulsen, B.A. Stress analysis of longwall top coal caving. *Int. J. Rock Mech. Min. Sci.* **2010**, *47*, 30–41. [CrossRef]
11. Si, G.; Jamnikar, S.; Lazar, J.; Shi, J.Q.; Durucan, S.; Korre, A.; Zavšek, S. Monitoring and modelling of gas dynamics in multi-level longwall top coal caving of ultra-thick coal seams, part I: Borehole measurements and a conceptual model for gas emission zones. *Int. J. Coal Geol.* **2015**, *144–145*, 98–110. [CrossRef]
12. Klishin, V.; Nikitenko, S.; Opruk, G. Longwall top coal caving (LTCC) mining technologies with roof softening by hydraulic fracturing method. *IOP Conf. Ser. Mater. Sci. Eng.* **2018**, *354*, 012015. [CrossRef]
13. Klishin, V.I.; Fryanov, V.N.; Pavlova, L.D.; Opruk, G.Y. Modeling top coal disintegration in thick seams in longwall top coal caving. *J. Min. Sci.* **2019**, *55*, 247–256. [CrossRef]
14. Ti, Z.; Li, J.; Wang, M.; Wang, K.; Jin, Z.; Tai, C. Fracture mechanism in overlying strata during longwall mining. *Shock. Vib.* **2021**, *2021*, 4764732. [CrossRef]
15. Zhang, N.; Liu, C.; Wu, X.; Ren, T. Dynamic Random Arching in the Flow Field of Top-Coal Caving Mining. *Energies* **2018**, *11*, 1106. [CrossRef]
16. Unver, B.; Yasitli, N.E. Modelling of strata movement with a special reference to caving mechanism in thick seam coal mining. *Int. J. Coal Geol.* **2006**, *66*, 227–252. [CrossRef]
17. Ghosh, A.K.; Gong, Y. Improving coal recovery from longwall top coal caving. *J. Mines Met. Fuels* **2014**, *62*, 51–64.
18. Klishin, V.I.; Klishin, S.V. Coal extraction from thick flat and steep beds. *J. Min. Sci.* **2010**, *46*, 149–159. [CrossRef]
19. Klishin, S.V.; Klishin, V.I.; Opruk, G.Y. Modeling coal discharge in mechanized steep and thick coal mining. *J. Min. Sci.* **2013**, *49*, 932–940. [CrossRef]
20. Yu, K.; Qiang, W. Application of ant colony clustering algorithm in coal mine gas accident analysis under the background of big data research. *J. Intell. Fuzzy Syst.* **2020**, *38*, 1381–1390. [CrossRef]
21. Ruilin, Z.; Lowndes, I. The application of a coupled artificial neural network and fault tree analysis model to predict coal and gas outbursts. *Int. J. Coal Geol.* **2010**, *84*, 141–152. [CrossRef]
22. Jiang, H.; Song, Q.; Gao, K.; Song, Q.; Zhao, X. Rule-based expert system to assess caving output ratio in top coal caving. *PLoS ONE* **2020**, *15*, e0238138. [CrossRef] [PubMed]
23. Fan, S. Study of the Safety Assessment Model of Coal Mine Based on BP Neural Network. In Proceedings of the International Conference on Intelligent Computation Technology & Automation IEEE, Changsha, China, 10–11 October 2009.
24. Meng, X.; Lu, P.; Wang, B. Coal mine safety warning system based on principal component method and neural network. In Proceedings of the 2017 6th Data Driven Control and Learning Systems (DDCLS) IEEE, Chongqing, China, 26–27 May 2017.
25. O'Rourke, D.; Ringer, A. The impact of sustainability information on consumer decision making. *J. Ind. Ecol.* **2015**, *20*, 882–892. [CrossRef]
26. Treichler, E.B.H.; William, D. Beyond shared decision-making: Collaboration in the age of recovery from serious mental illness. *Am. J. Orthopsychiatry* **2017**, *87*, 567–574. [CrossRef] [PubMed]
27. Li, W. Consumer Decision-Making Power Based on BP Neural Network and Fuzzy Mathematical Model. *Wirel. Commun. Mob. Comput.* **2021**, *2021*, 6387633. [CrossRef]
28. Joshi, Y.; Rahman, Z. Predictors of young consumer's green purchase behavior. *Manag. Environ. Qual. Int. J.* **2016**, *27*, 452–472. [CrossRef]
29. Hortaçsu, A.; Madanizadeh, S.A.; Puller, S. Power to choose? an analysis of consumer inertia in the residential electricity market. *Nber Work. Pap.* **2015**, *9*, 192–226.

30. Dong, L.-J.; Tang, Z.; Li, X.-B.; Chen, Y.-C.; Xue, J.-C. Discrimination of mining microseismic events and blasts using convolutional neural networks and original waveform. *J. Cent. South Univ.* **2020**, *27*, 3078–3089. [CrossRef]
31. Dong, L.; Zhou, Y.; Deng, S.; Wang, M.; Sun, D. Evaluation methods of man-machine-environment system for clean and safe production in phosphorus mines: A case study. *J. Cent. South Univ.* **2021**, *28*, 3856–3870. [CrossRef]

Article

Semi-Empirical Time-Dependent Parameter of Shear Strength for Traction Force between Deep-Sea Sediment and Tracked Miner

Wei Yi [2] and Feng Xu [1],*

[1] School of Mathematics and Physics, University of South China, Hengyang 421001, China
[2] School of Civil Engineering, Central South University, Changsha 410083, China; yi.wei@csu.edu.cn
* Correspondence: hsubong@usc.edu.cn

Abstract: Based on our direct shear creep experiment and the direct shear rheological constitutive model, a semi-empirical time-dependent parameter of the shear strength is obtained by Mohr–Coulomb shear strength theory, and different time-dependent traction force calculations between deep-sea sediment and a tracked miner are conducted by the work-energy principle. The time-dependent traction force calculation under its influencing factors, including the time, track shoe number, and grounding pressure, are analyzed and proved to be valid by the traction force experiment of a single-track shoe. The results show that the time-dependent cohesion force obtained by a semi-empirical way can be easily used to deduce the time-dependent traction force models under the different grounding pressure distributions and applied into deep-sea engineering application conveniently; the verified traction force models by the traction force experiment of a single-track shoe illustrate that traction force under the decrement grounding pressure distribution is the worst among the four kinds of grounding pressure distributions and suggested for evaluating the most unfavorable traction force and calculating the trafficability and stability of the deep-sea tracked miner.

Keywords: time-dependent cohesion; traction force; deep-sea sediment; tracked miner; rheology

Citation: Yi, W.; Xu, F. Semi-Empirical Time-Dependent Parameter of Shear Strength for Traction Force between Deep-Sea Sediment and Tracked Miner. *Sensors* **2022**, *22*, 1119. https://doi.org/10.3390/s22031119

Academic Editor: Konstantinos Kalpakis

Received: 31 December 2021
Accepted: 25 January 2022
Published: 1 February 2022

Publisher's Note: MDPI stays neutral with regard to jurisdictional claims in published maps and institutional affiliations.

1. Introduction

Many countries shift their attention from the land to the deep sea, which is abundant with mineral resources, due to the gradual lack of non-renewable land resources with the development of social economy. In the deep sea, there are nearly 1500 billion tons of mineral resources, including poly-metallic nodules, cobalt-rich shells, and poly-metallic sulfides [1–3]. Therefore, these countries target ocean exploitation as their strategic development direction. At present, there are three types of commonly used deep-sea mining systems: drag bucket mining system, continuous rope bucket mining system, and mining system with a tracked miner and pipeline lifting device (i.e., hydraulic lifting pipeline mining system). The tracked miner is one of the most critical pieces of equipment of the hydraulic lifting pipeline mining system, adopted mainly by China because of its low cost and high mining efficiency [4–6]. The tracked miner encounters extremely special deep-sea environments, such as more obvious rheological performance of sediment than land soil [7] and complex deep-sea topography (e.g., gullies, ditches, and slopes). The special environments inevitably result in the change of grounding pressure distribution and the lack of traction force for the tracked miner, and the tracked miner eventually fails the normal walk. It will cause severe turnover of the tracked miner due to the deep sinkage and breakdown of the whole mining system [8]. Therefore, it is significant for the safety and stability of the deep-sea mining system to study time-dependent characteristics of deep-sea sediment and the traction force of the tracked miner under different grounding pressure.

Currently, the research on time-dependent characteristics of the deep-sea sediment is mainly carried out from two aspects, i.e., the creep experiment and the rheological constitutive model. The research on the traction force mainly includes three methods, i.e., the traction force theory, the traction experiment, and the related simulation. In terms of the time-dependent characteristics of the deep-sea sediment, a direct shear creep experiment was conducted for analyzing the deep-sea simulant sediment with Burgers creep model and obtaining related rheological parameters of the direct shear rheological model [9]. A tri-axial compression creep experiment was studied for getting the creep curves of the deep-sea simulant sediment under the same confining pressure and different vertical compression, and then the creep curves were employed to identify the parameters by the different rheological models for determining the most proper compression rheological model of deep-sea sediment [10]. Xu et al. [7] discovered the compression-shear coupling effect between compression creep displacement and shear creep displacement by the compression-shear coupling creep experiment, and then a compression-shear coupling rheological model was deduced and proved to be reliable. In regard to the traction force of the tracked vehicle, the track-terrain interaction theory is often adopted mostly involving the Bekker's theory (for brittle soil) [11], Janosi–Hanamoto's theory (for plastic soil) [12] and Wong's theory (for both brittle and plastic soil) [13], and a little involvement of rheological constitutive model (for cohesive soil) [14]. For example, Zhao et al. [15] categorized the compression model of soil based on the soil theory and offroad vehicle-terrain theory and introduced an improved sinkage model of the brittle soil by analyzing the Bekker's model and ultimate balance theory. Wu et al. [16] established a traction force model of a tracked miner on the deep-sea soft sediment by studying the cohesive action between a grouser and the sediment, which revealed the influence of parameters of the sediment and sizes of the tracked miner structure on the traction force. Wang et al. [17] tested the applicability of two kinds of empirical models of shear stress-displacement to the deep seabed and promoted a new empirical model for saturated and plastic soil; Xu et al. [18] initially employed a compression-shear coupling rheological model into the analyses of the sinkage and thrusting force of a tracked miner and deduced a new turning traction force. Li et al. [19] obtains a relationship between the grouser height and the water jet based on elastic-plastic traction force model aiming at the cause of sticky soil shaped on the track. Experimentally, Xu et al. [20] discussed the relationship between the slippage and traction force and determined the optimal grouser height by analyzing the motion of simplified track shoes and the track shoe experiment of different grouser heights. Shin et al. [21] discussed the loss mechanism of the lateral thrusting force by a traction force experiment with different shape ratios. Baek et al. [22] evaluated the traction performance by testing the slipping sinkage of the track shoes. Furthermore, in terms of the simulation of traction force, Rubinstein et al. [23] studied a transporting tracked vehicle with dynamics software LMS-DADS and established a multi-body dynamic simulation model for calculating the traction force on different locations of the track shoes. Yang et al. [24] studied the influence of the shearing ratio and grouser size on the traction force based on the simulation model of track shoe and soil. Li et al. [25] built a 3-D simulation model with McKyes–Ali software to analyze the interaction between interval track shoes and the soil and verify the interaction laws. Summarily, the existing study on the track force is mainly about the elastic-plastic and rheological performance of the deep-sea sediment, seldom involving the time-dependent performance of the shear strength in deducing the time-dependent traction force under different grounding pressure distributions, which brings the inconvenience of applying the temporal traction force model into the deep-sea mining engineering.

In this paper, the direct shear creep experiment is conducted for obtaining the semiempirical time-dependent parameters in the shear strength based on the analyses of direct shear rheological model and Mohr–Coulomb shear strength theory. The time-dependent shear strength parameter will be employed into deducing the models of traction force based on the work-energy principle under different grounding pressure distributions (uniform, linear, and sine) and analyzing the influence of parameters such as the time, number of the track shoe, and grounding pressure distribution on the traction force. The research results could provide scientific basis for designing and optimizing the crawler as well as evaluating the trafficability of the tracked miner.

2. Time-Dependent Transformation of Shear Strength Parameter

2.1. Direct Shear Creep Experiment of Simulant Deep-Sea Sediment

The deep-sea soft sediment burdens a vertical pressure (e.g., grounding pressure) and a horizontal force (e.g., traction force) when the tracked miner walks. Therefore, it is necessary to study the direct shear creep effect of deep-sea soft sediment (provided that grounding pressure is not varying over time). The simulant sediment with the most similar physical and mechanical properties to deep-sea sediment (undisturbed sediment) is prepared to satisfy a large number of tests by mixing different betonies with water in various ratios. The main parameters measured by our research team are shown in Table 1 [9]. It can be seen that the main physical and mechanical parameters of simulant sediment and undisturbed sediment are close to each other and satisfied the requirement of the experiment [9].

Table 1. Main physical and mechanical parameters of sediment.

Physical and Mechanical Parameters	Simulant Sediment	Undisturbed Sediment
Wet density, $\rho/(\text{t·m}^{-3})$	1.315	1.250
Water content, $w/\%$	165.6	246.5
Liquid limit, $w_L/\%$	190.2	145.2
Cohesion force, $c/(\text{kPa})$	6.2	6.0
Friction angle, $\varphi/(°)$	1.72	3.1
Penetration resistance, $P_s/(\text{kPa})$	87	50–90

The deep-sea simulant sediment is sampled into a standard cylinder shape of 60 mm × 30 mm (Figure 1) [10] by a ring knife, considering that its water content (w = 165.6%) is between the plastic limit and the liquid limit. The sheer creep test of simulant sediment (with constant compressive stress) is conducted on a self-developed pressure-shear creep test device (Figure 2) [20] by our research team. The simulant sediment sample is placed in a shear box and burdens the effect of vertical weights (compressive stress σ) and horizontal weights (shear stress τ), i.e., the pressure-shear loading. Since the average grounding pressure σ_0 (σ_0 = 5 kPa) is the minimum standard for determining compressive stress and the average shear strength (τ_b = 6 kPa) is the maximum standard for determining shear stress, six different groups of constant compressive stress (σ = 5 kPa, 10 kPa, 15 kPa, 20 kPa, 25 kPa, 30 kPa) and six different groups of shear stress (τ = 1 kPa, 2 kPa, 3 kPa,4 kPa, 5 kPa, 6 kPa) are arranged and combined into 36 groups in total for direct shear creep experiment.

Figure 1. Standard cylinder sample of deep-sea simulant sediment.

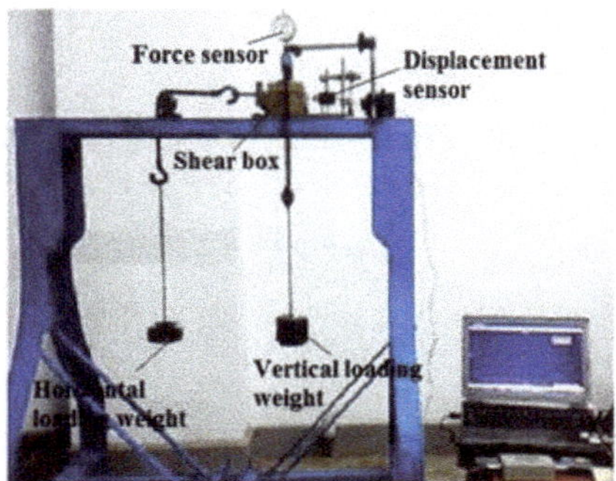

Figure 2. Direct shear creep apparatus of deep-sea simulant sediment.

For every group, the shear creep curves in the horizontal direction are obtained from a displacement test system comprising a NS-WY02 high-precision displacement sensor, a signal amplifier, and a display (computer). Noticeably, NS-YB data acquisition software is adopted for data storage due to the shortcoming of real-time data display [7].

2.2. Direct Shear Rheological Constitutive Model

Figures 3a and 4a show the typical shear creep curves of the simulant sediment under different constant compressive stresses σ (σ = 5 kPa [20] and 10 kPa) as examples, which are obtained under different constant shear stresses τ. Based on the characteristics analyses of the curves, Burgers rheological model, i.e., shear stress (τ)-displacement (s)-time (t) equation (Equation (1)) is adopted to fit these experimental creep curves (s-t) and the four rheological parameters (K_1, K_2, β_1 and β_2) in the model can be auto-fitted and determined by Sigma-plot software with fitted curved 3D surface consisting of experimental creep curves and are relevant with different constant compressive stresses (σ).

$$s(\tau,t) = \tau\left[\frac{1}{K_1} + \frac{t}{\beta_1} + \frac{1}{K_2}\left(1 - e^{-tK_2/\beta_2}\right)\right] \tag{1}$$

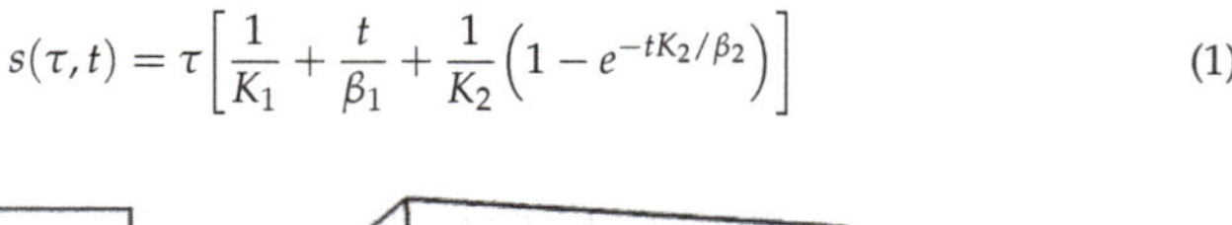

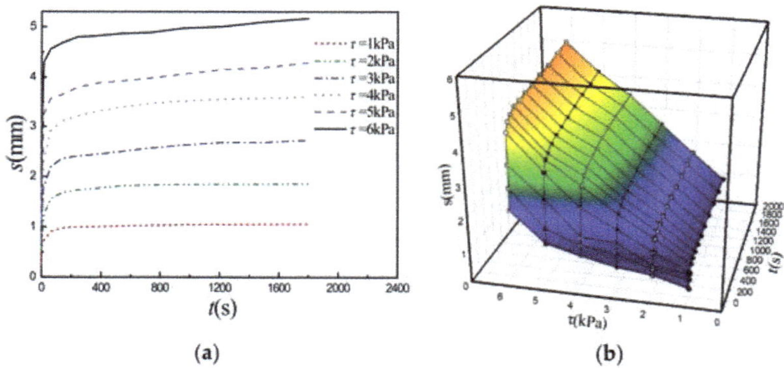

Figure 3. Shear creep curves and fitted 3D surface at σ = 5 kPa. (**a**) Shear creep curves under τ; (**b**) Fitted curved 3D surface.

As an example, Figures 3b and 4b show the curved 3D surface in τ-s-t space fitted by Equation (1) for the specific constant σ (σ = 5 kPa [20] and 10 kPa), where solid points are experimental data. By resorting to "Dynamic Fit Wizard" in the Sigma-plot software and inputting Equation (1) with user-defined function, the fitted Burgers rheological model parameters can be obtained (such as for the σ = 5 kPa, K_1 = 7.36 MPa, K_2 = 1.82 MPa, β_1 = 7380 MPa·s, β_2 = 22,900 MPa·s, coefficient of determination R-Square is 0.987). Table 2 lists all of the fitted shear creep parameters under different σ as well as R-Square and illustrates that the fitted shear creep parameters are the functions of σ and increase with σ. Therefore, the shear rheological Equation (1) can express the relationships between shear stress τ, shear displacement s, compressive stress σ, and time t, i.e., $\tau = \tau(s, \sigma, t)$ and can be rewritten as Equation (2), i.e., direct shear rheological constitutive model, where the expressions of $K_1(\sigma)$, $K_2(\sigma)$, $\beta_1(\sigma)$, and $\beta_2(\sigma)$ are obtained by fitting data in Table 2 [20] and described by Equation (3).

$$s(\tau, \sigma, t) = \tau \left[\frac{1}{K_1(\sigma)} + \frac{t}{\beta_1(\sigma)} + \frac{1}{K_2(\sigma)}\left(1 - e^{-tK_2(\sigma)/\beta_2(\sigma)}\right) \right] \tag{2}$$

$$\begin{cases} K_1(\sigma) = 0.002873\sigma^3 - 0.1312\sigma^2 + 2.373\sigma + 3.363 \\ K_2(\sigma) = 0.02327\sigma^{2.14} + 3.882 \\ \beta_1(\sigma) = 0.008\sigma^3 + 0.5629\sigma^2 - 4.483\sigma + 47.89 \\ \beta_2(\sigma) = 0.0011\sigma^2 - 0.00672\sigma + 0.0316 \end{cases} \tag{3}$$

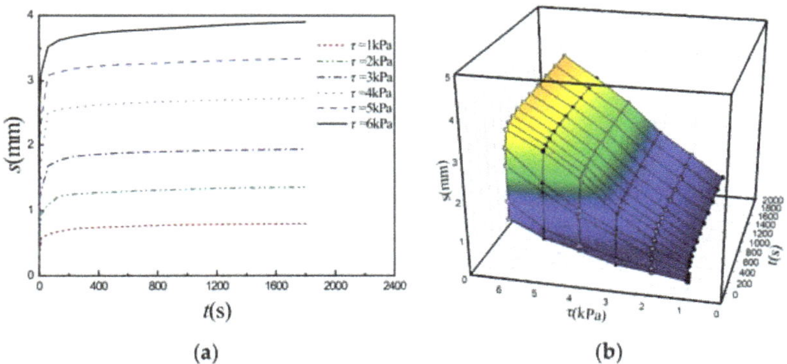

Figure 4. Shear creep curves and fitted 3D surface at σ = 10 kPa. (**a**) Shear creep curves under τ; (**b**) Fitted curved 3D surface.

Table 2. Fitted direct shear creep parameters.

σ/kPa	$K_1(\sigma)$/MPa	$K_2(\sigma)$/MPa	$\beta_1(\sigma)$/(MPa·s) $\times 10^3$	$\beta_2(\sigma)$/(MPa·s) $\times 10^3$	R-Square
5	7.36	1.82	7.38	22.90	0.987
10	10.27	2.25	10.24	27.50	0.992
15	11.20	2.95	11.26	24.73	0.984
20	11.22	5.51	11.41	304.04	0.995
25	13.29	8.09	13.90	421.70	0.984
30	19.24	9.87	19.15	466.84	0.983

2.3. Time-Dependent Parameters of Shear Strength

The shear stress (τ)-shear displacement (s) relationships (Equation (4)) are deduced by Janosi–Hanamoto based on Mohr–Coulomb shear strength theory and is often applied into deducing traction force of kinds of tracked vehicles [13]. For better deduction and analyses of traction force of different grounding pressure distributions under the deep-sea tracked miner, the direct shear rheological model is adopted to obtain the time-dependent parameters of Mohr–Coulomb shear strength and easily used to deduce traction force equation.

$$\tau = [c + \sigma \tan \varphi](1 - e^{-s/\kappa}) \tag{4}$$

where c is cohesion force, φ is friction angle and κ is shearing deformation module of the deep-sea simulant sediment. Since φ and κ are too small (φ is 1.72° and κ is 0.00424 mm), the two parameters can be regarded to be constants and not relevant with time.

Generally, the parameters in Equation (4) are determined based on the Figure 5, where the black dots represent the peak shear resistance and the straight line is fitted to the black dots. Obviously, the coordinate (σ, τ) of the intersection point between the straight line and τ-axial is $(0, c)$, i.e., $\tau(0) = c$. In order to obtain a time-dependent cohesion force c, let the σ in the Equation (2) be zero and shear stress τ will be the function of the shear displacement s and time t without the compressive stress σ, i.e., $c(s, t) = \tau(s, 0, t)$. Since shear displacement $s = vt$ and shear velocity v is constant, c is only the function of time t, i.e., $c = c(t)$.

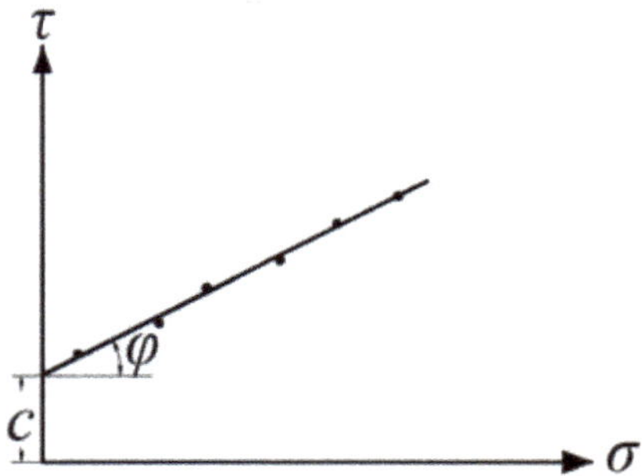

Figure 5. Determination of the cohesion force c and friction angle φ.

Hence, the parameters $K_1(\sigma)$, $K_2(\sigma)$, $\beta_1(\sigma)$ and $\beta_2(\sigma)$ become constant and the value of the parameters are as follows [20].

$$\begin{cases} K_1(0) = 3.363 \\ K_2(0) = 3.882 \\ \beta_1(0) = 47.89 \\ \beta_2(0) = 0.0316 \end{cases} \tag{5}$$

Accordingly, Equation (2) can be rewritten into Equation (6) given by

$$s = vt = c(t)\cdot\left[0.297 + \frac{t}{47.89} + \frac{1}{3.882}\left(1 - e^{-122.85t}\right)\right] \tag{6}$$

By means of modifying Equation (6) into a function of time t, the time-dependent cohesion force c is given by

$$c(t) = vt/\left[0.297 + \frac{t}{47.89} + \frac{1}{3.882}\left(1 - e^{-122.85t}\right)\right] \tag{7}$$

3. Time-Dependent Traction Force of Deep-Sea Miner Crawler

3.1. Traction Force Model

Figure 6 plots the simplified model of the miner's motion on the deep-sea sediment [26]. It can be seen that the whole crawler comprises several hinged track shoes with uniformed distribution. To simplify calculation of the traction force, it is assumed that every track shoe I $(i = 1, 2, \ldots, n)$ is a "T" type with a grouser thrusting and shearing the sediment and moves at the same horizontal displacement. When the tracked miner moves forwards straightly at a constant velocity v, every vertical grouser of the track shoe i develops a horizontal shear displacement s (i.e., slippage $i\Delta$). Choosing the whole crawler as an analysis object, as shown in Figure 7, the contacting force (i.e., grounding pressure $\sigma(x)$) on the track shoe is affected by the complex deep-sea topographies such as deep-sea mountains and ditches because of different contact conditions between track shoe surface and an idler. If every track shoe i moves under the action of traction force T_i and grounding pressure $\sigma(x)$, it is also assumed that the track shoes have the same horizontal displacement s but different sinkage z_i.

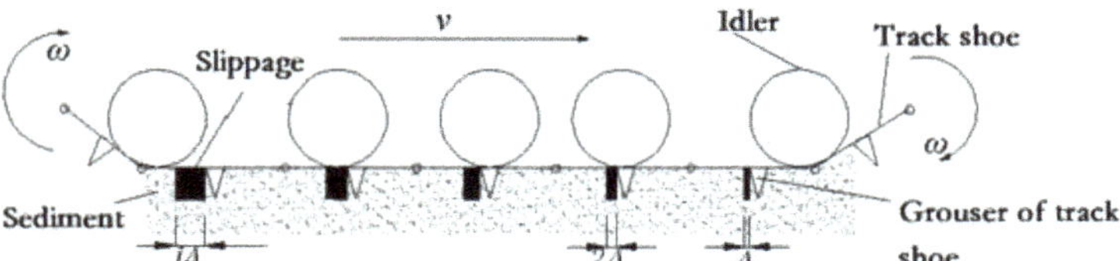

Figure 6. Simplified model of tracked miner's motion.

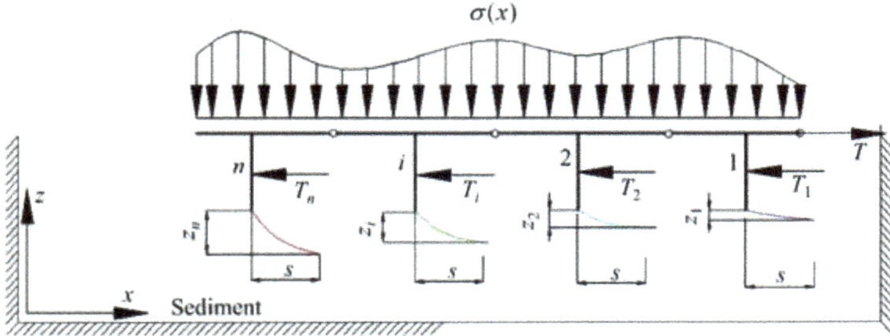

Figure 7. Simplified mechanic model of the crawler.

The traction force T of the whole crawler equals to the sum of the traction force T_i caused by the horizontal compression and shear of the grouser of every track shoe i and expressed by

$$T = \sum_{i=1}^{n} T_i \tag{8}$$

In order to calculate the traction force T_i of the grouser, it is assumed that every track shoe i has a length L, width B, grouser height h, and a location $x = Il\text{-}L/2$ in analyzing the kinetic process of a single-track shoe, as shown in Figure 8. When a single-track shoe moves from location I to location II under the action of the horizontal traction force T_i and vertical force $2BL\sigma(x)$, there is a horizontal shear displacement s and vertical sinkage z_i. Meanwhile, the horizontal shear stress τ_i changes from τ_{i0} to τ_{i1}, corresponding to disturbed sediment area A_{i1} and the vertical compressive stress ranges σ_i from σ_{i0} to σ_{i1} corresponding to disturbed sediment area A_{i2}. It can be known that the compressive stress σ_i equals to grounding pressure $\sigma(x)$ according to balance of the vertical forces. Hence, the work W_{i1} done by the shear stress τ_i and the work W_{i2} done by the compressive stress σ_i can be described by the following two equations, respectively:

$$W_{i1} = Bh \int_0^s \tau_i ds \tag{9}$$

$$W_{i2} = BL \int_0^z \sigma_i dz \tag{10}$$

Based on the word-energy principle and Wong's suggestion [27], it can be known that when a rigid track shoe, width B and contact length h, moves through a distance s, the work W_{i1}' to make the rut of area A_{i1}' can be assumed to be equal to the work W_{i2} necessary to compact the sediment of the area of A_{i2} corresponding to the contact part of track to the sinkage z_i. Therefore, the relationships between the two works can be given as follows:

$$W_{i2} = W_{i1}' \tag{11}$$

Hence, there are two kinds of works in the horizontal direction after the analysis above, and the sum work W_i ($W_i = T_i s$) of them equals to $W_{i1} + W_{i1}'$ or $W_{i1} + W_{i2}$. Eventually, the expression of traction force T_i under the consideration of vertical compressive stress is written as follows:

$$T_i = \frac{W_{i1} + W_{i2}}{s} = \frac{Bh\int_0^s \tau_i ds + BL\int_0^z \sigma_i dz}{s} \tag{12}$$

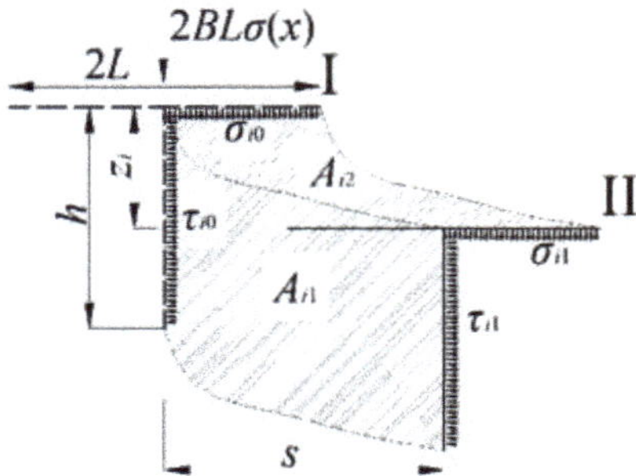

Figure 8. Kinetic process of single-track shoe.

3.2. Time-Dependent Traction Force T under Different Grounding Pressure Distributions

When the tracked miner walks on complex deep-sea topographies such as deep-sea mountains and sea ditches, different grounding pressure distributions develop under the crawler of the tracked miner. If the grounding pressure σ_i under every track shoe just changes with the horizontal location x but not with the time t, it is assumed that the types of grounding pressure distribution, relevant with the weight G of the tracked miner, can be simplified into four types: (a) uniform distribution with amplitude $\sigma_i(x) = G/BL$, (b) linear decrement distribution with amplitude $\sigma_i(x) = 2Gx/BL^2$, (c) linear increment distribution with amplitude $\sigma_i(x) = -2Gx/BL^2$, and (d) sine distribution with amplitude $\sigma_i(x) = (\pi G/2BL)\sin(\pi x/L)$, as shown in Figure 9.

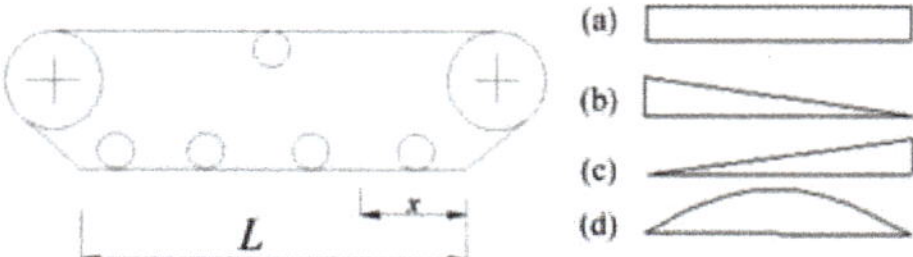

Figure 9. Four kinds of grounding pressure distribution under the crawler with length L [13].

Taking into account the time-dependent parameter $c(t)$ of shear strength and grounding pressure distribution $\sigma_i(x)$ to Equations (4) and (12), the sum of time-dependent traction force T_i can be obtained as follows:

(a) Uniform distribution, $\sigma_i(x) = G/BL$

$$T_{uniformed} = \sum_{i=1}^{n} T_i = \sum_{i=1}^{n} \frac{Bh\int_0^s \tau_i ds + BL\int_0^z \sigma_i dz}{s}$$
$$= \left\{ Bh\left[c(t) + \frac{G}{BL}\tan\varphi\right]\left(s + \kappa e^{-s/\kappa}\right) + \frac{Gz}{s} \right\} \frac{n^2 + n}{2} \tag{13}$$

(b) Linear decrement distribution, $\sigma_i(x) = 2Gx/BL^2$

$$T_{decrement} = \sum_{i=1}^{n} T_i = \sum_{i=1}^{n} \frac{Bh\int_0^s \tau_i ds + BL\int_0^z \sigma_i dz}{s}$$
$$= \left\{ Bh\left[c(t) + \frac{Gn^2}{BL}\tan\varphi\right]\left(s + \kappa e^{-s/\kappa}\right) + \frac{n^2 Gz}{s} \right\} \tag{14}$$

(c)   Linear increment distribution, $\sigma_i(x) = -2Gx/BL^2$

$$T_{increment} = \sum_{i=1}^{n} T_i = \sum_{i=1}^{n} \frac{Bh\int_0^s \tau_i ds + BL\int_0^z \sigma_i dz}{s}$$
$$= \left\{ Bh\left[c(t) - \frac{Gn^2}{BL}\tan\varphi\right]\left(s + \kappa e^{-s/\kappa}\right) - \frac{n^2 Gz}{s}\right\} \tag{15}$$

(d)   Sine distribution, $\sigma_i(x) = (\pi G/2BL)\sin(\pi x/L)$

$$T_{sine} = \sum_{i=1}^{n} T_i = \sum_{i=1}^{n} \frac{Bh\int_0^s \tau_i ds + BL\int_0^z \sigma_i dz}{s}$$
$$= \left\{ Bh\left[c(t) + \frac{\pi G}{2BL}\sin\frac{\pi n^2}{2}\tan\varphi\right]\left(s + \kappa e^{-s/\kappa}\right) + \frac{\pi n^2 LGz}{4s}\sin\frac{\pi n^2}{2}\right\} \tag{16}$$

4. Verification and Analysis of the Traction Force

4.1. Verification of the Traction Force Model

In order to verify the traction force model, the traction force experiment of a grouser is conducted on the apparatus designed by our research team as shown in Figure 10 [28]. It can be seen that simulant soil is prepared in the glass tank with a steel frame, and a grouser with width (4 cm) and height (4 cm) is fixed on a truck connecting with a tension sensor. Then, the constant compressive stress (5 kPa) is applied on the grouser (because there cannot be a distributed grounding pressure for only one grouser) with a constant velocity (v = 2 cm/s) from the motor controlled by a motor speed controller. When the truck with four pulleys moves along a rail, the traction force varying with time data from the tension sensor can be collected by the data collector and stored in the computer for analyses. Figure 11 shows a rut shaped by the grouser after the experiment [28].

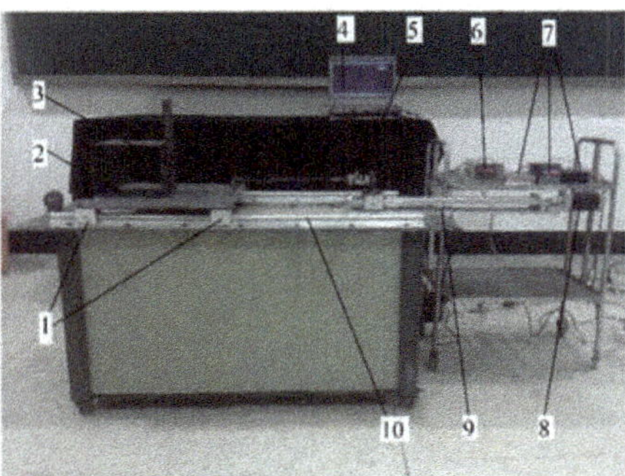

Figure 10. Traction force experiment apparatus of track shoe grouser. 1, Pulley; 2, Truck; 3, Grouser; 4, Computer; 5, NS-WL1 tension sensor; 6, Data collector; 7, Motor speed controller; 8, Motor; 9, Screw rod; 10, Rail.

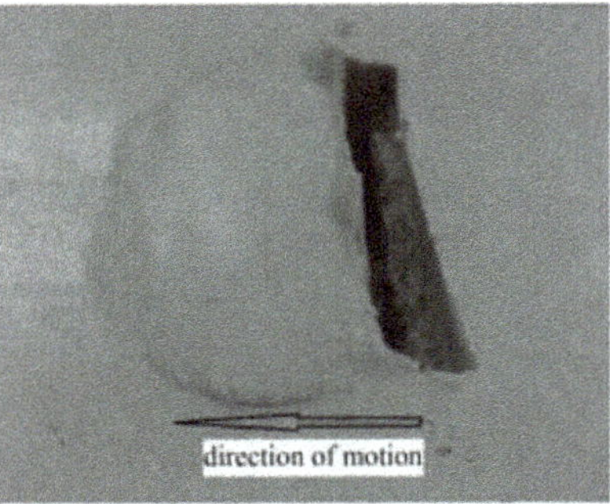

Figure 11. A rut shaped by the grouser (the grouser is removed).

Figure 12 illustrates the two curves of an experimental way [27] and theoretical way for comparison by adopting the parameters listed in Table 3 (not mentioned parameters above). Since the curve from the experimental way is from one track shoe grouser by interacting with the deep-sea simulant sediment, letting $n = 1$ (only one track shoe grouser) in the Equation (13) under the uniformed grounding pressure, the theoretical traction force curve under the same condition as the experimental way is obtained, and it can be seen that the two curves are close and have basically the same trend of change, which proves that the theoretical values of traction force are reliable. The two curves have the close peak values, and it is important for evaluating the traction force varying with time and proves the reliability of the traction force model.

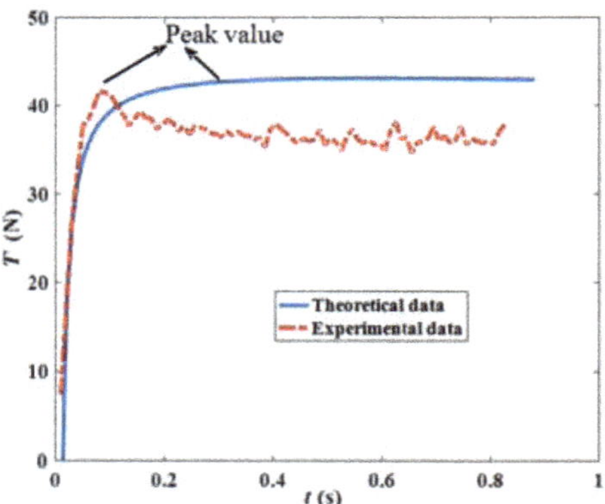

Figure 12. Comparison of the theoretical data and experimental data.

Table 3. Main size of crawler in time-dependent traction force model.

Length L (m)	Width B (m)	Weight G (kN)
6	1.7	110

4.2. Influence of Time and Track Shoe Number on the Traction Force under Different Grounding Pressure Distributions

4.2.1. Traction Force (T)- Time (t) Relationships

Figure 13 shows the relationships between traction force T and time t of the track shoes ($n = 10$). It can be seen that at the beginning of moving (i.e., $t = 0$ s), the instant traction forces are positive under uniformed distribution and linear decrement distribution conditions, but negative under linear increment distribution conditions. After the tracked miner moves (i.e., $t > 0$ s), the traction force changes greatly with different trends. It illustrates that the traction force is greatly influenced by the time. Moreover, during the period between 0 and 4 s, the traction forces decrease with time under uniformed distribution and linear decrement distribution conditions, but basically keep constant under sine distribution and increase with time under linear increment distribution. All the traction forces arrive at zero at about $t = 4$ s. It can be seen that when $t > 4$ s, the $T_{increment} > T_{sine} > T_{uniformed} > T_{decrement}$ during the same period. This is because the track shoes near the front of the crawler head are subject to the largest magnitude of the grounding pressure under the linear increment distribution condition, which leads to $c(t)$ getting the largest in a short time and helps increase the $T_{increment}$. For the sine distribution condition, the traction force under the different periodic grounding pressure is basically eliminated, but the sum of the traction force is not zero because of the effect of time on different track shoes with different locations. Under the uniformed and linear decrement distribution conditions, the $T_{uniformed}$ and $T_{decrement}$ are always negative between and 4 s and 10 s and signify that the $T_{uniformed}$ and $T_{decrement}$ becomes the resistance for the moving tracker miner, which is due to the smaller effect of temporal summation on the time-dependent parameter of the shear strength.

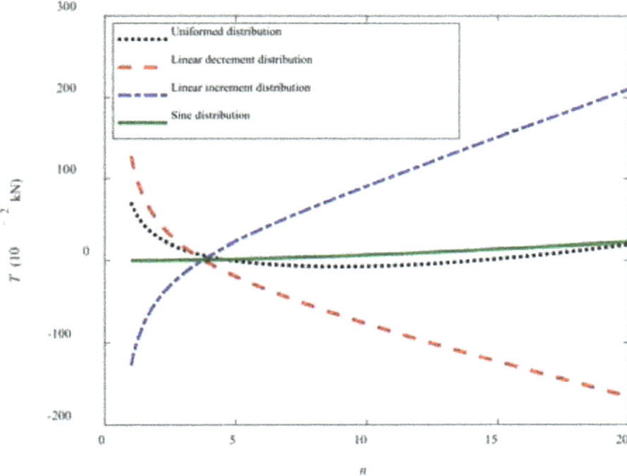

Figure 13. Curves of traction force vs. time.

Under the same grounding pressure distribution, the curve of the traction force under the linear increment and sine distribution conditions are increasing during the main analysis period and beneficial for the moving tracked miner. For other conditions, the corresponding curves become negative and resistance for the tracked miner.

4.2.2. Traction Force (T)- Number of Track Shoe (n) Relationships

Figure 14 illustrates the relationships between the traction force T and the number of track shoe n under the different grounding pressure distributions. It can be seen that within the 10 s, the magnitudes of T are increasing or decreasing obviously with n, except the uniformed condition. The cause is that the increase of n leads to more grounding pressure applied on the track shoes and develops more work or negative work to overcome the corresponding sinkage; for the uniformed increment condition, less contact time with sediment of larger grounding pressure weakens the magnitude and change of T and leads to slight traction force change.

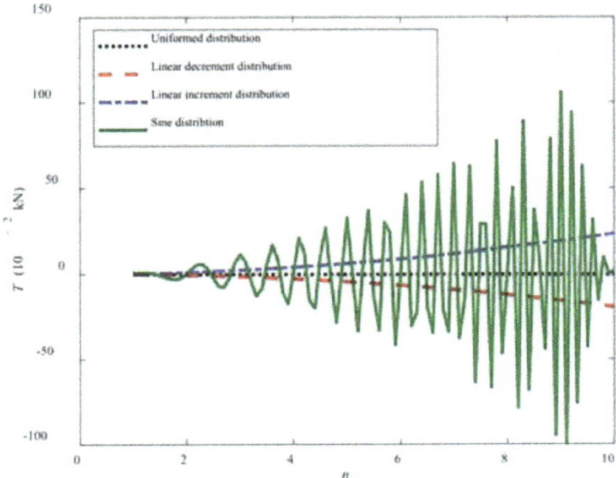

Figure 14. Curves of number of track shoe vs. traction force.

Moreover, under the same number of track shoe n and during the same period of time t, the absolute value relationships of T are $T_{sine} > T_{increment} > T_{decrement} > T_{uniform}$. For the sine condition, the curve of T is close to sine and gradually increases, but with an abrupt decrease as the number of track shoe becomes 10. This is because the increasing n in the traction force model may narrow the sine characteristics based on the fact that the maximum value of sine is one.

5. Conclusions

(1) A semi-empirical time-dependent shear strength (cohesion force) of describing the shear strength can be obtained by letting the compressive stress be zero in direct shear rheological constitutive model based on the analysis of Mohr–Coulomb shear strength theory and direct shear rheological experiment. Several time-dependent traction force models under the different grounding pressure distributions are deduced with the time-dependent cohesion force based on the work-energy principle. The models take the time, grounding pressure, and track shoe number into account and is used for conveniently analyzing the influence of kinds of key parameters on traction force of the deep-sea tracked miner.

(2) The traction force model is verified by a comparison between an experimental curve and a theoretical curve of a single-track shoe. By analyzing the influence of time and track shoe number on time-dependent traction force, it is found that $T_{increment} > T_{sine} > T_{uniformed} > T_{decrement}$ when $t > 4$ s under different grounding pressure distributions. The linear increment grounding pressure distribution is suggested for evaluating the most favorable traction force and the linear decrement grounding pressure distribution for calculating the worst traction force. Both grounding pressure distributions can better help the crawler design and optimization for better trafficability and stability of the deeps-sea tracked miner when adopting the time-dependent cohesion force.

(3) The traction force calculation is proved to be valid by the traction force experiment of a single-track shoe, and the influence of time, number of the track shoe, and grounding pressure distribution on the traction force can provide scientific basis for designing the crawler and evaluating the trafficability of tracked miner.

Author Contributions: Conceptualization, W.Y.; methodology, F.X.; software, W.Y.; validation, F.X.; formal analysis, F.X.; investigation, W.Y.; resources, F.X.; writing—original draft preparation, W.Y.; writing—review and editing, F.X.; visualization, W.Y.; supervision, F.X.; project administration, F.X.; funding acquisition, F.X. and W.Y. All authors have read and agreed to the published version of the manuscript.

Funding: This research was funded by the Natural Science Foundation of Hunan Province (2021JJ40448), and the Excellent Postdoctoral Innovative Talents Project of Hunan Province (2020RC2001).

Institutional Review Board Statement: Not applicable.

Informed Consent Statement: Not applicable.

Data Availability Statement: All data included in this study are available upon request by contact with the first author or corresponding author.

Conflicts of Interest: The authors declare no conflict of interest.

References

1. Jian, Q. Introduction on ocean mining technology of China. *Min. Technol.* **2001**, *1*, 9–11.
2. Chen, J.W.; Cao, C.; Ge, Y.Q.; Zhu, H.; Xu, C.; Sheng, Y.; Tian, L.; Zhang, H. Experimental research on data synchronous acquisition method of subsidence monitoring in submarine gas hydrate mining area. *Sensors* **2019**, *19*, 4319. [CrossRef] [PubMed]
3. Okamoto, N. Deep-Sea mineral potential in the south pacific region. *J. MMIJ* **2015**, *131*, 597–601. [CrossRef]
4. Liu, S.J.; Liu, C.; Dai, Y. Status and progress on researches and developments of deep ocean mining equipments. *J. Mech. Eng.* **2014**, *50*, 8–18. [CrossRef]

5. Filho, W.L.; Abubakar, I.R.; Nunes, C.; Platje, J.; Ozuyar, P.G.; Will, M.; Nagy, G.J.; Al-Amin, A.Q.; Hunt, J.D.; Li, C. Deep seabed mining: A note on some potentials and risks to the sustainable mineral extraction from the oceans. *J. Mar. Sci. Eng.* **2021**, *9*, 521. [CrossRef]

6. Deng, L.; Hu, Q.; Chen, J.; Kang, Y.; Liu, S. Particle distribution and motion in six-stage centrifugal pump by means of slurry experiment and CFD-DEM simulation. *J. Mar. Sci. Eng.* **2021**, *9*, 716. [CrossRef]

7. Xu, F.; Rao, Q.H.; Zhang, J.; Ma, W. Compression -shear coupling rheological constitutive model of the deep -sea sediment. *Mar. Georesour. Geotechnol.* **2018**, *36*, 288–296. [CrossRef]

8. Dai, Y.; Liu, H.; Zhang, T.; Liu, S.J. Research on traction trafficability of seafloor tracked mining vehicle. *China Sci. Pap.* **2015**, *10*, 1203–1208.

9. Ma, W.B.; Rao, Q.H.; Li, P.; Guo, S.C.; Feng, K. Shear creep parameters of simulative soil for deep-sea sediment. *J. Cent. South Univ.* **2014**, *21*, 4682–4689. [CrossRef]

10. Ma, W.B.; Rao, Q.H.; Feng, K.; Xu, F. Experimental study on tri-axial compressive creep model of simulative soil for deep-sea sediment. *J. Cent. South Univ.* **2014**, *12*, 4342–4347.

11. Evans, I. The sinkage of tracked vehicles on soft ground. *J. Terramech.* **1964**, *1*, 33–43. [CrossRef]

12. Mizukami, N.; Ishigami, G.; Yoshimitsu, T.; Kubota, T. Evaluation of the shear deformation model in the process of wheel sinking by the wheel experiment. *Trans. JSME* **2015**, *81*, 14-00514.

13. Wong, J.Y. *Theory of Ground Vehicles*, 3rd ed.; John Wiley & Sons, Inc.: New York, NY, USA, 2001.

14. Xu, F.; Rao, Q.H.; Ma, W.B. Track shoe structure optimization of deep-sea mining vehicle based on new rheological calculation formulae of sediment. *Mech. Based Des. Struct.* **2019**, *47*, 479–496. [CrossRef]

15. Zhao, J.F.; Wang, W.S.; Zhong, X.; Su, X. Improvement and verification of pressure-sinkage model in homogeneous soil. *Trans. Chin. Soc. Agric. Eng.* **2016**, *32*, 60–66.

16. Wu, H.Y.; Chen, X.M.; Liu, S.J. Influence of soft sediment adhered to track on adhesion performance of Seabed track vehicle. *Trans. Chin. Soc. Agric. Eng.* **2010**, *26*, 6–17.

17. Wang, M.; Wu, C.; Ge, T.; Gu, Z.M.; Sun, Y.H. Modeling, calibration and validation of tractive performance for seafloor tracked trencher. *J. Terramech.* **2016**, *66*, 13–25. [CrossRef]

18. Xu, F.; Rao, Q.H.; Ma, W.B. Turning traction force of tracked mining vehicle based on rheological property of deep-sea sediment. *Trans. Nonferr. Met. Soc. China* **2017**, *28*, 1233–1240. [CrossRef]

19. Li, J.W.; Yin, B.L.; Chen, B.Z. Study on reducing sediment adhesion of the tracks in deep-sea mining vehicle by submerged water jet. *Min. Res. Dev.* **2020**, *40*, 128–133.

20. Xu, F.; Rao, Q.H.; Ma, W.B. Predicting the sinkage of a moving tracked mining vehicle using a new rheological formulation for soft deep-sea sediment. *J. Oceanol. Limnol.* **2018**, *2*, 230–237. [CrossRef]

21. Shin, G.B.; Baek, S.H.; Park, K.H.; Chung, C.K. Investigation of the soil thrust interference effect for tracked unmanned ground vehicles (UGVs) using model track tests. *J. Terramech.* **2020**, *91*, 117–127. [CrossRef]

22. Baek, S.H.; Shin, G.B.; Lee, S.H.; Yoo, M.; Chung, C.-K. Evaluation of the slip sinkage and its effect on the compaction resistance of an off-road tracked vehicle. *Appl. Sci.* **2020**, *10*, 3175. [CrossRef]

23. Rubinstein, D.; Hitron, R. A detailed multi-body model for dynamic simulation of off-road tracked vehicles. *J. Terramech.* **2004**, *41*, 163–173. [CrossRef]

24. Yang, C.B.; Gu, L.; Li, Q. Finite element simulation of track shoe and ground adhesion. *Appl. Mech. Mater.* **2014**, *644–650*, 402–405. [CrossRef]

25. Li, J.; Bieerdebierke, W.Z.; Jin, S.C. Establishment of interval type track-terrain interaction model and experimental validation. *J. Acad. Armored Force Eng.* **2016**, *30*, 42–48.

26. Xu, F.; Rao, Q.H.; Liu, Z.L.; Ma, W.B. Traction force calculation method for mining vehicle based on rheological performance of deep-sea sediment and grounding pressure of crawler. *Chin. J. Nonferr. Met.* **2021**, *31*, 2817–2828.

27. Muro, T.; O'Brien, J. *Terramechanics: Land Locomotion Mechanics*, 1st ed.; Taylor & Francis: New York, NY, USA, 2004; p. 1.

28. Ma, W.B.; Rao, Q.H.; Xu, F. Experimental research on grouser traction of deep-sea mining machine. *Appl. Math. Mech.-Engl.* **2015**, *36*, 1243–1252. [CrossRef]

 *minerals*

Article

A Joint Experiment and Discussion for Strength Characteristics of Cemented Paste Backfill Considering Curing Conditions

Shunman Chen [1,2], Wei Wang [1,2,*], Rongfu Yan [3,*], Aixiang Wu [3], Yiming Wang [3] and Erol Yilmaz [4]

[1] State Key Laboratory of Mechanical Behavior and System Safety of Traffic Engineering Structures, Shijiazhuang Tiedao University, Shijiazhuang 050043, China; chenshunman@stdu.edu.cn

[2] School of Safety Engineering and Emergency Management, Shijiazhuang Tiedao University, Shijiazhuang 050043, China

[3] School of Civil and Resources Engineering, University of Science and Technology Beijing, Beijing 100083, China; wuaixiangustb@sina.cn (A.W.); wangyimingustb@sina.cn (Y.W.)

[4] Department of Civil Engineering, Geotechnical Division, Recep Tayyip Erdogan University, Fener, Rize TR53100, Turkey; erol.yilmaz@uqat.ca

* Correspondence: wangwei@stdu.edu.cn (W.W.); yan_rongfu@163.com (R.Y.)

Citation: Chen, S.; Wang, W.; Yan, R.; Wu, A.; Wang, Y.; Yilmaz, E. A Joint Experiment and Discussion for Strength Characteristics of Cemented Paste Backfill Considering Curing Conditions. *Minerals* **2022**, *12*, 211. https://doi.org/10.3390/min12020211

Academic Editors: Longjun Dong, Yanlin Zhao, Wenxue Chen and Carlito Tabelin

Received: 4 January 2022
Accepted: 2 February 2022
Published: 7 February 2022

Abstract: As lots of underground mines have been exploited in the past decades, many stope instability and surface subsidence problems are appeared in the underground mines, while the cemented paste backfill (CPB) technology has been applied for more than 40 years, and it can solve these problems. As it is shown that the effect of backfilling is mainly affected by the mechanical properties of the CPB, and there are lots of factors which can influence the strength of the CPB, but the coupled effects of curing conditions has not been reported. In this research, the coupled effects of curing conditions are importantly considered, and the uniaxial compressive strength (UCS) is adopted as the important evaluation index of CPB, then the evolution law of the UCS for CPB are analyzed, also the mathematical strength model of CPB is established. The findings suggest that the relationship between the UCS of CPB and curing stress develops the function of quadratic polynomial with one variable, while the UCS of the CPB shows the power function as the curing temperature increases. Moreover, the established mathematical strength model is verified on the basis of laboratory experiments, the error between the measured UCS and the prediction UCS is less than 15%. It shows that the established strength model of the CPB by considering the curing conditions can predict the UCS very well, it has great significance for the safety design of CPB.

Keywords: cemented paste backfill; curing conditions; mechanical properties; mathematical strength model

1. Introduction

As more and more mines have been excavated, many voids are produced in the underground mines, also a lot of tailings are disposed in the tailing dam, therefore, there are huge challenges we are facing, such as the surface subsidence, stope instability and safety of the tailing dams [1,2]. For the voids, they may cause the surface subsidence, stope collapse and other geological risks [3]. According to the statistics by the researchers, there are about 14.6 billion tons of tailings in China, most of them are disposed in the tailings or on the surface, it may cause the environment pollution or dam-failure accidents [4,5]. As the society has been rapidly developed, the environment protection is getting more and more attention, while the filling mining methods can dispose the surface tailings, which has been widely applied in many mines at home and abroad [6]. On the one hand, the CPB technology has the advantages of safety, economy and high efficiency; on the other hand, it also has the advantages of environmental protection [7]. Moreover, the total tailings are used in the CPB technology, so it can reduce the surface tailings deposit for 50~60% [8].

For the CPB technology, the safety of the backfilling stope is the main areas of concern, while the mechanical property is one of the main indexes to reflect the backfilling effect, also the study shows that the UCS is the main index to show the mechanical properties of CPB [9,10]. In terms of the UCS for CPB, many studies had been done by the scholars, Andrew and Fall's [11,12] study indicated that the UCS was significantly influenced by the curing temperature, because the hydration reaction rate of the cement was firstly affected. The correlation of the electrical properties, microstructure properties and mechanical properties of the CPB was investigated. Secondly, the laboratory tests aere done considering the curing time, sulphur contents and mineral admixtures by Jiang and Liu [13–15]. Some studies had revealed that the UCS was significantly influenced by the water type, water content and mixing time, also the microstructure of the CPB was affected by the mixing time [16,17]. Libos, Wang and Yilmaz [18–20] had studied the relationship between the cement types, cement content and mechanical properties of CPB, indicating that the hydration reaction rate and UCS of the CPB were mostly affect by them. Moreover, it was found that the specimen size, curing stress and tailings fineness can also influence the mechanical properties of the CPB [21–24]. In a short word, the majority of the above studies are concentrated on one influencing factor, while there are few studies about the coupled effects of curing conditions on the mechanical properties of CPB.

However, most of the studies were based on the laboratory tests, the costs of the laboratory experiments are high, many researchers had established the UCS prediction model to forecast the UCS of CPB when designing the mix proportion of CPB [25,26]. Ehsan et al. used the particle swarm optimization algorithm to optimize the multi-objective mixture design of CPB. Li and Zhao et al. [27–29]. optimized the admixture of CPB on the basis of the response -surface method, and it can provide technical references for the engineering design in the mines, while the fitting formula was complex and the cross term had no specific physical meaning. Moreover. the BP neural network and intelligent modelling framework were adopted to show the UCS with different conditions, but the model was based on the laboratory experiment, and the data should be large, it was difficult to applied in the engineering design [30–32]. Based on the laboratory experiments, Mitchell and Fu et al. [33–35] established the mathematical model to forecast the UCS, also the model was applied to the engineering design. According to the above studies, there were many prediction models which can predict the strength of CPB, but the prediction model can be established better [36–38], the influencing factors were incomplete, the studies showed that the UCS is affected by curing conditions, therefore, it is needed to establish the prediction model considering the curing conditions.

In this paper, the strength characteristics of CPB considering curing conditions is studied, and the UCS of the CPB is tested in the laboratory, then evolution role of the UCS for CPB considering curing conditions is analyzed, based on the results, the mathematical strength model of CPB considering different curing conditions is established, also the established mathematical strength model is verified based on the experiment results.

2. Experimental Materials and Methods

2.1. Materials

2.1.1. Tailings

As the tailings is one of the main components of CPB, therefore, the total tailings can be disposed, and it is selected from one copper mine. Firstly, the laser particle scanning analyzeris used to analyze the particle size distribution (PSD) of the total tailings, then the X-ray diffractometry (XRD)is adopted to obtain the chemical component of the tailings [39,40], therefore, the physical and chemical properties of the tailings are obtained, and the chemical characteristics of the tailings is shown in Table 1, it indicates that the chemical components of the total tailings mainly include MgO, Al_2O_3, SiO_2 and CaO.

Table 1. Chemical characteristics of the tailings.

Parameter	Al_2O_3 (%)	CaO (%)	FeO_3 (%)	MgO (%)	P_2O_5 (%)	SiO_2 (%)	TiO_2 (%)	Na_2O (%)	K_2O (%)
Tailings	1.04	44.05	2.89	5.02	0.18	43.55	0.35	0.10	2.82
OPC	5.37	66.18	3.36	3.01	0.11	21.12	0.20	0.2	0.45

As it is known that the compactness of the CPB is affected by the gradation of the total tailings [41]. Based on the experimental results, the PSD of the CPB is shown in Figure 1. The statistics shows that the diameter of cumulative volume with 20 μm, 37 μm and 74 μm are 29.60%, 47.33% and 77.52%, respectively. Based on the theoretical calculation formula, then the non-uniformity coefficient and curvature coefficient can be obtained, which are 8.05 and 1.31, respectively. In contrast to the standard values, the distribution of the total tailing is good. According to the tailings which have been studied before, the tailings studied in the copper mine have similar properties to the tailings which have been reported in the literature by other scholars.

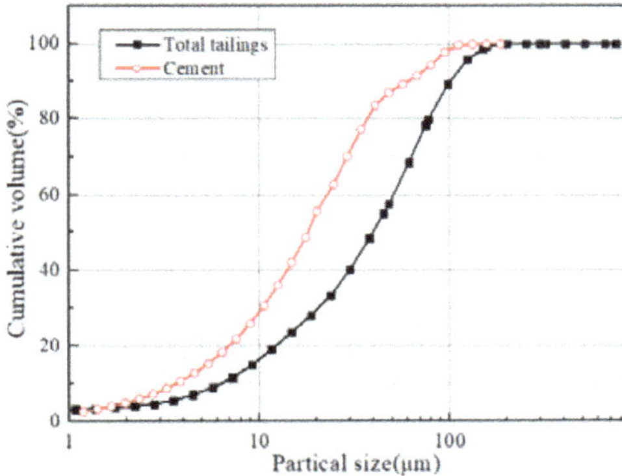

Figure 1. The curves of the cement and tailings between cumulative volume and particle size.

2.1.2. Binders

As the binder is also one of the important components of CPB, the Portland cement, fly ash, fiber and waste glass are included [42,43]. The type of Ordinary Portland cement (OPC) is adopted in this study, based on the tests by the laser particle scanning analyzer, the PSD of the cement is also obtained, which is shown in Figure 1, in contrast to the PSD of the tailings, the cement is finer. Moreover, the chemical properties of the cement is obtained based on the X-ray diffraction (XRD) tests, which is shown in Table 1, it also indicates that the SiO_2, CaO and other elements are included.

2.1.3. Water

As the CPB consists of cement, tailing and water, therefore, the water is the essential part of the CPB, for it can provide the hydration reaction medium for the CPB. The water is selected from the laboratory tap water, because it can reduce the experimental error. The acid and alkali stability experiment results show that the tap water in the laboratory is weakly alkaline, and the PH value of it is 7.5. Moreover, the Fe, Na, Ca, Si and SO_4^{2-} are included for the chemical properties of the water.

2.2. Preparation and Mix Proportion of CPB

In this study, the mixer is adopted to make the cement, tailing and water at the same time, and the curing conditions is mainly considered, therefore, the cement-tailing ratio is controlled to be 1:6, also the mass concentration of it is controlled to be 76%. As for the curing conditions, all of the specimens are controlled to be 20 °C for the initial temperature, and the relative humidity is 90 ± 2% RH, therefore, the curing temperature and curing stress are mainly considered for the curing conditions.

According to the studies before, the temperature of 20 °C, 35 °C and 50 °C are selected as the curing temperature schemes. The stress condition of the CPB under different stope heights is shown in Figure 2, as there are seven points ranging from the top to the bottom, the height of them is 0 m, 5 m, 10 m, 15 m, 20 m, 25 m and 30 m between each point to the bottom of the stope for point A, B, C, D, E, F and G. Moreover, the density of the CPB is selected as 2700 kg/m^3. Based on the CPB density and backfilling technology, the curing stress of 0 kPa, 90 kPa, 180 kPa, 270 kPa, 360 kPa, 450 kPa and 540 kPa are selected as the curing stress conditions. Firstly, the curing stress is applied to be a quarter of the total required curing stress as the curing time is 12 h; Secondly, the curing stress is applied to be a half of the total required curing stress as the curing time is 1 day; Finally, the curing stress is applied to be the total required curing stress as the curing time is 2 days. Previous studies showed that the curing stress had the appreciable impact on the consolidation of CPB in the first 48 h [44,45], therefore, the stress is applied within 2 days for each schemes. Moreover, as the specimens have been cured for 3 days, the demoulding is required for all the specimens, then all the specimens are in the standard curing conditions, the schemes of the curing stress application is shown in Figure 2.

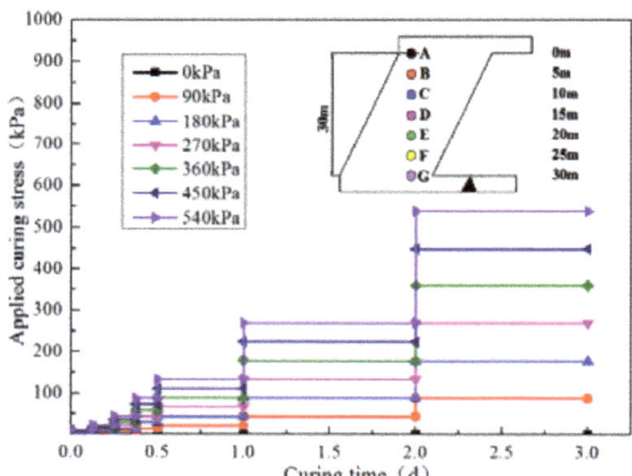

Figure 2. Schemes of the curing stress application.

2.3. UCS Tests

As it is known that the UCS is one of the main influencing indexes to reflect the mechanical properties of CPB, therefore, after all the specimens are cured for the required curing times, then all the specimens are tested to obtain the UCS of the CPB. Before the tests, it is needed to dispose all the specimens. the top and bottom of the specimen should be the plain surface to be parallel, also the standard size of the specimen is 50 mm and 100 mm for its diameter and height. In this study, the WDW-50 servo control uniaxial pressure testing machine is adopted to obtain the UCS of CPB, the control accuracy of it is ±0.2%.

3. Results and Discussion

3.1. Influence of Curing Stress on UCS

The experiment results are shown in the Figure 3, it can reflect the correlation between the curing stress and UCS of CPB, it has made clear that the UCS is significantly increased with the increase of the curing stress while other curing conditions are unchanged. The reason for the above phenomenon can be explained as follow: as the curing stress increases, resulting the faster hydration reaction rate, then more hydration products are produced, also pores between the tailing and cement are compressed by the CPB particles, then the space between the particles are decreased, and the pores are filling into the hydration products, therefore, the UCS of CPB is improved.

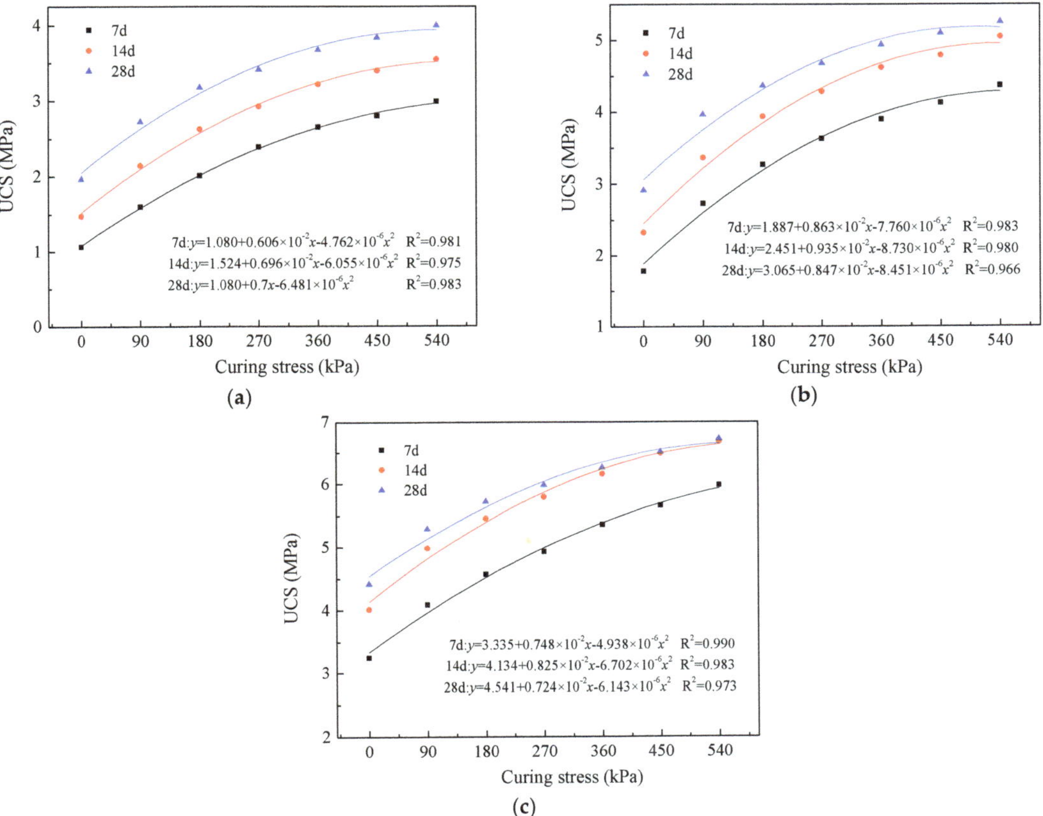

Figure 3. Relationship between the UCS and curing stress with different curing temperatures: (**a**) 20 °C; (**b**) 35 °C; (**c**) 50 °C.

In another aspect, the Table 2 shows the correlation between the curing stress and growth rate of UCS, it can reflect the evolution of the UCS growth rate under different curing conditions, the UCS growth rate is significantly decreased when the curing stress is increased. From Figure 3 and Table 2, it also shows that the UCS is increased when the curing time is increased regardless of other curing conditions, but the growth rate of UCS is trending downward. The reason for the above phenomenon is that the cementitious behavior of CPB is improved as the curing time is longer. In a word, the curing time can improve the UCS of CPB, but the growth property of it is certain.

Table 2. The distribution of the UCS growth rate for CPB.

Curing Time (d)	Curing Temperature (°C)	UCS Growth Rate (%)						
		0 kPa	90 kPa	180 kPa	270 kPa	360 kPa	450 kPa	540 kPa
7	20	0	49.53	25.63	18.91	10.88	5.66	6.79
	35	0	53.37	19.78	11.01	7.44	5.90	5.81
	50	0	25.85	11.74	7.88	8.72	5.60	5.65
14	20	0	45.27	22.33	11.41	9.90	5.59	4.41
	35	0	44.64	16.91	8.88	7.69	3.68	5.43
	50	0	24.13	9.42	6.23	6.21	5.36	2.93
28	20	0	38.58	16.48	7.55	7.60	4.35	4.17
	35	0	35.96	10.08	7.09	5.56	3.24	3.14
	50	0	19.91	8.11	4.54	4.51	3.99	3.38

By studying the correlation between the curing stress and UCS quantitatively, the linear, power, exponential, logarithm, and quadratic function of one variable are selected to fit them, the most appropriate function and its correlation coefficient (R^2) are shown in Figure 3, the p value of all the fitting equations is less than 0.05 [46], therefore, the function of quadratic polynomial with one variable can reflect the correlation between UCS and curing stress, which can be expressed by the Formula (1):

$$\sigma_c = a + bp + cp^2 \tag{1}$$

where σ_c is the UCS, p is the curing stress, a, b, and c are the fitting of constant.

For the Formula (1), as the correlation coefficient, all of the correlation coefficient are more than 0.95, which shows that the correlationship among the curing stress and UCS are high. Therefore, the function of quadratic polynomial with one variable is suitable to express the relationship between them.

3.2. Influence of Curing Temperature on UCS

The correlation between the curing temperature and UCS is clearly shown in Figure 4, the UCS of CPB is increased with the increase of curing temperature regardless of other curing conditions. Moreover, it indicates that curing temperature has greater effect on the UCS for the smaller curing stress than the larger curing stress. This phenomenon is caused by the following reasons: when the curing temperature becomes higher, more hydration products are produced, then most of them are filled into the pores between the tailing particles. Therefore, the pore spacing is decreased, and the compactness of the CPB is more dense, which resulting in the increase of UCS.

The Table 3 shows the evolution rule of the UCS growth rate for the CPB, as the curing stress is 0 kPa, the growth rate of UCS are 66.36% and 82.58% for the curing temperature of 35 °C and 50 °C under the curing time of 7 days; while the growth rate of UCS are 46.15% and 36.84% for the same curing temperature under the curing time of 28 days. Therefore, it shows the decline trend for the growth rate of UCS as the curing temperature is increased, and it also shows similar rule for other conditions. Moreover, the results indicate that the growth rate of UCS decreases with the longer curing time regardless of other curing conditions, and the curing temperature shows greater influence on the early UCS than the later UCS.

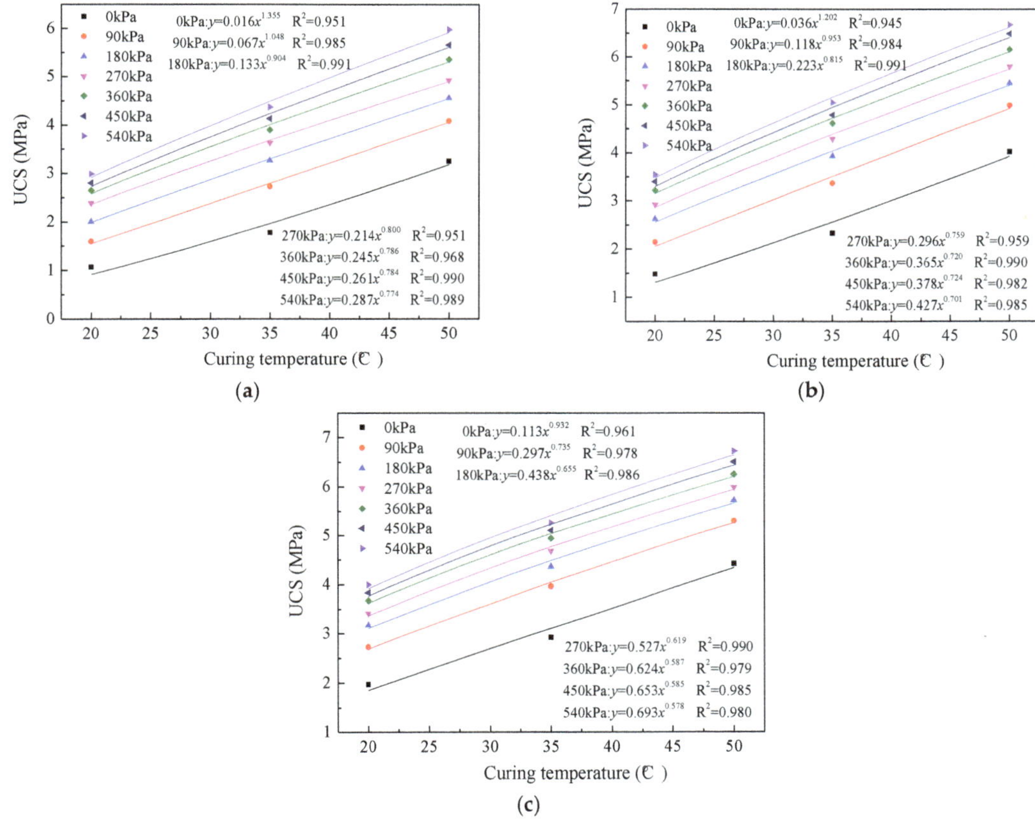

Figure 4. Relationship between UCS and curing temperature with different curing stress: (**a**) 7 d; (**b**) 14 d; (**c**) 28 d.

Table 3. The evolution rule of the UCS growth rate for the CPB.

Curing Time (d)	Curing Temperature (°C)	UCS Growth Rate (%)						
		0 kPa	90 kPa	180 kPa	270 kPa	360 kPa	450 kPa	540 kPa
7	20	0.00	0.00	0.00	0.00	0.00	0.00	0.00
	35	66.36	70.63	62.69	51.88	47.17	47.50	46.15
	50	82.58	49.82	39.76	35.81	37.44	37.05	36.84
14	20	0.00	0.00	0.00	0.00	0.00	0.00	0.00
	35	57.43	56.74	49.81	46.42	43.48	40.88	42.25
	50	72.53	48.07	38.58	35.20	33.33	35.49	32.28
28	20	0.00	0.00	0.00	0.00	0.00	0.00	0.00
	35	48.22	45.42	37.42	36.84	34.24	32.81	31.50
	50	51.37	33.50	31.12	27.99	26.72	27.65	27.95

By studying the evolution law of the correlation between the UCS and curing temperature quantitatively, the linear, power, exponential, logarithm, and quadratic function are adopted to fitting the relation between them. It can be seen that the power function showed the highest accuracy, that is to say the power function is fitted to show the correlation between the UCS and curing temperature, all of the correlation coefficient (R^2) are larger than 0.95, the p value of all the fitting equations is less than 0.05, therefore, the formula for the relationship between them can be shown bellow:

$$\sigma_c = dT^e \tag{2}$$

where σ_c is the UCS, T is the curing temperature, d and e are the fitting coefficient.

3.3. Coupled Effects of Curing Conditions on UCS

In order to find out the correlation between the coupled effects of curing conditions and UCS of CPB, based on the laboratory results, the correlation between them is given in Figure 5, it indicates that the UCS is significantly influenced by the coupled effects of curing conditions. As the curing stress decreases from 540 kPa to 360 kPa, also the curing temperature is increased from 20 °C to 50 °C, the UCS is increased sharply. While the curing temperature decreases from 50 °C to 20 °C, and the curing stress decreases from 360 kPa to 180 kPa, the UCS is decreased sharply. Therefore, it indicates that there may exist a complicated relation between them. Not only that, the trend of the three curves under different curing ages for the UCS are similar, indicating that there should be a regular rule for the UCS. According to the results, the formula considering the curing condition can be obtained.

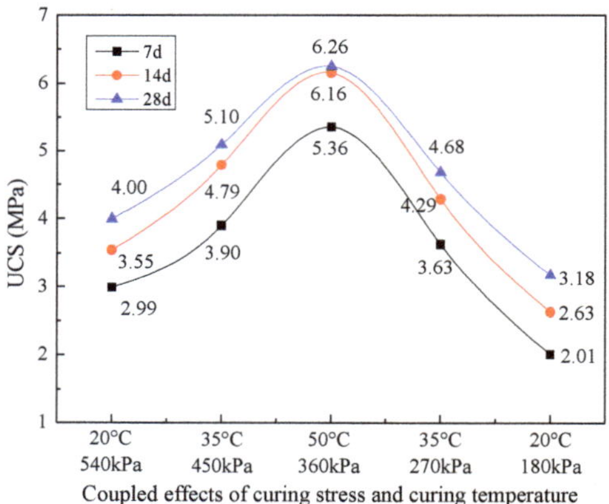

Figure 5. Correlation between the UCS and coupled effects of curing conditions.

4. Mathematical Strength Model

4.1. Establish of the Mathematical Strength Model

According to the results of the previous analysis, the curves for the coupled effects of curing conditions on the UCS develops a similar trend. Based on the above results, the mathematical strength model of CPB is established when considering different curing conditions, therefore the formula of it is as follows:

$$\sigma_c = f_c f_1(p) f_2(T) \tag{3}$$

$$\sigma_c = f_c \left(a + bp + cp^2 \right) dT^e \tag{4}$$

where σ_c is the UCS, f_c represents the UCS of CPB relation to the cement, $f_1(p)$ and $f_2(T)$ represent the UCS relation to the curing stress and curing temperature.

Assume that $\eta = adf_c$, $\xi = bdf_c$, $\lambda = cdf_c$, then the Formula (4) can be transformed to the Formula (5):

$$\sigma_c = (\eta + \xi p + \lambda p^2) T^e \tag{5}$$

where the η, ξ and λ are the fitting coefficients for the strength model.

Assume that the curing temperature of T is constant, and $a' = \eta T^e$, $b' = \xi T^e$, $c' = \lambda T^e$, then the Formula (5) can be simplified as the Formula (6):

$$\sigma_c = a' + b'p + c'p^2 \tag{6}$$

According to the relationship between the curing stress and UCS, the Formula (6) is exactly the same to the Formula (1).

Assume that the curing stress of p is constant, and the $d' = \eta + \xi p + \lambda p^2$, then the correlation between the UCS and curing temperature is shown below:

$$\sigma_c = d' T^e \tag{7}$$

According to the Formula (2), it indicates that the Formula (7) is exactly the same to it.

Based on the above analysis, it is evident that the mathematical strength model of CPB can be obtained by the Formula (5), and it contains the influencing factors of curing conditions, the parameters relation to the formula are clear. In the fact, in order to apply the established mathematical strength model, the parameters of η, ξ and λ can be determined by the laboratory results.

In order to verify the mathematical strength model considering the curing conditions, the experimental results of Figure 3 are adopted to analyze it. The parameters of the strength model relation to the UCS are obtained by the R programming language [47], also the fitting coefficients are obtained as shown in Table 4. It is evident that all of the fitting coefficient R^2 of the mathematical strength model under different curing ages are more than 0.95, also the p value of all the fitting equations is less than 0.05, indicating that the significance of the established mathematical strength model is good, it can verify the mathematical strength model.

Table 4. Parameters of the mathematical strength model.

Curing Time (d)	Parameters				R^2
	η	ξ	λ	e	
7	0.108	3.446×10^{-4}	-2.663×10^{-7}	0.847	0.977
14	0.175	4.854×10^{-4}	-7.603×10^{-7}	0.782	0.972
28	0.345	7.009×10^{-4}	-7.009×10^{-7}	0.637	0.970

Notice: The best fit parameters above can only be applied in the conditions of this experiment: (1) Curing stress of 0~540 kPa, (2) Curing temperature of 20~50 °C.

Based on the established mathematical strength model for CPB, the contours of CPB under different curing ages are shown in Figure 6, and they can reflect the correlation between the influencing factors and UCS. In the engineering design, if the curing conditions are given, then the UCS of CPB can be predicted based on the data in Figure 6. Moreover, if the design requirement is certain, in the actual engineering design, when the UCS is at a certain range with the curing time of 28 days, the contour plots of UCS considering the curing conditions can be applied to optimize the mix proportion.

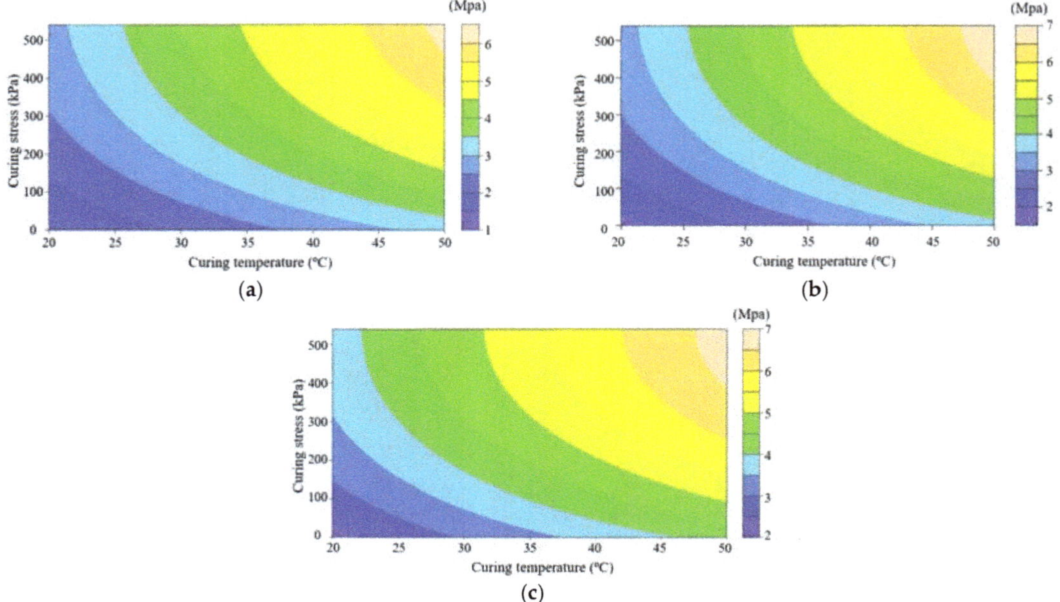

Figure 6. The theoretical model of the UCS contours for CPB: (**a**) 7 d; (**b**) 14 d; (**c**) 28 d.

4.2. Comparison of Experimental Results and Theoretical Model for UCS

As the mathematical strength model of CPB has been established above, the relationship between the model theoretical UCS and the measured UCS in the laboratory is shown in Figure 7, also the ideal fitting line and the prediction error within the 15% are listed in the Figure 7, and it indicates that the established mathematical strength model can predict the UCS of CPB very well. As shown in Figure 7a, most of the prediction error are less than 15%, only two specimen points are beyond the prediction error of 15%, moreover, all specimen points are in the range of prediction bands for 95%, indicating that the established mathematical strength model shows an important impact for predicting the UCS at early ages. The correlation between the model theoretical UCS and measured UCS under the curing time of 14 and 28 days are shown in Figure 7b,c, there are two specimen points which beyond the prediction error of 15% under the curing time of 14 and 28 days, but these two specimen points are near the ±15% prediction error line, respectively. Moreover, all of the specimen points are in the range of prediction bands for 95% under the curing time of 14 and 28 days, indicating that the prediction results are in consistent with the experiment results, and the established mathematical strength model is rational.

According to the above analysis, also the histogram of $UCS_{measured}/UCS_{predicted}$ based on the mathematical strength model are shown in Figure 8, it shows that the means and the median are all close to 1, which indicates that the predicted UCS of CPB is consistent with the experiment UCS. On the other hand, the skewness of the dataset under different curing ages are 0.0545, 0.0513 and 0.0958, respectively, it means that the frequency distribution of the $UCS_{measured}/UCS_{predicted}$ is positive partial, so the established mathematical strength model shows a slightly lower UCS than the experimental UCS under different curing ages, but this is good for the safety of the CPB when designing the mixture ratio of CPB.

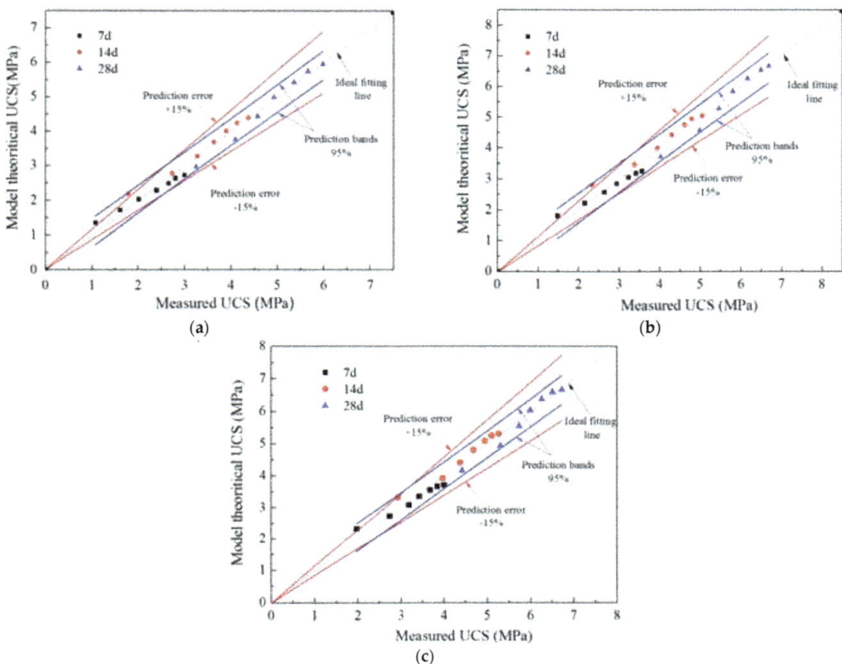

Figure 7. Relationship between the model theoretical UCS and measured UCS: (**a**) 7 d; (**b**) 14 d; (**c**) 28 d.

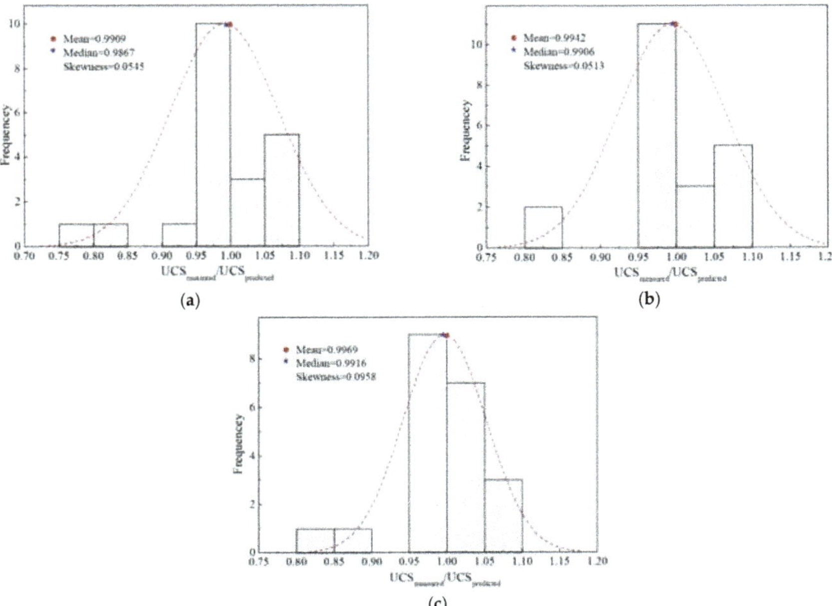

Figure 8. Histogram of $UCS_{measured}/UCS_{predicted}$ based on the mathematical strength model at: (**a**) 7 d; (**b**) 14 d; (**c**) 28 d.

4.3. Verification of the Mathematical Strength Model

The predicted UCS and experimental UCS have been analyzed above, in order to verify the rationality of the established mathematical strength model, the curing temperature is 35 °C, the curing temperature is 0 kPa, 90 kPa, 180 kPa, 270 kPa, 360 kPa, 450 kPa and 540 kPa, respectively, while other conditions and parameters are the same as the schemes before. Based on the studies above, the predicted UCS of CPB is computed by the established mathematical strength model which is shown in the Formula (5), then the relationship between the predicted UCS and the measured UCS is obtained, and it is shown in Figure 9. It can be observed that all the specimen points are in the range of the 15% prediction error, also all of the prediction UCS is slightly smaller than the measured UCS, indicating that the application of the established mathematical strength model to predict the UCS of CPB is good.

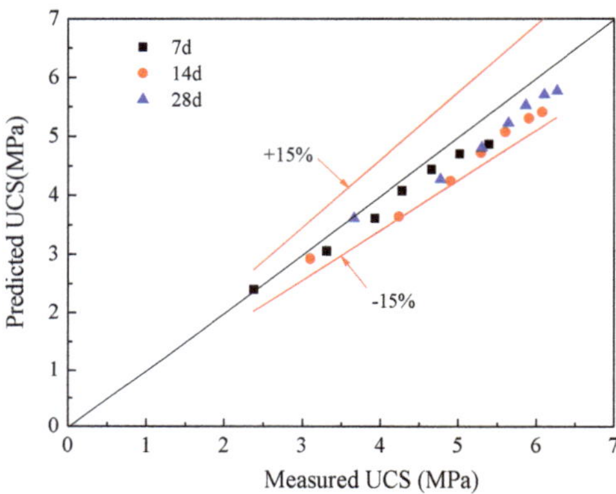

Figure 9. Correlation between the predicted UCS and measured UCS.

5. Conclusions

As the mechanical properties of the CPB considering curing conditions has been discussed, the strength testes are done in the laboratory, also the mathematical strength model is also established, and then the model is verified by the engineering example. According to the studies, some opinions can be concluded as follow:

(1) The results show that the function of quadratic polynomial with one variable can represent the correlation between the UCS and the curing stress well. Moreover, the correlation between the UCS and curing temperature shows the power function.

(2) It is found that the trend of the three curves under different curing ages for the UCS are extremely similar, indicating that there exists a regular rule for the UCS with different curing conditions. The established mathematical strength model for CPB considering the curing conditions is developed.

(3) The established mathematical strength model is verified by the engineering data, the prediction error between the measured UCS and the prediction UCS is less than 15%, all of the specimen points are in the range of prediction bands for 95%, indicating that the prediction results are in agreement with the experiment results under different curing ages, therefore, the established mathematical strength model is rational, which can provide some technical support for the mix proportion design of CPB.

Author Contributions: Conceptualization, S.C. and A.W.; methodology, S.C.; validation, R.Y.; formal analysis, S.C.; investigation, S.C.; resources, S.C.; data curation, E.Y.; writing—original draft preparation, S.C.; writing—review and editing, W.W.; visualization, Y.W.; supervision, Y.W. and S.C.; project administration, W.W.; funding acquisition, W.W. All authors have read and agreed to the published version of the manuscript.

Funding: This work is financially supported by the Hebei Natural Science Foundation (E2021210128) and the Joint Funds of the National Natural Science Foundation of China (U1967208).

Acknowledgments: We thank for the Chambishi copper mine for the help with the tailing sample and data collection.

Conflicts of Interest: The authors declare no conflict of interest. The authors declare that they have no known competing financial interests or personal relationships that could appear to have influenced the work reported in this paper.

References

1. Qi, C.; Fourie, A. Cemented paste backfill for mineral tailings management: Review and future perspectives. *Miner. Eng.* **2019**, *144*, 106025. [CrossRef]
2. Sun, Q.; Li, X.L.; Wei, X.; Mu, Q.W. Experimental study on the influence of mine water corrosion over filling paste strength. *Bull. Chin. Ceram. Soc.* **2015**, *34*, 1246–1251.
3. Wang, Q.; Shi, M.; Wang, D. Influence of elevated curing temperature on the properties of cement paste and concrete at the same hydration degree. *J. Wuhan Uni. Technol. Mater. Sci. Ed.* **2017**, *32*, 1344–1351. [CrossRef]
4. Wu, D.; Zhang, Y.; Liu, Y. Mechanical performance and ultrasonic properties of cemented gangue backfill with admixture of fly ash. *Ultrason* **2016**, *64*, 89–96. [CrossRef]
5. Hasan, E.; Atac, B. Influence of silica fume on mechanical property of cemented paste backfill. *Constr. Build. Mater.* **2022**, *317*, 1–9.
6. Lu, G.D.; Fall, M.; Cui, L. A multiphysics-viscoplastic cap model for simulating blast response of cemented tailings backfill. *J. Rock Mech. Geotech. Eng.* **2017**, *9*, 551–564. [CrossRef]
7. Kashani, A.; Nicolas, S.R.; Qiao, G.G.; Deventer, J.S.J.; Provis, J.L. Modelling the yield stress of ternary cement-slag-fly ash pastes based on particle size distribution. *Powder Technol.* **2014**, *266*, 203–209. [CrossRef]
8. Dong, Q.; Liang, B.; Jia, L.F.; Jiang, L.G. Effect of sulfide on the long-term strength of lead-zinc tailings cemented paste backfill. *Constr. Build. Mater.* **2019**, *200*, 436–446. [CrossRef]
9. Xiu, Z.G.; Wang, S.H.; Ji, Y.C.; Wang, F.L.; Ren, F.Y.; Wang, P.Y. An analytical model for the triaxial compressive Stress-strain relationships of Cemented Pasted Backfill (CPB) with different curing time. *Constr. Build. Mater.* **2021**, *313*, 1087–1102. [CrossRef]
10. Yilmaz, E.; Belem, T.; Benzaazoua, M.; Kesimal, A.; Ercikdi, B. Evaluation of the strength properties of deslimed tailings paste backfill. *Miner. Resour. Eng.* **2013**, *12*, 129–144.
11. Bull, A.J.; Fall, M. Curing temperature dependency of the release of arsenic from cemented paste backfill made with Portland cement. *J. Environ. Manag.* **2020**, *269*, 110772. [CrossRef] [PubMed]
12. Fall, M.; Samb, S.S. Influence of curing temperature on strength, deformation behaviour and pore structure of cemented paste backfill at early ages. *Constr. Build. Mater.* **2006**, *305*, 125–135. [CrossRef]
13. Jiang, H.Q.; Fall, M.; Yilmaz, E.; Li, Y.H.; Yang, L. Effect of mineral admixtures on flow properties of fresh cemented paste backfill: Assessment of time dependency and thixotropy. *Powder Technol.* **2020**, *372*, 258–266. [CrossRef]
14. Liu, L.; Xin, J.; Huan, C.; Qi, C.; Song, K.L. Pore and strength characteristics of cemented paste backfill using sulphide tailings: Effect of sulphur content. *Constr. Build. Mater.* **2020**, *237*, 117452. [CrossRef]
15. Tian, X.C.; Fall, M. Non-isothermal evolution of mechanical properties, pore structure and self-desiccation of cemented paste backfill. *Constr. Build. Mater.* **2021**, *297*, 123657. [CrossRef]
16. Yang, L.H.; Wang, H.J.; Wu, A.X.; Li, H.; Bruno, T.A.; Zhou, X.; Wang, X.T. Effect of mixing time on hydration kinetics and mechanical property of cemented paste backfill. *Constr. Build. Mater.* **2020**, *247*, 118516. [CrossRef]
17. Zhao, Y.; Taheri, A.; Karakus, M.; Chen, Z.W.; Deng, A. Effects of water content, water type and temperature on the rheological behaviour of slag-cement and fly ash-cement paste backfill. *Int. J. Min. Sci. Technol.* **2020**, *30*, 271–278. [CrossRef]
18. Libos, I.L.S.; Cui, L. Effects of curing time, cement content, and saturation state on mode-I fracture toughness of cemented paste backfill. *Eng. Fract. Mech.* **2020**, *235*, 107174. [CrossRef]
19. Wang, J.; Zhang, C.; Fu, J.X.; Song, W.D.; Zhang, Y.F. Effect of water saturation on mechanical characteristics and damage behavior of cemented paste backfill. *J. Mater. Res. Technol.* **2021**, *15*, 6624–6639. [CrossRef]
20. Yilmaz, E.; Belem, T.; Bussière, B.; Mbonimpa, M.; Benzaazoua, M. Curing time effect on consolidation behaviour of cemented paste backfill containing different cement types and contents. *Constr. Build. Mater.* **2015**, *75*, 99–111. [CrossRef]
21. Zhou, Y.B.; Fall, M.; Haruna, S. Flow ability of cemented paste backfill with chloride-free antifreeze additives in sub-zero environments. *Cem. Concr. Compos.* **2022**, *126*, 104359. [CrossRef]
22. Ghirian, A. Coupled Thermo-Hydro-Mechanical-Chemical (THMC) Processes in Cemented Tailings Backfill Structures and Implications for Their Engineering Design. Ph.D. Thesis, University of Ottawa, Ottawa, ON, Canada, 2016.
23. Ouyang, S.Y.; Huang, Y.L.; Zhou, N. Experiment on hydration exothermic characteristics and hydration mechanism of sand-based cemented paste backfill materials. *Constr. Build. Mater.* **2022**, *318*, 125870. [CrossRef]
24. Fall, M.; Benzaazoua, M.; Ouellet, S. Experimental characterization of the influence of tailings fineness and density on the quality of cemented paste backfill. *Miner. Eng.* **2005**, *18*, 41–44. [CrossRef]
25. Zhang, Q.L.; Li, X.P.; Yang, W. Optimization of filling slurry ratio in a mine based on back-propagation neural network. *J. Cent. South Univ. (Sci. Tech.)* **2013**, *44*, 2867–2874.

26. Huang, S.; Xia, K.W.; Qiao, L. Dynamic tests of cemented paste backfill: Effects of strain rate, curing time, and cement content on compressive strength. *J. Mater. Sci.* **2011**, *46*, 5165–5170. [CrossRef]
27. Ehsan, S.; Hakan, B.; Luo, G.H.; Ali, K.; Richard, D.; Andy, F.; Mohamed, E. Multi-objective mixture design of cemented paste backfill using particle swarm optimisation algorithm. *Miner. Eng.* **2020**, *153*, 106385.
28. Li, D.; Feng, G.R.; Guo, Y.X.; Qi, T.Y.; Jia, X.Q.; Feng, J.R.; Li, Z. Analysis on the strength increase law of filling material based on response surface method. *J. China Coal Soc.* **2016**, *41*, 392–398.
29. Zhao, G.Y.; Ma, J.; Peng, K.; Yang, Q.; Zhou, L. Mix ratio optimization of alpine mine backfill based on the response surface method. *Chin. J. Eng.* **2013**, *35*, 559–565.
30. Xu, M.F.; Gao, Y.T.; Jin, A.B.; Zhou, Y.; Guo, L.J.; Liu, G.S. Prediction of cemented backfill strength by ultrasonic pulse velocity and BP neural network. *Chin. J. Eng.* **2016**, *38*, 1059–1068.
31. Wei, W.; Gao, Q. Strength prediction of backfilling body based on modified BP neural network. *J. Harbin Inst. Technol.* **2013**, *45*, 90–95.
32. Qi, C.; Chen, Q.; Fourie, A.; Zhang, Q. An intelligent modelling framework for mechanical properties of cemented paste backfill. *Miner. Eng.* **2018**, *123*, 16–27. [CrossRef]
33. Mitchell, R.J.; Olsen, R.S.; Smith, J.D. Model studies on cemented tailings used in mine backfill. *Can. Geotech. J.* **2011**, *19*, 14–28. [CrossRef]
34. Fu, Z.G.; Qiao, D.P.; Guo, Z.L.; Li, K.G.; Xie, J.C.; Wang, J.X. A model for calculating strength of ultra-fine tailings cemented hydraulic fill and its application. *Rock Soil Mech.* **2018**, *39*, 3147–3156.
35. Qi, C.C.; Chen, Q.S.; Kim, S.S. Integrated and intelligent design framework for cemented paste backfill: A combination of robust machine learning modelling and multi-objective optimization. *Miner. Eng.* **2020**, *155*, 106422. [CrossRef]
36. Wei, X.; Li, C.; Zhang, L.; Zhou, X.; Hu, B. Strength prediction model of cemented backfill and inversion calculation of cement consumption. *Geotech. Geol. Eng.* **2018**, *36*, 649–656. [CrossRef]
37. Orejarena, L.; Fall, M. Artificial neural network based modeling of the coupled effect of sulphate and temperature on the strength of cemented paste backfill. *Can. J. Civ. Eng.* **2011**, *38*, 100–109. [CrossRef]
38. Cui, L.; Fall, M. Mathematical modelling of cemented tailings backfill: A review. *Int. J. Mining, Reclam. Environ.* **2018**, *33*, 389–408. [CrossRef]
39. Zhao, Y.L.; Wu, P.Q.; Qiu, J.P.; Guo, Z.B.; Tian, Y.S.; Sun, X.G.; Gu, X.W. Recycling hazardous steel slag after thermal treatment to produce a binder for cemented paste backfill. *Powder Technol.* **2022**, *395*, 652–662. [CrossRef]
40. Klein, K.; Simon, D. Effect of specimen composition on the strength development in cemented paste backfill. *Can. Geotech. J.* **2006**, *43*, 310–324. [CrossRef]
41. Cao, S.; Yilmaz, E.; Song, W.D. Fiber type effect on strength, toughness and microstructure of early age cemented tailings backfill. *Constr. Build. Mater.* **2019**, *223*, 44–54. [CrossRef]
42. Chen, S.J.; Du, Z.W.; Zhang, Z.; Zhang, H.W.; Xia, Z.G.; Feng, F. Effects of chloride on the early mechanical properties and microstructure of gangue-cemented paste backfill. *Constr. Build. Mater.* **2020**, *235*, 117504. [CrossRef]
43. Li, W.C.; Fall, M. Strength and self-desiccation of slag-cemented paste backfill at early ages: Link to initial sulphate concentration. *Cem. Concr. Compos.* **2018**, *89*, 160–168. [CrossRef]
44. Fahey, M.; Helinski, M.; Fourie, A. Development of specimen curing procedures that account for the influence of effective stress during curing on the strength of cemented mine backfill. *Geotech. Geol. Eng.* **2011**, *29*, 709–723. [CrossRef]
45. Ghirian, A.; Fall, M. Strength evolution and deformation behaviour of cemented paste backfill at early ages: Effect of curing stress, filling strategy and drainage. *Int. J. Min. Sci. Technol.* **2016**, *26*, 809–817. [CrossRef]
46. Qiu, J.P.; Guo, Z.B.; Yang, L.; Jiang, H.Q.; Zhao, Y.L. Effect of tailings fineness on flow, strength, ultrasonic and microstructure characteristics of cemented paste backfill. *Constr. Build. Mater.* **2020**, *263*, 120645. [CrossRef]
47. Zisi, C.; Pappa-Louisi, A.; Nikitas, P. Separation optimization in HPLC analysis implemented in R programming language. *J. Chromatogr. A.* **2020**, *1617*, 460823. [CrossRef] [PubMed]

 *sustainability*

Article

Experimental Study on the Hydroponics of Wetland Plants for the Treatment of Acid Mine Drainage

Aijing Wu [1], Yongbo Zhang [1,*], Xuehua Zhao [1], Jiamin Li [1], Guowei Zhang [1], Hong Shi [2], Lina Guo [1] and Shuyuan Xu [1]

[1] College of Water Resources Science and Engineering, Taiyuan University of Technology, Taiyuan 030024, China; wuaijing0077@link.tyut.edu.cn (A.W.); zhaoxuehua@tyut.edu.cn (X.Z.); lijiamin0404@link.tyut.edu.cn (J.L.); zhangguowei0421@link.tyut.edu.cn (G.Z.); guolina0069@link.tyut.edu.cn (L.G.); xushuyuan0062@link.tyut.edu.cn (S.X.)
[2] College of Environmental Science and Engineering, Taiyuan University of Technology, Taiyuan 030024, China; sx_hosh196@sina.com
* Correspondence: zhangyongbo2021@sina.com

Abstract: Acid Mine Drainage (AMD) has become an important issue due to its significant ecological pollution. In this paper, phytoremediation technology and mechanism for AMD were investigated by hydroponic experiments, using six wetland plants (*Phragmites australis, Typha orientalis, Cyperus glomeratus, Scirpus validus, Iris wilsonii, Juncus effusus*) as research objects. The results showed that (1) the removal of sulfate from AMD was highest for *Juncus effusus* (66.78%) and *Iris wilsonii* (40.74%) and the removal of Mn from AMD was highest for *Typha orientalis* (>99%) and *Phragmites australis* (>99%). In addition, considering the growth condition of the plants, *Juncus effusus, Iris wilsonii,* and *Phragmites australis* were finally selected as the dominant plants for the treatment of AMD. (2) The removal pathway of pollutants in AMD included two aspects: one part was absorbed by plants, and the other part was removed through hydrolysis and precipitation processes. Our findings provide a theoretical reference for phytoremediation technology for AMD.

Keywords: AMD; phytoremediation; sulfate; hydroponic experiment; wetland plants; ecological pollution

Citation: Wu, A.; Zhang, Y.; Zhao, X.; Li, J.; Zhang, G.; Shi, H.; Guo, L.; Xu, S. Experimental Study on the Hydroponics of Wetland Plants for the Treatment of Acid Mine Drainage. *Sustainability* **2022**, *14*, 2148. https://doi.org/10.3390/su14042148

Academic Editor: Christophe Waterlot

Received: 3 January 2022
Accepted: 11 February 2022
Published: 14 February 2022

Publisher's Note: MDPI stays neutral with regard to jurisdictional claims in published maps and institutional affiliations.

1. Introduction

Acid mine drainage (AMD) is formed when sulfide minerals are exposed to oxidizing conditions after mining and other excavation processes [1,2]. As AMD is highly acidic and contains a large number of heavy metals, sulfates, and other pollutants [3,4], when it is discharged to the ground, it will cause great pollution to the surrounding water bodies and soil, lower the pH value of surface water, inhibit the growth and reproduction of aquatic organisms, destroy the granular structure of the soil, make the soil caked, salinized, barren, which will lead to the withering and death of crops [5]. In addition, the contaminants in AMD can pose a risk to human health through the food chain [6,7]. Previous studies have shown that AMD has become a long-term source of pollution, as it can continue to be generated for hundreds of years even after mining activities have ceased [8,9]. Therefore, there is an urgent need to investigate economical and efficient treatment technologies to minimize the negative impacts of AMD in response to its serious pollution problem.

Various techniques which cut across physical, chemical, and biological processes have been used to remediate water, air, and soil contaminated by AMD. Traditionally, AMD has been treated by adding calcium carbonate, lime, hydrated lime, caustic soda, and soda ash to AMD to neutralize the acidity [10]. However, it was found that 107–640 g of limestone is required to neutralize 1 L of AMD, making the application of neutralization quite expensive and unsafe when treating large amounts of AMD produced in coal mines [11]. The use of constructed wetlands (CWs) for AMD treatment is a rapidly developing passive treatment technology that focuses mainly on metal and sulfate removal [12,13]. CW is a substrate-microbial-plant composite ecosystem with physical, chemical, and biological triple synergy [14], which is an economic, efficient, and environment-friendly AMD remediation technology.

Wetland plants are an important part of CWs, which can not only remove pollutants from water bodies, accelerate the recycling and reuse of nutrients, but also maintain and beautify the wetland environment, improve the regional climate and promote a virtuous cycle of the ecological environment [15,16]. Since the selection of wetland plants has a significant impact on pollutant removal efficiency, the ability to release oxygen, and the species and number of microorganisms in the wetland substrate [17], cultivating or selecting plants that meet the treatment requirements can enhance the purification capacity of CWs and achieve long-term stable operation of CWs.

Currently, phytoremediation technology for the treatment of heavy metals has been studied more extensively, including studies on the removal effect and removal mechanism of heavy metals by plants and the tolerance mechanism of plants to heavy metals. For example, Oyuela Leguizamo, et al. [18] studied the behavior of 41 native or endemic species towards heavy metal pollution and screened the plants of the dominant family in the process of heavy metal enrichment. Muthusaravanan, et al. [19] reviewed the methods, mechanisms, and enhancement processes of phytoremediation of heavy metals. Han, et al. [20] investigated the Pb tolerant mechanisms, plant physiological response, and Pb sub-cellular localization in the root cells of Iris halophila. Although relatively mature research results have been achieved in the phytoremediation technology of heavy metals, however, little research has been done on the phytoremediation technology of sulfate. Thus, the study of the mechanism of sulfate removal from AMD by wetland plants in this paper is necessary.

In this study, six acid-tolerant wetland plants commonly found in China were used as research objects. Using hydroponic experiments, the growth status of six plants under different concentrations of AMD stress, the removal effects of six plants on pollutants in AMD and the accumulation of pollutants in plants were studied, and the removal mechanism of pollutants in AMD was analyzed, while the optimal wetland plants suitable for treating AMD were screened, which provided a reference basis for the construction of CWs at a later stage.

2. Materials and Methods

2.1. Synthetic AMD Composition

The chemical composition of AMD varies from site to site. In this study, the AMD was formulated manually based on the types and concentrations of the main pollutant ions in the AMD outflow from the Shandi River basin (38°1′32″ N, 113°31′37″ E) in Yangquan City, Shanxi Province, China. Since the constructed wetland investigated in this study was, actually, the final step of the AMD treatment technology, and the first stage was via permeable reaction barriers (PRB) when Fe was well removed, while Mn, Zn, and Cd were not [21], Fe was not considered in this study. Measured amounts of Na_2SO_4, $MnCl_2$, $Zn(NO_3)_2$, and $Cd(NO_3)_2$ powders were added in distillate water to produce three sulfate concentrations of AMD shown in Table 1. Under acidic conditions, metals exist mainly in dissolved forms.

2.2. Wetland Plants

Phragmites australis (*P. australis*), *Typha orientalis* (*T. orientalis*), *Cyperus glomeratus* (*C. glomeratus*), *Scirpus validus* (*S. validus*), *Iris wilsonii* (*I. wilsonii*), *Juncus effusus* (*J. effusus*) were purchased from Anxin County, Baoding City, Hebei Province, China. As shown in Table 2, these six plants are all common acid-tolerant perennial wetland plants in China, which have some economic value and can therefore reduce the maintenance costs of CW systems.

Table 1. Chemical composition and target pollutant concentrations in AMD.

	Theoretical Concentration (mg/L)	Reagent Used	Amount (mg) Per L Water
SO_4^{2-}	500 (C1)	Na_2SO_4	739.5833
	2000 (C2)		2958.3333
	4000 (C3)		5916.6667
Mn	18	$MnCl_2$	41.1853
Zn	10	$Zn\,(NO_3)_2{\cdot}6H_2O$	45.7677
Cd	0.5	$Cd\,(NO_3)_2{\cdot}4H_2O$	1.3721
pH	4	HCl	

Table 2. Ecological habits and economic value of tested plants [22].

The Plant	Ecological Habits	Economic Value
P. australis	The perennial aquatic herb that grows along irrigation ditches, riverbank marshes, etc. It is found throughout the world and often forms contiguous reed colonies due to its rapidly expanding reproductive capacity.	It can be used for making medicine, paper, weaving, and construction, and has ornamental value.
T. orientalis	Perennial aquatic or marsh herb grows in lakes, ponds, ditches, rivers in slow-flowing shallow water, also seen in wetlands and swamps, can withstand low temperatures of $-30\,°C$.	It is a weaving material, can be used for making medicine, paper, food, and has ornamental value
C. glomeratus	A perennial herb of the Cyperaceae family, growing mostly in wet places or swamps.	It can be used for weaving and making medicinal
S. validus	Perennial emergent aquatic herb, produced in many provinces in China, growing in lakesides or shallow ponds, and can tolerate low temperatures.	It can be used for weaving and has ornamental value.
I. wilsonii	Perennial herb, with fibers of old leaves remaining at the base of the plant, born on mountain slopes, forest margins, and wetlands along riverside ditches, light-loving, also more shade-tolerant, cold-hardy.	It has great ornamental value and can also be used to make medicine.
J. effusus	Perennial herbaceous aquatic plants, suitable for growing by rivers, ponds, ditches, rice fields, grasslands, marshes.	It can be used to weave utensils and make medicines, and the pith of the stem can be used to make lamp wicks and pillow wicks, etc.

2.3. Experimental Operation

Each of the six weighed plants was put into a measuring cup (150 g of each plant in each cup) with a capacity of 2 L, and then 1 L of experimental water was added to each measuring cup. There were four groups of experimental water (Table 3), including the control group with no contaminants and the AMD treatment group with three sulfate concentrations (low, medium, and high). The water level at this point was marked as the initial water level. Then, 5 mL of 1/5 strength Hoagland solution [23] was added to the measuring cup each day, and then distilled water was added to bring the water level to the initial level (water was consumed due to evaporation and plant transpiration). The experiment was carried out for 60 days.

2.4. Water Sample Analysis

The water samples were measured and analyzed for each indicator every 10 days. The pH value was measured using a pH meter (PHS-3C, Rex, Shanghai, China), while Ec was measured using a digital conductivity meter (DDS-307A, Rex). Concentrations of metals were determined by flame (acetylene) ionization using an atomic absorption spectrophotometer (TAS-990, Persee, Beijing, China) after sample filtration with 0.45 μm filter, and the concentration of SO_4^{2-} in the water samples was determined by ion chromatography (883 Basic IC plus, Metrohm, Shanghai, China). All represented data are the average of three replicate values. The removal efficiency of pollutants (SO_4^{2-}, Mn, Zn, Cd) was calculated (Equation (1)):

$$\text{Removal rate } (\%) = \frac{C_i - C_e}{C_i} \times 100 \tag{1}$$

where C_i and C_e represent the initial water and effluent pollutants concentrations, respectively.

Table 3. Schemes for hydroponic experiments.

Code	Experimental Water	pH	SO_4^{2-}	Mn	Zn	Cd
			mg/L			
CK	Control group (distillate water)		0	0	0	0
C1	Low sulfate concentration AMD	4	500	18	10	0.5
C2	Medium sulfate concentration AMD		2000	18	10	0.5
C3	High sulfate concentration AMD		4000	18	10	0.5

2.5. Plant and Water Sediment Analysis

After the experiment, the plants were rinsed repeatedly with tap water and then with distilled water to remove surface impurities. The plants were dried in an oven at 8 °C for 24 h and weighed for dry biomass. The dried plants were ground to fine powder with a grinder and then one gram of the powder was digested using a tri-acid mixture (HNO_3, $HClO_4$, and H_2SO_4; 5:1:1) at 80 °C until the solution became clear. The obtained solution was filtered and its contaminant content was determined using the method of water sample analysis in the previous section.

The bioconcentration factor (BCF) is defined as the ratio of total metal content in plant tissues (Cp, mg·kg^{-1}) to total metal content in the surrounding environment (Cw, mg·L^{-1}). It is given by Equation (2) [24]:

$$\text{BCF} = \frac{C_p}{C_w} \tag{2}$$

The experimental water was filtered at the end of the experiment and the precipitates were collected from the filter paper. The mineral composition of the precipitates was determined by X-ray diffraction (XRD) (Bruker D8-Advance X-ray polycrystalline powder diffractometer, Germany). XRD spectra were recorded over an angular range of 10–90° with Cu Kα anode (wavelength = 0.154 nm, 40 mA, 40 kV) with a step size of 0.01°.

2.6. Statistical Analysis

A one-way ANOVA was performed to identify significant differences among treatments, and, when detected, a post hoc Duncan's Multiple Range Test was performed using the SPSS 26.0 statistical software. Differences between the two treatments were analyzed using a t-test (SPSS 26.0). The differences were considered significant when $p < 0.05$.

3. Results and Discussion

3.1. The Growth State of Plants

Figure 1 shows the comparison of the growth status of the six plants before and after the experiment. It could be seen that the plants grew well at the beginning of the experiment, however, the growth status of the six plants showed a large difference after 60 days. which was partly due to the impact of AMD and partly due to the addition of sodium in the configuration of AMD, and the high concentration of sodium would cause a series of osmotic and metabolic problems to the plants thus inhibiting their growth [25,26]. In this paper, the growth status of six plants at the end of the experiment was evaluated based on four indicators (Table 4): the number of new shoots, the state of old branches, the phenomenon of roots rotted, and pest infestation.

Figure 1. Comparison of the growth status of each plant before and after the experiment.

Table 4. The growth state of each plant at the end of the experiment.

The Plant	Large Number of New Shoots	Old Branches in Good Condition	No Root Rotted	No Pest Infestation	Aggregate
P. australis	×	√	√	√	3√ 1×
T. orientalis	×	×	×	×	0√ 4×
C. glomeratus	√	×	×	√	2√ 2×
S. validus	×	√	×	×	1√ 3×
I. wilsonii	×	√	×	√	2√ 2×
J. effusus	√	×	√	√	3√ 1×

The "√" indicates that the plant meets the growth status described in the table header, while the "×" does the opposite.

The growth of *J. effusus* and *P. australis* was in good condition due to the high number of new shoots of *J. effusus* and the good condition of old shoots of *P. australis*, and both were free from the phenomenon of roots rotted and pest infestation. *I. wilsonii* and *C. glomeratus* both passed two indicators; however, observation of Figure 1 shows that although *C. glomeratus* had more new shoots, its overall growth status was not as good as *I. wilsonii* due to its thinner branches and poor condition of old branches. The growth status of *T. orientalis* and *S. validus* was poor and both were infected with pests during the experiment, which were found to be difficult to eradicate after many repellent measures. Therefore, based on the overall growth of the plants under AMD stress, the plants with better growth were selected as *J. effusus*, *P. australis* and *I. wilsonii*.

3.2. Removal of Contaminants in AMD

Figure 2 shows the variation of metals removal rates in AMD with different sulfate concentrations over time. As shown in Figure 2a, the removal rate of Mn from AMD was the highest for both *P. australis* and *T. orientalis*, with the removal rate reaching more than 99% at 20 days of the experiment and remaining stable afterward. In addition, the removal rate of Mn by *J. effusus* was basically above 90%. However, the removal rates of Mn from AMD by *C. glomeratus*, *S. validus* and *I. wilsonii* were relatively low and decreased in the later stages of the experiment. This was mainly since some of the roots of these three plants had decayed in the later stages of the experiment, leading to the release of the Mn absorbed by the plants. Moreover, the release of Mn in C3 water samples was stronger than that in C1 and C2 water samples, which may be due to the saline stress on plants in C3 water samples. Studies have shown that high concentrations of soluble salts in the environment can cause damage to plant cells and affect the normal nutrient uptake of plants, which may eventually cause the stomata of plants to close and plants to wilt or even die [27], so the wilting and root rot of plants in C3 water samples were more serious.

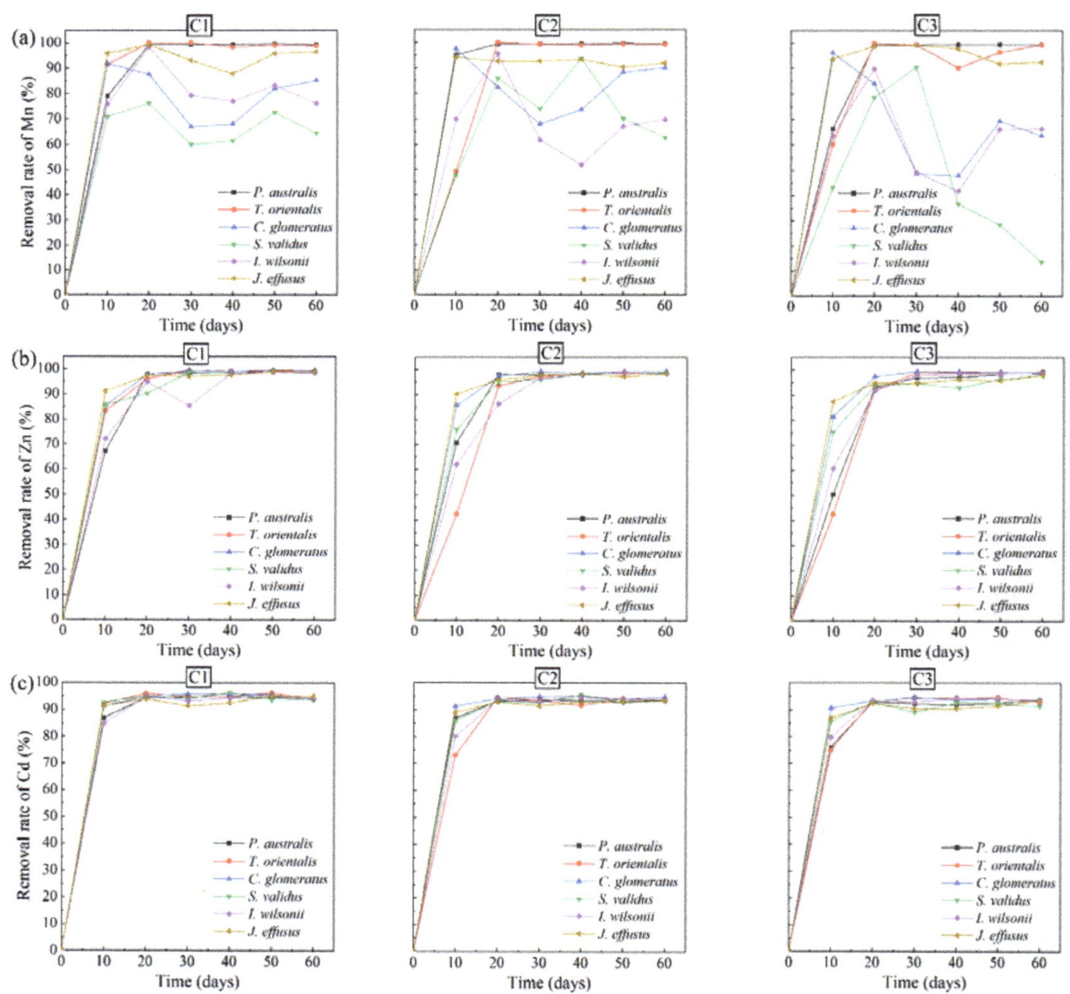

Figure 2. The removal rates of (**a**) Mn; (**b**) Zn; (**c**) Cd in AMD with three sulfate concentrations vs. time. C1, C2 and C3 represent AMD with different sulfate concentrations (Table 3).

As shown in Figure 2b,c, the trends of Zn and Cd removal rates in AMD by the six plants were all rising first and then stable, and the Zn and Cd removal rates reached more than 97% and 90%, respectively, at 60 days of the experiment. From the figure, it could be seen that the differences in the removal rates of Mn from AMD were greater among the different plants, while there were no significant differences in the removal rates of Zn and Cd, indicating that Zn and Cd in the water samples were more stable than Mn and not easily released by dissolution. Many earlier pieces of literature also reported the difficult removal of Mn [28,29], because, under anoxic conditions, Mn is often present in the form of Mn^{2+}, which is more soluble. Therefore, before treating AMD through CW systems, it needs to be pretreated to oxidize manganese to insoluble manganese dioxide or manganese hydroxide.

Figure 3 shows that sulfate concentrations in AMD decreased continuously with time. Among the six plants, the highest SO_4^{2-} removal was achieved by *J. effusus*. The removal of SO_4^{2-} from all three concentrations of AMD by *J. effusus* reached more than 50% after 60 days of the experiment with the highest removal rate of 66.78% (C1). The plant with the next highest SO_4^{2-} removal was *I. wilsonii*, which removed more than 35% of SO_4^{2-} from all three concentrations of AMD after 60 days of the experiment, with the highest removal rate of 40.74% (C2). This result is closely related to the growth status of the six plants because sulfur is an indispensable element for the growth and development of all plants, as well as a structural component element of plants, which is involved in many important biochemical reactions in plants [30]. Among the six plants, *J. effusus* had the highest number of new shoots and *I. wilsonii* had the best state of old branches, so they needed more water and nutrients, which explains the higher removal of SO_4^{2-} from AMD by *J. effusus* and *I. wilsonii*.

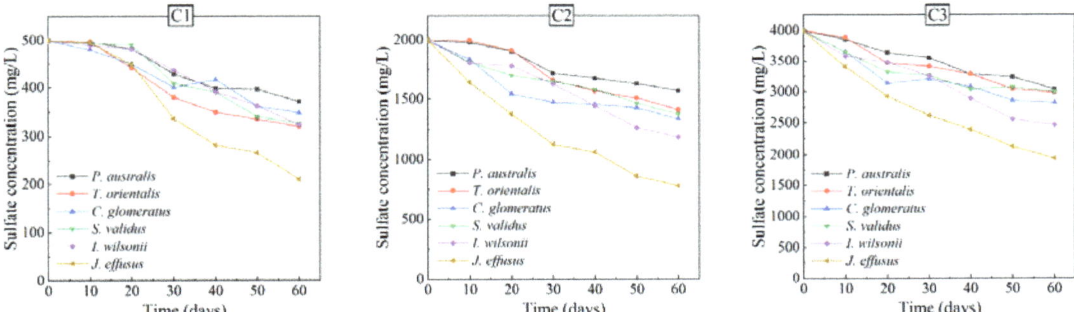

Figure 3. Sulfate concentrations in three concentrations of AMD vs. time. C1, C2 and C3 represent AMD with different sulfate concentrations (Table 3).

Studies have shown that water quality changes, plant growth, and microbial reproduction are all affected by the pH of the water [31], so continuous monitoring of pH in water is necessary. Figure 4 shows the variation profiles of pH of the experimental water over the operation time. It can be seen that the pH of the experimental water increased from 4 to more than 7 after 20 days of planting the plants in water, which means that the water changed from acidic to weakly alkaline, and the pH of the water was stabilized at 7–8 in the later stages of the experiment. This may be partly due to the ability of plants to regulate the pH of the water during growth, and partly due to the presence of microorganisms, such as sulfate-reducing bacteria, which can use organic matter in the water as electron donors to produce bicarbonate while reducing sulfate, leading to an increase in pH [32], which is important for the removal of metals from the water because the acidity of the solution allows the metals to be transported in the most soluble form.

The electrical conductivity (Ec) of the solution is an important indicator of its salt content, ionic content, impurity content, etc. Figure 5 shows the variation profiles of Ec of the experimental water over the operation time. In CK (control group), the Ec of the water samples of *J. effusus* decreased more and the Ec of the water samples of *P. australis* changed very little, while the Ec of the water samples of the remaining four plants increased to different degrees, which was mainly due to some root rot of these four plants in the late stage of the experiment, resulting in the dissolution and release of internal plant components. In C1 (low sulfate concentration AMD), the Ec of the water samples of *P. australis*, *T. orientalis* and *J. effusus* decreased, the Ec of the water samples of *S. validus* and *I. wilsonii* decreased and then increased slightly, while the Ec of the water samples of

C. glomeratus increased more, which was caused by the combination of its growth status and weaker purification effect on AMD. In C2 and C3 (medium and high sulfate concentration AMD), the Ec of the water samples of all six plants decreased, because the removal of pollutants from the water samples had a more significant effect on the Ec than the release of the internal plant components, with the largest decrease in the Ec of water samples of *J. effusus*, indicating that *J. effusus* had a greater effect on the removal of pollutants from the water samples, which was consistent with the previous experimental findings.

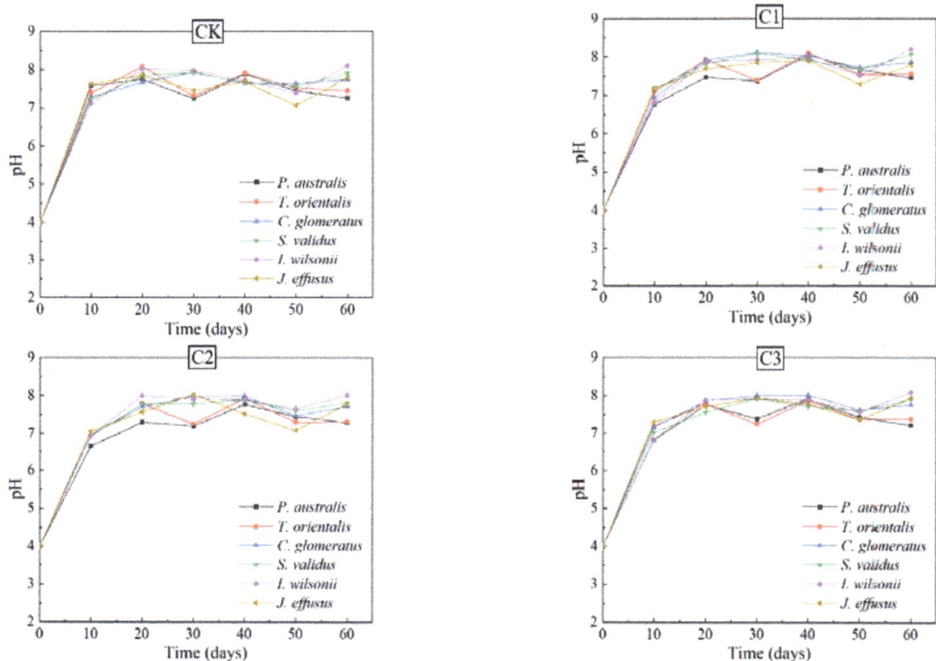

Figure 4. Variation profiles of pH of the experimental water over the operation time. CK represents distilled water, and C1, C2 and C3 represent AMD with different sulfate concentrations (Table 3).

3.3. Removal Mechanism of Pollutants in AMD

Table 5 shows the metal concentrations and bioconcentration factors (BCFs) of six plants in C3 water samples. The BCF indicates the ability of plants to enrich heavy metals from their surroundings [33], and it could be seen that the BCFs of six plants for each metal were greater than one, indicating that the accumulation of metals in plant tissues was greater than that in the growth medium, so all six plants could be used for phytoextraction of Mn, Zn, and Cd [34]. The order of BCF for most of the plants (except *P. australis*) was Mn > Cd > Zn, which indicated that Mn was more readily absorbed by plants, thus explaining the decrease in Mn removal in AMD due to the poor growth state of plants at the later stages of the experiment. In addition, the BCFs of plants for Mn followed the order: *I. wilsonii* > *S. validus* > *C. glomeratus* > *J. effusus* > *T. orientalis* > *P. australis*, the BCFs of plants for Zn followed the order: *I. wilsonii* > *J. effusus* > *S. validus* > *T. orientalis* > *C. glomeratus* > *P. australis*, the BCFs of plants for Cd followed the order: *I. wilsonii* > *J. effusus* > *S. validus* > *C. glomeratus* > *P. australis* > *T. orientalis*. This showed that *I. wilsonii* had the highest metal enrichment capacity, followed by *J. effusus* and *S. validus*.

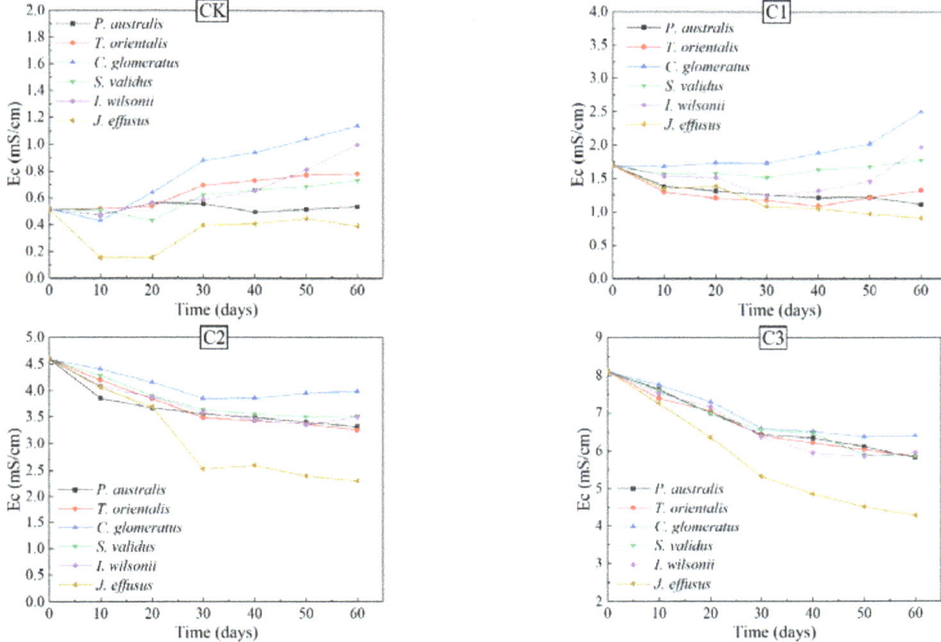

Figure 5. Variation profiles of electrical conductivity (Ec) of the experimental water over the operation time.

In this paper, the removal mechanism of pollutants in AMD was analyzed by taking the C3 treatment group as an example. The sulfur, Mn, Zn, and Cd contents of six plants in CK and C3 water samples at the end of the experiment were detected, the difference was taken as the sulfur, Mn, Zn, and Cd contents absorbed by each plant in C3 water samples, and the ratio of the difference to the initial content of pollutants in C3 water samples was taken as the proportion of sulfur, Mn, Zn, and Cd absorbed by plants in C3 water samples, as shown in Figure 6.

Table 5. Metal concentrations and bioconcentration factors (BCFs) of six plants in C3 water samples.

Parameters	The Plant	Mn	Zn	Cd
Concentration(mg/kg)	*P. australis*	171.75 ± 11.24	71.01 ± 5.68	5.67 ± 0.96
	T. orientalis	210.55 ± 16.69	101.76 ± 10.57	5.54 ± 0.57
	C. glomeratus	406.12 ± 20.25	86.32 ± 8.98	6.03 ± 0.84
	S. validus	450.23 ± 15.55	111.83 ± 13.54	7.12 ± 1.11
	I. wilsonii	503.89 ± 23.57	171.00 ± 15.14	11.48 ± 1.21
	J. effusus	393.27 ± 8.89	121.38 ± 9.63	9.11 ± 0.98
BCF	*P. australis*	9.54	7.10	11.35
	T. orientalis	11.70	10.18	11.08
	C. glomeratus	22.56	8.63	12.06
	S. validus	25.01	11.18	14.24
	I. wilsonii	27.99	17.10	22.96
	J. effusus	21.85	12.14	18.23

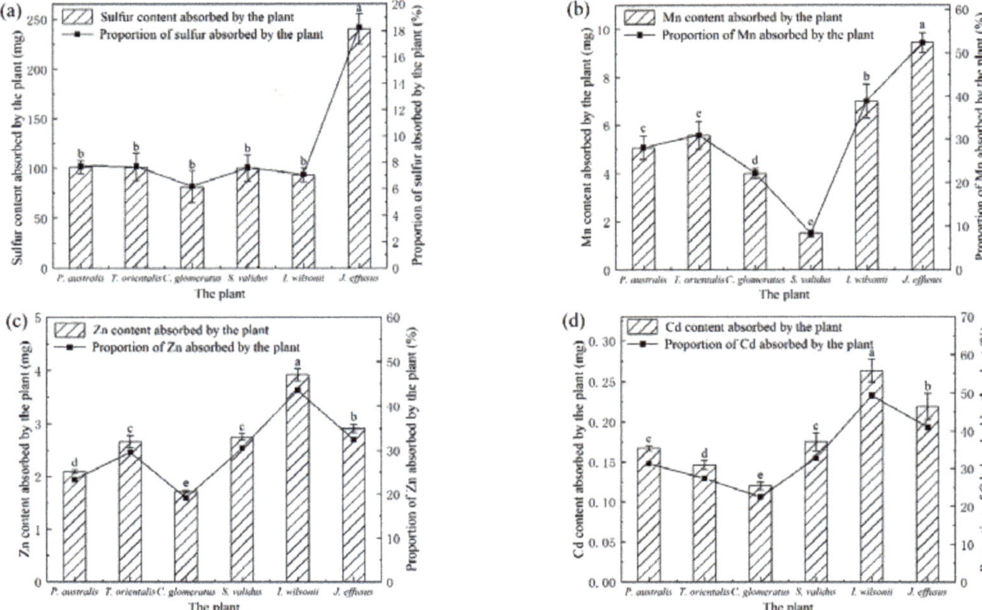

Figure 6. The content and proportion of (**a**) sulfur; (**b**) Mn; (**c**) Zn; (**d**) Cd absorbed by the plant in C3 water samples.

Bars and error bars represent the mean ± SD of three replicates. The same letter in the histogram of a certain plant represents no significant difference at the level of 0.05 (Duncan's Multiple Range Test). From Figure 6a, it can be seen that *J. effusus* absorbed the highest amount of sulfur, which accounted for 18.23% of sulfur in AMD, while the other five species absorbed a small percentage of sulfur, ranging from 6% to 8%. This is also the reason for the highest SO_4^{2-} removal rate in AMD by *J. effusus*. However, compared with Figure 3, it can be found that the proportion of sulfur absorbed by plants is smaller than the sulfur removal rate in AMD, indicating that only a part of sulfur in AMD is absorbed by plants. In addition, white crystals were observed to precipitate from the plant surface during the experiment, and the higher the sulfate concentration in AMD and the longer the experiment, the more white crystals were precipitated from the plant. Examination of the composition of the white crystals using X-ray diffraction (XRD) revealed that the main component was sodium sulfate (Figure 7), suggesting that the plant first absorbed the sodium sulfate into its body and then excreted the portion that could not be absorbed and used by its own tissues, thus allowing the removal of sulfate by harvesting.

Figure 7. Images and XRD patterns of precipitates on the plant surface.

As seen in Figure 6b–d, among the six plants, the metal uptake by *J. effusus* and *I. wilsonii* was higher than the other four plants. *J. effusus* absorbed 52.44%, 32.37%, and 41.02% of the Mn, Zn, and Cd contents in AMD, respectively, and *I. wilsonii* absorbed 38.89%, 43.47%, and 49.27% of the Mn, Zn, and Cd contents in AMD, respectively. However, compared with Figure 2a–c, it can be found that the proportion of metals absorbed by plants is smaller than the metal removal rate in AMD, which means that only part of the metals in AMD are absorbed by plants. Therefore, the precipitates in the water samples were examined for composition using an X-ray diffractometer (XRD) (Figure 8), and it was found that the precipitates could be $Mn(OH)_2$, $Zn(OH)_2$, $Cd(OH)_2$, $CdSO_4$, etc., indicating the existence of other ways (hydrolysis, precipitation, etc.) for the removal of heavy metals from AMD, which could be related to the change of pH in AMD.

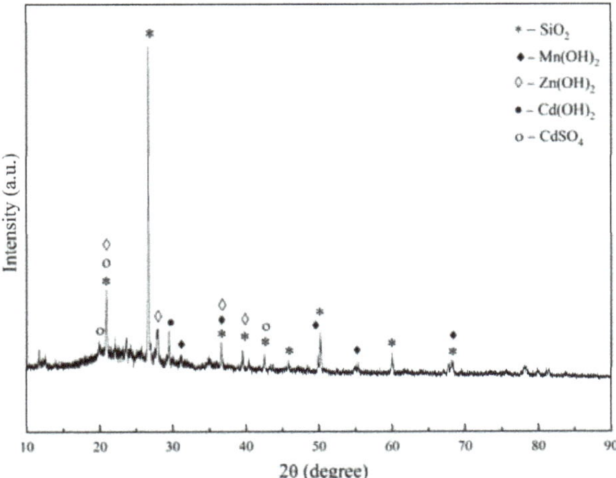

Figure 8. XRD patterns of precipitates in water samples.

4. Conclusions

In this paper, six wetland plants (*Phragmites australis, Typha orientalis, Cyperus glomeratus, Scirpus validus, Iris wilsonii, Juncus effusus*) were used as research objects to conduct an experimental hydroponic study of phytoremediation of AMD, and the following main conclusions were obtained:

(1)   There was no significant difference in the removal rates of Zn and Cd in AMD among the six plants, while the removal rates of SO_4^{2-} and Mn in AMD varied greatly.

Therefore, the six wetland plants were screened in terms of their growth status and the removal effects of the plants on pollutants in AMD, and *Juncus effusus*, *Iris wilsonii* and *Phragmites australis* were preferably finally selected as the dominant plants for the treatment of AMD.

(2) The analysis of the uptake of pollutants in plants and the precipitates in AMD showed that the removal pathway of pollutants in AMD consisted of two aspects: one part was absorbed by the plants, and the other part was removed by means of hydrolysis, precipitation, etc. It was noteworthy that the plants first absorbed sodium sulfate into their bodies and then excreted the part that could not be absorbed and utilized by their own tissues, which precipitated as white crystals on the plant surface; hence, sulfate could be removed by harvesting.

Author Contributions: Conceptualization, A.W. and Y.Z.; software, A.W. and L.G.; investigation, A.W., J.L. and G.Z.; writing—original draft preparation, A.W.; writing—review and editing, S.X. and X.Z.; supervision, Y.Z.; project administration, Y.Z. and H.S.; funding acquisition, Y.Z. and H.S. All authors have read and agreed to the published version of the manuscript.

Funding: This research was funded by key research programs of the Ministry of Science and Technology for water resource efficiency development and utilization project, grant number 2018YFC0406403.

Data Availability Statement: The datasets generated and/or analyzed during the current study are available from the corresponding author on reasonable request.

Conflicts of Interest: The authors declare no conflict of interest.

References

1. Fernando, W.A.M.; Ilankoon, I.M.S.K.; Syed, T.H.; Yellishetty, M. Challenges and opportunities in the removal of sulphate ions in contaminated mine water: A review. *Miner. Eng.* **2018**, *117*, 74–90. [CrossRef]
2. Singh, S.; Chakraborty, S. Performance of organic substrate amended constructed wetland treating acid mine drainage (AMD) of North-Eastern India. *J. Hazard. Mater.* **2020**, *397*, 122719. [CrossRef]
3. Guo, L.; Cutright, T.J. Remediation of AMD Contaminated Soil by Two Types of Reeds. *Int. J. Phytore Mediat.* **2015**, *17*, 391–403. [CrossRef]
4. Mang, K.C.; Ntushelo, K. Phytoextraction and phytostabilisation approaches of heavy metal remediation in acid mine drainage with case studies: A review. *Appl. Ecol. Environ. Res.* **2019**, *17*, 6129–6149. [CrossRef]
5. Cutright, T.J.; Senko, J.; Sivaram, S.; York, M. Evaluation of the Phytoextraction Potential at an Acid-Mine-Drainage-Impacted Site. *Soil Sediment Contam.* **2012**, *21*, 970–984. [CrossRef]
6. Karaca, O.; Cameselle, C.; Reddy, K.R. Mine tailing disposal sites: Contamination problems, remedial options and phytocaps for sustainable remediation. *Rev. Environ. Sci. Bio-Technol.* **2018**, *17*, 205–228. [CrossRef]
7. Nagy, A.; Magyar, T.; Juhasz, C.; Tamas, J. Phytoremediation of acid mine drainage using by-product of lysine fermentation. *Water Sci. Technol.* **2020**, *81*, 1507–1517. [CrossRef]
8. Younger, P.L. The longevity of minewater pollution: A basis for decision-making. *Sci. Total Environ.* **1997**, *194–195*, 457–466. [CrossRef]
9. Zheng, Q.; Zhang, Y.; Zhang, Z.; Li, H.; Wu, A.; Shi, H. Experimental research on various slags as a potential adsorbent for the removal of sulfate from acid mine drainage. *J. Environ. Manag.* **2020**, *270*, 110880. [CrossRef]
10. Tong, L.; Fan, R.; Yang, S.; Li, C. Development and Status of the Treatment Technology for Acid Mine Drainage. *Min. Metall. Explor.* **2021**, *38*, 315–327. [CrossRef]
11. Dutta, M.; Islam, N.; Rabha, S.; Narzary, B.; Bordoloi, M.; Saikia, D.; Silva, L.F.O.; Saikia, B.K. Acid mine drainage in an Indian high-sulfur coal mining area: Cytotoxicity assay and remediation study. *J. Hazard. Mater.* **2019**, *389*, 121851. [CrossRef]
12. Skousen, J.; Zipper, C.E.; Rose, A.; Ziemkiewicz, P.F.; Nairn, R.; McDonald, L.M.; Kleinmann, R.L. Review of Passive Systems for Acid Mine Drainage Treatment. *Mine Water Environ.* **2017**, *36*, 133–153. [CrossRef]
13. Blanco, I.; Sapsford, D.J.; Trumm, D.; Pope, J.; Kruse, N.; Cheong, Y.-W.; McLauchlan, H.; Sinclair, E.; Weber, P.; Olds, W. International Trials of Vertical Flow Reactors for Coal Mine Water Treatment. *Mine Water Environ.* **2018**, *37*, 4–17. [CrossRef]
14. Pat-Espadas, A.M.; Loredo Portales, R.; Amabilis-Sosa, L.E.; Gomez, G.; Vidal, G. Review of Constructed Wetlands for Acid Mine Drainage Treatment. *Water* **2018**, *10*, 1685. [CrossRef]
15. Palihakkara, C.R.; Dassanayake, S.; Jayawardena, C.; Senanayake, I.P. Floating Wetland Treatment of Acid Mine Drainage using Eichhornia crassipes (Water Hyacinth). *J. Health Pollut.* **2018**, *8*, 14–19. [CrossRef]
16. Yan, A.; Wang, Y.; Tan, S.N.; Mohd Yusof, M.L.; Ghosh, S.; Chen, Z. Phytoremediation: A Promising Approach for Revegetation of Heavy Metal-Polluted Land. *Front. Plant Sci.* **2020**, *11*, 359. [CrossRef]

17. Brisson, J.; Chazarenc, F. Maximizing pollutant removal in constructed wetlands: Should we pay more attention to macrophyte species selection? *Sci. Tech.* **2008**, *407*, 3923–3930. [CrossRef]
18. Oyuela Leguizamo, M.A.; Fernandez Gomez, W.D.; Sarmiento, M.C.G. Native herbaceous plant species with potential use in phytoremediation of heavy metals, spotlight on wetlands—A review. *Chemosphere* **2017**, *168*, 1230–1247. [CrossRef]
19. Muthusaravanan, S.; Sivarajasekar, N.; Vivek, J.S.; Paramasivan, T.; Naushad, M.; Prakashmaran, J.; Gayathri, V.; Al-Duaij, O.K. Phytoremediation of heavy metals: Mechanisms, methods and enhancements. *Environ. Chem. Lett.* **2018**, *16*, 1339–1359. [CrossRef]
20. Han, Y.; Zhang, L.; Yang, Y.; Yuan, H.; Zhao, J.; Gu, J.; Huang, S. Pb uptake and toxicity to Iris halophila tested on Pb mine tailing materials. *Environ. Pollut.* **2016**, *214*, 510–516. [CrossRef]
21. Zheng, Q. The Study on Mechanism of the Treatment of Acid Mine Drainage Using PRB with Loess-Steel Slag. Ph.D. Thesis, Taiyuan University of Technology, Taiyuan, China, 2020. [CrossRef]
22. Chinese Academy of Sciences, Editorial Committee of the Flora of China. *Flora of China*; Science Press: Beijing, China, 1993.
23. Park, J.H. Contrasting effects of Cr(III) and Cr(VI) on lettuce grown in hydroponics and soil: Chromium and manganese speciation. *Environ. Pollut.* **2020**, *266*, 115073. [CrossRef] [PubMed]
24. Dan, A.; Oka, M.; Fujii, Y.; Soda, S.; Ishigaki, T.; Machimura, T.; Ike, M. Removal of heavy metals from synthetic landfill leachate in lab-scale vertical flow constructed wetlands. *Sci. Total Environ.* **2017**, *584*, 742–750. [CrossRef]
25. Maser, P.; Eckelman, B.; Vaidyanathan, R.; Horie, T.; Fairbairn, D.J.; Kubo, M.; Yamagami, M.; Yamaguchi, K.; Nishimura, M.; Uozumi, N.; et al. Altered shoot/root Na+ distribution and bifurcating salt sensitivity in Arabidopsis by genetic disruption of the Na+ transporter AtHKT1. *FEBS Lett.* **2002**, *531*, 157–161. [CrossRef]
26. Wu, H. Plant salt tolerance and Na+ sensing and transport. *Crop. J.* **2018**, *6*, 215–225. [CrossRef]
27. Fang, S.; Hou, X.; Liang, X. Response Mechanisms of Plants Under Saline-Alkali Stress. *Front. Plant Sci.* **2021**, *12*, 1–20. [CrossRef]
28. Vymazal, J. Removal of heavy metals in a horizontal sub-surface flow constructed wetland. *J. Environ. Sci. Health* **2005**, *40*, 1369–1379. [CrossRef]
29. Lesage, E. *Behaviour of Heavy Metals in Constructed Treatment Wetlands*; Ghent University: Ghent, Belgium, 2006.
30. Wu, S.; Kuschk, P.; Wiessner, A.; Mueller, J.; Saad, R.A.B.; Dong, R. Sulphur transformations in constructed wetlands for wastewater treatment: A review. *Ecol. Eng.* **2013**, *52*, 278–289. [CrossRef]
31. Aoyagi, T.; Hamai, T.; Hori, T.; Sato, Y.; Kobayashi, M.; Sato, Y.; Inaba, T.; Ogata, A.; Habe, H.; Sakata, T. Hydraulic retention time and pH affect the performance and microbial communities of passive bioreactors for treatment of acid mine drainage. *AMB Express* **2017**, *7*, 142. [CrossRef]
32. Sanchez-Andrea, I.; Luis Sanz, J.; Bijmans, M.F.M.; Stams, A.J.M. Sulfate reduction at low pH to remediate acid mine drainage. *J. Hazard. Mater.* **2014**, *269*, 98–109. [CrossRef]
33. Mulenga, C.; Clarke, C.; Meincken, M. Bioaccumulation of Cu, Fe, Mn and Zn in native Brachystegia longifolia naturally growing in a copper mining environment of Mufulira, Zambia. *Environ. Monit. Assess.* **2022**, *194*, 13. [CrossRef]
34. Cruzado-Tafur, E.; Bierla, K.; Torro, L.; Szpunar, J. Accumulation of As, Ag, Cd, Cu, Pb, and Zn by Native Plants Growing in Soils Contaminated by Mining Environmental Liabilities in the Peruvian Andes. *Plants* **2021**, *10*, 241. [CrossRef] [PubMed]

sustainability

Article

Comprehensive Diagnosis Method of the Health of Tailings Dams Based on Dynamic Weight and Quantitative Index

Kai Dong [1,2], **Zhankuan Mi** [1,3,*] **and Dewei Yang** [1]

[1] Nanjing Hydraulic Research Institute, No. 223, Guangzhou Road, Nanjing 210029, China; kdong@stu.scu.edu.cn (K.D.); dwyang@nhri.cn (D.Y.)

[2] College of Water Resources & Hydropower, Sichuan University, No. 24 South Section 1, Yihuan Road, Chengdu 610065, China

[3] Key Laboratory of Failure Mechanism and Safety Control Techniques of Earth-Rock Dam of the Ministry of Water Resources of P.R., No. 34, Hujuguan Road, Nanjing 210024, China

* Correspondence: zkmi@nhri.cn

Abstract: As a dangerous source of man-made debris flow with high potential energy, tailings dams can cause huge losses to people's lives and property downstream once they break, and their safety control problem is particularly prominent. The health diagnosis of tailings dams is a complex and nonlinear problem full of uncertainty. At present, the health diagnosis of tailings dams is mostly qualitative evaluation or quantitative analysis aiming at a single index, so this study puts forward a comprehensive quantitative diagnosis method of tailings dam health based on dynamic weight. Slope stability, deformation stability and seepage stability are taken as project layers, and the diagnosis index system of the tailings dam is constructed. The quantitative methods of diagnosis indexes of project layers are proposed. For the dam slope stability project, the safety factor and the reliability index of tailings dams are determined based on the Monte Carlo method, which can consider the uncertainty of tailings material parameters. For the deformation stability project, the normal operation values of deformation rate and deformation amount are determined by analyzing the in situ observation data and combining them with the numerical simulation results. For the seepage stability project, through the analysis of seepage and stability, the relationship curve between the depth of saturation line and the safety factor of anti-sliding stability is established. The norms method is used to determine the quantitative standards for the diagnosis indexes of the basic layer. Based on the analytical hierarchy process method and the penalty variable weight method, the method of dynamic weight of the project layer index is proposed. The proposed methods are applied to a practical engineering project. The results show that the methods can accurately reflect the health status of tailings dams. This study provides a new method for evaluating the safety of tailings dams.

Keywords: tailings dam; safety factor; quantitative evaluation; dynamic weight; comprehensive diagnosis of health

Citation: Dong, K.; Mi, Z.; Yang, D. Comprehensive Diagnosis Method of the Health of Tailings Dams Based on Dynamic Weight and Quantitative Index. *Sustainability* **2022**, *14*, 3068. https://doi.org/10.3390/su14053068

Academic Editors: Longjun Dong, Yanlin Zhao and Wenxue Chen

Received: 18 January 2022
Accepted: 3 March 2022
Published: 6 March 2022

Publisher's Note: MDPI stays neutral with regard to jurisdictional claims in published maps and institutional affiliations.

1. Introduction

The tailings pond is a place for storing tailings, and the tailings dam is a dam structure around the tailings pond, which is a key project to ensure the normal operation of the tailings pond. At present, there are about 8869 non-coal mine tailings ponds in China, among which there are about 1112 "overhead tailings ponds", accounting for 14.3%. Tailings pond accidents rank 18th in the ranking of hidden dangers in the world, and their hazards are second only to nuclear radiation and nuclear explosions. [1,2]. On 8 September 2008, the tailings dam of Xinta Mining in Xiangfen, Shanxi Province, China collapsed, resulting in the deaths of at least 277 people and having an extremely bad social impact [3]. On 4 August 2014, the Mount Polley tailings dam in Canada broke due to design reasons, flooded large forests and lakes, and seriously damaged the ecological environment [4]. On 5 November 2015, the Samarco tailings dam in Brazil was liquefied and collapsed due to a small earthquake, which killed at least 19 people and polluted 600 km of rivers, causing the most serious environmental disaster in Brazilian history [5]. In order to ensure the safety of people's lives and properties, the Chinese government has paid more and more attention to the safe operation of tailings dams in recent years and has put forward higher requirements for the safe operation and risk control of tailings dams. Therefore, the health diagnosis method of tailings dams is put forward in order to assess its operating health state.

The health diagnosis of tailings dams refers to the evaluation of key performance indicators based on the design, operation and monitoring data of the tailings dam during the operation period, and the diagnosis of its health status and defects [6]. At present, very rich research results [7–10] have been obtained in the health diagnosis of tailings dams, which are mainly divided into qualitative diagnosis and quantitative diagnosis. Most studies [11–15] focus on qualitative diagnosis, represented by the safety checklist method, which lists inspection items according to the relevant laws and regulations, and then scores and summarizes the overall health status of tailings dams by experts. This method is convenient to operate and intuitive in its evaluation results, but it has strong subjective randomness, and its accuracy and credibility are insufficient. In recent years, scholars have gradually introduced new mathematical methods and theories, such as fuzzy theory [16–18] and uncertainty measurement theory [19], which have improved the accuracy of safety evaluation. However, the methods are mostly used to deal with the relationship between indexes in the diagnosis system, and the diagnosis of the basic indexes is still mainly qualitative. In terms of the quantitative diagnosis of tailings dam health, because the tailings dams are mostly constructed in stages and the tailings materials have obvious anisotropy, heterogeneity and temporal and spatial variability, it is extremely difficult to quantitatively diagnose their health status. Scholars mainly evaluate the health status of tailings dams by numerical simulation and mathematical models. For example, Wang [20] and Xu [21] used the limit equilibrium method to analyze the stability of tailings dams and diagnose whether the stability of tailings dams meets the requirements; Wang [22] and Li [23] calculated and analyzed the seepage field through numerical simulation and a theoretical model, respectively, and evaluated the seepage safety of tailings dams. Dong established the pre-alarm system based on monitoring data and numerical simulation for tailings dams, and verified the applicability and accuracy of the system [24]. Li applied the strength reduction method to analyze the overall stability of the tailings dam [25]. Dong summarized and compared three common tailings dam stability analysis methods: limit equilibrium method, numerical simulation method and uncertainty method, and analyzed their applicability [26]. These analytical methods are based on monitoring data and test data to diagnose the health status of tailings dams, which is more convincing and scientific than qualitative diagnosis methods. The research is mostly concentrated on a single working condition and a single index. However, there are many factors that affect the health status of tailings dams, and the multi-index and multi-factor evaluation method is more suitable for its safety evaluation. At present, there is little research on comprehensive quantitative diagnosis of tailings dam health. In the comprehensive diagnosis of tailings dam health, the index weights have an important influence on the accuracy of the diagnosis result. Generally, the deterministic weight is used, that is, the weight will not change as the indicator's health status deteriorates. The method will lead to a state of imbalance. When a certain index deteriorates while other indexes are still in a healthy state, the influence of the deterioration index may be ignored in the traditional diagnosis based on deterministic weight. Aiming at the existing problems in the comprehensive diagnosis of tailings dam health, a comprehensive quantitative diagnosis method of tailings dam health based on dynamic weight is proposed in this study, which provides a new method for reference to the health diagnosis of tailings dams.

The main contents of this paper are as follows: (a) The diagnosis index system is constructed and the health grade is determined. (b) Based on the analytic hierarchy process and the penalty variable weight method, the dynamic weight method of the diagnosis index is proposed. (c) The quantified methods and standards for the basic-layer diagnosis index such as the slope stability, the deformation stability and the seepage stability are proposed. (d) The method was applied to tailings dam I in Brazil, and its applicability and accuracy were verified.

2. Materials and Methods

2.1. Construction of Diagnosis Index System for Tailings Dam Health

2.1.1. Diagnosis Index

As the basis of comprehensive diagnosis, whether the diagnosis index is appropriate or not will directly affect whether the diagnosis result is reasonable and accurate. In this paper, the diagnosis index system is divided into a project layer, an effect-quantity layer and a basic layer. With reference to related literatures [27,28] and norms [29,30], combined with the tailings dam failure modes and hazard factors, the health diagnosis index system of tailings dam was established. As shown in Figure 1, the project layer of the index system includes three aspects: slope stability, deformation stability and seepage stability, and the effect quantity layer is obtained by further concretization of the project layer. The slope stability project includes the deterministic safety factor and the reliability index, and the reliability index is added to consider the randomness of physical and mechanical characteristics of tailings. As a non-linear material, tailings will deform significantly during the life cycle of the tailings dam. Normal consolidation deformation is beneficial to the stability of the tailings dam, but for tailings with strong cementation characteristics and structural properties, excessive deformation will cause its strength to be lost. The deformation stability project is divided into two indexes: the deformation rate and the total deformation. The tailings dam is a permeable structure, and the saturation line is the lifeline. Therefore, the depth of the saturation line is taken as an index of the seepage project. The basic layer is the bottom layer diagnosis index in the index system, including the in situ monitoring data of the tailings dam and the corresponding calculation results obtained by numerical analysis based on the structure of the tailings dam and the physical and mechanical properties of the tailings.

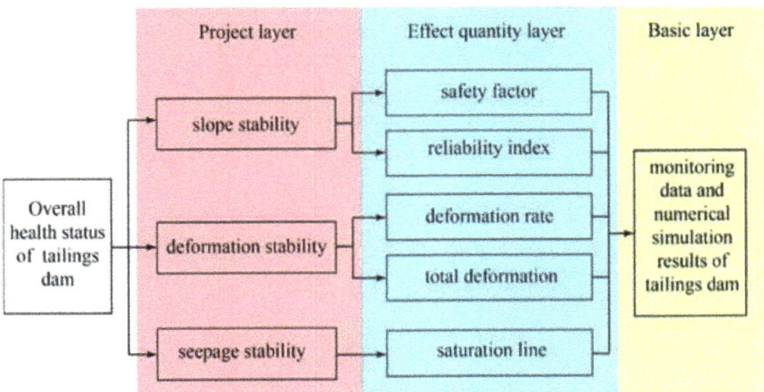

Figure 1. Diagnosis index system of tailings dam health.

2.1.2. Classification of Health Levels

Comprehensive health diagnosis needs to classify the diagnosis results and refer to relevant literatures and norms [29], and the classification is mainly based on the fourth and third levels. For example, the Chinese standard Tailing facilities design code [30] is divided into four levels according to the degree of safety: normal pond, disease pond, risk pond, and dangerous pond. Guidelines on the safe design and operating standards for tailings storage [31] issued by Australia is divided into three levels: high, significant and low. Because the definition of safety degree in the four levels is mainly qualitative description, the boundary between risk pond and dangerous pond is relatively fuzzy. Therefore, the study divides the health levels of tailings dams into three levels. Referring to medical comments on human health, the levels are determined as healthy, diseased, and dangerous.

2.1.3. Standardization Method of Health Value

The effect quantity of each diagnosis index is different, and it is not commensurable. Therefore, the index diagnosis result is processed to make it dimensionless, that is, it is converted into the form of health value, and the influence of index unit and numerical magnitude is excluded. The health value interval is set to [0, 1], in which the health value of 1 is the most ideal health state. The specific classification criteria are shown in Table 1.

Table 1. Classification criteria for health values.

Health Level	Healthy	Diseased	Dangerous
Health value	$[1.0, 0.67)$	$[0.67, 0.33)$	$[0.33, 0]$

The health value of the basic index is calculated by standardized Equation (1):

$$X = \begin{cases} 1 & U \in \left(-\infty, u_1^*\right) \\ 0.67 + 0.33\frac{u_1-U}{u_1-u_1^*} & U \in \left[u_1^*, u_1\right) \\ 0.33 + 0.34\frac{u_2-U}{u_2-u_1} & U \in \left[u_1, u_2\right) \\ 0.33\frac{u_2^*-U}{u_2^*-u_2} & U \in \left[u_2, u_2^*\right) \\ 0 & U \in \left[u_2^*, +\infty\right) \end{cases} \tag{1}$$

where X is the index health value; U is the diagnosis parameter of the basic index; u_1 is the threshold value of the healthy and diseased level; u_2 is the threshold value of the diseased and dangerous level;.u_1^*. and u_2^*. are the upper and lower limits of the diagnosis parameters, respectively, and the health values exceeding the limits are 1 and 0.

2.2. Determination of Dynamic Weight of Diagnosis Indexes

2.2.1. The Analytical Hierarchy Process

The analytic hierarchy process is a comprehensive evaluation method that combines qualitative and quantitative analysis by objectively describing subjective judgments [32]. It has a wide range of applications in comprehensive evaluation [33]. The steps of the analytic hierarchy process to determine the weight of index are as follows:

(1) Hierarchical structure reflects the relationship between indexes, and the proportion of each index in the same target layer is quantitatively analyzed by the judgment matrix $A = (a_{ij})_{n\times n}$. The judgment matrix is positive and the reciprocal matrix is constructed by comparing the factors in pairs, and is generally represented by the scale of 1–9.

(2) The constructed judgment matrix has a certain degree of inconsistency, so its rationality is checked for consistency. When the consistency index $CR < 0.1$, the consistency of the judgment matrix is considered acceptable, and the weight coefficients are allocated reasonably.

(3) The maximum eigenvalue $\lambda_{\max}$ and the corresponding eigenvector x are obtained by solving the judgment matrix $A = (a_{ij})_{n\times n}$, and the weights of indexes are obtained by normalizing eigenvector x.

2.2.2. The Penalty Variable Weight Method

The deterioration of each index in the project layer will have a significant impact on the health of the tailings dam. When a certain index deteriorates while other indexes are still in healthy state, the traditional fixed-weight diagnosis ignores the impact of the deterioration index. The penalty variable weight method can adjust the weight of the index according to the change of health value, thereby highlighting the diagnosis index that has deteriorated. The method increases the influence of the index with the lower health value in the comprehensive diagnosis, so as to diagnose the overall health status of the tailings dam more reasonably and accurately.

According to the definition of the penalty variable weight function [34], the three axioms of normalization, continuity and monotonicity need to be satisfied, and, combined with the characteristics of the tailings dam, the variable weight function is constructed as follows:

$$S(x) = \begin{cases} 1 & x > 0.67 \\ \ln(0.67/x)^{10} + 1 & 0.33 < x \leq 0.67 \\ \ln(0.67/x)^{20} - 7.08 & x \leq 0.33 \end{cases} \tag{2}$$

where $S(x)$ is the value of variable weight; x is the health value of the index. The variable weight function is composed of three sections: non-penalty function, penalty function and heavy penalty function, which respectively correspond to three health states: healthy, diseased, and dangerous. The variable weight value of each index is calculated by Equation (3):

$$w_i(x_i) = w_i^{(0)} S_i(x) / \sum_{j=1}^{n} w_j^{(0)} S_j(x) \tag{3}$$

where x_i is the health value of the ith index; n is the number of index; $w_i(x_i)$ is the variable weight of the ith index; $w_i^{(0)}$ is the fixed weight of the ith index.

2.3. Diagnosis Method of the Index of Effect Quantity Layer

The data dimensions of each index of the basic layer are different, and the reflections on the health status of the tailings dam are also different. Therefore, according to the characteristics of diagnosis indexes, the reasonable quantitative methods are selected to analyze the indexes, such as numerical simulation, statistical analysis and so on. Then, the norms method and the confidence interval method are used to determine the quantitative standard of the index, to complete the quantitative diagnosis of the indexes.

2.3.1. Slope Stability Project

The typical profile of the tailings dam is selected, and the stability of the tailings dam is calculated by the limit equilibrium method to obtain the slope safety factor corresponding to deterministic parameters. Considering the variability of tailings material and taking the mean and standard deviation of its physical and mechanical parameters as random variables, the reliability analysis is carried out by the Monte Carlo method. A large number of tailings parameter combinations are extracted and their anti-sliding stability safety factors are calculated respectively to determine the probability of the tailings dam break and the reliability index. Quantitative diagnosis is performed by combining the calculation results of the deterministic safety factor and reliability index. The quantitative standard adopts the norms method, and the specific quantitative standard is:

(1) Regarding the deterministic safety factor index, the minimum safety factor stipulated in the code is regarded as u_1, which is the threshold value of the healthy and diseased level. $u_1 * u_1$ is taken as u_1^* that is the upper limit value. The norm [35] describes that the tailings dams with a minimum safety factor of less than 0.95 times the stipulated value belong to the dangerous reservoir, so this value is taken as u_2, which is the threshold value of the diseased and dangerous level, and the safety factor of 1 is taken as u_2^*, which is the lower limit value.

(2) Regarding the reliability index, referring to the value in the norm [35], the specific values are shown in Table 2.

Table 2. Reliability index of hydraulic structure.

Structural Safety Level		1	2	3
Category of destruction	The first	3.7	3.2	2.7
	The second	4.2	3.7	3.2

The second category of destruction is suitable for sudden brittle destruction, and is mostly applied to concrete structures with higher requirements. Therefore, the reliability index of the tailings dam refers to the first category of destruction. According to the corresponding relationship between reliability index and failure probability, the hierarchical corresponding relationship between the structural safety level and the probability of tailings dam failure is shown in Table 3.

Table 3. Safety level and dam-break probability of the tailings dam.

Structural Safety Levels	Tailings Dam	Reliability Index	Dam Break Probability
1	1	3.7	1.08×10^{-4}
2	2,3	3.2	6.87×10^{-4}
3	4,5	2.7	3.47×10^{-3}

In the norms [29], the magnitude of the probability of dam failure is taken as the basis for the classification of risk significance, and the significance difference of different health levels is reflected by enlarging or reducing the failure probability by one magnitude. According to the reference manual, the value in Table 2 is taken as u_1, and the value is reduced by two orders of magnitude as u_1^*. The value in Table 2 is enlarged by one order of magnitude as u_2, and the u_2 is enlarged by one order of magnitude as u_2^*.

2.3.2. Deformation Stability Project
Deformation Rate

The normal operating value of deformation rate is determined by statistical analysis of historical monitoring data, such as curve fitting and statistical regression. The quantitative standard adopts the norms method, which stipulates that the yellow warning value is 1.3 times the normal operating value, the orange warning value is 2 times the normal operating value, and the red warning value is 3 times the normal operating value. Therefore, $u_1^*,u_1,u_2,$ and u_2^* are 1, 1.3, 2 and 3 times of the normal operating value respectively.

Total Deformation

After the tailings dam stops filling sub-dam, there is no new load on the upper part, and the deformation of the dam body is mainly the secondary consolidation deformation of tailings. For tailings with cementing properties, creep deformation may lead to loss of strength and eventually instability failure. The quantification of the total deformation index adopts the numerical analysis. Through forward and inverse analysis of the creep deformation of the tailings dam, the total deformation of the dam body can be calculated more accurately. Deformation failure is defined as a penetrating failure area formed by excessive deformation.

Taking the safety factor of deformation as the quantitative index, the greater the safety factor, the greater the safety margin of the total deformation index. When the safety factor is 1, the monitored deformation of the dam body reaches the destruction deformation, that is, the tailings dam is on the verge of instability. Therefore, u_2^* and u_2 take 1 and F_s, which is the slope safety factor specified in the norm. Referring to relevant norms and study, u_1 and u_1^* take 2 and $2 * F_s$.

2.3.3. Seepage Stability Project

The depth of saturation line of the tailings dam has an important influence on the stability of the slope, so the quantitative basis of the saturation line index is the coupling relationship between the depth of saturation line and the stability of the slope.

The distribution of the saturation line under different dry beach lengths is obtained by seepage calculation. Then, the stability safety factors of the dam slope corresponding to different saturation lines are calculated, so as to establish the coupling relationship between the depth of saturation line and the stability safety factor. According to the relationship curve, u_1 takes the depth of saturation line corresponding to F_s. u_1^* is twice of u_1, and u_2 is 0.95 times of F_s. u_2^* is the minimum depth of saturation line specified in the norm.

Finally, the flow chart of this paper can be shown in Figure 2.

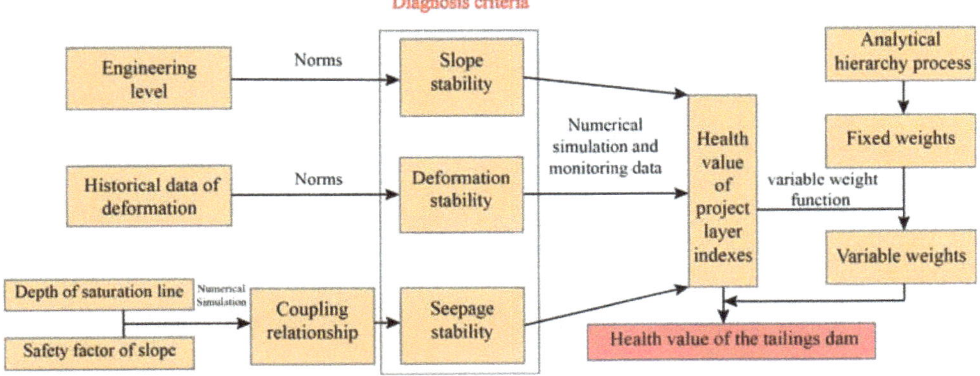

Figure 2. The flow chart of diagnosis process.

3. Case Study

The tailings dam B-I at Córrego do Feijão Iron Ore Mine (dam I) located in Brumadinho, Brazil, was constructed in 1976. The tailings discharge was stopped in July 2016. Before the dam break, the height of the tailings dam was 86 m and the storage capacity was 12.7 Mm^3. On 25 January 2019, the tailings dam broke, and about 9.7 Mm^3 tailings flowed out of the pond, killing 235 people and flooding an area of about 40 km^2 downstream [36]. Figure 3 is the satellite image taken before the dam break, in which ◆ is the monitoring point of the saturation line and ● is the monitoring point of the deformation.

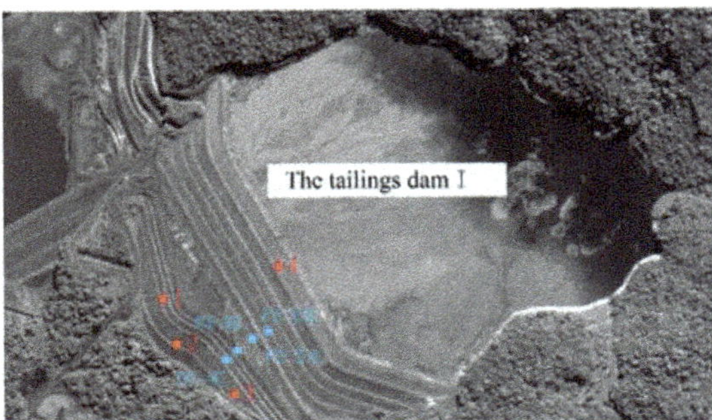

Figure 3. Satellite image of tailings dam I.

3.1. Diagnosis of the Slope Stability Project

Tailings dam I is a Class 3 dam, and the minimum safety factor specified in the norm [30] is 1.3. According to the quantitative method of the effect quantity index, the quantitative standards of deterministic safety factor index and reliability index are determined. Specific diagnostic criteria are shown in Table 4.

Table 4. Diagnosis criteria of the effect quantity indexes (tailings dam I).

Project Layer	Effect Quantity Layer		Healthy	Diseased	Dangerous
Dam slope stability	Safety fator	Normal operating conditions	[1.690, 1.300]	(1.300, 1.235]	(1.235, 1.000]
	Reliability	Reliability index	[4.348, 3.200]	(3.200, 2.464]	(2.464, 1.485]
		Probability of failure	$[6.87 \times 10^{-6}, 6.87 \times 10^{-4}]$	$(6.87 \times 10^{-4}, 6.87 \times 10^{-3}]$	$(6.87 \times 10^{-3}, 6.87 \times 10^{-2}]$
Deformationstability	Deformation rate(mm/d)	1	[0.244, 0.317)	[0.317, 0.488)	[0.488, 0.732]
		2	[0.203, 0.264)	[0.264, 0.406)	[0.406, 0.609]
		3	[0.222, 0.288)	[0.288, 0.443)	[0.443, 0.666]
		4	[0.240, 0.312)	[0.312, 0.480)	[0.480, 0.720]
	Total deformation	Deformation safety factor	[2.6, 2)	[2, 1.3)	[1.3, 1]
Seepagestability	Depth of the saturation line	PZ-4C	[13.60, 6.80)	[6.80, 4.40]	[4.40, 2.47]
		PZ-5C	[22.60, 11.30)	[11.30, 7.50)	[7.50, 2.87]
		PZ-24C	[28.92, 14.46)	[14.46, 9.21)	[9.21, 3.20]
		PZ-23C	[36.50, 18.25)	[18.25, 11.46)	[11.46, 3.53]

The typical section of the tailings dam is selected to establish a two-dimensional model, as shown in Figure 4. The iron content of the tailings is more than 50%, which makes the bulk density of the tailings very high, about 26 kN/m^3. The material parameters required in the stability calculation refer to the test data [37], as shown in Table 5.

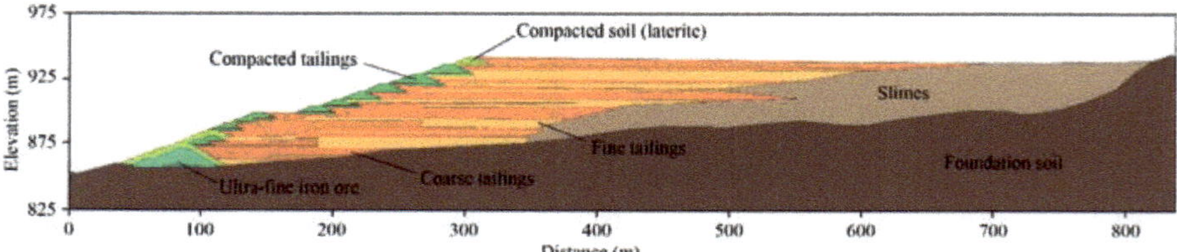

Figure 4. Typical profile of tailings dam I.

Table 5. Mechanical parameters of the materials.

Parameters	γ (KN/m^3)	c (kPa)	φ (°)	k_h (m/s)	k_v/k_h
Coarse tailings	26.5	0	33	5.00×10^{-6}	0.2
Fine tailings	26.0	0	32	1.00×10^{-7}	0.2
Compacted tailings	27.5	0	36	5.00×10^{-7}	0.2
Ultra-fine iron ore	25	0	35	1.20×10^{-6}	1
Compacted soil (laterite)	20	12	29	1.20×10^{-9}	1
Slimes	23	0	25	1.00×10^{-7}	0.2
Foundation soil	23	15	30	9.30×10^{-7}	1

The calculation results of stability and reliability are shown in Figure 5. The most dangerous sliding surface of the tailings dam is located between the downstream toe and the fourth sub-dam, and its deterministic safety factor is 1.307, which is basically consistent with the results of the accident investigation report. The failure probability is 1.45×10^{-3}, and the reliability index is 2.704. The diagnostic results determined by the quantified standard and the Equation (1) show that the health value of the safety factor index is 0.6759, and the health value of the reliability index is 0.4376.

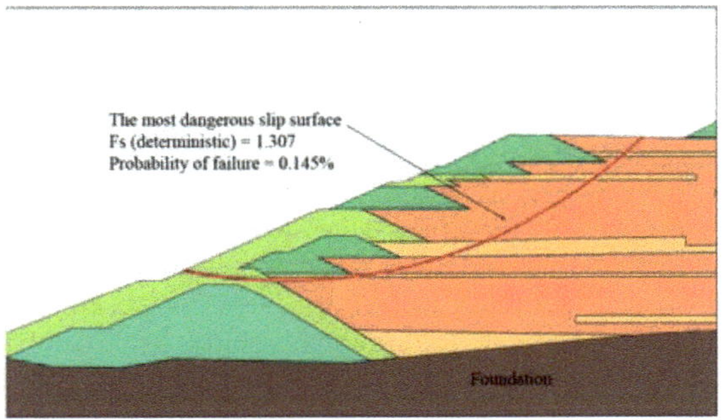

Figure 5. Calculation results of stability and reliability of tailings dam I.

3.2. Diagnosis of the Deformation Stability Project

3.2.1. Deformation Rate Index

The deformation monitoring data is derived from the radar monitoring data, including the vertical component and a small amount of the horizontal component. Because the horizontal component is close to the noise level, the vertical deformation is used for diagnosis of the deformation rate index. Figure 6 shows the vertical deformation curve of the points, with No.1, No.2 and No.3 measuring points located at the bottom of the dam and No.4 measuring point located at the top of the dam.

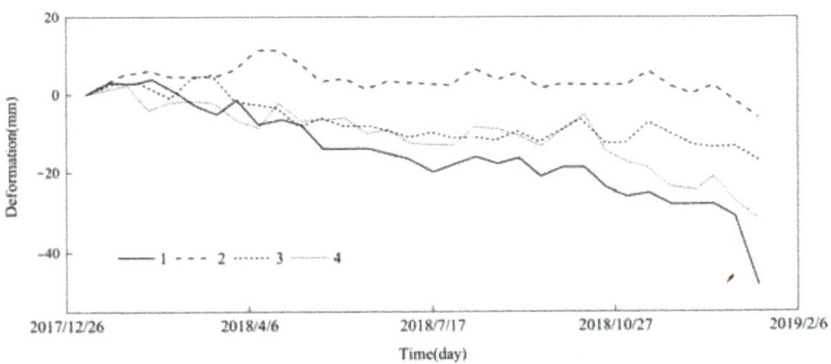

Figure 6. Vertical deformation curve of the measuring points.

Based on the monitoring data from 6 January 2018 to 20 December 2018, it can seen that the deformation in the historical period has no obvious acceleration stage, and the overall deformation is uniform. Therefore, the average deformation rate in the period is taken as the normal operation value. According to the determination method of quantification standard, the quantification standard of each point is obtained, which is shown in Table 4. The standard is used to diagnose the average deformation rate of each point within one month before the dam break, and the healthy value of the deformation rate index before the dam break is obtained. The specific results are shown in Table 6.

Table 6. Diagnostic results of the deformation rate indicator.

Measuring Points	1	2	3	4
Deformation rate (mm/d)	0.82	0.33	0.28	0.42
Health value	0.000	0.5066	0.7100	0.4479

3.2.2. Total Deformation Index

Because of the high iron content in the tailings, the oxidation of iron will lead to cementation between particles. The test results [37] also show that under the constant load, the continuous creep of the tailings may lead to obvious and rapid strength loss, so the strain-softening model is adopted in the simulation calculation.

The initial stress state is obtained by simulating the accumulation process of the tailings dam. Then the creep deformation is added to the calculation, and the deformation is increased step by step, and the stability of the tailings dam after creep is calculated step by step until the tailings dam is destroyed. The investigation report shows that when the added creep is consistent with the deformation indicated by the monitoring data, the strength loss of tailings will be caused and dam failure will occur. Combined with the diagnostic standard shown in Table 4, the safety factor of the total deformation is 1, and the total deformation index of the tailings dam is determined to be in a dangerous state and the health value is 0.

3.3. Diagnosis of the Seepage Stability Project

The seepage field of the dry beach with lengths of 50 m, 100 m, 125 m, 150 m and 200 m is calculated. The results show that the saturation lines all overflow from the drainage body between the fifth and sixth sub-dams, and then flow into the downstream channel through the drainage channel on the dam surface, which is consistent with the actual situation. By calculating the stability safety factor of the dam slope corresponding to the seepage field under various working conditions, the relationship curves between the depth of saturation line at PZ-4C, PZ-5C, PZ-24C and PZ-23C points and the stability safety factor of the dam slope are established, as shown in Figure 7.

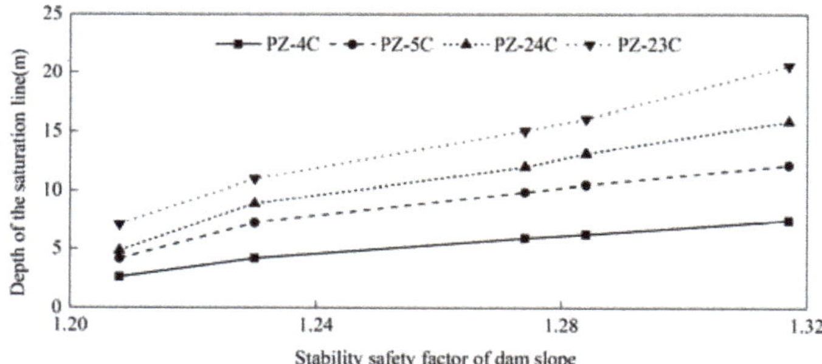

Figure 7. The relation curve between the depth of saturation line and the safety factor.

According to the curve, the quantification standard of each point is determined, as shown in Table 4. The diagnosis data adopts the average value of the depth of the saturation line within one month before the dam break, and the diagnosis results are shown in Table 7.

Table 7. Diagnosis results of the saturation line index.

Measuring Points	PZ-4C	PZ-5C	PZ-24C	PZ-23C
Depth of the saturation line (m)	14.29	21.48	15.55	10.47
Health value	1.000	0.9673	0.6949	0.2888

3.4. Comprehensive Health Diagnosis

The health value of the project-layer index is obtained by summarizing the health value of the effect quantity index, and the health values of the slope stability project, the deformation stability project and the seepage stability project are 0.5568, 0.2080 and 0.7377, respectively, as shown in Table 8. The slope stability project is in a diseased state, and the deformation stability project is in a dangerous state. Before the dam break, the seepage field of the tailings dam is in good condition, and the depth of saturation line has a downward trend year by year, and the seepage stability project is in healthy state.

Table 8. Health values of each layer index.

Project Layer		Effect Quantity Layer		Basic Layer	
Dam slope stability	0.5568	Safety fator	0.6759	Safety fator of dam slope	0.6759
		Reliability	0.4376	Reliability index	0.4376
Deformation stability	0.2080	Deformation rate	0.4161	1	0.0000
				2	0.5066
				3	0.7100
				4	0.4479
		Total deformation	0.0000	Deformation safety factor	0.0000
Seepage stability	0.7378	Depth of the saturation line	0.7378	PZ-4C	1.0000
				PZ-5C	0.9673
				PZ-24C	0.6949
				PZ-23C	0.2888

The analytical hierarchy process method was used to weight the project-layer indexes, and eight experts were invited to give judgment matrix considering the importance of each index to the overall health status of the tailings dam and the operation characteristics of the tailings dam.

The consistency indexes of the judgment matrices are all less than 0.1, and the consistency test meets the requirements. The weight of the project-layer index is obtained by solving the eigenvalue of the judgment matrices. As can be seen from Table 9, the opinions of experts are relatively unified, and they all think that the slope stability project is more important. Combining the opinions of eight experts by the arithmetic average method, the fixed weights of the project-layer index are $w = (0.4483, 0.2612, 0.2905)$. The health value of the slope stability project falls within the diseased state. Therefore, the penalty function should be adopted in Equation (2), and its value of variable weights should be determined to be 2.5813. Similarly, the values of variable weights of the project layer indexes obtained by Equation (2) are $S = (2.8513, 16.3090, 1.000)$, and the final variable weights calculated by Equation (3) are $w_0 = (0.2193, 0.7309, 0.0498)$. The health value of the tailings dam based on dynamic weight is 0.3109 calculated by the weighted average method, that is, the tailings dam before the dam break was in a dangerous state, which is consistent with the actual situation of the tailings dam. The result proves the accuracy and applicability of this method. The health value of the tailings dam based on fixed weight is 0.5283. The traditional diagnosis method based on fixed weight will lead to the distortion of diagnosis results, while the dynamic weight can effectively solve the problem of state imbalance and improve the reliability of diagnosis results.

Table 9. Weighting results of the project-layer indexes.

Experts		A1	A2	A3	A4	A5	A6	A7	A8	Results
project-layer	Slope stability	0.4934	0.5396	0.2402	0.6250	0.3874	0.1740	0.6337	0.4934	0.4483
	Deformation	0.3108	0.1634	0.5499	0.1365	0.1692	0.3715	0.1919	0.1958	0.2612
	Seepage	0.1958	0.2970	0.2098	0.2385	0.4434	0.4545	0.1744	0.3108	0.2905
Consistency index		0.0517	0.0088	0.0176	0.0157	0.0176	0.0166	0.0089	0.0516	-

4. Conclusions

(1)   The index system of comprehensive diagnosis of tailings dam health is established, and the dynamic change of index weight is realized based on the analytical hierarchy process and the penalty variable weight method, which increases the importance of the deterioration index in comprehensive diagnosis and makes the diagnosis result more accurate and reasonable.

(2)   Based on numerical simulation and the statistical analysis method, the diagnosis method of the indexes of effect quantity layer is put forward, and the quantitative

standard of each index is determined. The comprehensive diagnosis method of tailings dam health based on monitoring data is put forward, and the quantitative diagnosis of tailings dam health status is realized.

(3) This method was applied to tailing dam I, and the health value of 0.3109 indicates that the tailings dam is in a dangerous state before the dam failure, which is consistent with the actual situation and verifies the accuracy and applicability of the method.

The comprehensive diagnosis based on the dynamic weight of tailings dam is very valuable. This method overcomes the influence of artificial subjective judgment and provides a new method for evaluating the safety of tailings dams.

Author Contributions: Conceptualization, K.D. and Z.M.; methodology, K.D. and Z.M.; software, K.D.; validation, D.Y.; formal analysis, K.D.; investigation, K.D.; resources, K.D.; data curation, K.D. and D.Y.; writing—original draft preparation, K.D.; writing—review and editing, K.D. and D.Y.; visualization, K.D.; supervision, Z.M. and D.Y.; project administration, K.D.; funding acquisition, Z.M. All authors have read and agreed to the published version of the manuscript.

Funding: This research was funded by the National Key R&D Program of China (No. 2017YFC0804605), and the National Natural Science Foundation of China (51539006, 51909174).

Institutional Review Board Statement: Not applicable.

Informed Consent Statement: Not applicable.

Data Availability Statement: Not applicable.

Conflicts of Interest: The authors declare no conflict of interest.

References

1. McDermott, R. The Aznalcóllar Tailings Dam Accident—A Case Study. *Miner. Resour. Eng.* **2000**, *9*, 101–118. [CrossRef]
2. Fourie, A.B.; Blight, G.E.; Papageorgiou, G. Static liquefaction as a possible explanation for the Merriespruit tailings dam failure. *Can. Geotech. J.* **2001**, *38*, 707–719. [CrossRef]
3. Guodong, M. Quantitative Assessment Method Study Based on Weakness Theory of Dam Failure Risks in Tailings Dam. *Procedia Eng.* **2011**, *26*, 1827–1834. [CrossRef]
4. Byrne, P.; Hudson-Edwards, K.A.; Bird, G.; Macklin, M.G.; Brewer, P.; Williams, R.D.; Jamieson, H.E. Water quality impacts and river system recovery following the 2014 Mount Polley mine tailings dam spill, British Columbia, Canada. *Appl. Geochem.* **2018**, *91*, 64–74. [CrossRef]
5. Burritt, R.L.; Christ, K. Water risk in mining: Analysis of the Samarco dam failure. *J. Clean. Prod.* **2018**, *178*, 196–205. [CrossRef]
6. Su, H.Z.; Hu, J.; Wu, Z.R. A study of safety evaluation and early-warning method for dam global behavior. *Struct. Health Monit.* **2012**, *11*, 269–279.
7. Xu, Q.; Xu, K. Assessment of air quality using a cloud model method. *R. Soc. Open Sci.* **2018**, *5*, 171580. [CrossRef] [PubMed]
8. Wang, Y.; Jing, H.; Yu, L.; Su, H.; Luo, N. Set pair analysis for risk assessment of water inrush in karst tunnels. *Bull. Eng. Geol. Environ.* **2016**, *76*, 1199–1207. [CrossRef]
9. Zhu, L. Research and application of AHP-fuzzy comprehensive evaluation model. *Evol. Intell.* **2020**, *13*, 1–7. [CrossRef]
10. Do, T.; Laue, J.; Mattsson, H.; Jia, Q. Numerical Analysis of an Upstream Tailings Dam Subjected to Pond Filling Rates. *Appl. Sci.* **2021**, *11*, 6044. [CrossRef]
11. Li, W.; Ye, Y.; Hu, N.; Wang, X.; Wang, Q. Real-Time Warning and Risk Assessment of Tailings Dam Disaster Status Based on Dynamic Hierarchy-Grey Relation Analysis. *Complexity* **2019**, *2019*, 5873420. [CrossRef]
12. Yuan, L.W.; Li, X.M.; Li, S.M.; Chen, Y.M. Study on Risk Assessment Method for Tailings Pond Disaster Based on Improved Index Weight Method. *Adv. Mater. Res.* **2015**, *1092–1093*, 753–761. [CrossRef]
13. Krupskaya, L.; Zvereva, V.P.; Kirienko, O.A.; Imranova, E.L.; Volobueva, N.G. Using an comprehensive method for evaluation of environment state in a zone of influence of a tailing dam. *Russ. J. Gen. Chem.* **2016**, *86*, 3012–3014. [CrossRef]
14. Liu, C.; Shen, Z.; Gan, L.; Xu, L.; Zhang, K.; Jin, T. The Seepage and Stability Performance Assessment of a New Drainage System to Increase the Height of a Tailings Dam. *Appl. Sci.* **2018**, *8*, 1840. [CrossRef]
15. He, L.; Gao, B.; Luo, X.; Jiao, J.; Qin, H.; Zhang, C.; Dong, Y. Health Risk Assessment of Heavy Metals in Surface Water near a Uranium Tailing Pond in Jiangxi Province, South China. *Sustainability* **2018**, *10*, 1113. [CrossRef]
16. Pan, K.; Dong, Y. Risk Assessment of Tailings Pond Based on Triangular Fuzzy Theory. *Adv. Mater. Res.* **2012**, *524–527*, 515–519. [CrossRef]
17. Chu, H.; Xu, G.; Yasufuku, N.; Yu, Z.; Liu, P.; Wang, J. Risk assessment of water inrush in karst tunnels based on two-class fuzzy comprehensive evaluation method. *Arab. J. Geosci.* **2017**, *10*, 179. [CrossRef]

18. Gu, H.; Fu, X.; Zhu, Y.; Chen, Y.; Huang, L. Analysis of Social and Environmental Impact of Earth-Rock Dam Breaks Based on a Fuzzy Comprehensive Evaluation Method. *Sustainability* **2020**, *12*, 6239. [CrossRef]
19. Wang, G.; Tian, S.; Hu, B.; Chen, J.; Kong, X. Regional Hazard Degree Evaluation and Prediction for Disaster Induced by Discharged Tailings Flow from Dam Failure. *Geotech. Geol. Eng.* **2020**, *39*, 2051–2063. [CrossRef]
20. Wang, T.; Zhou, Y.; Lv, Q.; Zhu, Y.; Jiang, C. A safety assessment of the new Xiangyun phosphogypsum tailings pond. *Miner. Eng.* **2011**, *24*, 1084–1090. [CrossRef]
21. Xu, B.; Wang, Y. Stability analysis of the Lingshan gold mine tailings dam under conditions of a raised dam height. *Bull. Eng. Geol. Environ.* **2014**, *74*, 151–161. [CrossRef]
22. Wang, D.; Shen, Z.Z.; Tao, X.H. Three-dimensional finite element analysis and safety assessment for seepage field of a tailings dam. *J. Hohai Univ. (Nat.Sci.)* **2012**, *3*, 307–312. (In Chinese)
23. Li, Q.; Gao, S.; Niu, H.K.; Shang, Y.L. Analytical solution to saturation line of tailings pond and its applicability analysis. *Rock Soil Mechannics* **2020**, *41*, 9. (In Chinese)
24. Dong, L.; Shu, W.; Sun, D.; Li, X.; Zhang, L. Pre-Alarm System Based on Real-Time Monitoring and Numerical Simulation Using Internet of Things and Cloud Computing for Tailings Dam in Mines. *IEEE Access* **2017**, *5*, 21080–21089. [CrossRef]
25. Li, Q.M.; Yuan, H.N.; Zhong, M.H. Safety assessment of waste rock dump built on existing tailings ponds. *J. Cent. South Univ.* **2015**, *22*, 2707–2718. [CrossRef]
26. Dong, L.; Deng, S.; Wang, F. Some developments and new insights for environmental sustainability and disaster control of tailings dam. *J. Clean. Prod.* **2020**, *269*, 122270. [CrossRef]
27. Tan, Q.W.; Xin, B.Q.; Wan, L.; Dong, Y.; Du, S. Risk evaluation indexes and gradation method of major hazard installations for tailings pond. *J. Saf. Sci. Technol.* **2018**, *14*, 99–106. (In Chinese)
28. Jiang, F.; Wu, H.; Liu, Y.; Chen, G.; Guo, J.; Wang, Z. Comprehensive evaluation system for stability of multiple dams in a uranium tailings reservoir: Based on the TOPSIS model and bow tie model. *R. Soc. Open Sci.* **2020**, *7*, 191566. [CrossRef] [PubMed]
29. ANCOLD. *Guideline on Risk Assessment*; Australian National Committee on Large Dams: Sydney, Australia, 2003.
30. Ministry of Housing and Urban-Rural Development of the People's Republic of China. *Tailings Facility Design Specification*; GB50863–2013; China Planning Press: Beijing, China, 2013. (In Chinese)
31. Australia, Western. Guidelines on the Safe Design and Operating Standards for Tailings Storage. Department of Minerals and Energy Western Austrlia. 1999. Available online: https://www.scribd.com/document/235421799/Guidelines-Safe-Design-Operating-Standards-for-Tailings-Storage-Wa (accessed on 1 December 2020).
32. Ishizaka, A.; Labib, A. Review of the main developments in the analytic hierarchy process. *Expert Syst. Appl.* **2011**, *38*, 14336–14345. [CrossRef]
33. Jena, R.; Pradhan, B.; Beydoun, G.; Nizamuddin, N.; Ardiansyah; Sofyan, H.; Affan, M. Integrated model for earthquake risk assessment using neural network and analytic hierarchy process: Aceh province, Indonesia. *Geosci. Front.* **2019**, *11*, 613–634. [CrossRef]
34. Li, H.X. Factor space and mathematical frame of knowledge representation (VIII)—Variable weights analysis. *Fuzzy Syst. Math.* **1995**, *3*, 1–9. (In Chinese)
35. Ministry of Housing and Urban-Rural Development of the People's Republic of China. *Unified Standard for Reliability Design of Hydraulic Engineering Structures*; GB50199–2013; China Planning Press: Beijing, China, 2013. (In Chinese)
36. Palmer, J. Anatomy of a Tailings Dam Failure and a Caution for the Future. *Engineering* **2019**, *5*, 605–606. [CrossRef]
37. Robertson, P.K.; Melo, L.; Williams, D.J.; Wilson, G.W. *Report of the Expert Panel on the Technical Causes of the Failure of Feijão Dam I*. Australia. 2019. Available online: https://www.resolutionmineeis.us/documents/robertson-et-al-2019. (accessed on 1 December 2020).

Article

Study on Surface Subsidence Characteristics Based on Three-Dimensional Test Device for Simulating Rock Strata and Surface Movement

Xingyin Ma [1,2], Zhiyong Fu [3,*], Yurong Li [1,2,*], Pengfei Zhang [1,2], Yongqiang Zhao [4] and Guoping Ma [5]

[1] College of Energy and Mining Engineering, Shandong University of Science and Technology, Qingdao 266590, China; ma.xingyin@sdust.edu.cn (X.M.); pfzhang@sdust.edu.cn (P.Z.)

[2] State Key Laboratory of Mining Disaster Prevention and Control Co-Founded by Shandong Province and the Ministry of Science and Technology, Shandong University of Science and Technology, Qingdao 266590, China

[3] College of Mining, Liaoning Technical University, Fuxin 123000, China

[4] State Key Laboratory of Water Resource Protection and Utilization in Coal Mining, Beijing 102209, China; 20039429@chnenergy.com.cn

[5] Qianjiaying Mining Branch, Kailuan (Group) Co., Ltd., Tangshan 063000, China; maguoping@kailuan.com.cn

* Correspondence: fuzhiyonghappy@163.com (Z.F.); li.yurong@sdust.edu.cn (Y.L.)

Abstract: The main functions of a three-dimensional test device for simulating rock formations and surface movement affected by underground coal mining were described in detail, and a series of similar related tests were carried out. The device consisted of an outer frame, a pressurization unit, a pulling unit, and a coal seam simulation portion. Using this test device, supported by monitoring methods such as the three-dimensional laser scanner method, a model test study on the surface subsidence characteristics caused by coal seam mining was carried out. Combined with the field measurements, the transfer law of surface subsidence caused by coal seam mining was revealed, and the whole surface subsidence response process was analyzed. The experimental results show that the subsidence caused by mining disturbances below the coal seam accounts for 79.3% of the total subsidence, which is the dominant factor of the total surface subsidence. After long-term surface observations, surface subsidence can be divided into four stages after coal mining, and the settlement value of the obvious settlement stage accounts for more than 60% of the total settlement value. The above test results fully reflect the feasibility and practicality of the three-dimensional test device to simulate rock strata and surface movement and provide a new experimental research tool that can be used to further study the surface subsidence characteristics and control caused by coal mining.

Keywords: rock formations; surface subsidence law; surface subsidence process; 3D test device; 3D laser scanning

Citation: Ma, X.; Fu, Z.; Li, Y.; Zhang, P.; Zhao, Y.; Ma, G. Study on Surface Subsidence Characteristics Based on Three-Dimensional Test Device for Simulating Rock Strata and Surface Movement. *Energies* **2022**, *15*, 1927. https://doi.org/10.3390/en15051927

Academic Editor: Longjun Dong, Yanlin Zhao and Wenxue Chen

Received: 22 January 2022
Accepted: 3 March 2022
Published: 7 March 2022

Publisher's Note: MDPI stays neutral with regard to jurisdictional claims in published maps and institutional affiliations.

1. Introduction

With the transformation and upgrade of coal development and people's increased awareness of environmental protection issues, the vast majority of coal mines in China will encounter problems related to coal pressing to protect buildings, structures, water bodies and other protected bodies during the construction and production process, as well as mining problems that are influenced by protective bodies, that is, problems related to subsidence control and coal mining activities under special conditions, seriously restricting the production of coal mining enterprises [1–3]. Fundamentally, technical measures that reduce subsidence and control loss mainly include filling mining, partial mining, coordinated mining, etc. [4–7]. Filling mining is a method that has been proven to solve pressed coal problems. The use of this method supports the rock mass over the mined-out area, thereby alleviating surface subsidence, reducing damage to surface buildings, achieving the goals of efficiently mining coal mine resources and of controlling surface damage [8–13]. Controlling the deformation and destruction of structures, such as villages, railways and other structures is one of the problems that the mining industry urgently needs to solve. After mining, stress in the overlying rock mass is redistributed, causing local stress concentration in the surrounding rock, causing the top of the mined-out area to sink, become crushed, or fall [14–16]. As the stress in the support body changes, surface deformation will be induced within a certain range [17–20]. Therefore, studying the influence of strata movement and surface subsidence on the villages and buildings on the surface during and after coal mining is of great significance [21–23].

There are many theoretical methods and monitoring methods related to surface subsidence, and many achievements have been made [24,25]. For example, Sun proposed a theoretical method to predict surface subsidence caused by inclined coal seam mining [26]. Dong studied the influence of different factors on tomography [27]. Existing test devices for similar coal mining material simulations are mostly two-dimensional test benches that simulate the roof and rock formation movement in coal mines and represent mature technology, but there are certain surface movement limitations that must be accounted for during simulation [28–30]. The control process of the coal seam simulation components in three-dimensional test equipment is complex, and successful trial production is difficult or is limited by the bearing capacity, function and size, making it difficult to effectively combine these simulations with engineering practices to carry out model test research [31–33]. Therefore, in order to better study the strata and surface movement characteristics caused by coal mining, this paper adopts a method combining field measurements and the development of a test device to conduct simulation tests. Through the "three-dimensional test device to simulate the influence of underground coal mining on strata and surface movement", developed by the authors of this paper, combined with three-dimensional laser scanning technology, simulation tests determining surface subsidence after coal mining are carried out. Combined with long-term surface observations, the laws of strata and surface movement caused by coal mining are revealed.

2. Testing Device

2.1. The Overall Structure of the Test Device

This paper introduces a three-dimensional test device that simulates the impact of underground coal mining on rock formations and surface movement, as shown in Figure 1. This device belongs to independent research and development, and has obtained the Chinese utility model patent authorization. And entrust Qingdao local testing machine manufacturers to cooperate in manufacturing. It includes an outer frame, pressurization unit, pulling unit, and coal seam simulation portion. The coal seam simulation portion is located inside the outer frame, the upper surface of which is filled with similar m coal seam materials, and the coal seam simulation portion consists of multiple mining blocks and multiple reserved coal pillar assemblies. The pressurization unit is located on the top of the outer frame, and the pressurization unit is connected to the outer frame by the pressurizing position adjustment unit. The pulling unit is located at the bottom of the outer frame, and the pulling unit is connected to the outer frame by the pulling position adjustment unit. This device can be combined with the coal mining site to simulate the mining process and can simulate the variable mining height, controlled mining speed and ease of pressurization, and laying of similar materials.

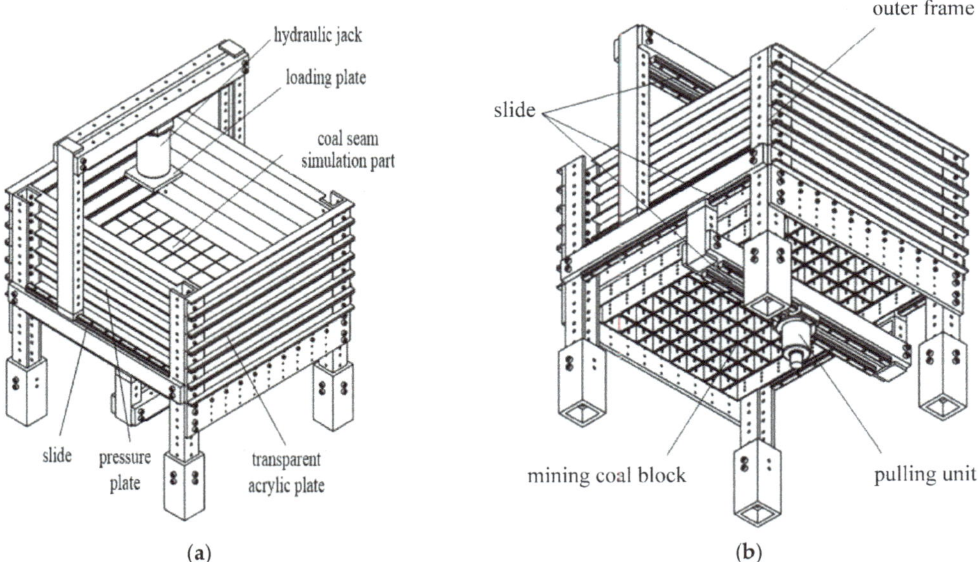

(a)　　　　　　**(b)**

Figure 1. Schematic diagram of the device structure: (**a**) the top structure and components of the device; (**b**) bottom structure and components of the device.

2.2. Introduction of Function and Test Method

Using a "three-dimensional test device for simulating surface movement in underground coal mining" that was developed by the authors independently to carry out the test, a certain degree of model simplification was carried out during the physical processing process of the test device. The specific size of the model is as follows: $x \times y \times z = 0.60$ m $\times$ 0.60 m $\times$ 0.80 m. A detailed image of the model and its size parameters is shown in Figure 2.

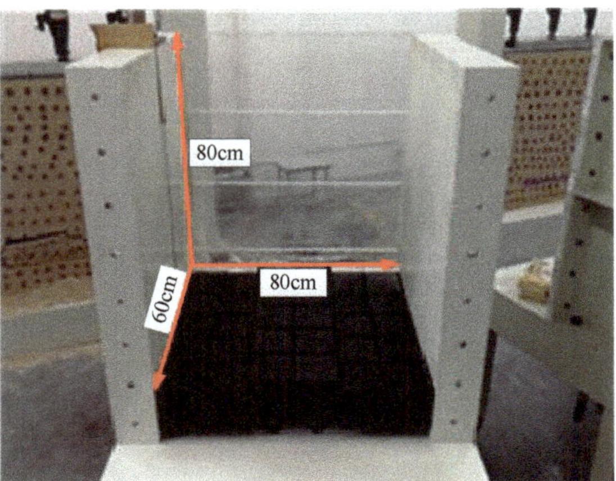

Figure 2. The three-dimensional test device for simulating surface movement in underground coal mining.

This device mainly addresses the technical problems that are present in the existing technology, thus providing a three-dimensional simulation test device and test method that can simulate different coal seam mining schemes and that can facilitate the observation of surface deformation characteristics, as shown in Figure 3. In order to achieve the design purpose of the device, it was determined that the overall structure of the device should mainly consist of the outer frame, pressurization unit, pulling unit, and coal seam simulation portion:

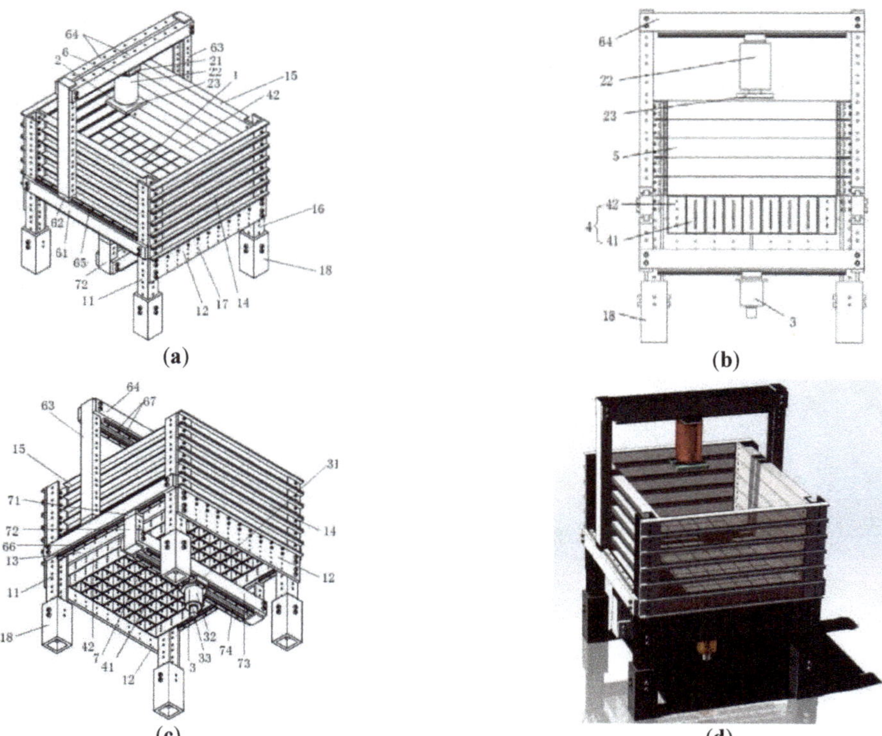

(a)

(b)

(c)

(d)

Figure 3. Structure diagram of the test device. The numbers in the diagrams can be described as follows: 1—outer frame, 11—column, 12—the first pressure plate, 13—the second pressure plate, 14—the first baffle, 15—second baffle, 16—threaded hole, 17—the first threaded hole, 18—bearing, 19—transparent acrylic plate, 2—pressurization unit, 21—the first ball slide, 22—hydraulic jack, 23—loading plate, 3—pulling unit, 31—third ball slide, 32—drawing instrument, 33—pull rod, 4—coal seam simulation part, 41—mining coal block, 42—reserved coal pillar assembly, 5—similar materials for coal rock formations, 6—pressurizing position adjustment unit, 61—central rail beam, 62—second ball slide, 63—upper track column, 64—upper rail beam, 65—upper slide, 66—lower slide, 67—first slide, 7—pulling position adjustment unit, 71—fourth ball slide, 72—lower track post, 73—lower track beam, 74—second slide. (a) direction view 1 (b) left view (c) direction view 2 (d) 3D schematic.

(1) The outer frame includes columns located in the four corners. Multiple threaded holes are spaced throughout the column. Four pressure plates are installed in four columns on all sides with bolts.

(2) The coal seam simulation portion is composed of mining coal blocks and reserved coal column components that are staggered and connected on the horizontal plane. The coal seam simulation portion is connected with four pressure plates, which are all around this part of the model.

(3) The pressurization unit is set at the top of the outer frame and is connected to the outer frame by the pressurizing position adjustment unit. The pressurization unit is used to pressurize the surface of coal rock formations that are composed of similar material; if surface deformation observations are required, then the test geometric similarity ratio can

be adjusted to make the whole range of the model correspond to the whole stratum. Then, there is no need to apply the surface pressure to the model.

(4) The pulling unit is located at the bottom of the outer frame and is connected by the pulling position adjustment unit to the outer frame. The pulling unit is used to pull down the mining coal block in order to simulate coal mining.

In order to coordinate the operation of the unit and to work closely with the various parts of the system, the detailed features of each component were designed so that the functions could be achieved without affecting the overall structure of the equipment: The reserved coal pillar assembly and the mining block comprise a rectangular steel body with a bottom opening, a waist through-hole is located on the inner four walls of the rectangular steel body, and the top also has a welded nut that extends inward and that is connected to the pulling unit. The coal seam simulation portion is bolted through the threaded hole to connect it to the four pressure plates; the surface is also filled with a similar coal seam material surrounded by baffles that are set on all sides, and the front baffle is fitted with a transparent acrylic plate to observe the overall deformation of the specimen during the experimental process. The pressurization unit consists of the first ball slide, hydraulic jack and load plate; the first ball slide is connected to the pressurizing position adjustment unit, the position of the loading plate corresponds to the surface position of the similar coal rock formation material, and the pressurization unit is connected to the outer frame by the pressurizing position adjustment unit. The pressurizing position adjustment unit consists of a central rail beam, a second ball slide, an upper rail column, and an upper rail beam.

By combining the overall structure and the other components, the device can effectively simulate the surface subsidence characteristics of coal seam mining. During the test, the HandyScan700 three-dimensional laser scanner was used to scan the surface of the model multiple times in order to obtain the deformation characteristics of the model. The HandyScan700 three-dimensional laser scanner includes a handheld scanner as well as the control host and control software VXelements, as shown in Figure 4. During operation, the scanner is able to calculate the shape of the object accurately based on the triangulation principle combined with the positioning spots on the object by projecting the laser mesh onto the object being tested, and the camera is used to capture the laser mesh shape. The following are the positive characteristics of this method: fast, non-contacting, high-density, high-precision, digital, automatic, etc.

Figure 4. HandyScan700 three-dimensional scanner and support equipment.

2.3. Test Steps

When using this device to simulate underground coal mining processes and to observe its surface movement features, the following steps should be followed:

(1) According to the actual situation of the project site, develop pre-simulation mining conditions (mining area, number of working faces, mining methods, etc.), assemble the test device;

(2) Obtain the on-site coal seam and rock formations parameters, and formulated similar materials indoors;

(3) Test to determine the parameters of the formulated similar materials;

(4) Place similar materials in the test device, arrange the sensors, and compact the layers;

(5) Place the three-dimensional scanners in the surface deformation observation positions of the test device and scan and store the surface and building deformation and movement in real-time;

(6) Conduct similar simulation tests according to the established mining simulation scheme: the non-slip fastening screws between the mining coal blocks will be loosened, and the pulling mechanism will be pulled down one by one to simulate coal mining. After the test is completed, according to the corresponding data and the processing steps, conduct the analysis.

3. Characteristics Analysis of Surface Subsidence in Coal Seam Mining

3.1. Test Scheme

During the field engineering measurement process, due to the influence of building surfaces and other factors, irregular monitoring points will be laid down according to the actual situation on the surface to obtain the subsidence law of the monitoring polyline on the surface. Two-dimensional planar or the three-dimensional spatial features of surface subsidence can only be studied by means of indoor experiments. In order to better analyze the dynamic spatial characteristics of surface subsidence caused by coal seam mining, a test device for simulating rock formations and surface movement can be used, and supplementary research and analysis of the temporal and spatial surface subsidence characteristics are conducted through the test results.

3.1.1. Formulation Ratio of Similar Materials

The test scheme is designed based on a similar material test design principle, combined with the characteristics of the overlying strata of Tangshan ore. To highlight the geological features of the thick, loose layers on the surface, the thickness of the loose layer on the surface is set to 300 m, and the parameters of the remaining rock formations are listed in Table 1.

Table 1. Formulation ratio of the test.

Layer No.	Lithology	Thickness/m	Model Thickness/cm	Unit Weight g/cm^3	Formulation Ratio	Total Weight	Amount of Material/kg			
							Sand	Calcium Carbonate	Gypsum	Water
R4	Loose Layer	300	33	0.95	673	182.40	156.34	18.24	7.82	20.27
R3	Bedrock Layer	300	33	1.60	537	230.40	192.00	11.52	26.88	25.60
R2	Basic Roof	100	11	1.70	755	73.44	64.26	4.59	4.59	8.16
R1	Immediate Roof	30	3	1.80	773	34.56	30.24	1.30	3.02	3.84
	Total	730	80			520.80	442.84	35.65	42.31	57.87

3.1.2. Test Steps

(1) Design the similarity ratio of the test according to the purpose, calculate the material formulation ratio based on the similarity ratio and rock formations, and determine the formulation scheme;

(2) Adjust the mining block according to the test scheme and raise the expected mined coal seam range, as shown in Figure 5a;

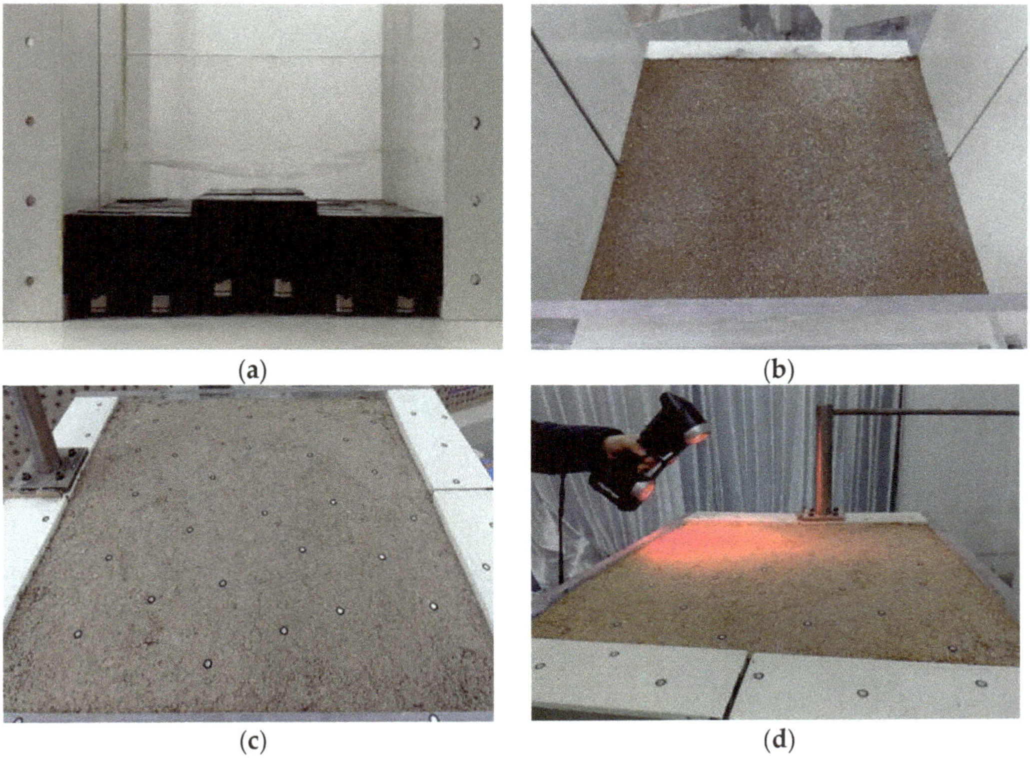

Figure 5. Operating steps of the test process: (**a**) adjusted mining blocks; (**b**) layered laying model; (**c**) placement location punctuation; (**d**) three-dimensional scanning monitoring.

(3) Formulate the similar material of each rock formation according to the test scheme, lay the model out, and compact each layer, as shown in Figure 5b;

(4) Lay out the three-dimensional scanning positions of the surface on the model, as shown in Figure 5c;

(5) Mine the model step-by-step and lower the pre-raised mining blocks to their original position while using a three-dimensional laser scanner to monitor the vertical deformation of the surface on the model, as shown in Figure 5d.

3.2. Analysis of Test Results

3.2.1. Surface Subsidence Characteristics of Coal Seam Mining

In order to compare the surface subsidence morphology after coal seam mining, a cloud map of the vertical displacement on the surface of the model and a schematic of the measuring line of the model are calculated by Geomagic Control X, as shown in Figures 6 and 7, respectively. After coal seam mining, the surface subsidence pattern is symmetrically distributed along the working surface. The sediment volume gradually decreases from the center to the edges of the mined-out area: With coal seam mining, the surface subsidence gradually radiates in the direction of the work surface, and the peak settlement position gradually shifts from the center to the back of the mined-out area. The early stage of mining has a greater impact on surface subsidence. During coal seam mining, partial positive vertical displacement occurs. At the beginning of coal seam mining, positive vertical displacement is concentrated at the edges of the mining area and at the edges of the model. In the mid to late stages of mining, positive vertical displacement is more distributed at the edges of the surface subsidence, which is because the stress is redistributed during the surface subsidence, resulting in the formation particles being squeezed and the upward vertical displacement occurring.

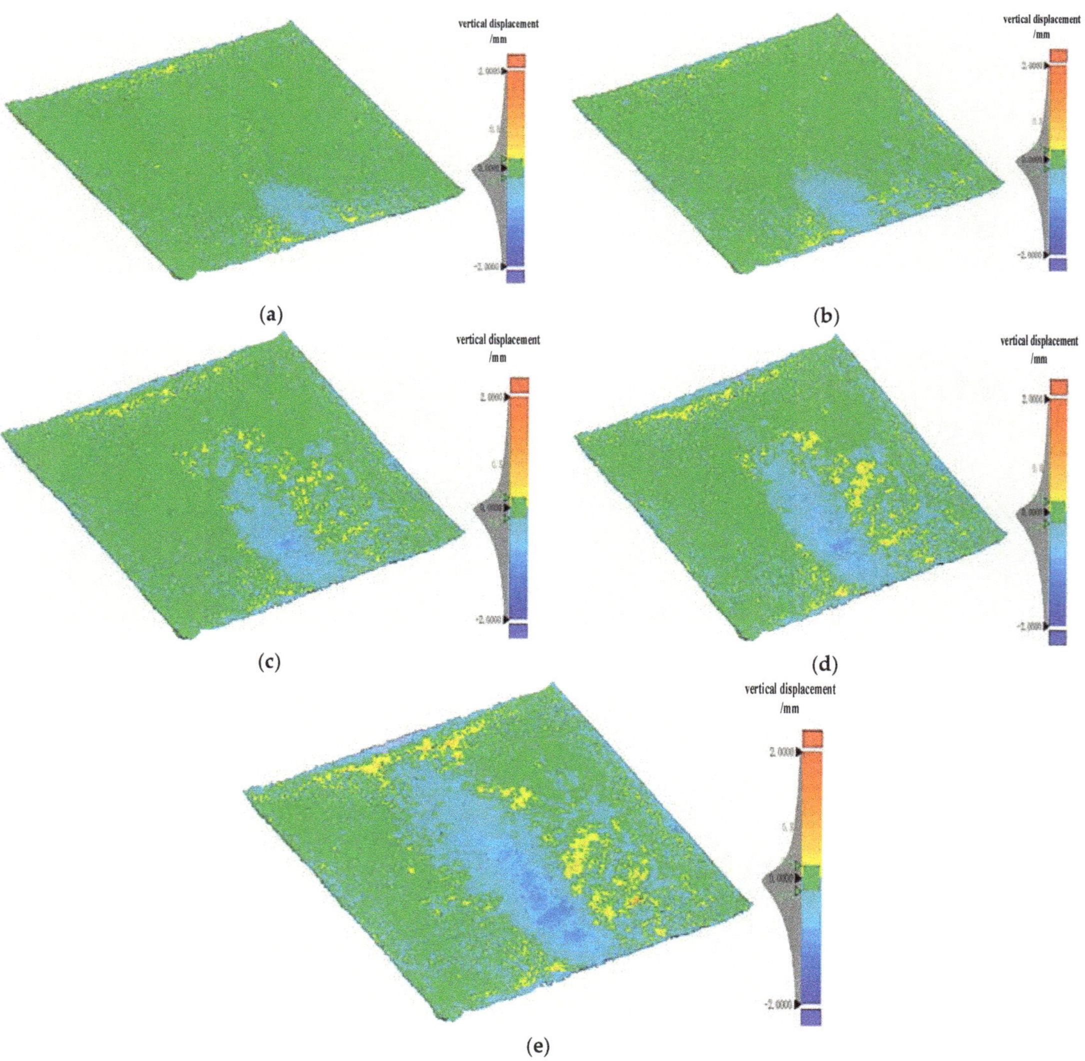

Figure 6. Cloud map of cumulative settlement of the model: (**a**) first mining; (**b**) second mining; (**c**) third mining; (**d**) fourth mining; (**e**) fifth mining.

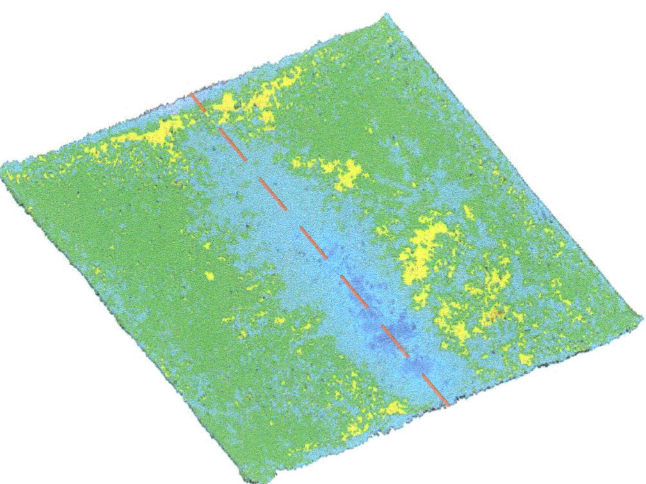

Figure 7. Schematic diagram of the model measuring line.

The statistical average and maximum sedimentation curves of the surface after coal seam mining are shown in Figure 8. As seen from the figure, the average and maximum surface settlement volume gradually increase with the work surface, and after the coal seam is mined out, the average surface settlement volume is about 0.837 mm, and the maximum settlement volume is about 1.841 mm.

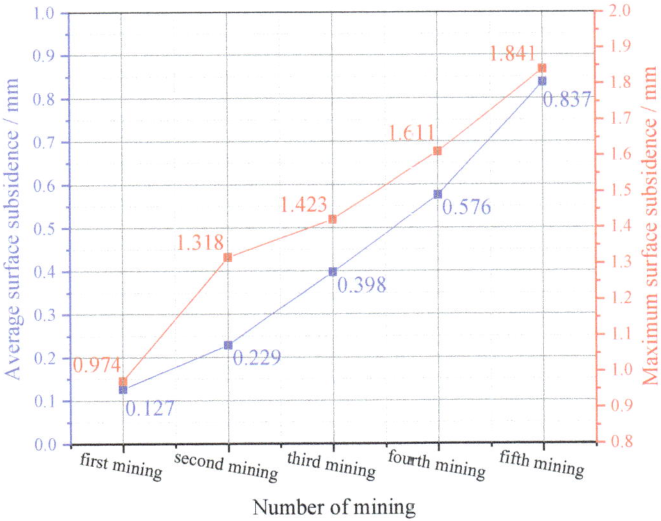

Figure 8. Statistical curve of surface settlement.

The sediment characteristics of the surface measuring line are shown in Figure 9, and the corresponding settlement monitoring curve is shown in Figure 10. According to Figure 9, after coal seam mining, the surface subsidence parallel to the mined-out area is presented as asymmetrical distribution, the settling pattern is similar to the spoon type, the central settlement of the mined-out area is larger, the edge settlement is smaller, the sedimentation slope is larger, the slope is steeper on the open-off cut side, the sedimentation slope is smaller, and the slope is relatively slower on the working face side.

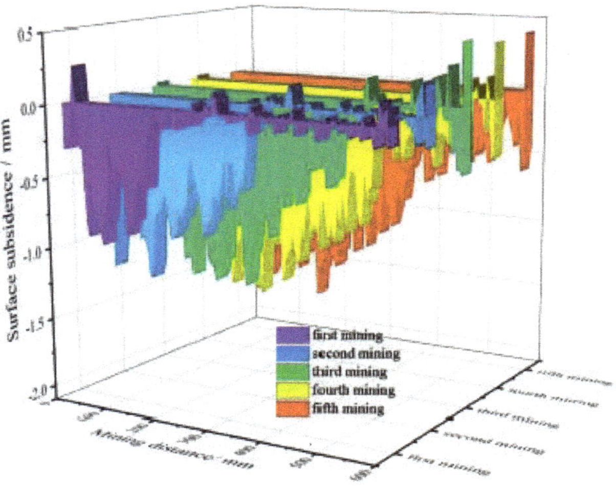

Figure 9. Three-dimensional strip diagram of surface measuring line settlement.

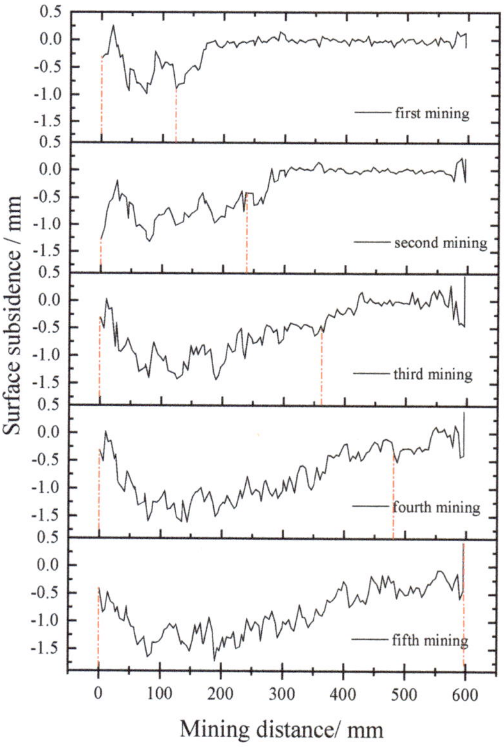

Figure 10. Monitoring curves of surface measuring line settlement.

3.2.2. Transfer Law of Surface Subsidence in Coal Seam Mining

In order to compare the impact of coal seam mining on the sediment at different areas of the surface, the surface range right above the five mining stages is divided into five areas to analyze the changes in the measuring line sediment in the five regions, as shown in Figure 11. The analysis shows that:

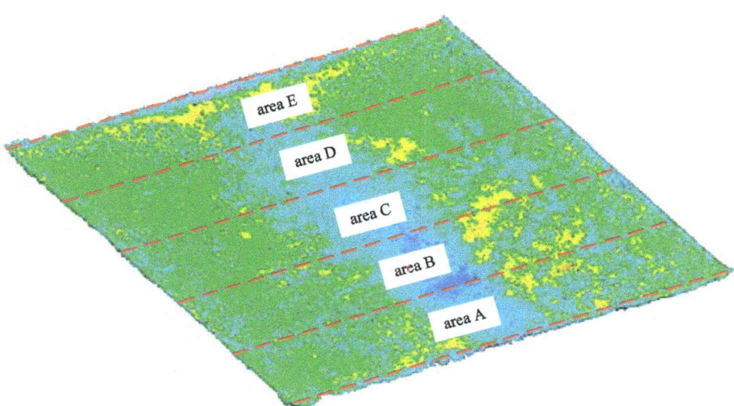

Figure 11. Region division for surface analysis.

(1) The changes in average sedimentation in different surface regions are shown in Figure 12. As seen from the figure, all the surface regions saw an increase in the amount of sedimentation, and the increase gradually decreased. After the stability of the strata, the total settlement is the largest in area II, with an average of 1.323 mm, and the smallest overall settlement is observed in area V, with an average of 0.334 mm. Therefore, the early middle stage of the surface in the mined-out area is the area with the largest amount of settlement, and measures need to be taken to focus on prevention and control.

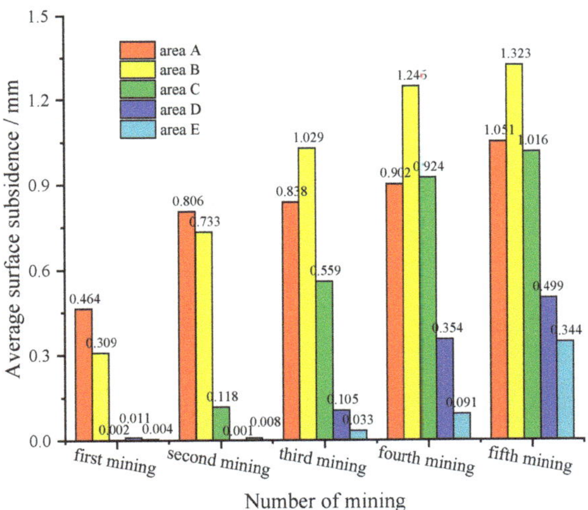

Figure 12. Histogram of average sediment volume change in different regions of the surface.

(2) The radar map of the average sedimentation for the different surface regions is shown in Figure 13. With the exception of area I, each area settles when the coal seam corresponds to the previously mined area, and the amount of advanced settlement caused by coal seam mining in areas II to V is 0.309 mm, 0.118 mm, 0.105 mm, 0.091 mm, respectively, with an overall decreasing trend being observed. At the same time, during the third mining operation, area V shows a mild response, indicating that as the range of the mined-out area increases, the degree of surface subsidence advance that is caused by the continuously advancing working surface is gradually reduced, but the advance impact range gradually increases.

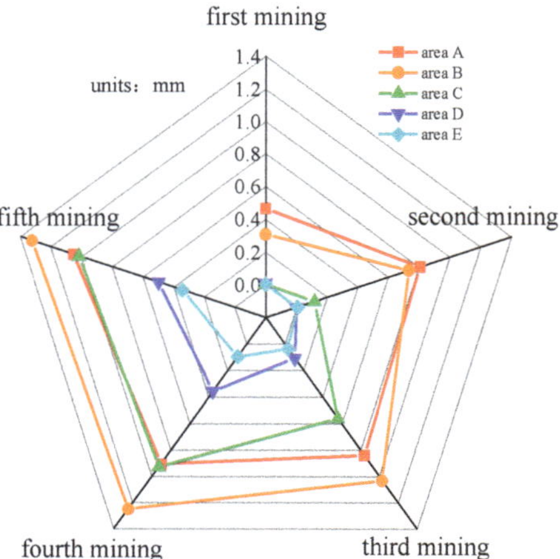

Figure 13. Radar map of average sediment volume change in different regions of the surface.

(3) Different coal seam mining stages have different degrees of influence on the corresponding surface. The settlement increments caused by coal seam mining right under the five areas are 0.464 mm, 0.424 mm, 0.441 mm, 0.249 mm, and 0.253 mm, and exceed the settlement increments caused by coal seam mining in other areas. Taking area III as an example for analysis, the surface subsidence caused by the first two, the third and fourth, and the fifth mining simulations accounted for 11.6%, 79.3%, 9.1% of the total subsidence, respectively. The analysis shows that the surface subsidence is the superposition of advanced settlement caused by coal seam mining, disturbance settlement caused by subsurface coal seam mining, and prolonged post-mining subsidence. The settlement caused by the disturbance of subsurface coal seam mining is the dominant factor in total surface subsidence.

4. Field Engineering Validation

In order to analyze the surface subsidence characteristics of coal seam filling mining in the Tangshan mine, a regional surface subsidence observatory was established in the corresponding surface area. The station layout is shown in Figure 14 and has a total of 87 observation points, the average observation point spacing is 30 m, the total length of the measuring line is 2700 m, and both ends of the measuring point distance from the T3292 working surface boundary are located about 750 m or so away [34].

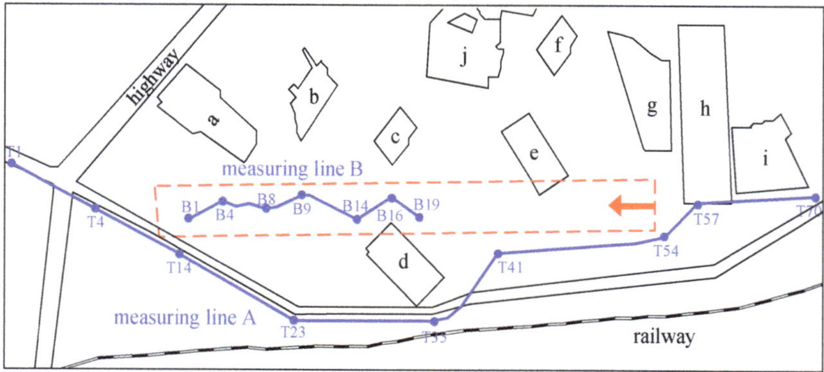

Figure 14. Measuring line layout.

Surface subsidence observation line B was established in the main section and was located on both sides of the main section of the working surface and had a total length of 470 m as well as a total of 19 observation stations from B1 to B19. The average distance between the observation positions was 26 m, and the surface subsidence caused by mining on the T3292 working surface was observed.

By studying the influence range of the surface subsidence of the working surface and analyzing the long-term observation results of each measuring point of measuring line B, the settlement vs. time curve of some observation positions could be obtained, as shown in Figure 15.

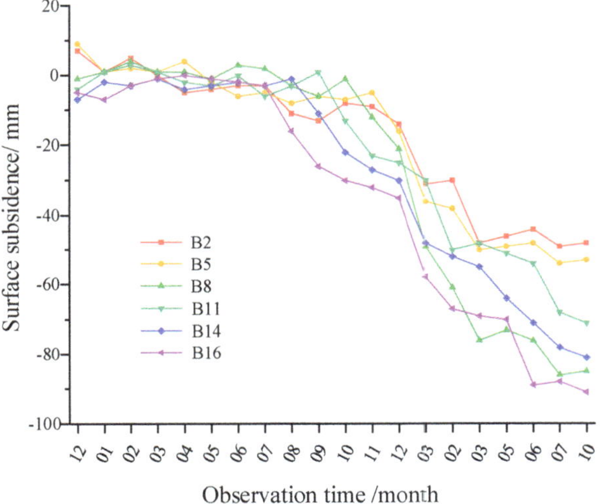

Figure 15. Settlement vs. time curve of measuring line B.

In order to study the surface subsidence response process, we take observation positions B11, B14s, and B16 as examples. The advanced influence surface subsidence process can be divided into four stages, as shown in Figure 16.

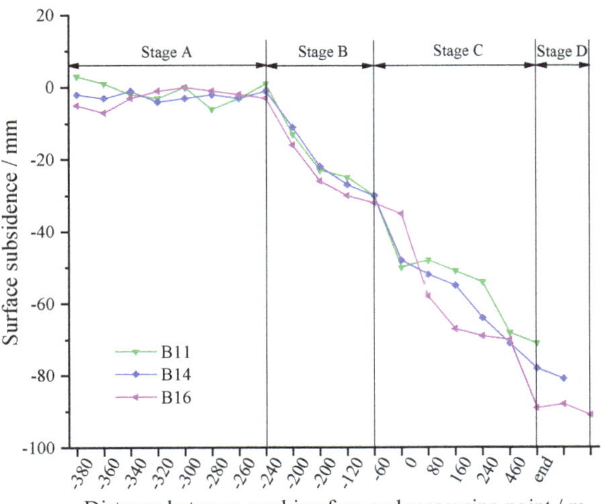

Figure 16. Advanced impact stage division.

Combined with field engineering tests, the field engineering situation and the analysis of the device test results are consistent. The study reveals the following:

(1) Stage A—No settlement stage. Before the working surface advances to 240 m, each measuring point settlement value is small, fluctuating around the threshold value, and the average value is lower than the surface subsidence threshold value (10 mm). Multiple observations on the surface lasting more than a dozen months showed small changes in the settlement, indicating that the observation area is far away from the mining area and that it is only slightly affected by the advance.

(2) Stage B—Slight settlement stage. When the working surface advances to the position of 240-60 m ahead of the measuring point, the curve slope increases, the settlement increased to 30 mm, exceeding the threshold of surface subsidence (10 mm), the surface of the observation area begins to be affected by the advance, and displacement settlement occurs.

(3) Stage C—Significant settlement stage. After the working surface advances to the position of 60 m ahead of the measuring point, the settlement exhibits a major increase as the working surface advances. After mining, the settlement continues to increase over time, with the final settlement of each measuring point increasing to more than 70 mm.

(4) Stage D—Residual settlement stage. After experiencing significant growth, the settlement of each observation point and the ground surface gradually reach a stable state; the stabilized subsidence value is affected by the location of the measuring point and the surface situation, which has a larger stabilized subsidence value near the strike of the working surface inclination and is close to the middle position.

5. Conclusions

(1) A "three-dimensional test device for simulating surface movement in underground coal mining" was self-designed and developed. The overall structure of the device consisted of an outer frame, pressurization unit, pulling unit, and coal seam simulation portion that can effectively simulate the law of surface subsidence caused by underground coal seam mining.

(2) The final surface subsidence state is the superposition of advance settlement caused by coal seam mining, disturbance settlement caused by subsurface coal seam mining, and prolonged post-mining subsidence. The surface subsidence caused by the three mining stages accounted for 11.6%, 79.3%, and 9.1% of the total surface subsidence, respectively. The settlement caused by the disturbance of subsurface coal seam mining is the dominant factor in the total surface subsidence.

(3) The device model test was effectively combined with actual engineering practices. The field engineering tests and model test results analysis are consistent, and a conclusion can be drawn: after coal seam mining, the surface subsidence comprised four stages, including a no settlement stage (ahead of 240 m), a slight settlement stage (ahead of 240~60 m), a significant settlement stage (ahead of 60 m ~ the end of mining), and a residual settlement stage (after the end of mining), with the settlement from the significant settlement stage accounting for more than 60% of the total settlement.

Author Contributions: Methodology, X.M.; software, Z.F.; validation, Y.L. and P.Z.; formal analysis, Y.Z.; investigation, X.M.; resources, Y.L. and Z.F.; data curation, Z.F.; writing—original draft preparation, X.M.; writing—review and editing, X.M.; visualization, P.Z.; supervision, G.M.; funding acquisition, Y.L. and Z.F. All authors have read and agreed to the published version of the manuscript.

Funding: This research was funded by the Shandong Provincial Natural Science Foundation, grant number ZR2020QE135, and by the State Key Laboratory of Water Resource Protection and Utilization in Coal Mining, grant number SHGF-16-25.

Institutional Review Board Statement: Not applicable.

Informed Consent Statement: Not applicable.

Data Availability Statement: Not applicable.

Acknowledgments: The authors acknowledge the Shandong Provincial Natural Science Foundation and the State Key Laboratory of Water Resource Protection and Utilization in Coal Mining.

Conflicts of Interest: The authors declare no conflict of interest.

References

1. Adhikary, D.; Khanal, M.; Jayasundara, C.; Balusu, R. Deficiencies in 2D simulation: A comparative study of 2D versus 3D simulation of multi-seam longwall mining. *Rock Mech. Rock Eng.* **2016**, *49*, 2181–2185. [CrossRef]
2. Ghabraie, B.; Ghabraie, K.; Ren, G.; Smith, J.V. Numerical modelling of multistage caving processes: Insights from multi-seam longwall mining-induced subsidence. *Int. J. Numer. Anal. Methods Geomech.* **2017**, *41*, 959–975. [CrossRef]
3. Whittles, D.N.; Reddish, D.J.; Lowndes, I.S. The development of a coal measure classification (CMC) and its use for prediction of geomechanical parameters. *Int. J. Rock Mech. Min. Sci.* **2007**, *44*, 496–513. [CrossRef]
4. Sun, X.K. Present situation and prospect of green backfill mining in mines. *Coal Sci. Technol.* **2020**, *48*, 48–55. (In Chinese)
5. Bell, F.G.; Stacey, T.R.; Genske, D.D. Mining subsidence and its effect on the environment: Some differing examples. *Environ. Geol.* **2000**, *40*, 135–152. [CrossRef]
6. Sun, G.J.; Wang, P.; Feng, T.; Yu, W.J.; Liu, J.H. Strata movement characteristics of the deep well gangue filling on the fully mechanized mining face. *J. Min. Saf. Eng.* **2020**, *37*, 562–570. (In Chinese)
7. Altun, A.A.; Yilmaz, I.; Yildirim, M. A short review on the surficial impacts of underground mining. *Sci. Res. Essays* **2010**, *5*, 3206–3212.
8. Barbato, J.; Hebblewhite, B.; Mitra, R.; Mills, K. Prediction of horizontal movement and strain at the surface due to longwall coal mining. *Int. J. Rock Mech. Min. Sci.* **2016**, *84*, 105–118. [CrossRef]
9. Vivanco, F.; Melo, F. The effect of rock decompaction on the interaction of movement zones in underground mining. *Int. J. Rock Mech. Min. Sci.* **2013**, *60*, 381–388. [CrossRef]
10. Soni, A.K.; Singh, K.K.K.; Prakash, A.; Singh, K.B.; Chakraboraty, A.K. Shallow cover over coal mining: A case study of subsidence at Kamptee Colliery, Nagpur, India. *Bull. Eng. Geol. Environ.* **2007**, *66*, 311–318. [CrossRef]
11. Zhao, J.Y. Study on Dynamic Subsidence Law of Mining Surface of Adjacent Working Face in Yushen Mining Area. Ph.D. Thesis, Xi'an University of Science and Technology, Xi'an, China, 2020. (In Chinese).
12. Gligoric, M.V.; Gligoric, Z.M.; Beljic, C.R.; Lutovac, S.M.; Damnjanovic, V.M. Long-term room and pillarmine production planning based on fuzzy 0-1 linear programing and multicriteria clustering algorithm with uncertainty. *Math. Probl. Eng.* **2019**, *2019*, 3078234. [CrossRef]
13. Sementsov, V.V.; Prokopenko, S.A. Room-and-pillar mining of thick coal seams in the conditions of high gas dynamic hazard in Kuzbass. In Proceedings of the Conference on Challenges for Development in Mining Science and Mining Industry, Novosibirsk, Russia, 1–5 October 2018; Volume 262, p. 12064.
14. Singh, R.; Mandal, P.K.; Singh, A.K.; Kumar, R.; Sinha, A. Optimal underground extraction of coal at shallow cover beneath surface//subsurface objects: Indian practices. *Rock Mech. Rock Eng.* **2008**, *41*, 421–444. [CrossRef]
15. Sasaoka, T.; Takamoto, H.; Shimada, H.; Oya, J.; Hamanaka, A.; Matsui, K. Surface subsidence due to underground mining operation under weak geological condition in Indonesia. *J. Rock Mech. Geotech. Eng.* **2015**, *7*, 337–344. [CrossRef]
16. Sivakugan, N.; Widisinghe, S.; Wang, V.Z. Vertical stress determination within backfilled mine stopes. *Int. J. Geomech.* **2014**, *14*, 06014011. [CrossRef]
17. Wang, F.; Jiang, B.; Chen, S.; Ren, M. Surface collapse control under thick unconsolidated layers by backfilling strip mining in coal mines. *Int. J. Rock Mech. Min. Sci.* **2019**, *113*, 268–277. [CrossRef]
18. Brent, G.F. Quantifying eco-efficiency within life cycle management using a process model of strip coal mining. *Int. J. Min. Reclam. Environ.* **2011**, *25*, 258–273. [CrossRef]
19. Zuo, J.P.; Sun, Y.J.; Wen, J.H.; Li, Z.D. Theoretical and mechanical models of rock strata movement and their prospects. *Coal Sci. Technol.* **2018**, *46*, 1–11. (In Chinese)
20. Cheng, J.W.; Zhao, G.; Sa, Z.Y.; Zheng, W.C.; Wang, Y.G.; Liu, J. Overlying strata movement and deformation calculation prediction models for underground coal mines. *J. Min. Strat. Control Eng.* **2020**, *2*, 20–29. (In Chinese)
21. Cai, Y.F.; Li, X.J.; Deng, W.N.; Xiao, W.; Zhang, W.K. Simulation of surface movement and deformation rules and detriment key parameters in high-strength mining. *J. Min. Strat. Control Eng.* **2020**, *2*, 46–54. (In Chinese)
22. Ren, B.W. Study on Surface Subsidence of Mining Area Based on Multi-source Monitoring Data. Ph.D. Thesis, Hebei University of Engineering, Handan, China, 2020.
23. Panchal, S.; Deb, D.; Sreenivas, T. Variability in rheology of cemented paste backfill with hydration age, binder and superplasticizer dosages. *Adv. Powder Technol.* **2018**, *29*, 2211–2220. [CrossRef]
24. Sun, Y.J.; Zuo, J.P.; Karakus, M.; Liu, L.; Zhou, H.W.; Yu, M.L. A new theoretical method to predict strata movement and surface subsidence due to inclined coal seam mining. *Rock Mech. Rock Eng.* **2021**, *54*, 2723–2740. [CrossRef]
25. Dong, L.J.; Tong, X.J.; Ma, J. Quantitative investigation of tomographic effects in abnormal regions of complex structures. *Engineering* **2021**, *7*, 1011–1022. [CrossRef]
26. Zhu, G.A.; Dou, L.M.; Wang, C.B.; Ding, Z.W.; Feng, Z.J.; Xue, F. Experimental study of rock burst in coal samples under overstress and true-triaxial unloading through passive velocity tomography. *Saf. Sci.* **2019**, *117*, 388–403. [CrossRef]

27.   Li, X.B.; Dong, L.J.; Zhao, G.Y.; Huang, M.; Liu, A.H.; Zeng, L.F.; Dong, L.; Chen, G.H. Stability analysis and comprehensive treatment methods of landslides under complex mining environment—A case study of Dahu landslide from Linbao Henan in China. *Saf. Sci.* **2012**, *50*, 695–704. [CrossRef]

28.   Hou, D.; Li, D.; Xu, G.; Zhang, Y. Superposition model for analyzing the dynamic ground subsidence in mining area of thick loose layer. *Int. J. Min. Sci. Technol.* **2018**, *28*, 663–668. [CrossRef]

29.   Thongprapha, T.; Fuenkajorn, K.; Daemen, J.J.K. Study of surface subsidence above an underground opening using a trap door apparatus. *Tunn. Undergr. Space Technol.* **2015**, *46*, 94–103. [CrossRef]

30.   Suchowerska Iwanec, A.M.; Carter, J.P.; Hambleton, J.P. Geomechanics of subsidence above single and multi-seam coal mining. *J. Rock Mech. Geotech. Eng.* **2016**, *8*, 304–313. [CrossRef]

31.   Fathi Salmi, E.; Nazem, M.; Karakus, M. Numerical analysis of a large landslide induced by coal mining subsidence. *Eng. Geol.* **2017**, *217*, 141–152. [CrossRef]

32.   Salmi, E.F.; Nazem, M.; Karakus, M. The effect of rock mass gradual deterioration on the mechanism of post-mining subsidence over shallow abandoned coal mines. *Int. J. Rock Mech. Geotech. Eng.* **2017**, *91*, 59–71. [CrossRef]

33.   Sun, Y.; Zuo, J.; Karakus, M.; Wang, J. Investigation of movement and damage of integral overburden during shallow coal seam mining. *Int. J. Rock Mech. Min. Sci.* **2019**, *117*, 63–75. [CrossRef]

34.   Yin, B.J.; Wang, A.L.; Zhang, P.F.; Fu, Z.Y.; Qiu, D.W. Analysis of the laws and control effect of surface subsidence of solid backfilling mining. *China Coal* **2018**, *44*, 81–86. (In Chinese)

Article

Efficient Graphical Algorithm of Sensor Distribution and Air Volume Reconstruction for a Smart Mine Ventilation Network

Yujiao Liu [1,2], Zeyi Liu [1,2], Ke Gao [1,2,*], Yuhan Huang [3] and Chengyao Zhu [1,2]

[1] College of Safety Science and Engineering, Liaoning Technical University, Huludao 125105, China; liuyujiao@lntu.edu.cn (Y.L.); liuzeyijy@163.com (Z.L.); zcy1339645348@163.com (C.Z.)
[2] Key Laboratory of Mine Thermo-Motive Disaster and Prevention, Ministry of Education, Huludao 125105, China
[3] Centre for Green Technology, School of Civil and Environmental Engineering, University of Technology Sydney, Sydney, NSW 2007, Australia; yuhan.huang@uts.edu.au
* Correspondence: gaoke@lntu.edu.cn; Tel.: +86-13591991287

Abstract: The accurate and reliable monitoring of ventilation parameters is key to intelligent ventilation systems. In order to realize the visualization of airflow, it is essential to solve the airflow reconstruction problem using few sensors. In this study, a new concept called independent cut set that depends on the structure of the underlying graph is presented to determine the minimum number and location of sensors. We evaluated its effectiveness in a coal mine owned by Jinmei Corporation Limited (Jinmei Co., Ltd., Shanghai, China). Our results indicated that fewer than 30% of tunnels needed to have wind speed sensors set up to reconstruct the well-posed airflow of all the tunnels (>200 in some mines). The results showed that the algorithm was feasible. The reconstructed air volume of the ventilation network using this algorithm was the same as the actual air volume. The algorithm provides theoretical support for flow reconstruction.

Keywords: mine ventilation network; wind speed sensors distribution; air volume reconstruction; independent cut set

Citation: Liu, Y.; Liu, Z.; Gao, K.; Huang, Y.; Zhu, C. Efficient Graphical Algorithm of Sensor Distribution and Air Volume Reconstruction for a Smart Mine Ventilation Network. *Sensors* **2022**, *22*, 2096. https://doi.org/10.3390/s22062096

Academic Editors: Longjun Dong, Yanlin Zhao and Wenxue Chen

Received: 16 February 2022
Accepted: 4 March 2022
Published: 8 March 2022

Publisher's Note: MDPI stays neutral with regard to jurisdictional claims in published maps and institutional affiliations.

1. Introduction

Coal is a primary energy source and plays a key role in economic development in many countries. In China, coal accounts for 40–45% of total carbon emissions [1]. Meanwhile, due to the unique working environment in coal mines, the requirements for safe operations are very high. The working environment affects the safety of those working in coal mines. Mining accidents could cause significant loss of life, as well as economic losses to a country. In 2020 alone, 573 Chinese miners died in 434 mining accidents such as leaks of poisonous gases, explosions of natural gases, and the collapsing of mine stopes, especially from underground coal mining [2]. An efficient and reliable ventilation system is essential for safe and efficient production of coal [3].

Intelligent ventilation is a development trend for coal mines and other types of mines. The Chinese National Development and Reform Commission and eight other ministries jointly issued *Guiding Opinions on Accelerating the Intelligent Development of Coal Mines* to improve the intelligence level of coal mines. A set of systems, including development design, geological guarantee, mining, transportation, and ventilation, are needed to achieve the goal of intelligent decision-making and automation systems operating collaboratively by 2025 [4,5]. Intelligent coal mining is to be fully realized by 2035 [6]. The reliable operation of mine ventilation systems is the basis for safe production in intelligent coal and noncoal mines. Intelligent mine ventilation is a new ventilation system that can be adjusted automatically on demand, which is the development trend of mine ventilation technology in China [7]. Proper monitoring of airflows for all tunnels plays a key role in the development of intelligent mine ventilation systems [8,9]. Airflow monitoring depends on sensors. Hu [10] studied the influence of wall roughness on wind speed distribution to improve the accuracy of monitoring. However, mine ventilation is a very complex network system involving hundreds or even thousands of tunnels. As shown in Figure 1, the Changcun coal mine in Shanxi province, China has 2061 tunnels. It may be impossible to install wind speed sensors in every tunnel due to the high installation and maintenance costs. Air monitoring was highlighted in the assessment of mine ventilation systems and air pollution to improve health and safety for miners [11]. Therefore, it is of great importance to economically determine the minimum number of wind speed sensors that can sample sufficient information to accurately reconstruct the whole network. Airflow reconstruction consists of well-posed reconstruction and underdetermined reconstruction. The so-called well-posed reconstruction is to reconstruct the airflows of the whole network with the minimum number of sensors when the number of sensors is sufficient. The underdetermined reconstruction is realized when the number of sensors is insufficient. The independent cut set algorithm is a well-posed airflow reconstruction algorithm.

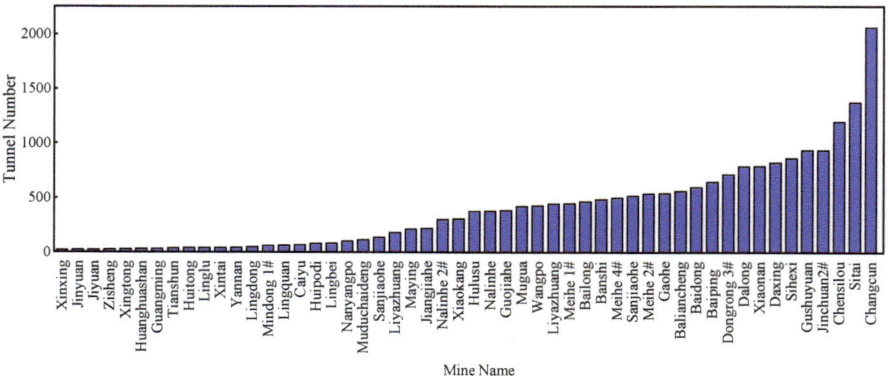

Figure 1. The number of tunnels in mines in China.

A fully visual environment of ventilation parameters is very important to realize intelligent ventilation of mines. In order to visualize the air volume, an air volume reconstruction algorithm is proposed in this paper. In past studies, optimization and reconstruction technologies of sensor networks were widely used in flow monitoring [12], for the development of intelligent systems [13], and for leak location diagnosis for water [14–16], oil [17,18], and gas [19–21] pipeline networks. Singh et al. [22] proposed a method to identify ideal monitoring sites for water quality. Huang et al. [23] established a PSO-LSTM model to realize the identification of sources with abnormal radon exhalation rates. Castillo et al. [24] proposed a method of flow reconstruction by constructing a matrix of constraints to give bounds on the number of sensors required. In this method, the construction of a constraint matrix depends on all possible paths of the network. For complex networks, the measurement process in this method will also be very complex. The method was optimized to reduce the impact of network complexity [25]. However, the construction of a constraint matrix depends on the network path; although it has been optimized, the construction process is still very complex. Rinaudo et al. [26] proposed a minimum number of sensors arrangement method to determine the temperature. Although this method reduces the cost of monitoring, the measurement process is too complex. Balaji et al. [27] proposed a method to maximize sensor lifetime by activating sensor covers one at a time to monitor the whole network. This method mainly depends on the effective usage of available resources. Li et al. [28] put forward a novel method based on deep learning techniques and transfer learning to deal with large-scale missing data problems. Ng [29] proposed a method to calculate the number of sensors without depending on the path. He [30] developed an efficient algorithm to determine the smallest subset of links in a traffic network for sensor installation. Muduli et al. [31] proposed a novel wireless sensor network deployment scheme for environmental monitoring in longwall coal mines, which provided the best area coverage and was very cost-effective. Automated ventilation system adjustment software has been developed by monitoring the air volume of the minimum remainder branch in the network, and all air volumes of the branches in the network can be obtained by inverse calculation of associated branches [32]. Song et al. [33] proposed a method for determining the changing trend of gas concentration in the goaf. Lyu et al. [34] proposed a gas concentration prediction method based on the ARMA model, the CHAOS model, and the encoder–decoder model (single-sensor and multisensor). Foorginezhad et al. [35] proposed advanced sensing systems utilized for relevant monitoring and recommendations for improving sensing accuracy.

Mine ventilation networks are similar to traffic networks on topology. Mine ventilation networks are high-order nonlinear systems [36]. In the literature, different approaches have been used for determining the minimum number of sensors needed to identify the location of a ventilation system's faults [37], such as the coverage of node flowrate method [38], the least full-coverage distribution method [39], minimum tree principle [40], tabu search (TS), Pareto ant colony algorithm (HPACA) [41], and the GA-DBPSO algorithm [42,43]. Changing the airflow in any one of the tunnels leads to changes in other associated tunnels. The sensitivity is used for measuring the airflow relationship between all the tunnels [44].

However, a mine ventilation system is a complex network, with different complexity levels in different areas. There are some problems with considering the accuracy to determine the minimum sensors' location and reconstruct the air volume of all tunnels in the mines. In this study, we aimed to solve the problems of sensor optimization and air flow reconstruction, and realize the air volume monitoring of the whole air network economically and effectively. In this paper, we propose an algorithm for the well-posed reconstruction of the airflow in a mine ventilation network. The algorithm is based on the structure of the underlying graph, and makes it possible to have a unique, optimized solution to the air volume of tunnels. Compared with other methods, this algorithm involves lower costs and a simpler process.

The rest of this paper is organized as follows. We state the problems we aimed to study, provide a theoretical analysis of our research, and explain some concepts that need to be used in Section 2. In Section 3, we introduce the methods of sensor location optimization and flow reconstruction. The above-mentioned methods are transformed into algorithms, and the specific calculation process is expressed by matrices and formulas, which are verified by taking a single-source and single-sink network as an example in Sections 4 and 5. In Section 6, we verify the algorithm by taking multisource and multisink network by the ventilation data of a coal mine. The main conclusions of this paper are given in Section 7.

2. Problem Statement

Our goal was to locate wind speed sensors so that the airflow of a ventilation network could be inferred from the measurements while minimizing the number of sensors used. We propose a well-posed flow reconstruction algorithm, which uses the flow conservation equations and breadth-first search algorithm to search the whole ventilation network, establish the equations, and then solve the airflow reconstruction problem.

2.1. Possibility Analysis on Well-Posed Flow Reconstruction

The correlation between the junction number and tunnel number was obtained, as shown in Figure 2. The junction number increases linearly with the tunnel number. This means that there is more mass conservation at the junctions as the tunnel number increases. For a mine ventilation network with a single sink and a single source, $m - 1$ mass conservation equations can be established for m junctions with a connecting sink and source as a virtual junction. There are n tunnels in the mine, and n unknown parameters, so $n - m + 1$ wind speed sensors are needed for the well-posed flow reconstruction if the sensors are set up accurately. Similarly, $n - m + k - 1$ wind speed sensors are used for multisinks and a multisource network (k denotes the total number of the sinks and sources). The above analysis indicates that the minimum number of sensors needed for a well-posed solution may be under 100 when the tunnel number is less than 380, while more than 300 sensors need to be installed when the tunnel number is over 1500; 100–300 sensors are needed if the range of tunnel numbers is 380–1500, as shown in Figure 3. This shows that the calculation complexity is also affected by the network structure.

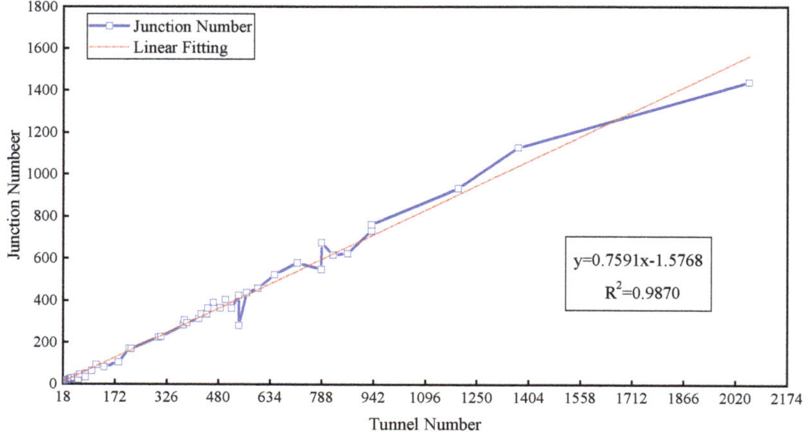

Figure 2. Correlation between the junction number and tunnel number.

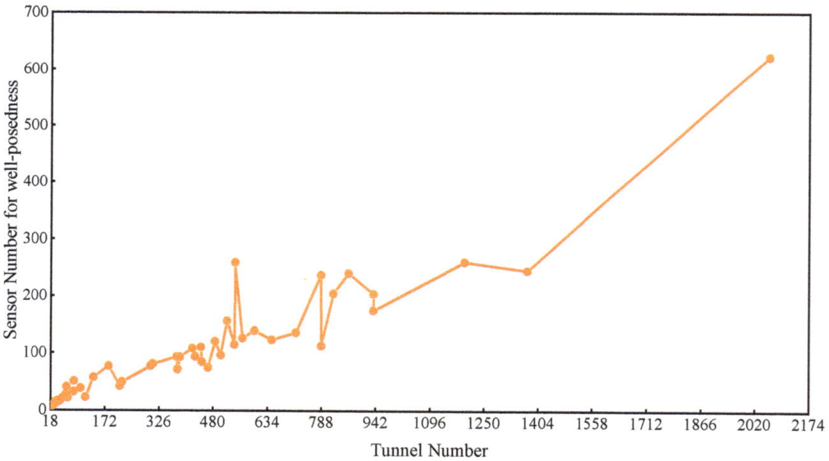

Figure 3. The number of sensors required for well-posed reconstruction.

Figure 4 shows that the ratio of sensor to tunnel number (RST) for well-posed reconstruction is over 30% and even up to 63% when there are fewer than 150 tunnels in the mine, while the RST is below 30% when the tunnels number more than 150. The above data demonstrated that sensors in some mines can be set up for well-posed flow reconstruction where the RST is less than 30% or there are few tunnels. However, there will be no solution or multiple solutions for the reconstruction if the locations of sensors are inaccurate. Therefore, the installation locations of sensors are critical for flow reconstruction.

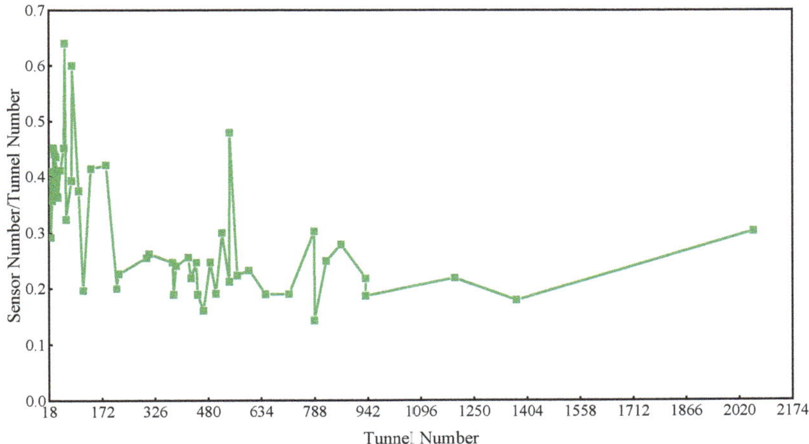

Figure 4. Ratio of sensor to tunnel number.

2.2. Flow Conservation Equation

Ventilation networks are represented by means of a directed graph:

$$G = (V, E), \tag{1}$$

where

$V = \{v_1, v_2, \cdots, v_m\}$ corresponds to the set of junctions,

m is the junction number,

$E = \{e_1, e_2, \cdots, e_n\}$ represents the set of tunnels, and

n is the tunnel number.

The out branches set is denoted as $E^+(v_i)$, $(v_i, v_j) \in E^+(v_i)$. The in branches set is denoted as $E^-(v_i)$, $(v_k, v_i) \in E^-(v_i)$. $\{s\}$, $\{t\}$ are the set of the source junctions and the sink junctions, respectively. For any junction $v \notin \{s, t\}$, inflow equals outflow:

$$\sum \rho_{ij} q_{ij} = \sum \rho_{ki} q_{ki}. \tag{2}$$

A generalized equation of mass can be derived from Equation (2) and the algebraic sum of the tunnels' flow mass is Q_t in any directed cut set of the network with sinks and sources:

$$SQ^T = \left(\sum_{j=1}^{n} s_{ij} q_j \right)_{s \times 1} = Q_t, \tag{3}$$

where

$S = \left(s_{ij} \right)_{s \times n}$ represents the directed cut set;

s is the cut set number;

Q_t is the total mass of the network.

2.3. Improved Breadth-First Search

An exact mathematical model (i.e., a direct solution) may not be found for some problems. In this case, search methods are generally used to solve the issues, among which a breadth-first search (BFS) is the simplest method.

A BFS is one of the graph algorithms. The process is to search for every possible edge by traditional BFS; each junction can only be visited once. The basic idea is to start from a junction v in the graph. In turn, the graph is traversed in breadth first from the unreachable adjacent junctions of v until the junctions in the graph are connected with the paths of v visited. If any junctions in the graph are not visited at this time, the breadth-first traversal is performed again from a junction that has not been visited until all the junctions in the graph have been accessed.

Figure 5 is a ventilation network diagram of single source and single sink. It shows a schematic diagram of an independent cut set. If we launch a BFS from junction v_2 (the following access order is not unique, and the second point can be either v_3 or v_4), we may get an access process as follows: $v_2 - v_3 - v_7 - v_8$. We then go back to v_3. We can get the access process as follows: $v_2 - v_3 - v_5 - v_6 - v_7 - v_8$. For the same reason, we can go back to v_2. We can then continue to search $v_2 - v_4 - v_6 - v_7 - v_8$. The search is finished when all junctions are visited. The result of each BFS must be a connected component of the graph.

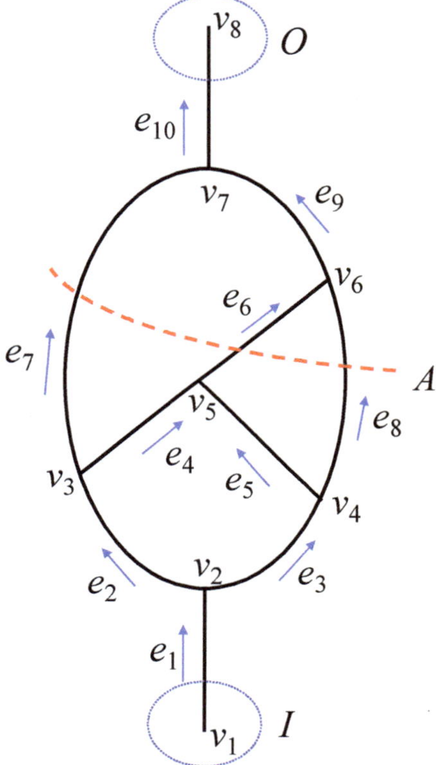

Figure 5. Schematic diagram of independent cut set. (Red line A represents cut line, and I and O represent source junction and sink junction respectively).

According to the analysis in Section 2.1, the spanning tree structure meets the requirements of well-posed reconstruction. Therefore, an improved breadth-first search (IBFS) method is proposed, based on the spanning tree structure instead of the whole network.

An IBFS is a search algorithm based on the spanning tree structure. The first step of an IBFS is to determine the spanning tree and sort the junctions and tunnels of the spanning tree. Then, all source and sink junctions of the spanning tree must be determined. The source junction is taken as the starting junction to start the search, and the incidence branches of the junction are searched. After searching all incidence branches, the search must be continued according to the previous step with the end junction of the incidence branch as the starting junction until the sink junction is searched. The previous junction must then be considered as the starting junction, and other branches are searched for. If there are no other branches, the retreat is continued until the starting junction with other branches is found and the search is maintained. When all branches of the spanning tree are searched, the search ends.

Take Figure 5 as an example; $e_1, e_2, e_3, e_4, e_6, e_9, e_{10}$ form a spanning tree. If we launch an IBFS from junction v_1, we may get an access process as follows: $v_1 - v_2 - v_3 - v_4$. Because v_4 is a sink of the spanning tree, we need to go back to v_2. The adjacent junction v_3 of junction v_2 has an incidence branch. The next search starts from v_3. The process is as follows: $v_3 - v_5 - v_6 - v_7 - v_8$. When all branches of the spanning tree have been searched, the search ends. The IBFS is the basis of our research on sensor optimization and air volume reconstruction. The following algorithms need to use IBFS.

2.4. Definition of Single Junction Cut Sets

In the directed graph $G = (V, E)$, $v_i \in v$, after deleting junction v_i, all tunnels associated with v_i from the graph cannot be connected. This is called a single junction cut set ($S = D_{v_i}$).

In Figure 5, $V = \{v_1, v_2, v_3, v_4, v_5, v_6, v_7 \cdot v_8\}$ and $m = 8$. In other words, there are eight single junction cut sets in the directed graph:

$$G = (V, E),$$
$$D = \{D_{v_1}, D_{v_2}, D_{v_3}, D_{v_4}, D_{v_5}, D_{v_6}, D_{v_7}, D_{v_8}\},$$
$$D_{v_1} = \{e_1\},$$
$$D_{v_2} = \{e_1, e_2, e_3\},$$
$$D_{v_3} = \{e_2, e_4, e_7\},$$
$$D_{v_4} = \{e_3, e_5, e_8\},$$
$$D_{v_5} = \{e_4, e_5, e_6\},$$
$$D_{v_6} = \{e_6, e_8, e_9\},$$
$$D_{v_7} = \{e_7, e_9, e_{10}\},$$
$$D_{v_8} = \{e_{10}\}.$$

2.5. Definition of Independent Cut Set

A new concept called independent cut set is proposed. The source–sink matrix describes the ventilation flow in the mine. The source junction set is I, and the sink junction set is O. In the mine ventilation network $G = (V, E)$, $I \subset V$, $O \subset V$, a cut $C = (S, T)$ partitions V into two subsets, S and T. The cut set A of a cut $C = (S, T)$ is the set $\{(u, v) \epsilon E | u \epsilon S, v \epsilon T, I \subset S, O \subset T\}$, which is called the independent cut set. The independent cut set can divide the input and output junctions into two disjointed parts, which means that the removal of all edges disconnects all paths from the input to output junctions, as shown in Figure 5.

2.6. Algorithm of Independent Cut Set

An independent cut set search algorithm for a single-source, single-sink ventilation network was studied. The graph is described as:

$$G = (V, E),$$
$$E = \{e_1, e_2, \ldots \ldots, e_n\},$$
$$n = |E|,$$
$$V = \{v_1, v_2, \ldots \ldots, v_m\}, \text{ and } m = |V|.$$

The sources set is denoted by $I \subset V$, $I = \{v_i | E^-(v_i) = 0\}$, where O represents the sink set, $O = \{v_i | E^+(v_i) = 0\} \subset V$. First, single-junction cut sets D_{v_2}, $i = 1, 2, \ldots, m$, were obtained, such as $D_{v_2} = \{e_1, e_2, e_3\}$ for the junction v_2. A spanning tree that connects all junctions in the mine network without forming a cycle was created, including some tunnels on demand. The spanning tree was then changed to an undirected graph. The breadth-first search starts from the source junction $S = D_{v_1}$. When it comes to junction v_b in the spanning tree, the independent cut set is calculated by $S = \left(S \cup D_{v_b}\right) - \left(S \cap D_{v_b}\right)$. It was followed until the searching of all the junctions of the spanning tree was completed (see Figure 6).

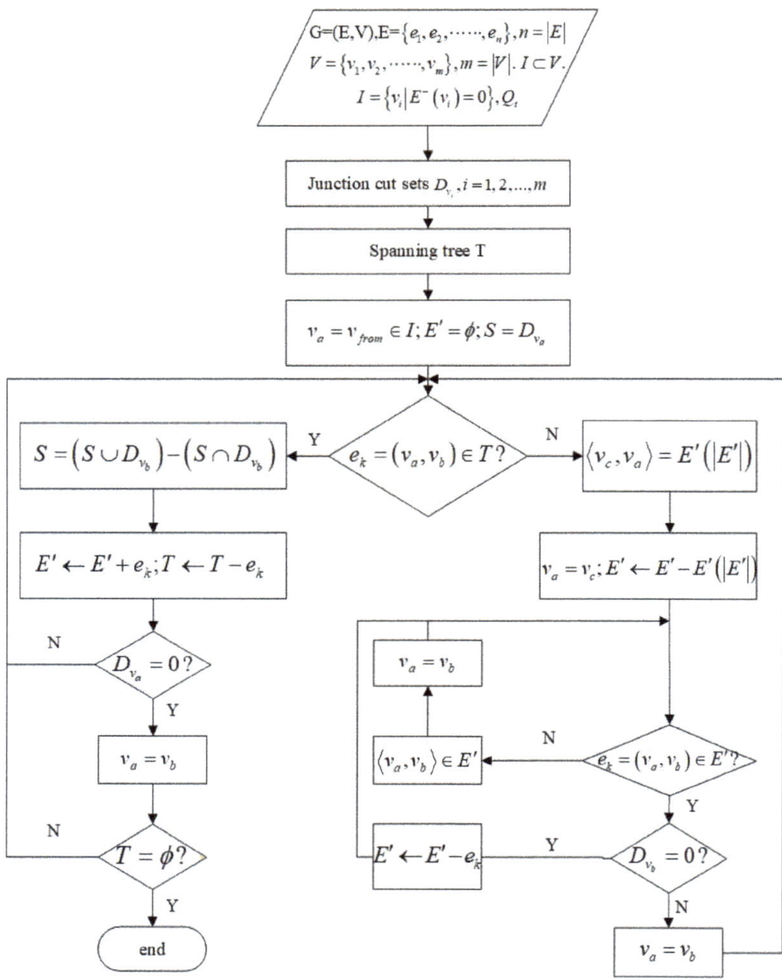

Figure 6. Flow diagram for the independent cut set.

Take the network shown in Figure 5, for example; the network graph is:
$G = (V, E)$,
$E = \{e_1, e_2, e_3, e_4, e_5, e_6, e_7, e_8, e_9, e_{10}\}, n = 10$,
$V = \{v_1, v_2, v_3, v_4, v_5, v_6, v_7, v_8\}, m = 8$,
$I = \{e_1\}$,
$O = \{e_{10}\}$.
The spanning tree is:
$T = \{e_1, e_2, e_3, e_4, e_6, e_9, e_{10}\}$.
The single junction cut sets for every junction are:
$D_{v1} = \{e_1\}$,
$D_{v2} = \{e_1, e_2, e_3\}$,
$D_{v3} = \{e_2, e_4, e_7\}$,
$D_{v4} = \{e_3, e_5, e_8\}$,
$D_{v5} = \{e_4, e_5, e_6\}$,
$D_{v6} = \{e_6, e_8, e_9\}$,
$D_{v7} = \{e_7, e_9, e_{10}\}$,

$$D_{v8} = \{e_{10}\}.$$

According to the flow diagram, the process of obtaining the independent cut set is as shown in Table 1.

Table 1. The independent cut set of the graph in Figure 5.

No.	v_a	$e_k \in \langle v_a, v_b \rangle \in T$	S	E'	Search	$\lvert D_{vb} \rvert$
1			$S_1 = D_{v1} = \{e_1\}$			0
2	v_1	$e_1 \in \langle v_1, v_2 \rangle \in T$	$S_2 = \{e_2, e_3\}$	e_1	$v_1 \xrightarrow{e_1} v_2$	0
3	v_2	$e_2 \in \langle v_2, v_3 \rangle \in T$	$S_3 = \{e_3, e_4, e_7\}$	e_1, e_2	$v_2 \xrightarrow{e_2} v_3$	1
4	v_2	$e_3 \in \langle v_2, v_4 \rangle \in T$	$S_4 = \{e_4, e_7, e_5, e_8\}$	e_1, e_2, e_3	$v_2 \xrightarrow{e_3} v_4$	0
5	v_3	$e_4 \in \langle v_3, v_5 \rangle \in T$	$S_5 = \{e_7, e_6, e_8\}$	e_1, e_2, e_3, e_4	$v_3 \xrightarrow{e_4} v_5$	0
6	v_5	$e_6 \in \langle v_5, v_6 \rangle \in T$	$S_6 = \{e_7, e_9\}$	e_1, e_2, e_3, e_4, e_6	$v_5 \xrightarrow{e_6} v_6$	0
7	v_6	$e_9 \in \langle v_6, v_7 \rangle \in T$	$S_7 = \{e_{10}\}$	$e_1, e_2, e_3, e_4, e_6, e_9$	$v_6 \xrightarrow{e_9} v_7$	0
8	v_7	$e_{10} \in \langle v_7, v_8 \rangle \in T$				0

3. Methods of Location Optimization and Flow Reconstruction

3.1. Location Optimization of Wind Speed Sensors

In order to study the minimum number and location of sensors that can completely recover the airflow in a mine network, an algorithm based on the topological structure of the network and the information was developed. The algorithm creates spanning trees and uses the IBFS, a well-known graph traversal algorithm, to search. The inputs of the algorithm on the location of wind speed sensors are the directed graph $G = (V, E)$ and single junction cut set, while the output is the independent cut set. The procedures of the algorithm are as follows:

1. Change all junctions in V with a single junction and find the single-junction cut set of all junctions.
2. Initialize $V_a = V_{from}$.
3. Use IBFS over spanning tree starting at v_a. Denote e_k and v_b as the visited edges and junctions, respectively.
4. Obtain the independent cut set S while $e_k \epsilon T$: end if $T\epsilon\theta$; otherwise, search the next spanning tree. To avoid repeated searches, the edges and junctions passing by should be recorded for every independent cut set obtained.
5. If any junctions in the graph are not visited at this time, the breadth-first traversal must be performed again from a junction that has not been visited until all the junctions in the graph have been accessed.
6. Put the known air volume into Equation (3) to solve for the unknown air volume.

3.2. Flow Reconstruction Method by the Wind Speed Sensors

According to the Chinese industry standards GB/T51272-2018 and AQ2031-2011, it is compulsory to install wind speed sensors in main return tunnels, the return tunnels of all mining areas, and the return tunnels of all sublevels. These tunnels are seen as cotrees for the spanning tree, which can be generated according to the cotrees by the Kruskal algorithm.

As shown in Figure 6, the independent cut sets can be searched by the spanning tree. There is a flow equation, Equation (3), for every independent cut set. Taking Table 1 as an example, seven flow equations can be set. Q_t is the air volume of the main fan that must be monitored. It is assumed that $Q_t = 15$, $Q_5 = 4$, $Q_7 = 5$, and $Q_8 = 2$.

$$\begin{cases} Q_1 = Q_t \\ Q_2 + Q_3 = Q_t \\ Q_3 + Q_4 + Q_7 = Q_t \\ Q_4 + Q_7 + Q_5 + Q_8 = Q_t \\ Q_6 + Q_7 + Q_8 = Q_t \\ Q_7 + Q_9 = Q_t \\ Q_{10} = Q_t \end{cases} \tag{4}$$

Equation (4) is solved by the Gaussian elimination method. The air volume of the tunnels in the spanning tree is $Q_1 = 15$, $Q_2 = 9$, $Q_3 = 6$, $Q_4 = 4$, $Q_6 = 8$, $Q_9 = 10$, and $Q_{10} = 15$.

4. Algorithm Optimization of Sensor Location Problem

In this paper, we propose an independent cut set algorithm for well-posed reconstruction of the airflow in a mine ventilation network; however, its process is too complex. For a large-scale ventilation network, the calculation amount and complexity are very large. In order to simplify the calculation process, the independent cut set algorithm can be optimized by matrixing the calculation process, i.e., all equations in the calculation process can be expressed by matrix. With the help of matrix characteristics, the sensor location problem can be calculated more quickly and intuitively. In addition, there is a mature calculation program for the matrix problem. It is more convenient to solve the matrix problem by computer than to directly solve the newly defined algorithm. Especially for large-scale network problems, faster and more accurate calculations can be made.

4.1. Optimization of Single Junction Cut Set

In order to more clearly represent the single junction cut sets, we can use the complete incidence matrix to express the single junction cut sets of each junction. The complete incidence matrix is a 0-1-(-1) matrix that describes the mine ventilation network structure through the spatial relationships between the tunnels and junctions of the network. Each column element of the matrix represents each tunnel of the mine ventilation network. Each row element denotes the junction. The matrix can be expressed as follows:

$$L_{m \times n} = \begin{pmatrix} l_{11} & \cdots & l_{1k} & \cdots & l_{1n} \\ \vdots & \ddots & \vdots & \ddots & \vdots \\ l_{j1} & \cdots & l_{jk} & \cdots & l_{jn} \\ \vdots & \ddots & \vdots & \ddots & \vdots \\ l_{m1} & \cdots & l_{mk} & \cdots & l_{mn} \end{pmatrix}, \tag{5}$$

where $L_{m \times n}$ is the complete incidence matrix of the network of dimension $(m \times n)$ and l_{ij} is 0, 1, or -1. $l_{ij} = 1$ indicates that e_j and junction v_i are associated, and e_j is the out branch of v_i. $l_{ij} = -1$ indicates that e_j is the in branch of v_i. Through the definition of single junction cut sets, if branch e_j and junction v_i are associated, then e_j belongs to D_{v_i}. Therefore, e_j is an element of D_{v_i} when $|l_{ij}| = 1$. On the contrary, $l_{ij} = 0$ indicates that e_j and junction v_i are not related.

The matrix can be represented through a set of column vectors. $L_{m \times n}$ denotes the complete incidence matrix with m junctions and n tunnels, and can be expressed as follows:

$$L_{m \times n} = \begin{bmatrix} L_1 L_2 \cdots L_j \cdots L_m \end{bmatrix}^T, \tag{6}$$

where L_j is the j_{th} row vector of dimension $(1 \times n)$. $D_{v_j} = \{e_k | e_k = E(v_j)\}$, i.e., $D_{v_j} = \{e_k | l_{jk} \neq 0\}$ $(j = 1, 2, \cdots, m; \ k = 1, 2, \cdots, n)$. e_k, corresponding to all nonzero entries l_{jk} in L_j, is the single-junction cut set corresponding to junction v_j; that is, e_k is all the elements of D_{v_j}.

4.2. Optimization of Independent Cut Set Algorithm

In Section 2.6, it is revealed that the independent cut set is calculated by $S = \left(S \cup D_{v_b}\right) - \left(S \cap D_{v_b}\right)$, which is a equation related to the single junction cut set and independent cut set. From Section 4.1, we know the matrix form of the single junction cut sets D_{v_j}, i.e., complete incidence matrix (see Equations (5) and (6)). Therefore, we can convert the solution process of the independent cut set into matrix form. An independent cut set–branch incidence matrix can be established to obtain the independent cut set of the ventilation network. The matrix can be expressed as follows:

$$
S_{m \times n} = \begin{pmatrix}
s_{11} & \cdots & s_{1k} & \cdots & s_{1n} \\
\vdots & \ddots & \vdots & \ddots & \vdots \\
s_{j1} & \cdots & s_{jk} & \cdots & s_{jn} \\
\vdots & \ddots & \vdots & \ddots & \vdots \\
s_{m1} & \cdots & s_{mk} & \cdots & s_{mn}
\end{pmatrix},
\tag{7}
$$

where $S_{m \times n}$ is the independent cut set–branch incidence matrix of the network of dimension $(m \times n)$ and s_{ij} is 0 or 1. $s_{ij} = 1$ indicates that e_j is a subset of the independent cut set S_i at junction v_i. $s_{ij} = 0$ denotes that e_j does not belong to the independent cut set S_i at junction v_i. Compared with the complete incidence matrix, each column element of the new matrix still represents each branch of the ventilation network. However, row elements are expressed by independent cut sets S. In Section 4.1, the complete incidence matrix (see Equation (6)) is expressed as follows:

$$
L_{m \times n} = \left[L_1 L_2 \cdots L_j \cdots L_m\right]^T.
$$

Through the calculation formula of independent cut sets $S = \left(S \cup D_{v_b}\right) - \left(S \cap D_{v_b}\right)$, we can obtain the matrix expression:

$$
\begin{cases}
S_{j+1} = S_j + L_{j+1} \\
S_1 = L_1
\end{cases} (j = 1, 2, \cdots, m - 1),
\tag{8}
$$

where e_k corresponds to all nonzero entries and s_{jk} in S_j is the independent cut sets. According to Equation (8), we can obtain the matrix form of the independent cut set.

5. Sensor Location and Flow Reconstruction Based on Algorithm Optimization

5.1. Sensor Location Based on Algorithm Optimization

Take the network shown in Figure 5 as an example; according to Equation (1), the network graph is:

$$
G = (V, E);
$$
$$
E = \{e_1, e_2, e_3, e_4, e_5, e_6, e_7, e_8, e_9, e_{10}\}, n = 10;
$$
$$
V = \{v_1, v_2, v_3, v_4, v_5, v_6, v_7, v_8\}, m = 8.
$$

The junctions and tunnels have been sorted in Figure 5. The whole calculation process is carried out according to the IBFS. According to Equations (5) and (6), we can obtain the single junction cut set expressed by the matrix, that is, the complete incidence matrix of ventilation network $G = (V, E)$. In order to facilitate the presentation of the row vector and column vector corresponding to each element of the complete incidence matrix, we show the complete incidence matrix corresponding to Figure 5 in the form of Table 2.

Table 2. The complete incidence matrix (corresponding to Figure 5).

Link Junction	e_1	e_2	e_3	e_4	e_5	e_6	e_7	e_8	e_9	e_{10}
v_1	1	0	0	0	0	0	0	0	0	0
v_2	−1	1	1	0	0	0	0	0	0	0
v_3	0	−1	0	1	0	0	1	0	0	0
v_4	0	0	−1	0	1	0	0	1	0	0
v_5	0	0	0	−1	−1	1	0	0	0	0
v_6	0	0	0	0	0	−1	0	−1	1	0
v_7	0	0	0	0	0	0	−1	0	−1	1
v_8	0	0	0	0	0	0	0	0	0	−1

The matrix can be expressed as $L_{8\times10}$. According to Equation (6), the form of its column vector is $L_{8\times10} = \begin{bmatrix} L_1 & L_2 & \cdots & L_j & \cdots & L_8 \end{bmatrix}^T$. All nonzero entries in L_1 are the result of D_{v_1}, that is, $L_1 = \{1\,0\,0\,0\,0\,0\,0\,0\,0\,0\}$; the first single junction cut set is $D_{v_1} = \{e_1\}$. Similarly, all nonzero entries in L_2 are the result of D_{v_2}, that is, $L_2 = \{-1\,1\,1\,0\,0\,0\,0\,0\,0\,0\}$, and the second single junction cut set is $D_{v_2} = \{e_1, e_2, e_3\}$, etc. Through the matrix, we can easily obtain the single-junction cut sets of each junction, and the result conforms to the definition of single-junction cut sets.

In order to obtain the independent cut set, $L_{8\times10}$ is substituted into Equation (8). The results are shown in Table 3.

Table 3. Independent cut set–branch incidence matrix (corresponding to Figure 5).

s_i \ e_j	e_1	e_2	e_3	e_4	e_5	e_6	e_7	e_8	e_9	e_{10}
s_1	1	0	0	0	0	0	0	0	0	0
s_2	0	1	1	0	0	0	0	0	0	0
s_3	0	0	1	1	0	0	1	0	0	0
s_4	0	0	0	1	1	0	1	1	0	0
s_5	0	0	0	0	0	1	1	1	0	0
s_6	0	0	0	0	0	0	1	0	1	0
s_7	0	0	0	0	0	0	0	0	0	1
s_8	0	0	0	0	0	0	0	0	0	0

The matrix can be expressed as $S_{8\times10}$. The form of its column vector is $S_{8\times10} = \begin{bmatrix} s_1 & s_2 & \cdots & s_j & \cdots & s_8 \end{bmatrix}^T$. All nonzero entries in s_j are the results of the independent cut set. For example, $s_1 = \{1\,0\,0\,0\,0\,0\,0\,0\,0\,0\}$ and the first independent cut set is $S_1 = \{e_1\}$; $s_2 = \{0\,1\,1\,0\,0\,0\,0\,0\,0\,0\}$ and the second independent cut set is $S_2 = \{e_2, e_3\}$; $s_3 = \{0\,0\,1\,1\,0\,0\,1\,0\,0\,0\}$ and the first independent cut set is $S_1 = \{e_3, e_4, e_7\}$, etc.

Through comparison, it can be seen that the results obtained through the matrix in Table 3 are consistent with those in Table 1.

5.2. Flow Reconstruction Based on Algorithm Optimization

In order to simplify the calculation process, the solution of the independent cut set is obtained by a matrix. Therefore, we can also express the calculation of Equation (3) by a matrix. According to Equation (3), the air volume in each tunnel can be denoted by air volume set Q, and its expression is as follows:

$$Q^T = \{Q_1 \; Q_2 \; \cdots \; Q_n\}, \tag{9}$$

where n denotes the tunnel number and Q_n represents the air volume of the nth tunnel. According to Equation (3), the following equation can be obtained:

$$SQ = Q_{t_{m\times1}}, \tag{10}$$

where $Q_{t_{m \times 1}}$ is the column vector of dimension $(m \times 1)$, $Q_{t_{m \times 1}} = \{Q_t \ Q_t \ \cdots \ Q_t\}^T$.

In Figure 5, $S = S_{8 \times 10} = [s_1 s_2 \ \cdots \ s_8]^T$ and $Q = \{Q_1 \ Q_2 \ \cdots \ Q_8\}^T$. It is assumed that $Q_t = 15$, $Q_5 = 4$, $Q_7 = 5$, and $Q_8 = 2$. From Equation (10),

$$
\begin{Bmatrix}
\begin{pmatrix}
1 & 0 & 0 & 0 & 0 & 0 & 0 & 0 & 0 & 0 \\
0 & 1 & 1 & 0 & 0 & 0 & 0 & 0 & 0 & 0 \\
0 & 0 & 1 & 1 & 0 & 0 & 1 & 0 & 0 & 0 \\
0 & 0 & 0 & 1 & 1 & 0 & 1 & 1 & 0 & 0 \\
0 & 0 & 0 & 0 & 0 & 1 & 1 & 1 & 0 & 0 \\
0 & 0 & 0 & 0 & 0 & 0 & 1 & 0 & 1 & 0 \\
0 & 0 & 0 & 0 & 0 & 0 & 0 & 0 & 0 & 1 \\
0 & 0 & 0 & 0 & 0 & 0 & 0 & 0 & 0 & 0
\end{pmatrix}
\end{Bmatrix}
\begin{Bmatrix}
Q_1 \\ Q_2 \\ Q_3 \\ Q_4 \\ Q_5 \\ Q_6 \\ Q_7 \\ Q_8
\end{Bmatrix}
=
\begin{Bmatrix}
Q_t \\ Q_t \\ Q_t \\ Q_t \\ Q_t \\ Q_t \\ Q_t \\ Q_t
\end{Bmatrix}
$$

After simplifying the above formula, the following equation can be obtained:

$$
\begin{aligned}
Q_1 &= Q_t \\
Q_2 + Q_3 &= Q_t \\
Q_3 + Q_4 + Q_7 &= Q_t \\
Q_4 + Q_7 + Q_5 + Q_8 &= Q_t \ . \\
Q_6 + Q_7 + Q_8 &= Q_t \\
Q_7 + Q_9 &= Q_t \\
Q_{10} &= Q_t
\end{aligned}
$$

The air volume of the tunnels in the spanning tree is $Q_1 = 15$, $Q_2 = 9$, $Q_3 = 6$, $Q_4 = 4$, $Q_6 = 8$, $Q_9 = 10$, and $Q_{10} = 15$.

6. Case Study

According to the standards AQ1028-2006 and GB/T 10178, the ventilation network is a fluid network formed by conveying air flow pipelines; various regulating facilities, power facilities, and air flow pipelines are connected. Figure 7 is the ventilation network of a coal mine run by Jinmei Co., Ltd. $G = (V, E)$, where the tunnel number is $|E| = 100$, and the junction number is $|V| = 71$. Table 4 shows the whole tunnel's reconstructed air volume and simulated air volume. The test air volume [45] is obtained by a mine ventilation simulation system (MVSS) [46]. The principle of solving the network by MVSS is to establish the mathematical equations for the ventilation resistance law, the air volume balance law, the air pressure balance law, the known total air volume of the fans, and the ventilation resistance coefficient, and then solve them. The main fans are installed in e_9, e_{39}, and e_{78}, and their characteristic equations are shown in Equation (11). For a multisource and multisink ventilation network, all sources and sinks should be connected to become a single-source and single-sink network. In a ventilation network, if the spanning tree is removed, the remaining part is called the cotree. As shown in Figure 8, 33 wind speed sensors are used for the network, including 30 cotrees and 3 wind shafts with fans. The tunnels that must be set up with wind speed sensors are e_3, e_9, e_{22}, e_{28}, e_{33}, e_{39}, e_{55}, e_{68}, e_{72}, e_{76}, e_{78}, e_{84}, e_{92}, and e_{94}. A spanning tree is generated that does not include the above tunnels (the magenta in Figure 8). Considering the difficulty of testing the lower wind speed, most of the cotrees should not contain the tunnels with structures such as throttles and confined walls. Based on the flow diagram in Figure 5, the air volume of the spanning trees can be calculated by the linear flow equation system of the independent cut sets.

$$
\left\{
\begin{aligned}
H(q_9) &= 1932.25 + 44.84 q_9 - 0.64 q_9^2 \\
H(q_{39}) &= 1932.25 + 44.84 q_{39} - 0.64 q_{39}^2 \\
H(q_{78}) &= 1932.25 + 44.84 q_{78} - 0.64 q_{78}^2
\end{aligned}
\right.
\tag{11}
$$

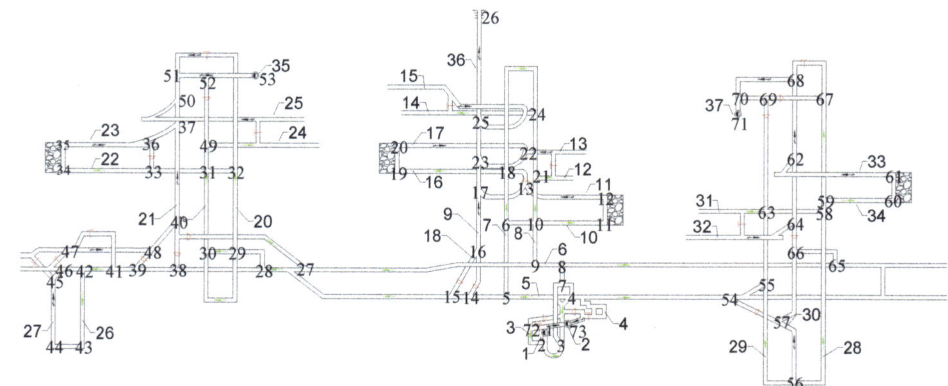

Figure 7. A coal mine ventilation system used by Jinmei Co., Ltd. (Red indicates the position of air doors or wind windows, and black arrows indicate the exhaust air flow. Green is the incoming air flow as well as the fresh air flow. Each number indicates the junction number and tunnel number).

Table 4. Ventilation system of Jinmei coal mine.

Edge	Test (m³/s)	Reconstruction (m³/s)	Edge	Test (m³/s)	Reconstruction (m³/s)	Edge	Test (m³/s)	Reconstruction (m³/s)
e_1	285.85	285.85	e_2	188.83	188.83	e_3	49.42	49.42
e_4	181.97	181.97	e_5	103.88	103.88	e_6	157.3	157.3
e_7	15.69	15.69	e_8	24.67	24.67	e_9	65.11	65.11
e_{10}	61.53	61.53	e_{11}	95.77	95.77	e_{12}	7.25	7.25
e_{13}	68.78	68.78	e_{14}	292.71	292.71	e_{15}	252.27	252.27
e_{16}	40.44	40.44	e_{17}	117.50	117.50	e_{18}	134.77	134.77
e_{19}	56.09	56.09	e_{20}	71.42	71.42	e_{21}	56.09	56.09
e_{22}	56.09	56.09	e_{23}	57.69	57.69	e_{24}	1.59	1.59
e_{25}	90.89	90.89	e_{26}	4.89	4.89	e_{27}	4.78	4.78
e_{28}	9.67	9.67	e_{29}	67.35	67.35	e_{30}	71.03	71.03
e_{31}	−3.84	−3.84	e_{32}	71.03	71.03	e_{33}	71.03	71.03
e_{34}	31.78	31.78	e_{35}	39.64	39.64	e_{36}	102.80	102.80
e_{37}	170.15	170.15	e_{38}	35.80	35.80	e_{39}	205.96	205.96
e_{40}	86.11	86.11	e_{41}	200.46	200.46	e_{42}	3.14	3.14
e_{43}	154.60	154.60	e_{44}	45.87	45.87	e_{45}	72.78	72.78
e_{46}	81.82	81.82	e_{47}	72.78	72.78	e_{48}	−1.89	−1.89
e_{49}	18.69	18.69	e_{50}	52.20	52.20	e_{51}	83.71	83.71
e_{52}	17.29	17.29	e_{53}	1.41	1.41	e_{54}	17.29	17.29
e_{55}	17.29	17.29	e_{56}	18.69	18.69	e_{57}	49.01	49.01

The reconstructed air volume in Table 4 was obtained by the independent cut set algorithm. Similarly, the results obtained by the independent cut set algorithm followed the three basic laws of air volume distribution. Their basic principles are similar. We also used part of the test air volume as a known quantity for the calculation. Therefore, their calculation results should be the same; otherwise, they will prove that the independent cut set algorithm is incorrect. It can be seen from Table 4 that their results were exactly the same. It is feasible to use the independent cut set algorithm to determine the minimum number and location of sensors through the case study. By placing sensors on the cotrees, the air volume of the spanning trees can be obtained. Due to the uniqueness of the well-posed solution, the error of flow reconstruction is zero (see Table 4). There will be no errors or other solutions.

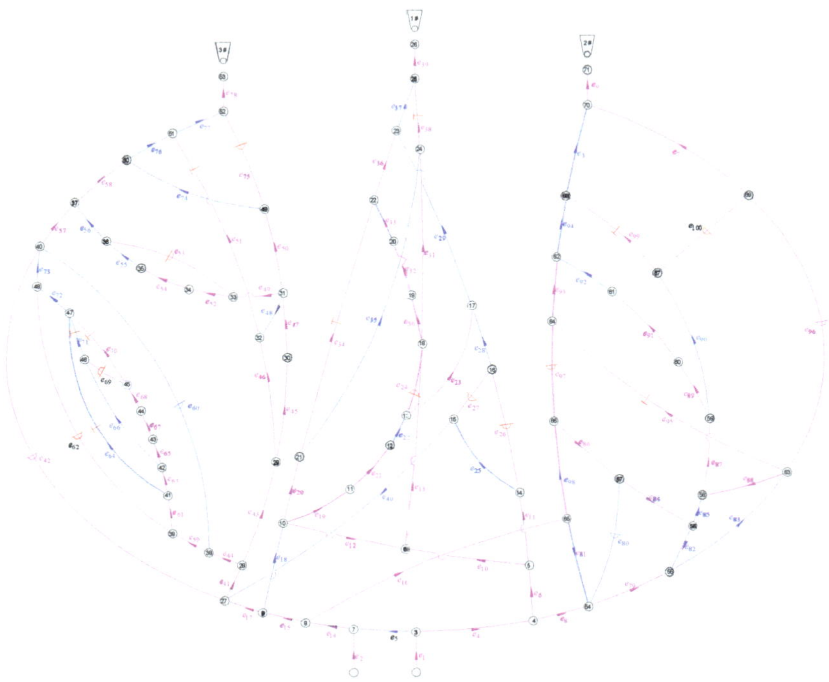

Figure 8. Coal mine ventilation used by Jinmei Co., Ltd. The magenta represents spanning trees and the blue represents cotrees.

Figure 5 is an example of a simple network with a single source and single sink, and Figure 8 is an example of a complex ventilation network with multiple sources and multiple sinks. Through the calculation of these two examples, it was found that the independent cut-set algorithm was suitable for both of the examples. Therefore, the independent cut set algorithm can be used to solve the sensor optimization and air volume reconstruction problems of all types of ventilation networks. In line with previous research, the calculation complexity was also affected by the network structure. Due to the uniqueness of the well-posed solution, the error of the result can only come from the measurement accuracy of the sensor and will not be affected by other factors. Moreover, the existing solution is unique and has no errors.

7. Conclusions

In this research, the wind speed sensor location problem of the mine ventilation network was solved by using the independent cut set algorithm, which is a new concept based on the structure of the underlying graph. Firstly, we found the problems: the sensor number was $n - m + k - 1$, and we located flow sensors in the cotrees that must be set up with sensors according to the Chinese standards. The calculation complexity was affected by the network structure. Secondly, we discussed the possibility of well-posed reconstruction based on the mine ventilation networks. Fewer than 30% of tunnels need to be set up with wind speed sensors to achieve a well-posed reconstruction of the airflow in all the tunnels if the mines have over 200 tunnels. For mines with fewer than 200 tunnels, more than 30% of tunnels should be installed with sensors. Due to the uniqueness of the well-posed solution, there is no error in the results unless the sensor accuracy does not meet the standard. Lastly, the algorithm of independent cut sets was presented for the well-posed reconstructions. The flow reconstructions were shown to be computationally efficient for the coal mine ventilation system used by Jinmei Co., Ltd. This algorithm works on velocity sensor distribution rather than sensors of hazardous and combustible gas. This algorithm is a well-posed reconstruction algorithm. It can only be used to determine the minimum number and location of sensors and cannot solve the underdetermined reconstruction of a given number of sensors. Therefore, the premise of using this algorithm is that the number of sensors will be sufficient. This study provides researchers with a sensor optimization scheme when the number of sensors is insufficient and provides the possibility for air volume reconstruction. At the same time, it reduces the cost for the air volume monitoring of the whole air network and provides some theoretical support for the realization of intelligent ventilation. In the following research, we will discuss the sensor location under the condition of underdetermined reconstruction.

Author Contributions: Conceptualization, K.G.; writing—original draft preparation, K.G. and Z.L.; writing—review and editing, K.G., Z.L., Y.L., Y.H. and C.Z.; supervision, K.G., Z.L. and Y.L. All authors have read and agreed to the published version of the manuscript.

Funding: The work is supported by National Natural Science Foundation of China (No.52104194 and 52074148).

Institutional Review Board Statement: Not applicable.

Informed Consent Statement: Not applicable.

Data Availability Statement: The datasets generated and analyzed during the current study are available from the corresponding author upon reasonable request.

Conflicts of Interest: The authors declare no conflict of interest.

References

1. Zhang, W.H.; Yan, Q.Y.; Yuan, J.H.; He, G.; Teng, T.-L.; Zhang, M.J.; Zeng, Y. A realistic pathway for coal-fired power in China from 2020 to 2030. *J. Clean. Prod.* **2020**, *275*, 122859. [CrossRef]
2. Zhao, Y.H.; Zheng, X.Z.; Hu, H.J.; Wang, S.F.; Lu, N. Press Conference of the State Administration of Mine Safety. Beijing. 2021. Available online: https://www.chinamine-safety.gov.cn/xw/xwfbh/ (accessed on 10 February 2022).
3. Zhang, P.; Lan, H.Q.; Yu, M. Reliability evaluation for ventilation system of gas tunnel based on Bayesian network. *Tunn. Undergr. Space Technol.* **2021**, *112*, 103882. [CrossRef]
4. Dong, L.J.; Hu, Q.C.; Tong, X.J.; Liu, Y.F. Velocity-Free MS/AE Source Location Method for Three-Dimensional Hole-Containing Structures. *Engineering* **2020**, *6*, 827–834. [CrossRef]
5. Dong, L.J.; Tong, X.J.; Ma, J. Quantitative Investigation of Tomographic Effects in Abnormal Regions of Complex Structures. *Engineering* **2021**, *7*, 1011–1022. [CrossRef]
6. Wang, G.F.; Xu, Y.X.; Ren, H.W. Intelligent and ecological coal mining as well as clean utilization technology in China: Review and prospects. *Int. J. Min. Sci. Technol.* **2019**, *29*, 161–169. [CrossRef]
7. Wang, K.; Jiang, S.G.; Zhang, W.Q.; Wu, Z.Y.; Shao, H.; Kou, L.W. Destruction mechanism of gas explosion to ventilation facilities and automatic recovery technology. *Int. J. Min. Sci. Technol.* **2012**, *22*, 417–422. [CrossRef]
8. Huang, D.; Liu, J.; Deng, L.J. A hybrid-encoding adaptive evolutionary strategy algorithm for windage alteration fault diagnosis. *Process Saf. Environ. Prot.* **2020**, *136*, 242–252. [CrossRef]
9. Gao, K.; Deng, L.J.; Liu, J.; Wen, L.X.; Wong, D.; Liu, Z.Y. Study on Mine Ventilation Resistance Coefficient Inversion Based on Genetic Algorithm. *Arch. Min. Sci.* **2018**, *63*, 813–826.
10. Hu, J.H.; Zhao, Y.; Zhou, T.; Ma, S.W.; Wang, X.L.; Zhao, L. Multi-factor influence of cross-sectional airflow distribution in roadway with rough roof. *J. Cent. South Univ.* **2021**, *28*, 2067–2078. [CrossRef]
11. Mayala, L.P.; Veiga, M.M.; Khorzoughi, M.B. Assessment of mine ventilation systems and air pollution impacts on artisanal tanzanite miners at Merelani, Tanzania. *J. Clean. Prod.* **2016**, *116*, 118–124. [CrossRef]
12. Liang, P.Y.; Han, Y.; Zhang, Y.Y.; Wen, Y.T.; Gao, Q.F.; Meng, J. Novel non-destructive testing method using a two-electrode planar capacitive sensor based on measured normalized capacitance values. *Measurement* **2021**, *167*, 108455. [CrossRef]
13. Rodriguez-Vega, M.; Canudas-de-Wit, C.; Fourati, H. Location of turning ratio and flow sensors for flow reconstruction in large traffic networks. *Transp. Res. Part B Methodol.* **2019**, *121*, 21–40. [CrossRef]
14. Hu, X.; Han, Y.M.; Yu, B.; Geng, Z.Q.; Fan, J.Z. Novel leakage detection and water loss management of urban water supply network using multiscale neural networks. *J. Clean. Prod.* **2021**, *278*, 123611. [CrossRef]
15. Lau, P.-W.; Cheung, B.-Y.; Lai, W.-L.; Sham, J.-C. Characterizing pipe leakage with a combination of GPR wave velocity algorithms. *Tunn. Undergr. Space Technol.* **2021**, *109*, 103740.
16. Liang, L.P.; Xu, K.J.; Wang, X.F.; Zhang, Z.; Yang, S.L.; Zhang, R. Statistical modeling and signal reconstruction processing method of EMF for slurry flow measurement. *Measurement* **2014**, *54*, 1–13. [CrossRef]

17. Lu, H.F.; Iseley, T.; Behbahani, S.; Fu, L.D. Leakage detection techniques for oil and gas pipelines: State-of-the-art. *Tunn. Undergr. Space Technol.* **2020**, *98*, 103249. [CrossRef]
18. Aida-zade, K.R.; Ashrafova, E.R. Localization of the points of leakage in an oil main pipeline under nonstationary conditions. *J. Eng. Phys. Thermophys.* **2012**, *85*, 1148–1156. [CrossRef]
19. Zhang, Z.W.; Hou, L.F.; Yuan, M.Q.; Fu, M.; Qian, X.M.; Duanmu, W.; Li, Y.Z. Optimization monitoring distribution method for gas pipeline leakage detection in underground spaces. *Tunn. Undergr. Space Technol.* **2020**, *104*, 103545. [CrossRef]
20. Santos, R.B.; de Sousa, E.O.; da Silva, F.V.; da Cruz, S.L.; Fileti, A.M.F. Detection and on-line prediction of leak magnitude in a gas pipeline using an acoustic method and neural network data processing. *Braz. J. Chem. Eng.* **2014**, *31*, 145–153. [CrossRef]
21. Li, J.; Li, Y.L.; Huang, X.J.; Ren, J.H.; Feng, H.; Zhang, Y.; Yang, X.X. High-sensitivity gas leak detection sensor based on a compact microphone array. *Measurement* **2021**, *174*, 109017. [CrossRef]
22. Singh, K.R.; Dutta, R.; Kalamdhad, A.S.; Kumar, B. An investigation on water quality variability and identification of ideal monitoring locations by using entropy based disorder indices. *Sci. Total Environ.* **2019**, *647*, 1444–1455. [CrossRef] [PubMed]
23. Huang, D.; Liu, Y.; Liu, Y.H.; Song, Y.; Hong, C.S.; Li, X.Y. Identification of sources with abnormal radon exhalation rates based on radon concentrations in underground environments. *Sci. Total Environ.* **2022**, *807*, 150800. [CrossRef] [PubMed]
24. Castillo, E.; Jiménez, P.; Menendez, J.M.; Conejo, A.J. The Observability Problem in Traffic Models: Algebraic and Topological Methods. *IEEE Trans. Intell. Transp. Syst.* **2008**, *9*, 275–287. [CrossRef]
25. Castillo, E.; Calviño, A.; Lo, H.K.; Menéndez, J.M.; Grande, Z. Non-planar hole-generated networks and link flow observability based on link counters. *Transp. Res. Part B Methodol.* **2014**, *68*, 239–261. [CrossRef]
26. Rinaudo, P.; Paya-Zaforteza, I.; Calderón, P.A. Improving tunnel resilience against fires: A new methodology based on temperature monitoring. *Tunn. Undergr. Space Technol.* **2016**, *52*, 71–84. [CrossRef]
27. Balaji, S.; Anitha, M.; Rekha, D.; Arivudainambi, D. Energy efficient target coverage for a wireless sensor network. *Measurement* **2020**, *165*, 108167. [CrossRef]
28. Li, Y.T.; Bao, T.F.; Chen, H.; Zhang, K.; Shu, X.S.; Chen, Z.X.; Hu, Y.H. A large-scale sensor missing data imputation framework for dams using deep learning and transfer learning strategy. *Measurement* **2021**, *178*, 109377. [CrossRef]
29. Ng, M. Synergistic sensor location for link flow inference without path enumeration: A node-based approach. *Transp. Res. Part B Methodol.* **2012**, *46*, 781–788. [CrossRef]
30. He, S.X. A graphical approach to identify sensor locations for link flow inference. *Transp. Res. Part B Methodol.* **2013**, *51*, 65–76. [CrossRef]
31. Muduli, L.; Jana, P.K.; Mishra, D.P. A novel wireless sensor network deployment scheme for environmental monitoring in longwall coal mines. *Process Saf. Environ. Prot.* **2017**, *109*, 564–576. [CrossRef]
32. Wang, K.; Jiang, S.G.; Wu, Z.Y.; Shao, H.; Zhang, W.Q.; Pei, X.D.; Cui, C.B. Intelligent safety adjustment of branch airflow volume during ventilation-on-demand changes in coal mines. *Process Saf. Environ. Prot.* **2017**, *111*, 491–506. [CrossRef]
33. Song, Y.W.; Yang, S.Q.; Hu, X.C.; Song, W.X.; Sang, N.W.; Cai, J.W.; Xu, Q. Prediction of gas and coal spontaneous combustion coexisting disaster through the chaotic characteristic analysis of gas indexes in goaf gas extraction. *Process Saf. Environ. Prot.* **2019**, *129*, 8–16. [CrossRef]
34. Lyu, P.Y.; Chen, N.; Mao, S.J.; Li, M. LSTM based encoder-decoder for short-term predictions of gas concentration using multi-sensor fusion. *Process Saf. Environ. Prot.* **2020**, *137*, 93–105. [CrossRef]
35. Foorginezhad, S.; Mohseni-Dargah, M.; Firoozirad, K.; Aryai, V.; Razmjou, A.; Abbassi, R.; Garaniya, V.; Beheshti, A.; Asadnia, M. Recent Advances in Sensing and Assessment of Corrosion in Sewage Pipelines. *Process Saf. Environ. Prot.* **2021**, *147*, 192–213. [CrossRef]
36. Hu, Y.N.; Koroleva, O.I.; Krstić, M. Nonlinear control of mine ventilation networks. *Syst. Control Lett.* **2003**, *49*, 239–254. [CrossRef]
37. Khan, K.S.; Tariq, M. Accurate Monitoring and Fault Detection in Wind Measuring Devices through Wireless Sensor Networks. *Sensors* **2014**, *14*, 22140–22158. [CrossRef]
38. Sun, J.P.; Tang, L.; Li, C.S.; Zhu, N.; Zhang, B. Application of air-volume Proportion rule in optimal placement of gas sensor in mine. *J. China Coal Soc.* **2008**, *33*, 1126–1130. (In Chinese)
39. Zhao, D.; Liu, J.; Pan, J.T.; Li, Z.X. Application study of air velocity fault source diagnosis technology for ventilation system in Daming Mine. *Chin. J. Saf. Environ.* **2012**, *12*, 204–207. (In Chinese)
40. Dong, X.L.; Chen, S.; Zhao, D.; Pan, J.T. Study on Application of Minimum Tree Principle in Layout of Wind Speed Sensor in Mine. *Chin. World Sci-Tech R D* **2015**, *37*, 680–683. (In Chinese)
41. Liang, S.H.; He, J.; Zheng, H.; Sun, R.H. Research on the HPACA Algorithm to Solve Alternative Covering Location Model for Methane Sensors. *Procedia Comput. Sci.* **2018**, *139*, 464–472. [CrossRef]
42. Zhao, D.; Zhang, H.; Pan, J.T. Solving Optimization of A Mine Gas Sensor Layout Based on A Hybrid GA-DBPSO Algorithm. *IEEE Sens. J.* **2019**, *19*, 6400–6409. [CrossRef]
43. Wu, C.Q.; Wang, L. On Efficient Deployment of Wireless Sensors for Coverage and Connectivity in Constrained 3D Space. *Sensors* **2017**, *17*, 2304. [CrossRef] [PubMed]
44. Semin, M.A.; Levin, L.Y. Stability of air flows in mine ventilation networks. *Process Saf. Environ. Prot.* **2019**, *124*, 167–171. [CrossRef]
45. Liu, J.; Jiang, Q.H.; Liu, L.; Wang, D.; Huang, D.; Deng, L.J.; Zhou, Q.C. Resistance variant fault diagnosis of mine ventilation system and position optimization of wind speed sensor. *J. China Coal Soc.* **2021**, *46*, 1907–1914. (In Chinese)
46. Jia, J.Z.; Liu, J.; Geng, X.W. Mathematical model of mine ventilation simulation system. *J. Liaoning Tech. Univ.* **2003**, *22*, 88–90. (In Chinese)

Article

Prediction Method for Surface Subsidence of Coal Seam Mining in Loess Donga Based on the Probability Integration Model

Bingchao Zhao [1,2], Yaxin Guo [1,2,*], Xuwei Mao [1,2], Di Zhai [1,2], Defu Zhu [2,3], Yuming Huo [4], Zedong Sun [3,5] and Jingbin Wang [1,2]

[1] Energy College, Xi'an University of Science and Technology, Xi'an 710000, China; zhaobc913@163.com (B.Z.); maoxuweivip@sina.com (X.M.); 21103077021@stu.xust.edu.cn (D.Z.); 20103077008@stu.xust.edu.cn (J.W.)
[2] State Key Laboratory of Coal Resources in Western China, Xi'an University of Science and Technology, Xi'an 710000, China; zhudefu@tyut.edu.cn
[3] Key Laboratory of In-Situ Property-Improving Mining of Ministry of Education, Taiyuan University of Technology, Taiyuan 030000, China; sunzedong0117@link.tyut.edu.cn
[4] College of Mining Engineering, Taiyuan University of Technology, Taiyuan 030000, China; huoyuming@tyut.edu.cn
[5] College of Coal Engineering, Shanxi Datong University, Datong 037000, China
* Correspondence: 19103077012@stu.xust.edu.cn; Tel.: +86-139-9465-2248

Citation: Zhao, B.; Guo, Y.; Mao, X.; Zhai, D.; Zhu, D.; Huo, Y.; Sun, Z.; Wang, J. Prediction Method for Surface Subsidence of Coal Seam Mining in Loess Donga Based on the Probability Integration Model. *Energies* **2022**, *15*, 2282. https://doi.org/10.3390/en15062282

Academic Editors: Nikolaos Koukouzas and Adam Smoliński

Received: 15 February 2022
Accepted: 19 March 2022
Published: 21 March 2022

Publisher's Note: MDPI stays neutral with regard to jurisdictional claims in published maps and institutional affiliations.

Abstract: The accurate prediction of surface subsidence is a significant foundation for the damage assessment of coal seam mining and ecological environment reclamation in loess donga. However, conventional models are very problematic, and the reliability of prediction is usually low. Therefore, we propose a method for predicting surface subsidence of coal seam mining in loess donga that is based on the probability integration model, combined with the movement principle of rock and soil layers in the respective study area, and considering the influence of slope stability and additional mining slip on mining subsidence. The feasibility of our new method was verified by a case study in the N1114 working face of the Ningtiaota coal mine (China) that is situated in an area with abundant loess dongas. The results show that slope slippage is the source of error in the prediction of subsidence in loess donga. The prediction idea of "dividing the surface of loess donga into horizontal strata area and slope sub-area, and predicting the subsidence value of the two areas, respectively" is put forward. A method for predicting the subsidence value of two regions is established. First, based on the theory of probability integral and rock formation movement, the probability integral parameters of the horizontal stratum area are determined, and the subsidence basins in the area are superimposed and calculated. Secondly, according to the slope stability and slip principle, the additional displacement of subsidence in the slope area with mining instability coefficient $G_{cs} > 0.87$ is calculated. Finally, combined with the subsidence prediction results of the strata area and the slope sub-area, and the position of the slope, the accurate prediction of the surface subsidence in loess donga is realized. Our results show that the agreement between the curves predicted from our calculations and from the measured data are between 88.7–97.8%. The calculated error of the additional displacement of slope mining slip is between 1.0–9.8%. The excellent correlation between the modelled and measured data documents that our method provides, demonstrated a new efficient and valuable tool for the precise prediction of damages induced by mining of underground coal seams in loess donga.

Keywords: surface subsidence; probability integration; loess donga; superimposed calculation; additional displacement of slope mining slip

1. Introduction

The Yushenfu coalfield [1] is the largest coalfield in China and one of the seven largest coalfields in the world. Loess dongas are widely distributed in this area. Major coal mining activity in this area has caused an increasing amount of subsidence areas. The subsidence has generated major damages to the cultivated land, water bodies, and buildings to varying degrees within the area affected by the mining, and also caused serious environmental problems [2], such as land desertification and soil erosion. An accurate prediction of surface subsidence is a prerequisite to reduce or even avoid such environmental problems. At present, among many surface subsidence prediction models, the probability integration model [3] is one of the most widely used methods for surface subsidence prediction in mining areas.

Although the model is sufficient several limitations exist: on the one hand, the parameter selection is the key to the probability integration method. However, the specific geological structure of loess donga may lead to large errors [4] in the predicted parameters, influencing the effective constraints of predicted parameters. On the other hand, the stability of the slope and the additional displacement [5] will deviate from the predicted results. We propose a prediction method of surface coal seam mining subsidence in loess donga by considering the essential causes of surface subsidence, i.e., the movement of underground rock layers and the stability of the slope [6], as important factors affecting the prediction accuracy, and combining the probability integral method model, the movement principles of rock and soil layers, and the influence of slope stability on mining subsidence.

The probability integration model [7] is a mining subsidence prediction method based on the random medium theory, with the predicted parameters as the key. Litwiniszyn [8] proposed the model in the 1950s and it has been further developed into a mature application model since then. The effective predicted parameters of the horizontal surface can be obtained in various ways. Therefore, the accuracy and reliability of the predicted results of the probability integral model are assured. However, concerning the prediction of the surface subsidence in donga areas, the model is affected by the probability integral parameters and the slope sliding effect [9], which leads to a decline of the application effect. Several studies have been conducted to solve this problem.

Field measurements [10] are a common method to obtain detailed and reliable predicted surface parameters. However, an accurate measurement is very time-consuming and requires a long period (at least 1 or 2 years), which squanders a lot of manpower and material resources. Moreover, the obtained parameters are only applicable to working faces under similar geological conditions, and the scope of application is limited [11]. Similarity simulation [12] is an effective method to acquire predicted parameters. A formation model, resembling the actual project, is constructed in the laboratory, according to the similarity principle, and the predicted parameters are inferred by monitoring the changes of the model. This method has major advantages, such as intuition, simplicity, and short experimental period. However, the complexity of mining geological conditions, material strength, and human factors have a significant influence on the experimental results [13]. Currently, predicted parameters can be acquired at low cost through numerical simulation [14,15] due to the rapid progress of the computer technology. However, the results are more random because of the influence of the simulation unit and value parameters of rock formation [16]. In addition, based on a large amount of measured field data, several further methods have been proposed to determine predicted parameters, including neural network method [17], support vector machine [18], and genetic algorithm [19]. These methods provide additional ideas for determining predicted parameters, but none of them consider the essential causes of mining subsidence, i.e., the movement of the subterranean strata, and the influence of the strata distribution on the predicted parameters in the donga areas. The stratum control theory [20] predicts that the surface subsidence will change periodically with the periodic breaking of the main key stratum in the overlying rocks. As the complex stratum distribution conditions, such as burial depth and soil layer thickness on the surface, vary in the donga area, a significant change of the parameters is expected [21]. Thus, studying the influence of the rock formation movement and formation distribution on the predicted parameters is of major significance for determining reasonable probability integral parameters in loess donga.

Mining operations below loess donga areas can easily induce geological disasters, such as loess slope slippage and collapse, which not only aggravates the destruction of the ecological environment, but also increases the difficulty of mining subsidence prediction and environmental disaster assessment. According to the topographic features in the special areas, Guo et al. [22] subdivided mining subsidence disasters into three types: collapse disasters, slump disasters, and landslide disasters. Stead, D et al. [23] used the theory of fracture mechanics and plastic mechanics to study the mechanical mechanism of landslides caused by mining. Wang et al. [24] applied numerical simulations to study the stability evaluation model of the mining slope body and to refine the influence principles of the slope stability. Luo et al. [25] established a mathematical model of slope stability under the influence of longwall mining subsidence and constrained a value of 1.5 as the critical value of slope stability. Instead of studying the influence of the slope stability in the loess donga on the prediction of mining subsidence, the previous studies mainly focused on the stability of the mining slope. However, the additional mining slip caused by slope instability in the loess donga is the main source for uncertainties in the subsidence prediction.

Considering findings from previous studies, we propose a method for predicting surface subsidence in loess donga based on a probability integration model, which resolves the deficiencies of the conventional probability integration model in predicting subsidence. The method combines the probability integral model with the influence of rock movement and stratum distribution on estimated parameters and considers the slope stability and the influence of the slip additional displacement on the subsidence results. The plausibility of the method is evaluated by a field test.

2. Prediction Method for Surface Subsidence of Coal Seam Mining in Loess Donga

Various subsidence prediction models have been proposed to predict surface subsidence, including the probability integration model [8], the Weibull distribution model [21], and the influence function model [7]. With the rapid development of computer technology, numerical simulation technology [14–16] became progressively used in mining subsidence prediction. The probability integration model is the numerical simulation with the longest application time and the widest application range. Moreover, the probability integration method is numerical simulation of the subsidence prediction models with the longest application time and the broadest scope of reasonable application. Unfortunately, it is not suitable for the application in loess donga and unsatisfactory results are expected due to the influence of strata distribution and topography. Considering these problems, we therefore improved the application of the traditional probability integral model in the donga area. Our study will expand the application range of the probability integral model and also has important significance for solving the limitations of the model.

2.1. The Probability Integration Model

The basic principle of the probability integration model is to superimpose the subsidence probability of countless mining units (ds) to form the surface subsidence curve $w(x)$ and the horizontal movement curve $u(x)$. The integration model of the coal seam unit mining is shown in Figure 1 [21].

Once the coal seam inclination attains full mining, the values of the subsidence and of the horizontal movement of any point on the main section in the strike direction can be calculated by Equation (1) [7].

$$w(x) = \frac{m\eta}{2}\left[\text{erf}\left(\frac{\sqrt{\pi}x}{r_0}\right) - \text{erf}\left(\frac{\sqrt{\pi}x - l + 2d}{r_0}\right)\right]$$

$$u(x) = bm\eta\left[e^{-\pi\frac{x^2}{r_0^2}} - e^{-\pi\frac{(x-l+2d)^2}{r_0^2}}\right] \tag{1}$$

m is the mining height, η the subsidence coefficient, erf the error integral function, r_0 the mining influence radius, $r_0 = h/\tan\beta$, h the buried depth of the coal seam, β the comprehensive moving angle, *l* the mining size, *d* the deviation of inflection point, and *b* is the horizontal movement coefficient.

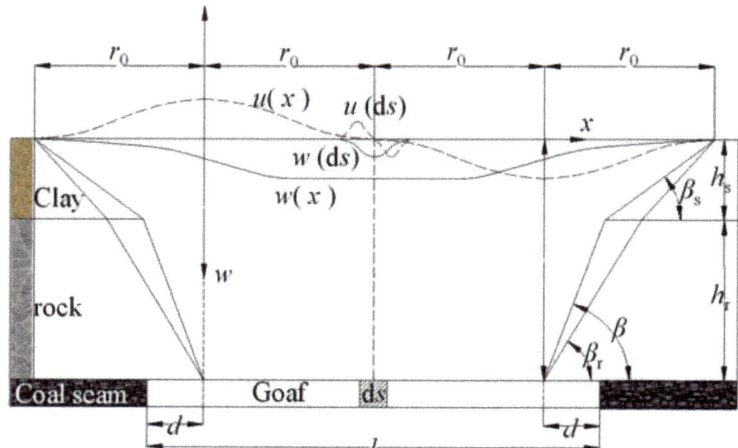

Figure 1. Coal seam unit mining integral model.

Figure 1 and Equation (1) show that the subsidence coefficient, the horizontal movement coefficient, the comprehensive movement angle, and the deviation of inflection point are the four predicted parameters of the probability integration model. According to the calculation principle of Figure 1, the one-dimensional mining parameters are subjected to two-dimensional normalization processing. In order to simplify the calculation, based on the principle of two-dimensional lattice calculation, two-dimensional lattice calculation software of probability integration has been independently developed [26]. By inputting the predicted parameters and mining parameters into the calculation software, the predicted results of the subsidence basins of the traditional horizontal surface can be obtained. However, in loess donga, influenced by stratum structural factors, including burial depth, soil layer thickness, slope angle, and slope height, the predicted parameters of different areas of the surface are not only different, but also difficult to obtain. Furthermore, the predicted results often have large deviations, as they are affected by slope slippage.

2.2. Prediction Process of Surface Subsidence in Loess Donga

Considering the characteristics of the probability integration model and the limitations of prediction process of surface subsidence in loess donga, we propose a prediction method of surface mining subsidence based on the probability integration model. Figure 2 illustrates the application of the method. At first, the loess donga is subdivided into several horizontal stratigraphic regions with the same characteristics, in light of the stratigraphic distribution, according to the stratigraphic histogram and the surface contour map. Among the horizontal stratigraphic area, slope sub-areas were separated according to the features of the slope (angle and height). The parameters of the stratigraphic structure determine the values of the predicted parameters. The movement of the subterranean strata is the critical cause for the movement of the surface. According to the stratum distribution and the movement principle of rock-soil layers, we calculated the probability integration parameters of different horizontal stratum regions. Based on the superposition calculation principle of the probability integration model, we subsequently calculated the subsidence basins formed by different horizontal strata regions. Finally, the subsidence basin formed by the superposition of the horizontal area is corrected according to additional displacement of slope sub-region mining slip.

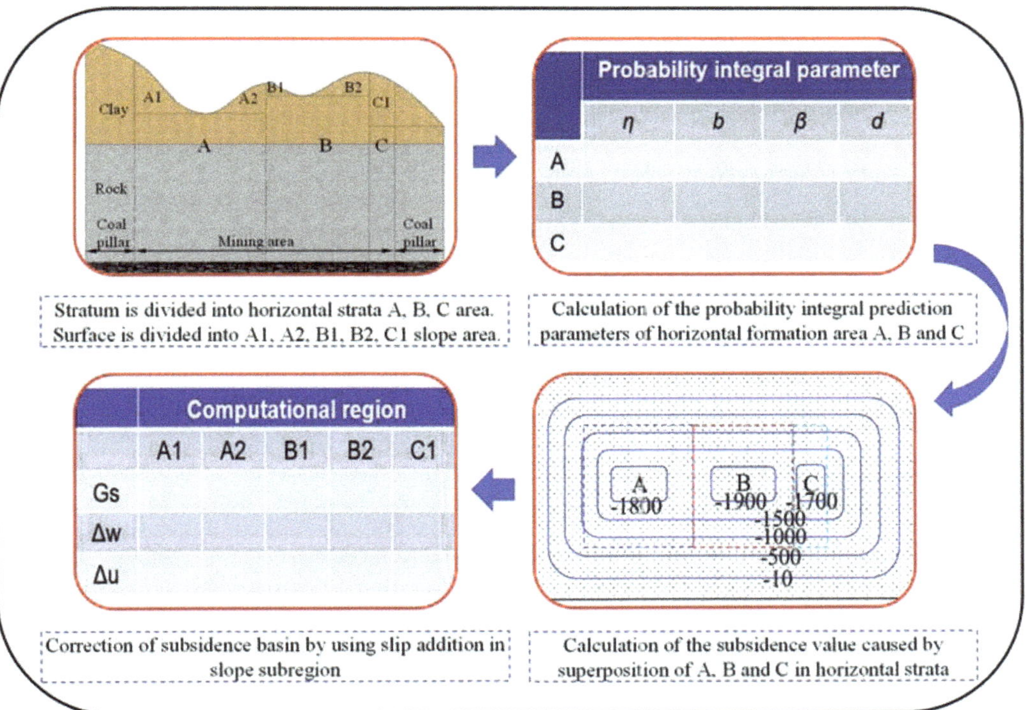

Stratum is divided into horizontal strata A, B, C area. Surface is divided into A1, A2, B1, B2, C1 slope area.

Calculation of the probability integral prediction parameters of horizontal formation area A, B and C

Correction of subsidence basin by using slip addition in slope subregion

Calculation of the subsidence value caused by superposition of A, B and C in horizontal strata

Figure 2. LDS application process.

3. Determination of Probability Integration Parameters in LDS

Various strata distribution and rock-soil movement principles lead to variations in the estimated subsidence parameters in differing regions. The subsidence coefficient is the key parameter of the probability integration model. The comprehensive movement angles, deviation of inflection points, and the horizontal movement coefficient control the influence range of the sinking basin and the horizontal movement value of the basin. The determination of the predicted parameters of the different regions is the main challenge of the coal seam mining in loess donga (LDS) prediction method. Previous studies have shown that the distribution and movement principles of rock-soil layers are closely related to the probability integration parameters [6]. Therefore, we studied the influences of the distribution and movement principles of the rock-soil layers on the probability integration parameters.

3.1. Calculation of Subsidence Coefficient

The key stratum theory in rock formation control implies that the movement of the underground rock formation is the critical cause of mining subsidence, and the mechanical state of the main key formation in the formation directly determines the migration principles of the overlying rock [20]. The main key stratum is located in the caving zone, the fracture zone or the bending subsidence zone of the mining overburden. The main stratum can be determined from the formation borehole histogram and the key stratum identification method [27]. The height of the caving zone (h_c) and the height of the fracture zone (h_f) of the mining overburden can be determined by various methods, including on-site measurement, empirical formula, engineering analogy, and theoretical calculations [28]. Based on the rock movement theory, we therefore propose a method for calculating the subsidence coefficient, according to the position of the main key stratum in the mining overburden.

Continuous deformation characteristics of the surface movement are the premise of predicting subsidence basins. If the height of the caving zone h_c is larger than the height of the main key stratum Σh ($h_c \geq \Sigma h$), the main key stratum is located in the caving zone. Figure 3 shows the discontinuous deformation characteristics of the surface when the main key layer is located in the caving zone. Under the influence of underground mining, the main key is completely broken, resulting in the loss of the support of the main key layer to the overlying rock and soil layer. Upward cracks formed by the rupture of rock and soil strata develop through the main key layers to the surface. Multiple vertical cracks cut the continuously deformed surface, showing the feature of stepped subsidence.

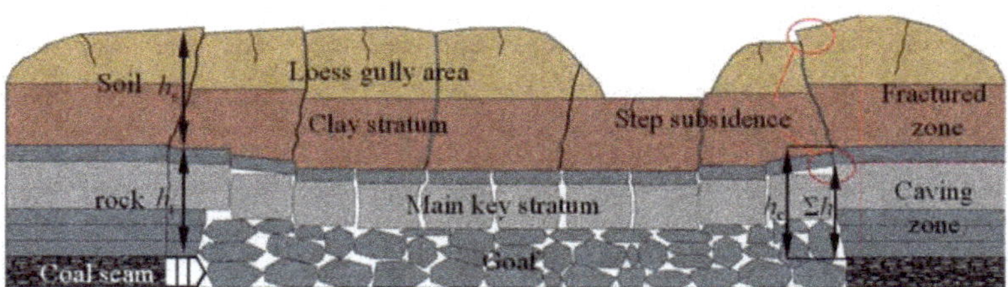

Figure 3. The main key stratum is located in the caving zone.

Therefore, we focus on the calculation method of the surface subsidence coefficient of continuous deformation for the condition that the main key layer is located in the curved subsidence zone or the fracture zone. Figure 4 illustrates the overlying strata movement principles of the continuous land surface deformation. In either of those cases, the main key stratum did not break, or was broken to form a stable masonry beam structure, which continued to undertake the supporting role of the overlying rock and soil strata. The upward cracks formed by the rupture of the rock and soil strata did not develop on the surface, and the surface curved and subsided to form a continuous subsidence basin. If the fracture zone height is smaller than h ($h_f \leq \Sigma h$), the main key stratum is in the bending subsidence zone (Figure 4a). In this case, the development height of the overburden mining fracture zone is below the main key stratum, not affecting the soil layer, and the overlying soil layer has no effect on the crack development. Based on the data of field measurements, Geweltzmann established Equation (2), which describes the link between the height of the fracture zone and the subsidence coefficient for mining a single horizontal coal seam by fallen method [29].

$$h_f^2 = \frac{7.25m\eta}{(\cot \beta_r + \cot \psi)^2 K_r} \tag{2}$$

ψ is the full mining angle, which is related to the full mining degree (usually 55–59°) and K_r is the limit curvature of the top rock formation. The value of K_r is related to the ratio A of the clay rock (mudstone and argillaceous sandstone) in the rock formation [30]. According to the ratio of the thickness of the bedrock constituted by the two rock types, K_r is calculated by Equation (3).

$$K_r = 0.002 + 0.04A \tag{3}$$

If $h_c < \Sigma h < h_f \leq h$, the main key stratum is located in the fracture zone. Figure 4b shows that the fracture zone is well developed through the bedrock and continues upward. The development height of the fracture zone is influenced by the thickness and properties of the soil layer. Since the clay layer thickness is not considered in Equation (2), the ultimate curvature K_0 ($K_0 = K_r + \Delta K_r$) of the clay layer is introduced into Equation (4) to replace K_r in Equation (2), and the soil layer movement angle is used to calculate the height of crack development [31].

$$h_f^2 = \frac{7.25m\eta}{(\cot \beta_s + \cot \psi)^2 (K_r + \Delta K_r)} \tag{4}$$

In Equation (4), ΔK_r is the ultimate curvature increment of the soil layer, $\Delta K_r = 0.1\,h_s$.

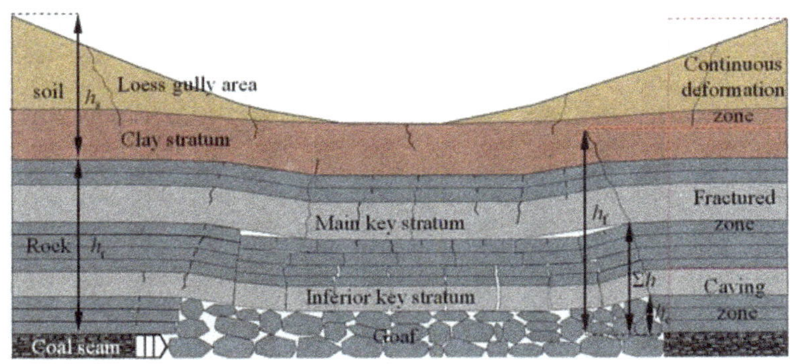

(a)

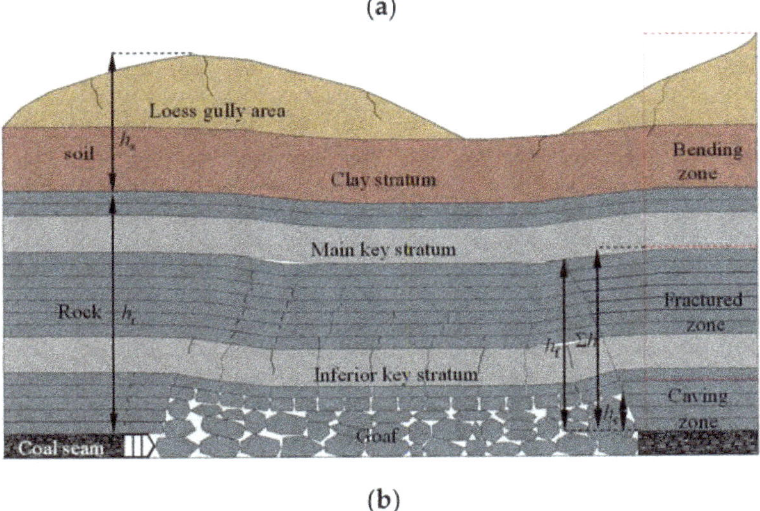

(b)

Figure 4. Migration principles of overlying rock under continuous deformation: (**a**) the main key stratum is located in the curved subsidence zone; and (**b**) the main key stratum is located in the bending zone.

The calculation method of the subsidence coefficient, derived from Equation (2) and Equation (4), is shown in Equation (5) in Table 1 for the scenario when the main key stratum is located at different positions.

Table 1. The subsidence coefficient of the main key stratum at different positions.

Analyzing Conditions	Main Key Stratum Location	Subsidence Coefficient Calculation Formula	
$h_c \geq \Sigma h$	curved subsidence zone	$\eta = \dfrac{h_f^2(\cot\beta_r + \cot\psi)^2 K_r}{7.25m}$	(5)
$h_c < \Sigma h < h_f \leq h$	Fracture zone	$\eta = \dfrac{(h_f - h_r)^2(\cot\beta_s + \cot\psi)^2(K_r + \Delta K_r)}{7.25m}$	

3.2. Comprehensive Movement Angle, Horizontal Movement Coefficient, and Deviation of Inflection Point

The size of the comprehensive movement angle is mainly related to the thickness of the rock (soil) layer and the rock movement parameters (Figure 1). The comprehensive moving angle is calculated from Equation (6):

$$\beta = \arctan\left(\frac{h\tan\beta_s\tan\beta_r}{h_s\tan\beta_r + h_r\tan\beta_s}\right) \tag{6}$$

with h_s as the thickness of the soil layer, h_r as the thickness of the rock layer, β_s as the displacement angle of the soil layer, and β_r as the displacement angle of the rock layer. β_s and β_r can be selected according to Tables 2 and 3 [21].

Table 2. Displacement angle (β_s) of loose layer.

Loose Layer Thickness (m)	Loose Layer Features		
	Dry, Water-Free	Strong Water Content	Quick-Curing Ground Content
<40	50°	45°	30°
40–60	55°	50°	35°
>60	60°	55°	40°

Table 3. Displacement angle (β_r) of strata.

	Average Rock Firmness Coefficient f		
	$f < 3$	$3 \leq f < 6$	$6 \leq f$
Rock displacement angle	65°	70°	75°

For the horizontal movement, coefficient b values of 0.2–0.3 are usually applied for China's coal mining fields. By relating the ratio of the thickness of the clay layer to the mining depth, b is calculated from Equation (7) [21].

$$b = 0.3 - 0.1\frac{h_s}{h} \tag{7}$$

The deviation of inflection point d is related to f, l, and h, d of the near-horizontal ore seam is calculated by Equation (8) in Table 4 [21].

Table 4. Displacement angle (β_s) of loose layer.

f	d	
$f > 6$	$d = \left(0.29 - 0.36\lg\frac{l}{h}\right)h$ if $l/h > 2.2$, $l/h = 2.2$	
$3 < f < 6$	$d = \left(0.19 - 0.35\lg\frac{l}{h}\right)h$ if $l/h > 1.4$, $l/h = 1.4$	(8)
$f < 3$	$d = \left(0.14 - 0.40\lg\frac{l}{h}\right)h$ if $l/h > 0.9$, $l/h = 0.9$	

Combined with the regional strata occurrence characteristics and mining rock movement principles (development height of caving zone and fracture zone) in loess donga, probabilistic integration parameters can be calculated from Equations (5)–(8) for each horizontal stratigraphic region. Subsequently, input of the constrained block parameters and predicted parameters of the subdivided horizontal stratigraphic areas into a self-development probability integration calculation software [26] enables the recognition of subsidence basins in the horizontal stratigraphic area after calculation.

4. Slope Stability and Slip Principles in Loess Donga

The subsidence basin calculated according to the superposition of the regional probability integration parameters of different horizontal strata only considers the influence of the horizontal strata in dongas, but excludes the effect of the mining-slip of the slope body. The main characteristics of the Yushenfu mining field (China): a plenitude of nearly horizontal coal seams, thick loose layers, high mining height, thin rock layers, and widespread distribution on the land surface of loess dongas. In the loess donga, the height of the slope body is usually 10–20 m, and the angle 10–30° [1]. Under specific circumstances, coal seam mining can easily lead to mining slip and instability of slopes, resulting in the different principles between mining subsidence and horizontal surface [6]. We studied the stability of the slope body and the additional displacement of mining slip to decipher the principles of the influence of the slope body on the subsidence in the loess gully area.

4.1. Analysis of Slope Stability in Loess Donga

Currently, the limit equilibrium method, the basic method for stability analysis of soil slope, is realized by calculating the safety factor of the slope to analyze the stability of the slope. Based on the Swedish strip method of Fellenius, we extended a series of other methods, including the Bishop method, the Janbu method, the Morgenster-Price method, and the Spencer method for the calculations [22,23,32].

We combined the analysis of the stability of the slope in the loess donga with the characteristics of the strip method. The slope is regarded as a uniform soil slope and its shear strength follows the Mohr-Coulomb failure criterion. The influence of the force between the soil strips on the stability of the soil slope is excluded. To simplify the calculation, we assume that the slip line of the soil slope is an arc. Figure 5 shows the mechanical model of the loess donga slope.

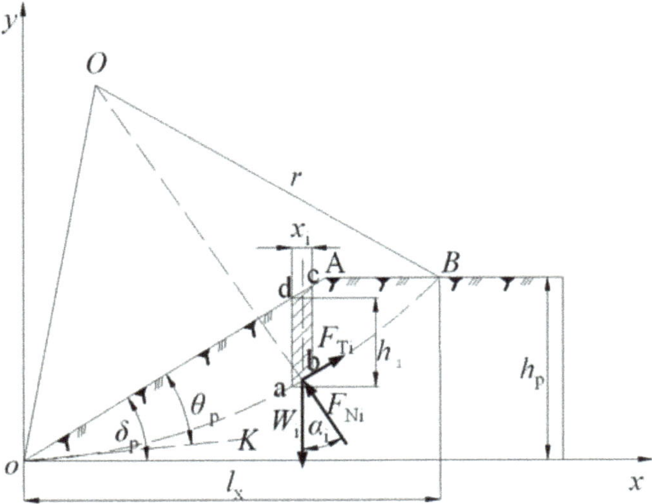

Figure 5. Mechanical model of slope body in loess dongas.

Figure 5 shows that the sliding slope oAB is approximately divided into parallelogram abcd vertical soil strips with number n. The safety factor of the slope in loess donga is set as K_S, which is the ratio of the sliding moment M_S to the anti-sliding moment M_T. The sliding moment is caused by the shear component induced by the gravity of the soil strip. The anti-slip strength of the soil strip generates anti-slip moments. According to the Mohr-Coulomb strength theory, K_S can be calculated from Equation (9) [32].

$$K_S = \sum_{i=1}^{n} \frac{M_{Si}}{M_{Ti}} = \sum_{i=1}^{n} \frac{r\left(\gamma_{si} x_i h_{pi} \tan \phi_{s_i} \cos \alpha_i + c_{si} l_i\right)}{r\gamma_{si} x_i h_{pi} \sin \alpha_i} = \sum_{i=1}^{n} \frac{\gamma_{si} x_i h_i \tan \phi_{s_i} \cos \alpha_i + c_{si} l_i}{\gamma_{si} x_i h_{pi} \sin \alpha_i} \tag{9}$$

with n as the number of soil strips, r as the radius of the slip line, γ_{si} as the bulk density of the soil strip, x_i as the width of the soil strip, h_{pi} as the height of the soil strip, α_i as the angle between the direction of the gravity of the soil strip and the normal force F_{Ni}, ϕ_{si} as the internal friction angle of the soil strip, c_{si} as the cohesive force of the soil strip, and l_i as the arc length of the soil strip slip line. Since the soil strip element is $x_i \approx l_i$ and the soil slope is a homogeneous soil mass, the soil strip parameters γ_{si}, h_{pi}, α_i, ϕ_{si}, and c_{si} can be converted into the bulk density γ_s of the loess slope, the height of the loess slope h_p, the angle of the loess slope δ_p, the internal friction angle ϕ_s, and soil cohesion c_s. Therefore, Equation (9) is further simplified, and Equation (10) is obtained:

$$K_S = \frac{2\left(\gamma_s h_p \cos^2 \delta_p \tan \phi_s + c_s\right)}{\gamma_s h_p \sin 2\delta_p} \tag{10}$$

Equation (10) documents for a height of the slope, a bulk density of the slope or an angle of the slope of 0, the denominator becomes 0 and the equation meaningless. To analyze the relationship between slope stability and influencing factors in the loess donga, the slope instability coefficient G_s was constrained in the experiment [6]. An increase in G_s reduces the stability of the slope, and the risk of slippage caused by the instability of the slope increases. G_s is calculated from Equation (11).

$$G_S = \frac{1}{K_S} = \frac{\gamma_s h_p \sin 2\delta_p}{2\left(\gamma_s h_p \cos^2 \delta_p \tan \phi_p + c_p\right)} \tag{11}$$

Equation (11) shows that the slope height, slope angle, soil bulk density, soil cohesion and soil internal friction angle are the main influencing factors of slope stability. To analyze the extent of the influence of every factor on the slope stability, the five factors are assumed to be independent of each other. Previous studies have shown that for $h_p > 30$ m or $\delta_p > 50°$, the slope will show collapse-type failure [32]. Therefore, we studied a range of the slope height from 0–30 m and a range of slope angle from 0–50°. The test results show that the bulk density of the loess layer ranges from 16,300–18,600 N/m^3, the cohesive force from 38,000–101,000 Pa, and the internal friction angle from 27.9–33.8° [29]. We studied the relationship between those factors and the stability of the slope by controlling the variable range of a single factor.

Figure 6a–c shows fitting degrees of 0.9860 for the G-h_p power index function curve, of 0.9988 for the G-δ_p proportional curve, and of 0.9992 for the G-γ_s linear curve. G is positively correlated with h_p, δ_p, and γ_s. With the increase in slope height, the slope angle, and the bulk density, the increasing speed of the slope sliding component force exceeds that of the anti-sliding component force, and the degree of the slope instability increases continuously. Figure 6d,e shows a fitting degree of 0.9998 for the G-c_s linear curve, of 0.9994 for the G-ϕ_s linear, and negative correlation of G with c_s and ϕ_s. With the increase in soil cohesion and internal friction angle, the shear strength of the soil slope, the anti-sliding component, and the stability of the slope increase. The results of grey correlation degree and orthogonal test analysis show that the sensitivity of the parameters affecting the slope stability in the loess donga is ranked from large to small, and the order is $\delta_p > h_p > \gamma_s > \phi_s > c_s$ [1]. The main reason is that the slope angle and slope height in this area have a large variation range, while the fluctuation range of the soil physical and mechanical parameters is small, which has little influence on the slope stability [22].

In conclusion, the slope instability coefficient in loess donga is positively correlated with slope height, slope angle, and soil bulk density, but is negatively correlated with soil cohesion and internal friction angle. According to Equation (11), the extent of instability of the slope body in donga areas can be determined prior to the impact of the mining activity.

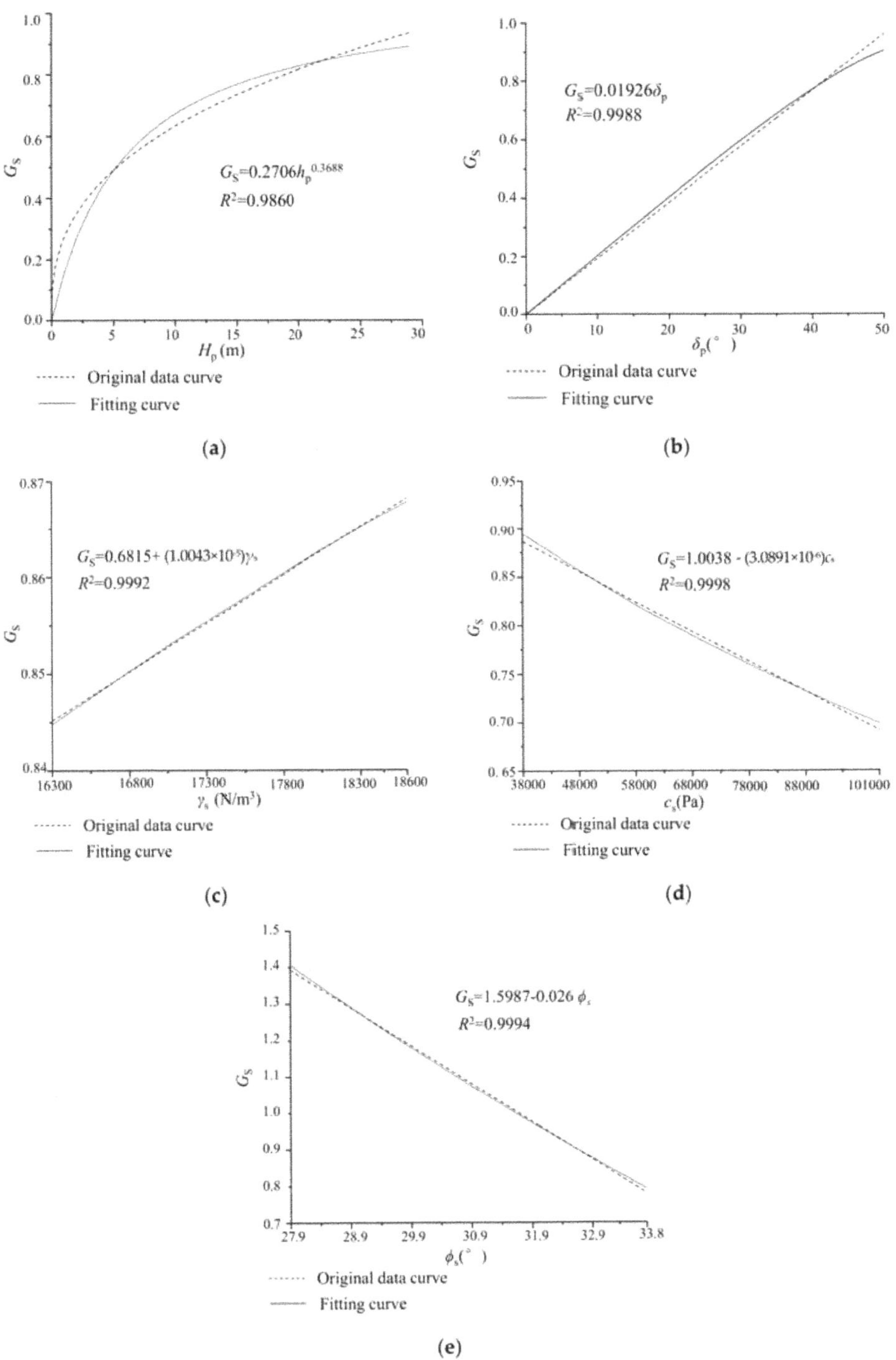

Figure 6. Relationship between different influencing factors and slope instability coefficient: (**a**) Slope height (h_p), (**b**) Slope angle (δ_p), (**c**) Soil bulk density (γ_s), (**d**) Soil cohesion (c_s), and (**e**) Internal friction angle of soil (ϕ_s).

4.2. Additional Displacement of Slope Mining Slip

The reliable understanding of striking differences in the movement principles of the surface and the horizontal surface is critical for mining in donga areas. The main movement and deformation characteristics of surface mining subsidence are subsidence, horizontal movement, and the influence range of the mining activity. The model of the movement and deformation of the inclined surface in the horizontal mining seam is shown in Figure 7 [6].

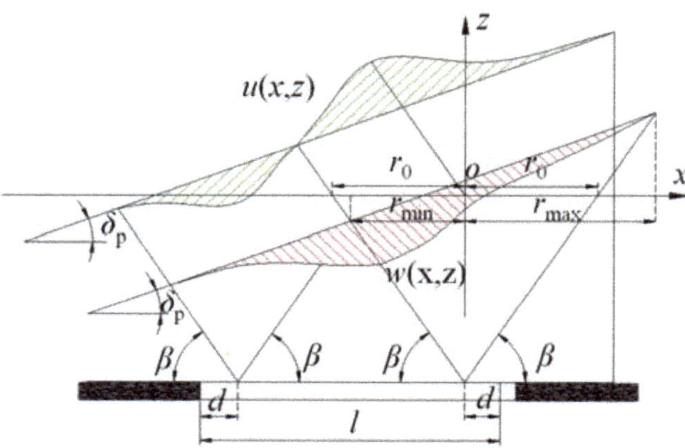

Figure 7. Deformation model of inclined surface in horizontal seam mining.

Figure 7 indicates an asymmetry of the inclined surface subsidence curve $w\,(x, z)$ and the horizontal movement curve $u\,(x, z)$ for a horizontal mine with the size l. The surface movement deformation curve is related to the stability of the slope, but also to the mining direction of the coal seam [21]. Therefore, the research progress should be combined with the influence of the stability of the mining slope in loess donga on the surface movement and deformation. Assuming that the slope slip is $R(x)$, the additional subsidence slip $\Delta w(x)$ and the additional horizontal displacement slip $\Delta u(x)$ caused by the slope slip can be expressed in Equation (12).

$$\begin{aligned} \Delta w(x) &= R(x) \sin \delta_p \\ \Delta u(x) &= R(x) \cos \delta_p \end{aligned} \tag{12}$$

The slippage of the slope is related to the mining instability degree of the slope. If the slope is unstable, it tends to slide. The higher the degree of instability, the higher the affinity for sliding. The slope can be subdivided into a first-class slope, a second-class slope, and a third-class slope according to the hazard degree of the slope body after sliding damage. The normative standards for the safety factor of slope stability are summarized in Table 5 [32].

Table 5. Slope stability safety factor.

Security Level	First-Class Slope	Second-Class Slope	Third-Class Slope
General condition safety factor	1.35	1.30	1.25
Temporary condition safety factor	1.25	1.20	1.15
Earthquake condition safety factor	1.15	1.10	1.05

The loess donga is a highly fragile ecological environment, which is attributed to the first-class slope, and mining in this area is regarded as an earthquake condition. According to Table 5, the safety factor K_S in this area is 1.15, and $G_S = 0.87$ as calculated by Equation (11).

The results of the numerical simulation and the field test show that, apart from the stability of the slope itself, mining has a larger impact on the stability of the slope. Compared with mining along the slope, the slope body is more prone to instability if mining occurs on the reverse slope. Taking K_C as the influence coefficient of the slope mining direction on the slope instability coefficient for mining along the slope K_C is 1.5, but for mining against the slope K_C becomes 2.0 [24]. The mining instability coefficient G_{CS} of the slope is expressed by Equation (13).

$$G_{CS} = K_C G_S \tag{13}$$

To sum up, a value of 0.87 is constrained as the critical coefficient that defines the slip instability in loess donga. At $G_{CS} \leq 0.87$ the slope is stable, and no slip occurs. At $0.87 < G_{CS}$, the slope is slippery and unstable, and the slip displacement $R(x)$ is calculated from Equation (14).

$$R(x) = G_{CS}\left[w(x_0)\sin\delta_p + u(x_0)\cos\delta_p\right] \tag{14}$$

with $w(x_0)$ and $u(x_0)$ as the initial subsidence and initial horizontal movement, respectively. From Equations (12) and (14), Equation (15) for the additional displacement of slope slip is derived.

$$\begin{aligned}
w(x) &= w(x_0) + G_{CS}\left[w(x_0)\sin\delta_p + u(x_0)\cos\delta_p\right]\sin\delta_p \\
u(x) &= u(x_0) + G_{CS}\left[w(x_0)\sin\delta_p + u(x_0)\cos\delta_p\right]\cos\delta_p
\end{aligned} \tag{15}$$

To summarize, the mining instability coefficient of the slope is calculated according to the slope parameters and the mining direction of the coal seam in loess donga. For conditions with $0.87 < G_{CS}$ the slope is slippery and unstable, and additional mining slip will be generated as the slope subsides. First, the predicted basin of horizontal formation subsidence is obtained according to the superposition calculation of the software, and the initial subsidence curve and initial horizontal movement curve of the surface are extracted from it. Secondly, combined with the initial subsidence value and horizontal movement value of the slope area that will generate slip, as well as δ_p and G_{CS}, the additional displacement of slope mining slip is calculated by Equation (14). Finally, the movement and deformation curve of the initial horizontal stratum area is corrected by the additional displacement mining slip, and the prediction of the surface subsidence in the loess donga is finally realized.

5. Case Study

We selected the N1114 working face as a case study to verify the predicted effect of LDS on surface mining subsidence in loess donga. The methods of LDS and the field test were used to predict the subsidence basin.

5.1. Engineering Case

The Ningtiaota Coal Mine is a typical mine in the loess donga of the Yushenfu mining field in China. The N1114 working face mines 1–2 near-horizontal coal seams with an average thickness of 1.85 m. The working face has a width of 245 m and a recoverable length of 1922 m. The working face is limited to north and south by the unmined N1116 working face and the N1112 working face, respectively, and to the west by the 135 m stop-line coal pillar and to the east by the 60 m mine field boundary coal pillar. An overview of the strata occurrence of N1114 working face is summarized in Table 6, that is constrained by combining drilling data with physical and mechanical parameters [33]. The key layers of the overlying rock are determined from the key layer theory and the data in Table 6. A 13.70 m thick horizon of fine sandstone is the main key layer and an 8 m horizon of fine sandstone is the sub-key layer. A 16.10 m layer of fine sandstone is a thick hard rock stratum, which is broken along with the breaking of the main key layer.

To fulfill the complete mining demand, the actual measurement and LDS subsidence comparison area is the section of N1114 working face from west to east, 300 m distant from the mining stop line. Full extraction has been achieved prior to the mining of the target area. Therefore, we added an additional 150 m computing area for the LDS to simulate actual mining on site. According to the contour distribution of the N1114 working face in Figure 8a, the trend direction with obvious slope distribution in the study area is selected for research. The profile A–A is constructed along the direction of the center of the working face, as shown in Figure 8b.

Table 6. Stratigraphy of the N1114 working face with physical and mechanical parameters.

Strata	Thickness (m)	Physical and Mechanical Parameters						
		σ_c (MPa)	σ_t (Mpa)	c (Mpa)	E (Mpa)	γ (N/m^3)	μ	φ (°)
Loess	30.00–60.00	0.29	0.03	0.059	33.42	18,600	0.32	28.2
Fine sandstone	16.10	29.6	0.50	1.50	1258	22,700	0.29	42.0
Sandy mudstone	7.40	34.7	0.54	0.26	2400	25,600	0.24	38.5
Fine sandstone	13.70	45.6	0.708	2.20	2113	23,000	0.27	41.5
Sandy mudstone	4.70	35.3	0.56	0.27	2415	26,180	0.24	38.8
Medium mudstone	3.40	40.6	0.56	1.50	1949	23,300	0.28	44.0
Silty sandstone	2.20	36	0.234	0.90	995	24,200	0.30	40.0
Fine sandstone	8.00	29.8	0.60	1.57	2024	23,050	0.27	39.2
Sandy mudstone	5.20	36.2	0.56	0.30	2423	25,780	0.24	38.5
Silty sandstone	2.00	32.9	0.27	0.94	979	22,700	0.27	39.3
1^{-2} Coal	1.85	15.7	0.29	1.10	845	12,900	0.28	37.5
Silty sandstone	4.90	36.4	0.28	1.01	977	23,030	0.27	40.2
Fine sandstone	12.70	34.1	0.52	1.55	1320	23,160	0.27	40.4

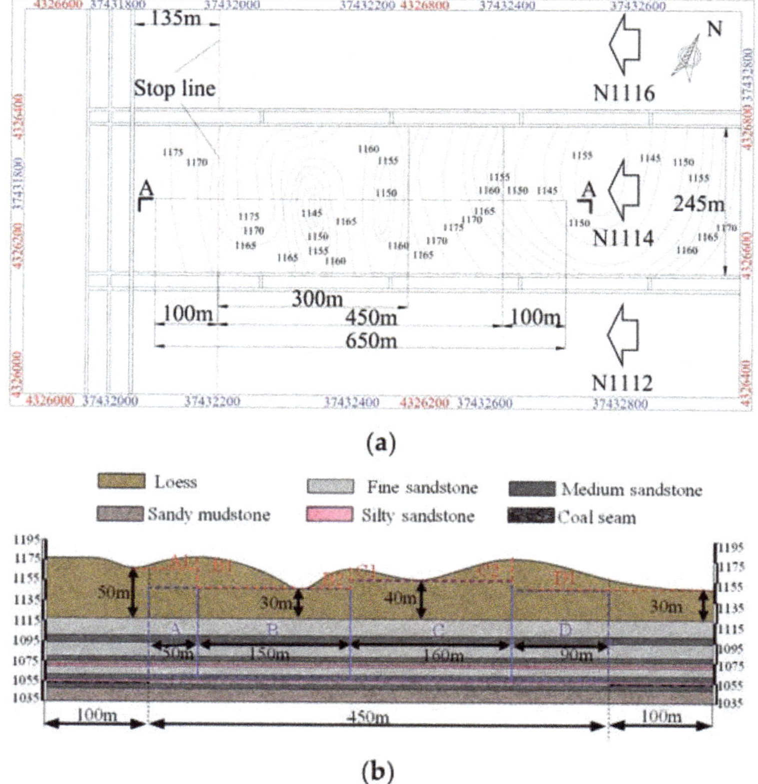

Figure 8. Overview of the N1114 working face: (**a**) work surface layout; and (**b**) A–A stratigraphic profile.

5.2. LDS Prediction

We modelled the surface of loess donga according to the LDS application process that is subdivided into following steps:

Step 1: Figure 8b shows that the coal seam mining area and overlying strata are subdivided into four horizontal areas (A, B, C, and D), according to the distribution characteristics of geotechnical layers. According to the slope characteristics of dongas, the slope is further divided into six slope sub-areas: A1, B1, B2, C1, C2, and D1.

Step 2: The probability integration parameters of the A, B, C, and D horizontal formation regions are calculated, respectively.

Field measurement shows that h_c is 7.5 m, h_f is 24.5 m, and 24.5 m = $h_f < \Delta h$ = 25.5 m [33]. The main key layer is in the curved subsidence zone and the surface forms a continuously deformed subsidence basin. From Figure 8 and the data of Table 6 it follows that the ratio of clay rock to the rock layer is 0.236 and the average firmness coefficient of the rock layer is 3.56. The mining height m of the working face is 1.85 m, the width is 245 m, and the calculated length is 450 m. The calculated length of the working face is larger than 1.4 h and the full extraction angle is taken is 55°. The results of the probability integral parameters of the individual horizontal formation areas, calculated from Equations (5)–(8), are summarized in Table 7.

Table 7. Calculated probability integration parameters of the individual areas.

		Calculation Areas			
		A	**B**	**C**	**D**
Parameters	h_s (m)	50.0	30.0	40.0	30.0
	h_r (m)	62.7	62.7	62.7	62.7
	β_s (°)	50.0	45.0	50.0	45.0
	β_r (°)	70.0	70.0	70.0	70.0
Calculated results	η	0.58	0.58	0.58	0.58
	δ (°)	60.1	64.9	63.4	64.9
	b	0.256	0.273	0.265	0.273
	d (m)	15.65	15.65	15.65	15.65

Step 3: Based on the probability integration model, we insert the parameters of the different regions and the coordinates of the inflection point of the block in the self-developed two-dimensional lattice probability integration calculation software [26] and calculated the value of the subsidence basin according to the mining order. The initial subsidence isoline cloud and horizontal movement cloud chart for the N1114 working face is shown in Figure 9.

Step 4: Subsequently, we calculated the slope instability coefficients of the six slope sub-regions (A1, B1, B2, C1, C2, and D1). According to Table 6 and Figure 8, the soil cohesion is 59 kPa, the friction angle in the soil is 28.2°, and the bulk density of the soil is 18,600 N/m3. The results of the calculations are shown in Table 8.

Table 8. Slope instability coefficient of slope sub-regions.

		Calculation Areas					
		A1	**B1**	**B2**	**C1**	**C2**	**D1**
Parameters	h_p (m)	10.0	30.0	20.0	10.0	20.0	30.0
	δ_p (°)	10.1	17.6	21.8	8.5	12.9	10.3
	mining direction	downward slope	reverse slope	downward slope	reverse slope	downward slope	reverse slope
Calculated results	G_{CS}	0.31	0.97	0.83	0.35	0.49	0.56

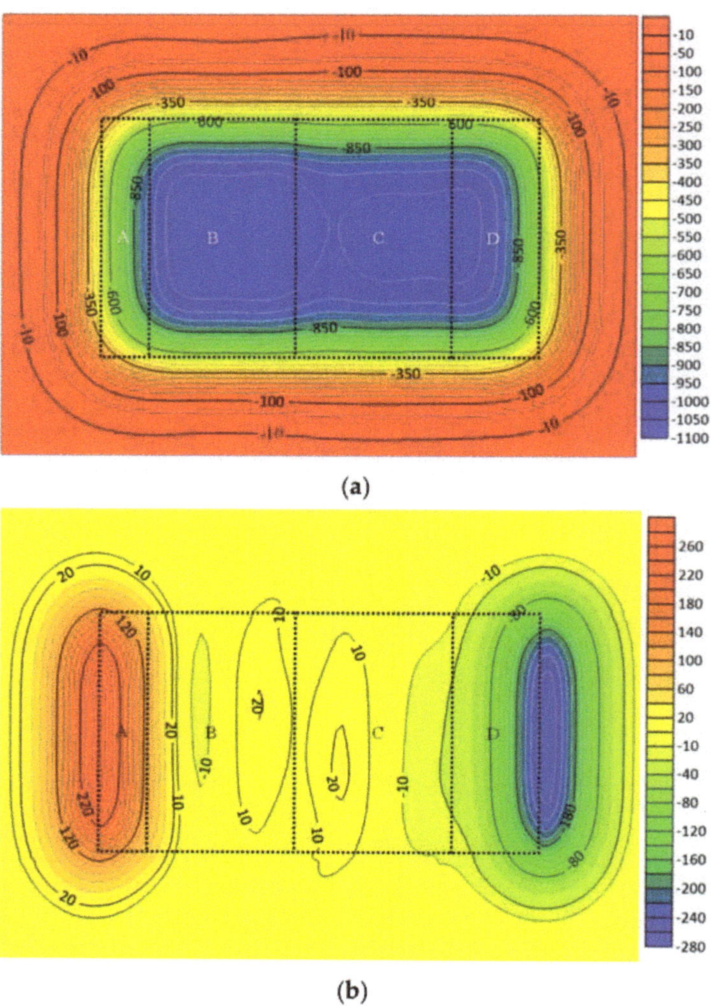

(a)

(b)

Figure 9. Probability integration calculation result: (**a**) subsidence isoline cloud; and (**b**) horizontal movement isoline cloud.

The $G_{CS\text{-}B1}$ relationship of the B1 slope is $0.87 < G_{CS\text{-}B1} = 0.97$ (Table 8) indicating that the B1 slope will generate instability and slip during the coal seam mining process. Therefore, we constructed a cross section along the strike direction in the center of the working face (Figure 9) and the surface curve of dongas is obtained in combination with Figure 9 Subsequently, we refined the sinking and horizontal movement curves according to the calculated data in Table 8, using Equation (15). Figure 10 shows the predicted surface movement curve in the strike direction of the N1114 working face.

Figure 10 shows that the trend of the surface subsidence and horizontal movement curve in loess donga essentially conforms to the principles of surface mining movement and deformation. The maximum subsidence value reaches 1096.23 mm, close to the middle of the goaf. The minimum value of the horizontal movement of the surface is in the center of the goaf. The surface subsidence values on both sides of the goaf are 505.86 mm and 566.81 mm, respectively, hence almost half the maximum subsidence. The maximum values of the horizontal movement of the ground surface are 270.25 mm and 274.09 mm, which are located above both sides of the goaf. Affected by loess dongas, the surface subsidence value in the center of the goaf ranges between 1039.62 mm to 1096.23 mm. The horizontal movement in the center of the goaf fluctuates around 0, and the floating range is 14.45 mm to 20.51 mm. The subsidence and horizontal movement correction curves of the dongas area show that the slope at B1 induces different degrees of additional slip. The additional displacement of subsidence and horizontal movement slip generated by the B1 slope are 16.97 mm to 33.88 mm and 0.64 mm to 5.97 mm, respectively.

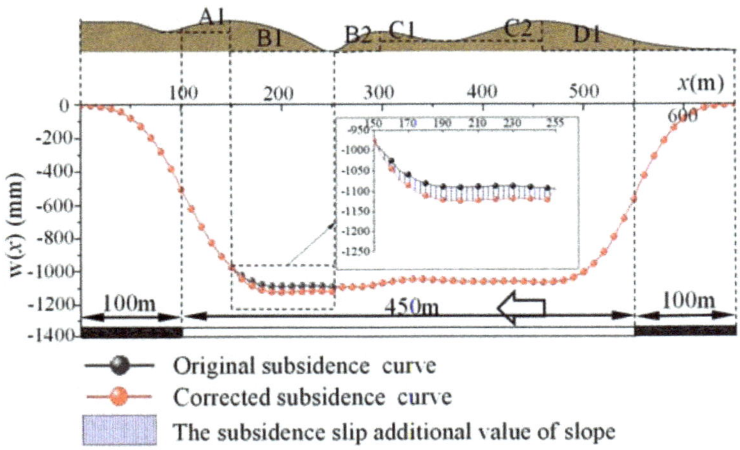

(a)

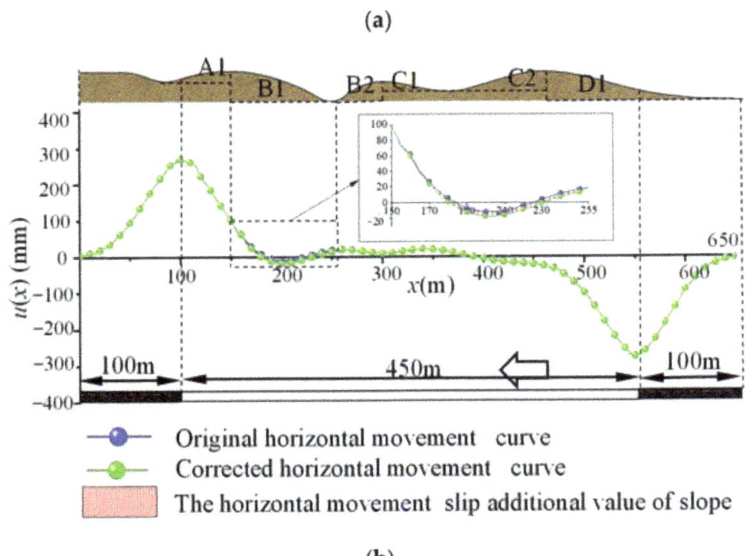

(b)

Figure 10. Surface movement prediction curve in dongas: (**a**) sinking curve; and (**b**) horizontal displacement curve.

5.3. The Field Test

A high-precision total station was used in the target research area for the field test to evaluate the application effect of LDS and refine the method. At first, two measurement control points were installed, and subsequently, the survey line A was arranged along the direction of the research area. From the west side, it was arranged 100 m away from the mining stop line of N1114 working face, including the 300 m research area. A total of 41 measuring points were arranged in the area. The distance between the measuring points was 10 m and the monitoring distance was 400 m. Since the mining speed of the working face was 20 m/d [33], until the mining of the working face was completed, the monitoring was performed five times once a week. After the mining stop of the working face, according to the change rate of the subsidence value, the monitoring time was 1 to 3 months and the monitoring time was five times until the surface movement stopped. The overall monitoring time was 1a, with a total of 10 monitoring times [21]. Figure 11 summarizes the measured curve of line A and the predicted curve of LDS in the target area.

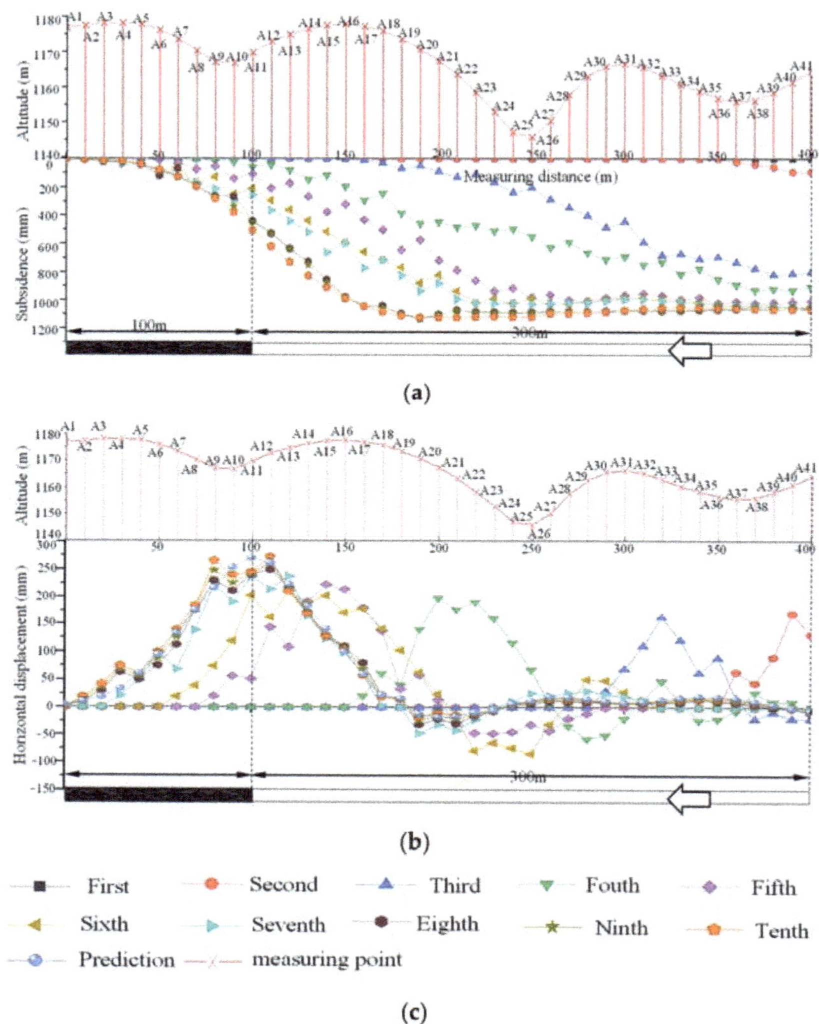

Figure 11. Surface movement deformation curve of mining under loess donga: (**a**) sinking curve, (**b**) horizontal movement curve, and (**c**) legend.

Figure 11a shows that the interval between the 8th and 10th monitoring is more than 6 months. The fluctuation range of the subsidence value between the measuring points is 0.96–28.56 mm. The surface mobile basin is considered to be stable if the subsidence increment is less than 30 mm for six consecutive months [21]. The 10 monitoring results basically conform to the principles of surface movement and deformation, indicating that the monitoring is reasonable and effective. Comparing the measurement point data A11–A41 (Figure 11a) shows that the theoretically predicted subsidence value in the study area deviates from the actual subsidence value by 0.74–77.07 mm, with an average of 7.11 mm. The relative deviation is 0.1–12.9%, with an average of 1.6%. The deviation between the theoretical predicted horizontal movement value and the actual horizontal movement value is 0.14–17.07 mm (Figure 11b), with an average of 2.96 mm. The relative deviation ranged from 0.8% to 22.6%, with an average of 11.3%. Excluding the measurement errors of individual discrete points shows a high extent of agreement between the predicted curve and the actual subsidence curve. Hence, the requirements of the engineering are fulfilled.

The theoretically predicted additional subsidence slip in the study area A17–A25 is 16.97–33.88 mm and the theoretically predicted additional displacement of horizontal movement slip is 0.64–5.97 mm. Figure 11a shows that, based on the average maximum subsidence of 1075.26 mm in the center of the goaf, the actual subsidence value at A17–A25 increases between 3.51–71.16 mm, with an average of 30.59 mm, which is in line with the expected range. The relative deviation between the theoretically predicted subsidence curve and the actual subsidence curve is 0.2–4.4%, with an average of 1.0%. The actual sinking horizontal movement slip increment cannot be calculated since a benchmark for the horizontal movement is lacking, Figure 11b shows that the relative deviation between the theoretical predicted horizontal shift curve and the actual horizontal shift curve is 2.7–22.6% at A17–A25, with an average of 9.8%. The data document that LDS can effectively predict the subsidence and horizontal movement of the slope due to slip in loess donga.

6. Discussion

The study of surface subsidence in loess donga of coal mines is of significance for ecological reclamation and mining damage assessment. However, the probability integral model, which is widely used in the prediction of horizontal surface subsidence, is unsuitable because of the influence of the surface of the donga area. To acquire probability integration parameters with a high reliability is the main challenge. However, additional displacement of slip, caused by the instability in loess donga, reduces the prediction accuracy of the probability integration model. Therefore, previous studies have focused on the probability integral parameters of the horizontal surface and on the mining stability of the slope in the donga area. Implementing findings from previous studies, we here propose a new method for predicting coal seam mining subsidence in loess donga from the probability integration model that is based on the distribution and movement principles of rock-soil layers and considers the mining-slip effect of the slope. This predicting method is suitable for subsidence forecasting for mining of surface-level ore seams in most donga areas. Due to the thickness of the sloping ore seam, the mining depth, and the offset caused by the sloping ore seam, the surface mining movement deformation basin is asymmetrical. Under the superposition of the surface in the donga area, the LDS prediction method is no longer applicable. In the future, we will study the impact of sloping seam mining on LDS prediction that will further refine the LDS subsidence prediction method.

7. Conclusions

We improved the subsidence prediction process of the probability integration model in loess donga areas to solve the problem that surface mining subsidence in such areas is difficult to predict. The following main conclusions are drawn from our new data:

(1) Slope slip can easily lead to large deviations in the prediction of subsidence in loess donga. In order to solve this long-standing problem, a solution of "regional subsidence prediction in loess donga" is put forward, and the subsidence prediction method for horizontal strata and slope area in loess donga is established.

(2) Determine the probability integration parameters of the horizontal stratigraphic region based on the position of the main key stratum and the stratigraphic distribution. With

the help of the probability integration software, the subsidence basin formed by the superposition of multiple horizontal stratigraphic regions can easily be obtained.

(3) When the mining instability coefficient of the slope $G_{cs} > 0.87$, the slope produces slip instability. The additional displacement of subsidence generated in the landslide area is calculated by the mining instability coefficient.

(4) Combined with the subsidence prediction results of the two regions in loess donga, the accuracy of the prediction of surface subsidence in the area is improved. Correlation of the calculated data with results of the field test document a 98.4% fit for the LDS subsidence curve and an 88.7% fit for the horizontal movement curve. In case of mining-slip instability of the slope, the estimated error of the subsidence curve is 0.96%, and the estimated error of the horizontal movement curve is 9.8%, thus fulfilling the requirements of engineering.

Author Contributions: The present article is addressed by the eight authors mentioned, each of whom were responsible for various aspects of the work. Conceptualization, Y.G. and B.Z.; methodology, Y.G. and B.Z.; software, B.Z.; validation, Y.G. and B.Z.; formal analysis, Y.G., Y.H., and B.Z.; investigation, Y.G., B.Z., X.M., D.Z. (Di Zhai) and J.W.; resources, B.Z.; data curation, Y.G., D.Z. (Defu Zhu) and B.Z.; writing—original draft preparation, Y.G. and B.Z.; writing—review and editing, Y.G., B.Z., X.M., D.Z. (Di Zhai), D.Z. (Defu Zhu), Y.H., Z.S. and J.W.; visualization, Y.G.; supervision, B.Z., X.M., D.Z. (Di Zhai), D.Z. (Defu Zhu), Y.H., Z.S. and J.W.; project administration, B.Z. and Y.G.; funding acquisition, B.Z., Y.G., D.Z. (Di Zhai), Y.H. and J.W. All authors have read and agreed to the published version of the manuscript.

Funding: This research was funded by the National Natural Science Foundation of China through (Grant No. 51874230 and 52074208). And Shanxi Applied Basic Research Programs, Science and Technology Foundation for Youths (Grant No. 202103021223073).

Data Availability Statement: Not applicable.

Conflicts of Interest: The authors declare no conflict of interest.

References

1. Song, S.; Zhao, X.; Xie, J.; Guan, Y. Grey Correlation Analysis and Regression Estimation of Mining Subsidence in Yu-Shen-Fu Mining Area. *Procedia Environ. Sci.* **2011**, *10*, 1747–1752. [CrossRef]

2. Jianjun, S.; Chunjian, H.; Ping, L.; Junwei, Z.; Deyuan, L.; Minde, J.; Lin, Z.; Jingkai, Z.; Jianying, S. Quantitative prediction of mining subsidence and its impact on the environment. *Int. J. Min. Sci. Technol.* **2012**, *22*, 69–73. [CrossRef]

3. Fan, H.; Gu, W.; Qin, Y.; Xue, J.; Chen, B. A model for extracting large deformation mining subsidence using D-InSAR technique and probability integral method. *Trans. Nonferrous Met. Soc. China* **2014**, *24*, 1242–1247. [CrossRef]

4. Tang, F.-Q. Mining subsidence prediction model in western thick loess layer mining areas. *Meitan Xuebao J. China Coal Soc.* **2011**, *36*, 74–78. [CrossRef]

5. Curtaz, M.; Ferrero, A.M.; Roncella, R.; Segalini, A.; Umili, G. Terrestrial Photogrammetry and Numerical Modelling for the Stability Analysis of Rock Slopes in High Mountain Areas: Aiguilles Marbrées case. *Rock Mech. Rock Eng.* **2014**, *47*, 605–620. [CrossRef]

6. Zhao, B.; Tong, C.; Liu, Z.; Liu, L.; Xueyi, Y.U. Characteristics of mining-induced surface damage in western ecological fragile region. *J. Cent. South Univ. Technol.* **2017**, *48*, 2990–2997.

7. Luo, Y. An improved influence function method for predicting subsidence caused by longwall mining operations in inclined coal seams. *Int. J. Coal Sci. Technol.* **2015**, *3*, 1–7. [CrossRef]

8. Guo, Y.; Zhu, G.; Jiang, X.; Dou, M. Study of Dynamic Coordinates Time Function in Mine Subsidence. *Adv. Mater. Res.* **2012**, *594–597*, 56–60. [CrossRef]

9. Peng, S.S. Topical areas of research needs in ground control—A state of the art review on coal mine ground control. *Int. J. Min. Sci. Technol.* **2015**, *25*, 1–6. [CrossRef]

10. Deguchi, T.; Kato, M.; Akcin, H.; Kutoglu, H.S. Automatic processing of interferometric SAR and accuracy of surface deformation measurement. *SPIE Eur. Remote Sens.* **2006**, *6363*, 47–54. [CrossRef]

11. Saeidi, A.; Deck, O.; Marwan, A.; Verdel, T.; Rouleau, A. Adjusting the Influence Function Method for Subsidence Prediction. *Key Eng. Mater.* **2013**, *553*, 59–66. [CrossRef]

12. Yanli, H.; Jixiong, Z.; Baifu, A.; Qiang, Z. Overlying strata movement law in fully mechanized coal mining and backfilling longwall face by similar physical simulation. *J. Min. Sci.* **2011**, *47*, 618–627. [CrossRef]

13. Zheng, R.; Kan, W.; Ru, L. Study on Overburden's Destructive Rules Based on Similar Material Simulation. *Int. J. Mod. Educ. Comput. Sci.* **2011**, *3*, 54–60. [CrossRef]

14. Guo, W.; Hou, Q.; Zou, Y. Relationship between surface subsidence factor and mining depth of strip pillar mining. *Trans. Nonferrous Met. Soc. China* **2011**, *21*, s594–s598. [CrossRef]
15. Zhang, B.; Ye, J.; Zhang, Z.; Xu, L.; Xu, N. A Comprehensive Method for Subsidence Prediction on Two-Seam Longwall Mining. *Energies* **2019**, *12*, 3139. [CrossRef]
16. Zhang, X.; Yu, H.; Dong, J.; Liu, S.; Huang, Z.; Wang, J.; Wong, H. A physical and numerical model-based research on the subsidence features of overlying strata caused by coal mining in Henan, China. *Environ. Earth Sci.* **2017**, *76*, 705.1–705.11. [CrossRef]
17. Kim, K.D.; Lee, S.; Oh, H.J. Erratum: Prediction of ground subsidence in Samcheok City, Korea using artificial neural networks and GIS. *Environ. Geol.* **2009**, *58*, 61–70. [CrossRef]
18. Tan, Z.; Li, P.; Yan, L.; Deng, K. Study of the method to calculate subsidence coefficient based on SVM. *Proced. Earth Planet. Sci.* **2009**, *1*, 970–976. [CrossRef]
19. Xiaofang, L.; Hui, S. Intelligent Planning of Tourism Scenic Routes Based on Genetic Algorithm in Coal WANBEI Mining Subsidence. *TELKOMNIKA Telecommun. Comput. Electron. Control* **2016**, *14*, 69. [CrossRef]
20. Xu, J.L.; Lian, G.M.; Zhu, W.B.; Qian, M.G. Influence of the key strata in deep mining to mining subsidence. *J. China Coal Soc.* **2007**, *32*, 686–690. [CrossRef]
21. Yu, X.; Zhang, E. *Mining Damage Science*; China Coal Industry Publishing House: Beijing, China, 2010.
22. Guo, W.; Yu, X.; Liu, Z. Analysis of Mining Subsidence Disasters in Loess Gully Areas and its Control Countermeasures. *Appl. Mech. Mater.* **2015**, *737*, 456–460. [CrossRef]
23. Stead, D.; Coggan, J.S.; Eberhardt, E. Realistic simulation of rock slope failure mechanisms: The need to incorporate principles of fracture mechanics. *Int. J. Rock Mech. Min. Sci.* **2004**, *41*, 563–568. [CrossRef]
24. Wang, F.; Shi, C.; Chen, K.; Li, D.; Han, K. Numerical Simulation Analysis on the Collapse of Slope in Iron Mine. *Adv. Mater. Res.* **2014**, *926–930*, 593–596. [CrossRef]
25. Luo, Y.; Cheng, J. An influence function method based subsidence prediction program for longwall mining operations in inclined coal seams. *Min. Sci. Technol. China* **2009**, *19*, 592–598. [CrossRef]
26. Zhao, B.C.; Xue-Yi, Y.U.; Zhao, J.Z. Visual study on prediction evaluation system of mining damage. *Coal Geol. Explor.* **2006**, *34*, 61–64. [CrossRef]
27. Changchun, H.; Xu, J. Subsidence Prediction of Overburden Strata and Surface Based on the Voussoir Beam Structure Theory. *Adv. Civ. Eng.* **2018**, *2018*, 2606108. [CrossRef]
28. Zhang, C.; Tu, S.; Zhao, Y. Compaction characteristics of the caving zone in a longwall goaf: A review. *Environ. Earth Sci.* **2019**, *78*, 27. [CrossRef]
29. Huang, Q. Research on roof control of water conservation mining in shallow seam. *J. China Coal Soc.* **2017**, *42*, 50–55. [CrossRef]
30. He, M.; Zhang, Z.; Zhu, J.; Li, N. Correlation Between the Constant mi of Hoek–Brown Criterion and Porosity of Intact Rock. *Rock Mech. Rock Eng.* **2022**, *55*, 923–936. [CrossRef]
31. Xie, X.; Hou, E.; Wang, S.; Sun, X.; Hou, P.; Wang, S.; Xie, Y.; Huang, Y. Formation Mechanism and the Height of the Water-Conducting Fractured Zone Induced by Middle Deep Coal Seam Mining in a Sandy Region: A Case Study from the Xiaobaodang Coal Mine. *Adv. Civ. Eng.* **2021**, *2021*, 6684202. [CrossRef]
32. Li, L.I.; Wang, Y.X.; Wang, W.B. Effects of Mining Subsidence on Physical and Chemical Properties of Soil in Slope Land in Hilly-gully Region of Loess Plateau. *Chin. J. Soil Sci.* **2010**, *41*, 1237–1240. [CrossRef]
33. Li, K. *Research on the Roof Structure and Support Resistance of the Working Face under Close Coal Seam Goaf in Shallow Buried*; Xi'an University of Science and Technology: Xian, China, 2020.

Article

Automatic Events Recognition in Low SNR Microseismic Signals of Coal Mine Based on Wavelet Scattering Transform and SVM

Xin Fan [1,*], **Jianyuan Cheng** [2], **Yunhong Wang** [2], **Sheng Li** [2], **Bin Yan** [2] and **Qingqing Zhang** [2]

[1] College of Geology and Environment, Xi'an University of Science and Technology, Xi'an 710054, China
[2] Xi'an Research Institute Co., Ltd., China Coal Technology and Engineering Group Corp., Xi'an 710077, China; cjy6608@163.com (J.C.); wangyunhong@cctegxian.com (Y.W.); lisheng_yj@163.com (S.L.); yanbin@cctegxian.com (B.Y.); zhangqingqing@cctegxian.com (Q.Z.)
* Correspondence: fanxin@stu.xust.edu.cn; Tel.: +86-029-81778060

Abstract: The technology of microseismic monitoring, the first step of which is event recognition, provides an effective method for giving early warning of dynamic disasters in coal mines, especially mining water hazards, while signals with a low signal-to-noise ratio (SNR) usually cannot be recognized effectively by systematic methods. This paper proposes a wavelet scattering decomposition (WSD) transform and support vector machine (SVM) algorithm for discriminating events of microseismic signals with a low SNR. Firstly, a method of signal feature extraction based on WSD transform is presented by studying the matrix constructed by the scattering decomposition coefficients. Secondly, the microseismic events intelligent recognition model built by operating a WSD coefficients calculation for the acquired raw vibration signals, shaping a feature vector matrix of them, is outlined. Finally, a comparative analysis of the microseismic events and noise signals in the experiment verifies that the discriminative features of the two can accurately be expressed by using wavelet scattering coefficients. The artificial intelligence recognition model developed based on both SVM and WSD not only provides a fast method with a high classification accuracy rate, but it also fits the online feature extraction of microseismic monitoring signals. We establish that the proposed method improves the efficiency and the accuracy of microseismic signals processing for monitoring rock instability and seismicity.

Keywords: mining water hazard; microseismic monitoring; intelligent recognition; feature extraction; support vector machine; classification model

Citation: Fan, X.; Cheng, J.; Wang, Y.; Li, S.; Yan, B.; Zhang, Q. Automatic Events Recognition in Low SNR Microseismic Signals of Coal Mine Based on Wavelet Scattering Transform and SVM. *Energies* **2022**, *15*, 2326. https://doi.org/10.3390/en15072326

Academic Editors: Manoj Khandelwal, Longjun Dong, Yanlin Zhao and Wenxue Chen

Received: 4 January 2022
Accepted: 22 March 2022
Published: 23 March 2022

Publisher's Note: MDPI stays neutral with regard to jurisdictional claims in published maps and institutional affiliations.

1. Introduction

Intelligent mining is the only way to achieve the safe and efficient production of coal in mines [1]. With the depth of mining, multifactorial compound disasters, such as the mining water hazards and others, become more frequent under high ground stress and some other conditions. At the same time, the rapid development of intelligent mining technology has put forward a new development opportunity for coal geological guarantee technology to be used to avoid more hazards. Coal geological guarantee technology runs through the whole cycle of coal mine production [2,3] and plays an important role in water disaster prevention and intelligent mining, especially the exploration and treatment of hidden disaster-causing geological factors in coal mines [4].

Generally, microseismic monitoring technology is one of the important technologies adopted for the dynamic monitoring of mine geological information, which can monitor the rock rupture phenomenon in real time and has a large monitoring range. To detect and explain the interior of the working surface by using microseismic monitoring technology, the online monitoring of the top and bottom plate damage of the working surface and the description of the whole process of the water guide channel from gestation and development to the final instability can be realized. We use the theory of the interference stress distribution of the surrounding rock to reveal the construction change law of the stress field, playing a key factor in water prevention and control, decreasing water and many other coal mining hazards [5,6].

A single microseismic event with a very short duration lasts tens of milliseconds. The highly accurate recognition of such an event requires careful discrimination between the microseismic and noise events [7]. However, because of the influence of extractive perturbation and the increased amount of monitoring data, the traditional methods of microseismic monitoring data collection and processing are slow and have a low level of accuracy. It is acknowledged that microseismic events and noise can be discriminated easily by the human senses, however, it is extremely difficult to do using automatic recognition methods [8]. Usually, the monitoring station is terribly disturbed by the surrounding noise, and sometimes microseismic events are even be submerged into noise. Along with the properties of microseismic signals, different researchers have proposed some discrimination methods in previous studies.

Mainly, methods based on both the sliding window and the threshold value are considered to be traditional events recognition algorithms. Some commonly used methods are the STA/LTA (the short-term average to long-term average ratio) algorithm [9–11], as well as multi-window techniques [12] and the modified energy ratio method [13]. This method, with an operation speed that is extremely fast, is an ordinary discrimination process for the detection of the first arrival of a seismic phase [7]. However, the obstruction signal is considered active, that is, the noise resistance characteristics of the process are invalid. The AR-AIC algorithm is another method used to calculate an autoregressive model of two signals combined in different time windows that use the Akaike Information Criterion (AIC). When the AIC value reaches its minimum, a pick of one microseismic event can be declared [14–16].

In these algorithms, because of an increased sensitivity to amplitude mutation, it is a common shortcoming that noise and its energy are portrayed as much larger than microseismic events. This is even more likely when the noise has a frequency content similar to that of a microseismic event. Recently, within the workings of the proposed calculation methods, some intelligent algorithms have been principally applied to the recognition of microseismic events, resulting in a lower efficiency of the processing of collected data, the discordance of recognition standards, and misjudgment.

Otherwise, spectral analyses of the different types of seismic waveforms, such as reflection and refraction tomography, have been adopted to provide more information concerning the source [17,18]. Almost all of these methods are achieved through the Fourier transform theory [19], a theory using orthogonal basis functions having perfect localization in frequency but infinite extent in time. The antileakage least-squares spectral analysis method, a method regularizing irregularly spaced data series, is an iterative one that estimates the statistically significant spectral peaks in the spectrum [20,21]. Because the frequency content is quite time-dependent, this may not be an appropriate way to process seismic signals. To address this issue, an approach called time–frequency transforms, such as wavelet transform, has been widely used in geophysical data processing and interpretations [22–24]. Features in the time–frequency domain are also applied for the automatic processing of microseismic signals [25]. However, it is easily changed by the time changing and can miss signal features, so it is not suitable for the analysis of time-varying non-stationary signals and the construction of a feature matrix.

The wavelet scattering decomposition (WSD) transform theory is mainly used to perform an analysis of the complexity of a signal sequence, achieving a nonlinearity analysis because of its high robustness as a rapid and common algorithm, which makes the analysis of time sequences more functional. For example, Mallat and Bruna [26] enabled the identification of audio signals, handwritten text, and image textures by constructing a wavelet scattering decomposition transformation network. Anden and Mallat [27,28] extracted the effective feature information through the wavelet scattering decomposition transformation network from the classical music data set GTZAN and the voice call data set TIMIT, achieving good classification results and applying the same method to the analysis of arrhythmia data in the same year. Based on the properties of wavelet scattering, Wiatowski et al. [29] demonstrated the superiority of this method by a process of rigorous mathematical derivation and generalization, achieving good results in different wavelet frameworks. Wang et al. [30] used a wavelet scattering transformation network to extract the features of synthetic aperture radar images, effectively identifying mobile and fixed targets. Li et al. [31] proposed an algorithm for cardiac tone signal classification, using the wavelet scattering transformation network to obtain cardiac tone signal characteristics, which were able to effectively express the feature information corresponding to the signal, and then obtained the feature matrix of the signal used for support vector machine classification. Recently, artificial intelligence algorithms have been widely used in the research involving the recognition of microseismic events in order to improve the efficiency and accuracy of microseismic signal processing for the monitoring of rock instability and seismicity [32,33]. The powerful artificial intelligence classification algorithm of the support vector machine (SVM) constructs the hyperplane with the largest margin in multi-dimensional space, separating different cases of each category label [34,35]. The SVM algorithm is explicitly designed to perform binary (two cluster) classifications and is an influential supervised machine learning algorithm that is widely used in image recognition, text detection, and protein classification. Here, we have successfully adapted the SVM algorithm to intelligently discriminate microseismic signals into microseismic events and noise ones with a higher degree of accuracy.

In this study, not only was an intelligent recognition method for microseismic events based on the support vector machine classification algorithm proposed, but wavelet scattering decomposition transform theory was also introduced into the field, used in performing a study of the influence of quality factors based on the characteristics of the collected data. The feature extraction method was performed based on the microseismic signals' features of the wavelet scattering decomposition transform. Combined with the SVM algorithm, we built the recognition model fitting low signal-to-noise ratio signals. The historical monitoring sample signals applied to experimental verification were determined in order to confirm the effectiveness and instantaneity of this model. Our results suggest that WSD is able to explain the different characteristics of the two classes of signals; that the established WSD-SVM model is able to discriminate microseismic events from noise has been identified. Overall, these studies taken together have revealed the significant discovery that the speed of the calculation process of this model is faster and more useful for real-time online recognition.

The rest of this paper is organized as follows. In Section 2, the effective microseismic signals classification model we proposed is presented. Then, the results of both testing and genuine signals are presented in Section 3. In Section 4, a comparison with other existing methods is presented and analyzed. Finally, our conclusions are given in Section 5.

2. Methods and Model Training

2.1. Methods

2.1.1. Wavelet Scattering Decomposition Theory

Wavelet transformation is an effective tool for time-varying non-stationary signal analysis [36]. Because of its scale variability and multi-resolution, it can describe both the time and frequency domain characteristics of the signal, so the local analysis of the signal has good results [37]. For signals in continuous finite time, the wavelet transform is defined as:

$$W(a, t) = \frac{1}{\sqrt{a}} \int_{-\infty}^{\infty} y(t)\psi * \left(\frac{t-b}{a}\right) dt \tag{1}$$

where a is a scale factor or frequency factor, and b is a translation factor or time factor, and the movement of the main wave is along t. Judging from the above formula, wavelet transformation does not have translation invariance.

The actual collected microseismic signals are usually much disturbed; even if overall there is no qualitative change, local changes will disturb the extracted signal features, thus affecting the analysis and recognition of the signal. Therefore, a signal analysis and feature extraction method with both translation invariance and local deformation stability is exactly what is needed.

With a module operation included, the operator $|W_m|$, which removes the complex phase of all wavelet coefficients, can be obtained. Convolution with the input signal yields a non-linear wavelet modulus:

$$|W|x = (x * \phi, |x * \psi_j|) \tag{2}$$

where ϕ refers to the low pass filter, so $S_m(x) = x * \phi$ refers to a local translation invariant descriptor of the signal x, the scattering coefficients, and the input signal with translation invariance, extracting the low-frequency information of the input signal and removing all high-frequency information. ψ_j represents a high-frequency wavelet. High-frequency information is recovered by the modulus transformation $U_j(x) = |x * \psi_j(x)|$, which represents the high-frequency information on scale j and obtains deformation stability by module operation on the nonlinear wavelet transform. Therefore, the low-frequency information (scattering coefficients) and high-frequency information of the wavelet scattering transformation of order 0 are as follows:

$$\begin{aligned} S_0(x) &= x * \phi \\ U_1(x) &= |x * \psi_{j_1}| \end{aligned} \tag{3}$$

The 0-order high-frequency information section $U_1(x)$ is used as input for the first order scattering transformation; this can be denoted as follows:

$$|W_1||x * \psi_{j1}| = (|x * \psi_{j_1}| * \phi, ||x * \psi_{j_1}| * \psi_{j_2}|). \tag{4}$$

Then, the first order scattering coefficients are indicated as follows:

$$S_1(x) = |x * \psi_{j_1}| * \phi \tag{5}$$

and so on; repeating the iterative procedure above can be done to obtain a scattering coefficient of an arbitrary order.

For arbitrary $j \geq 1$, the wavelet module transformation convolution of the signal can be expressed as follows:

$$U_j x = |||x * \psi_{j_1}| * \ldots| * \psi_{j_n}|. \tag{6}$$

As the next order input, $U_j x$ is low pass filtered to obtain the order m scattering coefficient:

$$S_m x = |||x * \psi_{j_1}| * \ldots| * \psi_{j_n}| * \phi = U_j x * \phi \tag{7}$$

Applying $|W_{m+1}|$ to $U_j x$, both $S_m x$ and $U_{j+1} x$ can be computed simultaneously. This can be expressed as:

$$|W_{m+1}|U_j x = (S_m x, U_{j+1} x). \tag{8}$$

The highest-order l of the scattering decomposition can be defined by initializing $U_0 x = x$, when $0 \leq m \leq l$ and $1 \leq j \leq n$, with the iteration of Equations (1)–(8).

Eventually, a feature vector is formed by the scattering coefficients on $0 \leq m \leq l$: $Sx = \{S_0 x, S_1 x, \ldots, S_m x\}$, known as $Sx = \{S_0 x, S_1 x, \ldots, S_m x\}$.

In conclusion, the process of wavelet scattering transformation can be described as a scattering transform iteration on the wavelet module operator $|W_m|$; convolution calculates the wavelet model transform $U_j x$ a value of m times and outputs the scattering coefficients $S_m x$ after low-pass filtering (Figure 1).

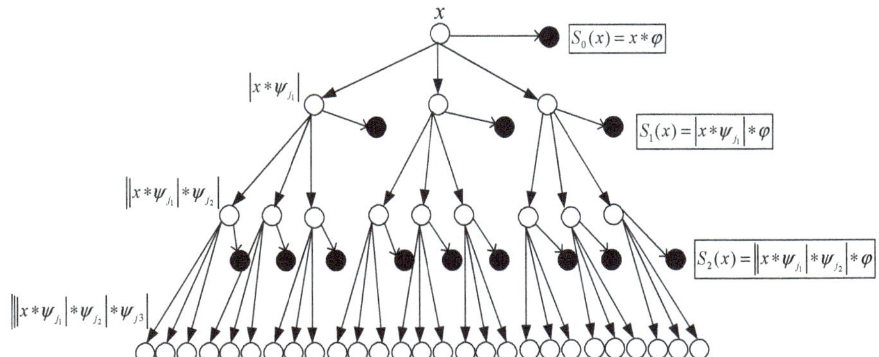

Figure 1. Wavelet Scattering Decomposition Structure.

2.1.2. Support Vector Machine Theory

In recent years, as one of the small sample algorithms based on supervised machine learning theory mainly adopted in image identification, text detection, and other fields, support vector machine theory (SVM) and possesses great advantages in the solving of nonlinear, high dimensional and small sample pattern discrimination problems and is becoming an effective classification algorithm. Usually, SVM employs an iterative training algorithm, where an optimal hyperplane with the maximum margin in multi-dimensional space can be constructed and applied to minimize an error function, as seen in Figure 2 [32,33]. In our case, we define the feature extraction of microseismic signals as a binary and nonlinear classification problem, which is an extremely significant step in the proposed algorithm for judging whether a vibration signal is a microseismic event or not. In this study, the given training vectors $x_j \in R, j = 1, \ldots, N$ in two classes and a label vector including *Microseismic events* (defined as M) and *Noise* (defined as N) are used and a quadratic optimization problem is solved by this model:

$$\min_{\beta, b, \xi} \left(\frac{1}{2} \beta' \beta + C \sum_{j=1}^{N} \xi_j \right) \tag{9}$$

which is subject to the constraints:

$$y_j \left(\beta' \phi(x_j) + b \right) \geq 1 - \xi_j$$
$$\xi_j \geq 0, j = 1, \ldots, N \tag{10}$$

where β is the normal vector to the hyperplane, b represents a constant. To avoid over-fitting, the penalty parameter C is defined on the training error. Note that ξ_j is the smallest non-negative number satisfying $y_j \left(\beta' \phi(x_j) + b \right) \geq 1 - \xi_j$. With the kernel ϕ adapted to convert the input data into the feature space, the kernel function $G(x_1, x_2) = \phi(x_1) \cdot \phi(x_2)$ is supposed to be a dot product of the input data, mapping into the higher dimensional feature space by the process of transformation ϕ.

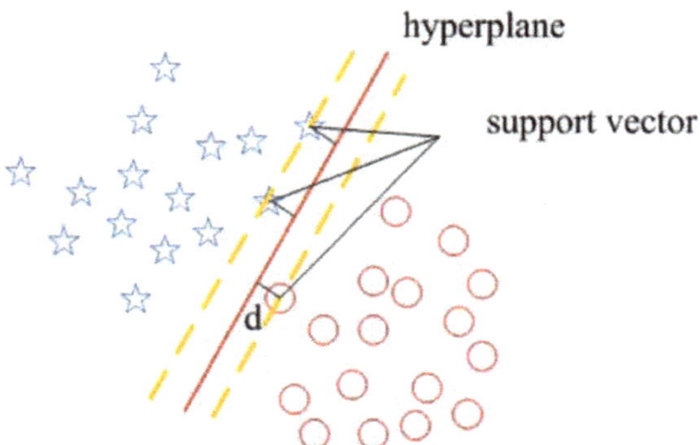

Figure 2. Schematic diagram of SVM. "d" means the largest diatance between the support vector and hyperplane.

2.2. Data Preparation and Model Training

Firstly, we selected equal numbers of microseismic events and noise sequences, operating the calculation of the wavelet scattering coefficients on the two signals and extracting the feature vectors of each one to form the feature matrix called the training set. Making use of the software package for SVM, the classifier, which assists in the classification of the testing signal samples, was built by training the set of selected sample signals. The workflow framework for the best performance and the establishing of our intelligent recognition model, as well as the iterative training and optimization of the predictive model was designed and is shown in Figure 3.

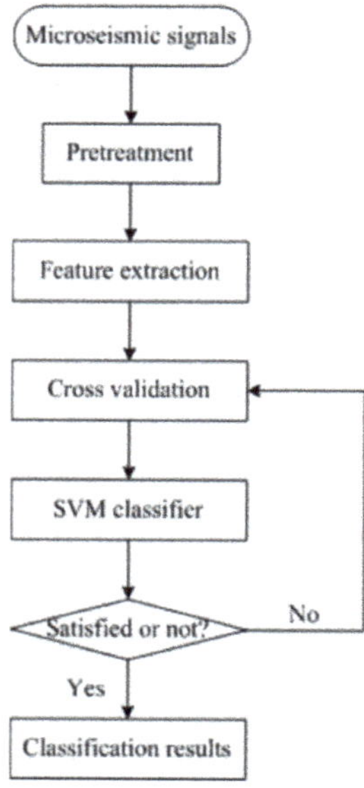

Figure 3. Workflow of the training and optimization of the SVM classification model.

To ensure that the classification model is able to differentiate microseismic events from noise in a low SNR environment, appropriate historical samples were selected to compose a strong data set. Because of the complexity of the microseismic monitoring environment in coal mines, the selected historical samples for training needed to meet the following criteria:

1. Samples selected for training the model should be a series of microseismic vibration signals, achieving clear waveform and obvious jumping.
2. An equal number of noise samples easily expressed as microseismic events should be selected in order to describe the noise features precisely.

The intelligent recognition algorithm based on the WSD and SVM of microseismic signals is as follows:

Input: the data set S, time invariance scale, transform times and quality factor
Output: the classification results.

1. Step 1: Sample selection. Because of the above-mentioned criteria, the input data set S for training can be made up of n ($n \geq 50$) samples. The data set S is composed of the same percent (50%) of the two types of signals.
2. Step 2: Feature extraction. The feature matrix of S is obtained by the calculation of scattering coefficients taking into account the certain number of the time invariance scale, the transform times and the quality factor.
3. Step 3: Cross validation. The k-fold cross validation method can be used to avoid over-fitting, evaluate classifier performance, and estimate the error rate or loss. Taking the level of computational efficiency into consideration, k in this study is 5.
4. Step 4: SVM classification. In this step, we fit a one-vs-one SVM to the training data only and then use the trained model to make predictions concerning the 30% of the data withheld for testing.

Large amounts of continuous microseismic signals were collected by stations and geophones working in environments with a high level of noise. It is a formidable task to discriminate the microseismic events contained in those signals with precision using previous methods. Many events submerged in the noise cannot help with source location and other processes. The purpose for the construction of the WSD-SVM model used for the processing of monitoring vibration signals obtained from certain stations is to improve the recognition accuracy of microseismic events through processes so that the data can be identified precisely.

3. Results

3.1. Testing Results

The data samples designed to fit the experiment were obtained from the *KJ959* microseismic monitoring system, which has a sampling frequency designated as 1 kHz, a standard widely adopted in coal mine inrush water hazards prediction and prevention. These samples provided an effective series of microseismic vibration signals. In addition, we chose single component detection sensors with a frequency response range from 10 Hz to 1 kHz as the geophones. A total number of 108 raw signals regarded as data set S were collected using automatic pick-up technology. For ease of analysis, 108 signals were interpreted as segments of equal length, with each segment consisting of 7000 sampling points. The data set S was split into S1, containing 54 microseismic events with an obvious jump, and S2, comprising 54 noise signals. After that, they were categorized into M (for microseismic events)and N (for noise). For convenience, one segment from S1 and another from S2 were picked for analysis; they are shown is Figure 4.

The time invariance scale $i = 6$, transform times $t = 3$, the quality factor $q = 3, 2, 1$ and the calculation of scattering coefficients for the two signals are shown in Figure 5, showing the distinctive differences between events and noise. The features of all the signal segments in data set S can be expressed by the feature matrix consisting of the scattering coefficients using the proposed method.

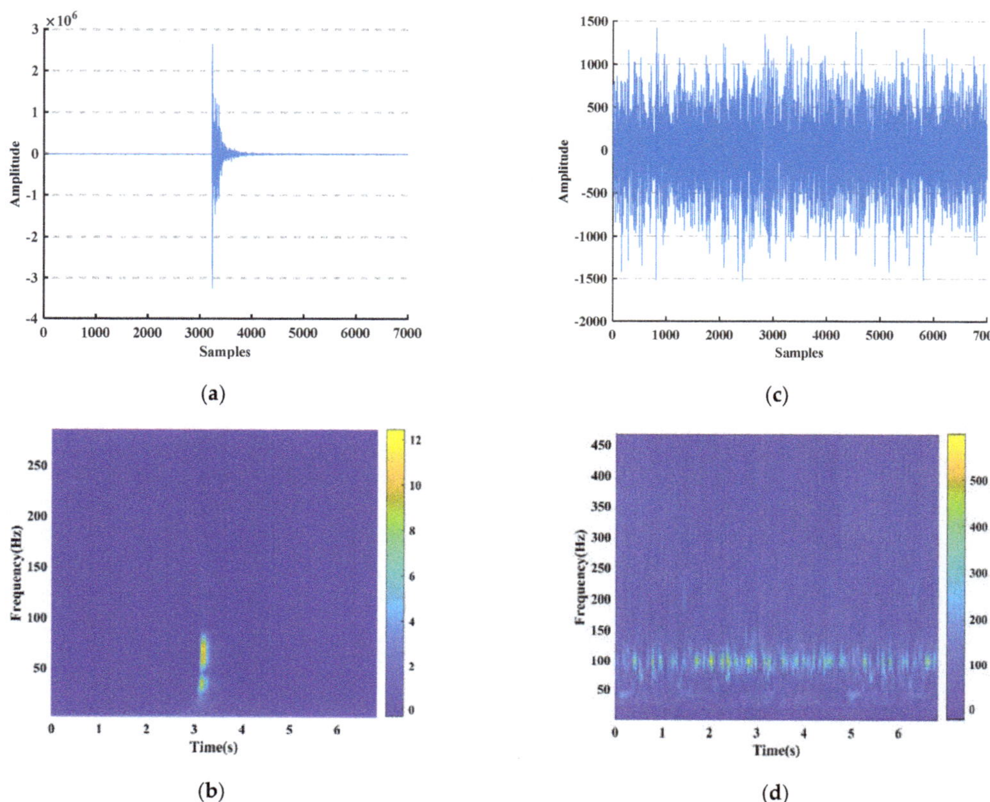

Figure 4. (**a**) microseismic signal, (**b**) scattergram time-frequency analysis of (**a**). (**c**) noise signal, (**d**) scattergram time-frequency analysis of (**c**).

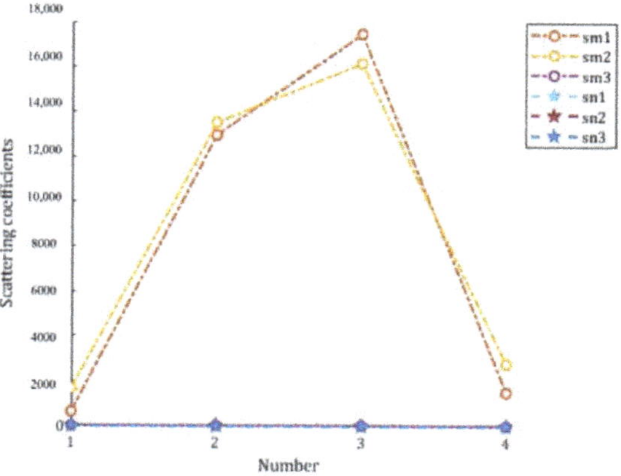

Figure 5. The distinctive scattering coefficients of microseismic events and noise.

To achieve better performance in the defined WSD-SVM model, 70 percent of the data in each class were randomly devoted to the formation of the training set *STr* which was trained in order to obtain the SVM classifier. Meanwhile, the remaining 30 percent was withheld for testing and assigned to the test set *STe*. As is known, the performance of a supervised machine learning algorithm is largely dependent on the training percent of the data set. The above process was repeated with different training percents, and the corresponding classification accuracy rates were calculated as shown in Figure 6.

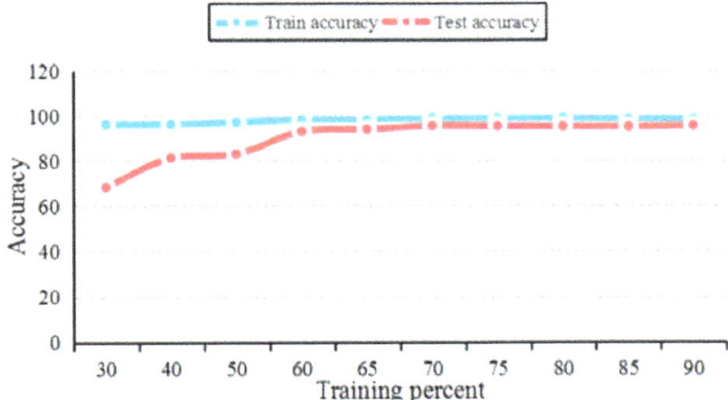

Figure 6. Recognition accuracy rates for different training percents.

Microseismic events and noise can be classified by the WSD-SVM algorithm effectively, as shown in Figure 6. As the following ten experiments illustrate, the recognition accuracy rates increased as the training percent became larger, reaching 99.6% in five experiments.

3.2. Application in Genuine Signals

We sought to verify the validity of the above algorithm, so a continuous microseismic signal with a duration of 56 s was selected for the experiment. Data were obtained using the monitoring equipment installed in a coal mine in northwestern China; results are shown in Figure 7.

Because the monitoring station is disturbed by ambient noise, the signal segment in Figure 7 shows a low SNR. There are 8 microseismic events in total in the sequence. Judging by the software, 3 (E2, E5, E7) of these have a clear waveform and can be verified directly, and another 5 (E1, E3, E4, E6, E8) events are covered by the noise. All 8 events are designed to be detected by the theory of STA/LTA and our trained model. Both the results of the detected event numbers and the corresponding time consumptions of the two methods are recorded in Table 1.

Only four microseismic events (E1, E4, E5, E7) were able to be recognized by the STA/LTA method with a lower threshold, while our proposed algorithm could recognize all eight events effectively. Taking the time consumption of the two algorithms into consideration, it took 2.488 s for the WSD-SVM model to recognize all eight events, irrespective of the training time, which is a little slow for calculation. In contrast, the method we proposed was able to recognize low SNR microseismic events accurately with little sacrifice in calculation time.

Table 1. Comparison of the method performance.

Method	Number of Detected Events	Time Consumption/s
STA/LTA	4	1.358
Algorithm in this paper	8	2.488

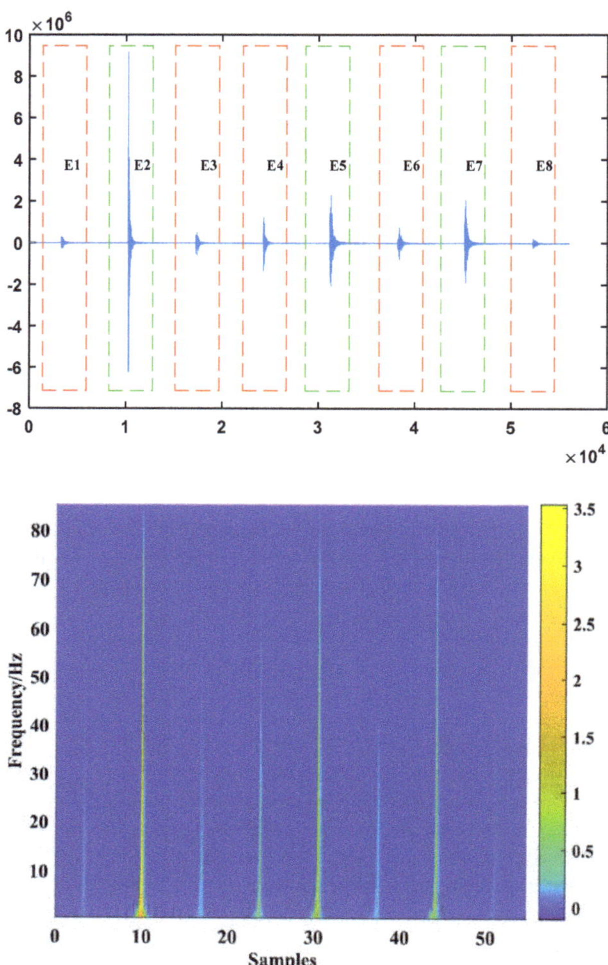

Figure 7. Microseismic signal for experiments and its scattergram of time-frequency analysis.

4. Discussion

4.1. The Recognition Ability of WSD-SVM

In order to further confirm the universality of our proposed method, ten microseismic signals were achieved from four unique monitoring stations as experimental sample data and were analyzed using the STA/LTA method and the method employed in this paper. Using professional software, the time–frequency analysis of the ten signals was observed and 28 events were concluded. The number of microseismic events successfully recognized is presented in Table 2, which shows that 28 microseismic events were detected from the selected samples for experiments, the recognition accuracy rate was 92.86%, and the recognition accuracy was better than that of the STA/LTA method.

4.2. The Influence of the Transform Times on the Classification Results

Whenever the WSD-SVM algorithm is used to recognize microseismic events, the selection of the appropriate transform times is a critical step, determining the level of classification accuracy. To work out the influence of the transform times on the classification results, 54 event samples and 54 noise samples from Section 3.1 were selected, as well as the WSD-SVM algorithm when $i = 6$ and $q = 1, 2, 3, 4, 5$.

Table 2. Comparison of the two methods.

No.	Duration of Signals/s	Number of Microseismic Events	Number of Events Recognized by the STA/LTA Method	Number of Events Recognized by the Proposed Method
1	15	2	0	2
2	23	3	1	3
3	10	1	1	1
4	26	2	1	2
5	36	5	3	4
6	20	3	2	3
7	12	1	0	1
8	29	4	2	3
9	32	4	3	4
10	30	3	1	3
Total	233	28	14	26

As we can see from Table 3, when the transform time is 1 or 2, it takes less time to complete the process of classification with a lower accuracy. However, when the transform time is greater than 4, it results in a higher level of accuracy with an extremely high level of time consumption because of the complexity of calsulation. A small number of transform times is not adequate to express the complexity of the samples, though it takes less time, but an excessive number introduces large time consumption. Therefore, according to these results, the best, most acceptable number of transform times is 3.

Table 3. Relationship between transform times and classification accuracy.

Transform Times	Classification Accuracy	Time Consumption/s
1	58.50	0.95
2	58.71	1.57
3	99.21	3.1
4	99.21	19.968
5	99.80	476.206

5. Conclusions

To conquer the noise problems in the microseismic monitoring data, a novel intelligent recognition method for microseismic signals with a low SNR was proposed in detail, consisting of the use of a support vector machine classifier in combination with the feature extraction method of wavelet scattering decomposition transform. Though the selected signals are expected to be further processed by the algorithm in this paper, the validity of and the favorable results for the WSD-SVM model have already been demonstrated by the accurate discrimination of genuine microseismic events from noise events. In addition, the scattering coefficients for each signal are shown to be useful as features for training the distinctive model. The recognition accuracy rate of the samples for experiments using the model reached 92.86%, showing that the model could be applied to recognize the microseismic events in the monitoring area. The increased utilization of a smaller feature matrix and an effective feature extraction method is the future direction of microseismic event classification.

Author Contributions: Writing—original draft, X.F. and S.L.; writing—review and editing, X.F., J.C. and S.L.; supervision, J.C. and B.Y.; funding acquisition, J.C. and Y.W.; data curation, B.Y. and Q.Z. All authors have read and agreed to the published version of the manuscript.

Funding: This research was funded by the National Natural Science Foundation of China (grant number 42074175).

Institutional Review Board Statement: Not applicable.

Informed Consent Statement: Not applicable.

Data Availability Statement: Not applicable.

Acknowledgments: The authors wish to thank the experiment platform of CCTEG Xi'an Research Institute and Xi'an University of Science and Technology as well as our families for their patience and support.

Conflicts of Interest: The authors declare no conflict of interest.

References

1. Wang, G.F. Speeding up intelligent construction of coal mine and promoting high-quality development of coal industry. *China Coal* **2021**, *47*, 2–10.
2. Wang, X.Z. Review and prospect of China coal mine safety production in the past fifty years. *Saf. Coal Mines* **2020**, *51*, 1–4.
3. Wu, J.W.; Zhao, Z.G. Development stages of coalmine mining geological works in China. *Coal Geol. China* **2010**, *22*, 26–28.
4. Wang, T.; Shao, L.; Xia, Y.; Fu, X.; Sun, Y.; Sun, Y.; Ju, Y.; Bi, Y.; Yu, J.; Xie, Z.; et al. Major achievements and future research directions of the coal geology in China. *Geol. China* **2017**, *44*, 242–262.
5. Jiang, Y.D.; Pan, Y.S.; Jiang, F.X.; Dou, L.M.; Ju, Y. State of the art review on mechanism and prevention of coal bumps in China. *J. China Coal Soc.* **2014**, *39*, 205–213.
6. Lu, C.P.; Liu, G.J.; Liu, Y. Micro-seismic multi-parameter characteristics of rockburst hazard induced by hard roof fall and high stress concentration. *Int. J. Rock Mech. Min. Sci.* **2015**, *76*, 18–32. [CrossRef]
7. Jia, R.S.; Sun, H.M.; Peng, Y.J.; Liang, Y.Q.; Lu, X.M. Automatic event detection in low SNR micro-seismic signals based on multi-scale permutation entropy and a support vector machine. *J. Seismol.* **2017**, *21*, 735–748. [CrossRef]
8. Zhao, Z.; Gross, L. Using supervised machine learning to distinguish micro-seismic from noise events. In *SEG Technical Program Expanded Abstracts*; Society of Exploration Geophysicists: Tulsa, OK, USA, 2017; pp. 2918–2923.
9. Allen, R. Automatic earthquake recognition and timing from single trace. *Bull. Seism. Soc. Am.* **1978**, *68*, 1521–1532. [CrossRef]
10. Baer, M.; Kradolfer, U. An automatic phase picker for local and teleseismic events. *Bull. Seism. Soc. Am.* **1987**, *77*, 1437–1445. [CrossRef]
11. Earle, P.S.; Shearer, P.M. Characterization of global seismograms using an automatic-picking algorithm. *Bull. Seism. Soc. Am.* **1994**, *84*, 366–376. [CrossRef]
12. Chen, Z.; Stewart, R.R. A multi-window algorithm for real-time automatic detection and picking of P-phases of microseismic events. *CREWES Res. Rep.* **2006**, *18*, 1–9.
13. Akram, J. Automatic P-wave arrival time picking method for seismic and micro-seismic data. In *CSPG CSEG CWLS Convention*; CSEG: Calgary, AB, Canada, 2011.
14. Sleeman, R.; van Eck, T. Robust automatic P-phase picking: An on-line implementation in the analysis of broadband seismogram recordings. *Phys. Earth Planet. Inter.* **1999**, *113*, 265–275. [CrossRef]
15. Leonard, M. Comparison of manual and automatic onset time picking. *Bull. Seismol. Soc. Am.* **2000**, *90*, 1384–1390. [CrossRef]
16. St-Onge, A. Akaike information criterion applied to detecting first arrival times on micro-seismic data. In Proceedings of the 81th Annual International Meeting, Yokohama, Japan, 14–17 April 2022; pp. 1658–1662.
17. Karaman, A.; Karadayilar, T. Identification of karst features using seismic P-wave tomography and resistivity anisotropy measurements. *Environ. Geol.* **2004**, *45*, 957. [CrossRef]
18. Hiltunen, D.R.; Cramer, B.J. Application of Seismic Refraction Tomography in Karst Terrane. *J. Geotech. Geoenviron. Eng.* **2008**, *134*, 938–948. [CrossRef]
19. Yang, Y.-S.; Li, Y.-Y.; Cui, D.-H. Identification of karst features with spectral analysis on the seismic reflection data. *Environ. Earth Sci.* **2014**, *71*, 753–761. [CrossRef]
20. Ghaderpour, E.; Liao, W.; Lamoureux, M.P. Antileakage least-squares spectral analysis for seismic data regularization and random noise attenuation. *Geophysics* **2018**, *83*, V157–V170. [CrossRef]
21. Ghaderpour, E. Multichannel antileakage least-squares spectral analysis for seismic data regularization beyond aliasing. *Acta Geophys.* **2019**, *67*, 1349–1363. [CrossRef]
22. Pinnegar, C.R.; Eaton, D.W. Application of the S transform to prestack noise attenuation filtering. *J. Geophys. Res. Earth Surf.* **2003**, *108*, 2422. [CrossRef]
23. Vallée, M.A.; Keating, P.; Smith, R.S.; St-Hilaire, C. Estimating depth and model type using the continuous wavelet transform of magnetic data. *Geophysics* **2004**, *69*, 191–199. [CrossRef]
24. Sinha, S.; Routh, P.S.; Anno, P.D.; Castagna, J.P. Spectral decomposition of seismic data with continuous-wavelet transform. *Geophysics* **2005**, *70*, P19–P25. [CrossRef]
25. Zhang, H.; Ma, C.; Pazzi, V.; Zou, Y.; Casagli, N. Microseismic Signal Denoising and Separation Based on Fully Convolutional Encoder–Decoder Network. *Appl. Sci.* **2020**, *10*, 6621. [CrossRef]
26. Bruna, J.; Mallat, S. Audio texture synthesis with scattering moments. *arXiv* **2013**, arXiv:1311.0407.
27. Anden, J.; Mallat, S. Deep Scattering Spectrum. *IEEE Trans. Signal Process.* **2014**, *62*, 4114–4128. [CrossRef]

28. Chudáček, V.; Anden, J.; Mallat, S.; Abry, P.; Doret, M. Scattering Transform for Intrapartum Fetal Heart Rate Variability Fractal Analysis: A Case-Control Study. *IEEE Trans. Biomed. Eng.* **2013**, *61*, 1100–1108. [CrossRef]

29. Wiatowski, T.; Bolcskei, H. A Mathematical Theory of Deep Convolutional Neural Networks for Feature Extraction. *IEEE Trans. Inf. Theory* **2018**, *64*, 1845–1866. [CrossRef]

30. Wang, H.; Li, S.; Zhou, Y.; Chen, S. SAR Automatic Target Recognition Using a Roto-Translational Invariant Wavelet-Scattering Convolution Network. *Remote Sens.* **2018**, *10*, 501. [CrossRef]

31. Li, J.; Ke, L.; Du, Q.; Ding, X.; Chen, X.; Wang, D. Heart Sound Signal Classification Algorithm: A Combination of Wavelet Scattering Transform and Twin Support Vector Machine. *IEEE Access* **2019**, *7*, 179339–179348. [CrossRef]

32. Pu, Y.; Apel, D.B.; Hall, R. Using machine learning approach for microseismic events recognition in underground excavations: Comparison of ten frequently-used models. *Eng. Geol.* **2020**, *268*, 105519. [CrossRef]

33. Dong, L.J.; Tang, Z.; Li, X.B.; Chen, Y.C.; Xue, J.C. Discrimination of mining microseismic events and blasts using convolutional neural networks and original waveform. *J. Cent. South Univ.* **2020**, *27*, 3078–3089. [CrossRef]

34. Boser, B.E.; Guyon, I.M.; Vapnik, V.N. A training algorithm for optimal margin classifiers. In Proceedings of the Fifth Annual Workshop on Computational Learning Theory, Pittsburgh, PA, USA, 27–29 July 1992; pp. 144–152. [CrossRef]

35. Cortes, C.; Vapnik, V. Support-vector networks. *Mach. Learn.* **1995**, *20*, 273–297. [CrossRef]

36. Olschneider, M.; Kronland-Martinet, R.; Morlet, J.; Tchamitchian, P. A real-time algorithm for signal analysis with the help of the wavelet transform. In *Wavelets*; Springer: Berlin/Heidelberg, Germany, 1990; pp. 286–297.

37. Daubechies, I. The wavelet transform, time-frequency localization and signal analysis. *IEEE Trans. Inf. Theory* **1990**, *36*, 961–1005. [CrossRef]

Article

Effect of Freeze–Thaw Cycles on Mechanical and Microstructural Properties of Tailings Reinforced with Cement-Based Material

Pengchu Ding [1,2,*], Yunbing Hou [2], Dong Han [2,3], Xing Zhang [2,4], Shuxiong Cao [2] and Chunqing Li [5]

1 School of Architecture and Civil Engineering, Zhongyuan University of Technology, Zhengzhou 450007, China
2 School of Energy and Mining Engineering, China University of Mining and Technology, Beijing 100083, China; houyunbing2000@163.com (Y.H.); handong9933@163.com (D.H.); tsp1600101008@student.cumtb.edu.cn (X.Z.); tsp1600101006@student.cumtb.edu.cn (S.C.)
3 Hydrologic Environment Center, Northwest Geological Exploration Institute, China Metallurgical Geology Bureau, Xi'an 710119, China
4 School of Civil Engineering, Wuhan University, Wuhan 430072, China
5 School of Engineering, RMIT University, Melbourne 3001, Australia; chunqing.li@rmit.edu.au
* Correspondence: dpc@zut.edu.cn

Abstract: In China, more than 10,000 Tailings storage facilities (TSF) have been created on the ground surface through mineral mining processes, these TSF occupy a large amount of land. The strength of the tailings is too low to be able to stand on its own without strengthening. In order to save land resources and alleviate the damage to the environment caused by mineral mining, it is necessary to reinforce the TSF so that they can store more tailings. China is one of the countries with the largest area of permafrost and seasonal frozen regions, accounting for about 75% of the country's total land area. The problem can be exacerbated in these regions where the freeze–thaw effect can further degrade the strength of tailings. A review of the literature suggests that there is little research on the mechanical and microstructural properties of tailings reinforced with cement-based materials under freeze–thaw conditions, especially when the tailings are to be discharged to land for sustainable development. This study investigates the effect of freeze–thaw cycles on the mechanical properties and microstructural changes of tailings reinforced with cement-based materials to mitigate environmental hazards. Unconfined compressive strength (UCS) tests, scanning electron microscopic images, X-Ray Diffraction tests, thermogravimetry tests and mercury intrusion porosimetry tests were conducted on samples of tailings. The results from this study show that freeze–thaw cycles reduce the UCS of all the tested samples eventually, but the frozen temperature does not significantly affect the UCS. The larger number of freeze–thaw cycles, the more damage is to the surface morphology and the matrix of the tailings. The results presented in the paper can help engineers and managers to effectively transport the TSF to other locations to minimize environmental hazards to achieve sustainable production of mineral mining processes.

Keywords: freeze–thaw cycles; tailings; mechanical behavior; SEM; MIP

Citation: Ding, P.; Hou, Y.; Han, D.; Zhang, X.; Cao, S.; Li, C. Effect of Freeze–Thaw Cycles on Mechanical and Microstructural Properties of Tailings Reinforced with Cement-Based Material. *Minerals* **2022**, *12*, 413. https://doi.org/10.3390/min12040413

Academic Editors: Longjun Dong, Yanlin Zhao, Wenxue Chen and Carlito Tabelin

Received: 30 January 2022
Accepted: 24 March 2022
Published: 27 March 2022

Publisher's Note: MDPI stays neutral with regard to jurisdictional claims in published maps and institutional affiliations.

1. Introduction

The excavation and removal of ore mass from the ground create a large volume of surface voids or ground subsidence, which poses environmental hazards [1,2]. Run-of-mine ores are initially processed through standard mineral processing operations, such as crushing, magnetic separation, gravity separation, dense medium separation and flotation to increase their grade for subsequent extraction of metals, the waste generated is called tailings. In China, more than 10,000 tailing storage facilities (TSF) have been generated on the ground due to mining [3], by the end of 2020, the reserves of tailings reservoirs reached 22.26 billion tons [4]. The type of tailings accumulated on the surface needs to be disposed of at a proper location to mitigate the environmental impact of the mineral process. If not handled well, the TSF most likely leads to serious geotechnical dangers (e.g., tailings dam failure) and land contamination (e.g., heavy metal leaching) [5,6]. Tailings generated from metal-sulfide mines usually contain non-valuable sulfide minerals, such as pyrite (FeS_2), pyrrhotite ($Fe_{1-x}S$, where $0 < x < 0.2$), and arsenopyrite ($FeAsS$) that generate acid mine drainage when exposed in the environment to oxygen (O_2) and water [7]. In addition, tailings can turn into mud in wet weather, e.g., rain, and into dust in dry weather, e.g., by wind. This creates severe pollution of the environment.

To solve these problems, some researchers [8–11] suggested that the tailings be used to backfill mined-out areas, which has a good effect on the control of geological subsidence. Backfilling is a good idea, but if the tailings used are rich in sulfide minerals, it can cause other problems, especially the enhanced generation of acid mine drainage. Another promising approach is to simultaneously repurpose and treat tailings via alkali activation in a process called geopolymerization. Some mine tailings are ideal geopolymer materials because they contain high clay and aluminosilicate minerals, which are essential components for the geopolymer matrix to form. In this process, aluminosilicates are dissolved in highly-concentrated alkali hydroxide or silicate solution to form a structurally stable material composed of amorphous, interconnected Si-O-Al polymeric matrices via a combination of diffusion, coagulation and polycondensation. They are potentially suitable raw materials for geopolymeric products with considerably high compressive strength and long-term durability [12]. Other researchers [13] proposed to add some cement-based material to the tailings to form a consolidated body with certain mechanical strength and discharge them into valleys or ground subsidence areas through transport by pipes or belts. Either way, the tailings should be strengthened before they can be transported to another location. In this way, it provides a method to support mining in its transition into an economically sustainable, socially responsible, and environmentally sensitive industry. This approach can simultaneously reduce waste generation, provide additional economic benefits to stakeholders, empower host communities, and improve rehabilitation programs [14]. In fact, a leaching test, such as TCLP showing that the slag and tailing samples are non-hazardous should also be provided as the greatest challenge in the repurposing of mining wastes is their potential to release hazardous elements to the environment. Alkali-activated cementitious materials can store harmful elements in the consolidated body, making them less prone to leaching [15].

The area of permafrost and seasonal frozen regions accounts for about 23% of the world's total land area [16]. It is, therefore, of significant importance to consider the influences of the cold weather on the strength of tailings reinforced with cement-based materials. In the past decades, researchers focused their attention on the internal microstructure in their studies of cement-based materials [17]. This is because the microstructure determines the strength and durability of the tailings. However, freeze–thaw cycles can destroy the reinforced body of tailings since frozen water, i.e., ice, generates crystallization pressure on the capillary and pore walls in the tailings, and the expanded volume leads to an increase in stress [18]. The ice then melts to water, causing a collapse of some pores and extensions into larger pore structures. As a result, the porosity of the tailings is increased. Clearly continuous cycles of freeze–thaw will degrade the mechanical properties of cement-based materials [19].

A number of researchers [20–23] investigated the degradation of compressive strength of concrete due to freeze–thaw cycles. Tang [24] applied the artificial freezing method in subway tunnel construction and found that the value of freezing temperature had a slight influence on the dynamic elastic modulus of soil, but freeze–thaw action can reduce the dynamic elastic modulus. Xie [25] studied the interaction between desertification and permafrost and found that both the cohesion values and the compressive strength of the permafrost samples that experienced freeze–thaw cycles were decreased compared to the unfrozen samples. Some researchers [26] observed that crystallization pressure by ice was the most important source of stress during freeze–thaw cycles for cement-based composites. The interfacial energy between the porous media played a vital role in the crystallization pressure, which is the stress that acted on the walls of pores due to the growing ice crystal [27–30]. Therefore, it is imperative to accurately measure the pore size distribution and micromorphology of the tailings reinforced with cement-based materials before and after freeze–thaw cycles.

Further review of the literature suggests that current research focused more on the mechanical and microstructural properties of concrete or soil only [31–34]. There is little or no research on tailings reinforced with cement-based materials under freeze–thaw conditions, especially when the tailings are to be discharged to land. Clearly knowledge on the degradation of mechanical and microstructural properties of tailings reinforced with cement-based materials under freeze–thaw cycles can help prevent failures of tailing dams and mitigate their environmental hazards. This can achieve not only sustainable production in the mining process but also bring about significant social, economic and environmental benefits. Therefore, there is a well-justified need to study the degradation of mechanical strength of tailings reinforced with cement-based materials under freeze–thaw cycles.

The main objective of this paper is to experimentally study the mechanical and microstructural properties of tailings reinforced with cement-based materials before and after freeze–thaw cycles and to identify the causes of changes in mechanical and microstructural properties. To achieve this objective, samples of tailings are prepared with different curing periods, namely 3, 7 and 28-days and frozen at different temperatures, namely $-5\,^{\circ}\mathrm{C}$, $-10\,^{\circ}\mathrm{C}$, and $-15\,^{\circ}\mathrm{C}$ for different numbers of cycles, namely 0, 3, 5, 7, 10, 12, 15, 20. Then, uniaxial compressive (UCS) tests are conducted on these samples. This w followed by the X-Ray Diffraction (XRD) and thermogravimetry (TG) tests to identify and assessment of the hydration products of cement-based materials before and after freeze–thaw cycles. Scanning electron microscopy (SEM) is performed on these samples to analyze the composition and microcosmic morphology of the hydration products. Finally, the microscope pore structures of the samples are studied through the mercury intrusion porosimetry (MIP) experiment. The paper focuses on the effects of freeze–thaw cycles on the mechanical and microscopic properties of the tailings to provide a theoretical and experimental basis for their wider application in cold regions to achieve environmental sustainability and cleaner production in the mining process.

2. Design of Experiment

2.1. Geology of the Mine and Test Materials

Lilou Iron Mine is located in Huoqiu County, western Anhui Province. The ore body has an elevation of $-520{\sim}-862$ m, an average thickness of 48.2 m, and an inclination angle of $65{\sim}85^{\circ}$. The roof of the deposit and its surrounding rocks are gneiss and dolomite marble. The hardness coefficient of specular hematite in the ore-bearing belt is 8 to 12, but the local extrusion and crushing have a broken structure. The lithology of the ore body floor is mainly dolomite marble, with no fissures and karst caves developed. The shape of the ore body within the ore deposit is layered, the ore body tends to be nearly north-south, and the ore body is inclined to the west. At present, the mining capacity of the mine is 7.5 million tons/year, which is the largest iron ore underground mining in China.

The tailings (from the Anhui Lilou Iron Mine, the main mineralogical composition is specular hematite) mixed with a new type of cement-based material were used as the test material. Although Ordinary Portland Cement is one of the most commonly used binders in backfilling and discharging [35], its costs account for almost 75% of the discharge expenditure [36]. Thus, a cheaper new type of cement-based material (denoted NCM) is developed to replace Ordinary Portland Cement. The composition of the NCM is clinker, lime, gypsum, blast furnace slag with a proportion of 14:6:10:70 (the main mineralogical compositions of lime, gypsum, blast furnace slag is calcium hydroxide, calcium sulfate and melilite, the main mineralogical compositions of clinker are tricalcium silicate, dicalcium silicate, tricalcium aluminate and tetracalcium ferroaluminate) [37]. The grain size distribution of NCM is given in Table 1 where d_{10}, d_{30}, d_{50}, d_{60}, d_{90} represent the cumulative content on the particle composition curve, with corresponding particle sizes of 10%, 30%, 50%, 60%, 90% of the volume, respectively. The main chemical compositions of NCM are given in Table 2; 0.4% of admixture was added to NCM. The admixture consists of sodium sulfate, alum, sodium fluorosilicate with a proportion of 2:1:1. Previous test results have proved [37] that the UCS of the tailings with NCM at the ages of 3, 7 and 28-days are 2.4, 2.4 and 1.7 times higher than that with OPC, respectively.

Table 1. Grain size distribution of NCM and tailings.

Element Unit	$d_{10}/\mu m$	$d_{30}/\mu m$	$d_{50}/\mu m$	$d_{60}/\mu m$	$d_{90}/\mu m$	Cu	Cc
slag	9.32	18.07	31.23	50.68	226.28	2.80	5.44
clinker	7.48	12.26	20.39	25.31	29.62	2.06	4.59
gypsum	2.47	9.19	18.73	24.86	74.81	2.71	3.09
lime	3.13	5.91	10.82	21.74	33.32	6.95	1.42
Tailings	14.55	26.61	38.32	54.27	82.33	3.73	0.89

Table 2. Main chemical compositions of NCM and tailings.

Element Unit	MgO (wt.%)	Al_2O_3 (wt.%)	SiO_2 (wt.%)	CaO (wt.%)	SO_3 (wt.%)	Fe_2O_3 (wt.%)	Total
slag	8.38	14.79	33.81	36.95	0.28	0.89	95.09
clinker	2.45	4.47	22.01	64.31	2.45	3.45	99.14
gypsum	2.14	0.12	0.98	45.85	42.45	0.11	91.66
lime	0.56	0.23	0.38	72.29	0.13	0.26	73.84
Tailings	2.41	3.85	82.05	2.46	0.18	8.01	98.96

Note: The experimental data are provided by the Key Laboratory of Orogenic Belt and Crustal Evolution of the Ministry of Education, Peking University.

Tap water was used to mix the binders, i.e., NCM with tailings. The grain size distribution of tailings is shown in Table 1. The main mineralogical compositions of the tailings are listed in Table 2. The grain size distribution of the tailings is shown in Figure 1.

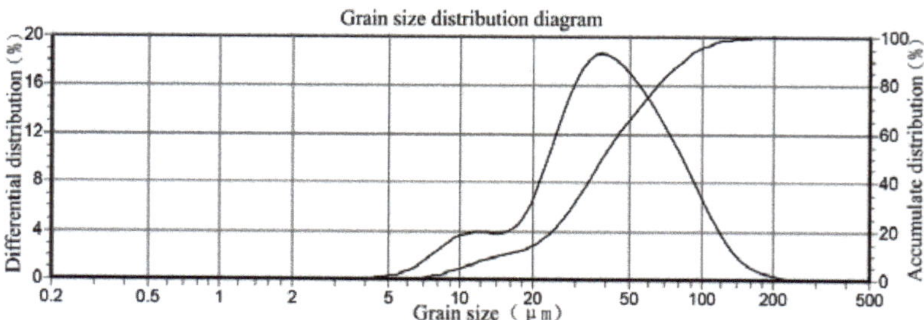

Figure 1. Grain size distribution of tailings.

2.2. Test Specimens

The water to binder (w/NCM) ratio of the mix proportion adopted in this study is 5.6. The binder (NCM) content of the tailings is 4.5% by weight. The preparation of the specimens basically follows the standard ASTMC39 [38] and includes the following steps: Firstly, the required amount of tailings, NCM and water were determined according to the experimental scheme, and the tailings mixtures were produced by mixing the required quantity of tailings, NCM and water for 7 min until a homogeneous paste was obtained. The prepared mortar was then poured into plastic cylinders, whose diameter and height are 50 mm and 100 mm, respectively. Then, the prepared samples of tailings mixtures were sealed with a plastic cover to prevent moisture loss. They were cured in an environmental chamber with a controlled temperature of $20 \pm 1\,°C$ and a minimum of 95% relative humidity for different curing times, i.e., 3, 7 and 28-days. To assure comparable results in the current measurements, identical experimental conditions are maintained for all the tests. After curing, tailings samples were frozen at three temperatures namely $-5\,°C$, $-10\,°C$ and $-15\,°C$, and then thawed in the environmental chamber with different numbers of cycles, namely, 0, 3, 5, 7, 10, 12, 15, 20 for each freezing temperature. The freezing and thawing time are both 12 h, according to the period of a natural cycle. To cover a range of designated values for different test variables of curing time, freezing temperature and number of freezing-thawing cycles, 230 samples of tailings mixture were prepared in this study.

A detailed test plan for this study is shown in Table 3. It should be noted that only water and binder are added to the XRD and TG tests, other operations are the same as above.

Table 3. Test plan for experiment.

Tests	Cured Time/Days	Cured Temperature/°C	Freeze–Thaw Times	Number of Samples
UCS	3, 7, 28	$-5, -10, -15$	0, 3, 5, 7, 10, 12, 15, 20	198
XRD	7, 28	-10	0, 20	4
TG	7, 28	$-5, -10, -15$	0, 20	8
SEM	3, 7, 28	-10	0, 5, 10, 20	12
MIP	7, 28	$-5, -10, -15$	0, 20	8

2.3. Tests on Specimens

Tests undertaken on tailings samples included mechanical tests, XRD, TG, SEM and MIP measurement. In accordance with ASTMC39 [38], the mechanical strength of tailings samples is represented by the UCS [39], which is one of the most important indicators to measure the macroscopic mechanical damage characteristics on samples. It is relatively intuitive to find the relationship between the damage and the number of freeze–thaw cycles. The UCS tests are performed on the tailings samples after they are cured for 3-days, 7-days, 28-days (referred to as 3-, 7-, 28-day samples hereafter), and experienced 0, 3, 5, 7, 10, 12, 15, 20 freeze–thaw cycles.

XRD is a common measurement for crystal phase structure identification in cement-based materials [40]. The tests were performed on NCM samples that are cured for 7-days and 28-days and experienced 20 freeze–thaw cycles at a freezing temperature of $-10\,°C$.

TG test is to assess the amount of hydration products on the microstructural development of the cement-based materials. The tests were performed on NCM samples that were cured for 7-days and 28 days and undergone different freeze–thaw cycles at different freezing temperatures.

SEM images are used to evaluate the influence of hydration products on tailings samples [41]. The tests were conducted on tailings samples that were cured for 3-days, 7-days, 28-days and had undergone freeze–thaw cycles of 0, 5, 10, 20 times at the freezing temperature of $-10\,°C$.

MIP is a high-precision method to analyze the micropore structure of materials [42,43]. Non-immersion liquid cannot penetrate the porous area unless external pressure is introduced. The tests were conducted on tailings samples cured for 7 and 28-days, after the 20th freeze–thaw cycle at the temperatures of $-5\,°C$, $-10\,°C$, $-15\,°C$ and $20\,°C$.

The relationship between the pore diameter d (m) and the pressure P (MPa) can be described by the well-known Ishburn equation [44]:

$$d = \frac{-4\sigma \cos\theta}{P} \tag{1}$$

where the applied pressure P is inversely proportional to pore diameter d, σ is the surface tension (N/m) and θ is the contact angle between mercury and the pore wall. The contact angle is believed in the range of $120°$ to $140°$ [45].

3. Results and Analysis

3.1. Effects of Freeze–Thaw Cycles on Strength of Tailings Samples

Figure 2 shows the UCS variation under different freeze–thaw cycles (0, 3, 5, 7, 10, 12, 15 and 20) and freezing temperatures ($-5\,°C$, $-10\,°C$, $-15\,°C$). It can be clearly observed that the number of freeze–thaw cycles and freezing temperature has a significant effect on the strength variation of the tailing samples.

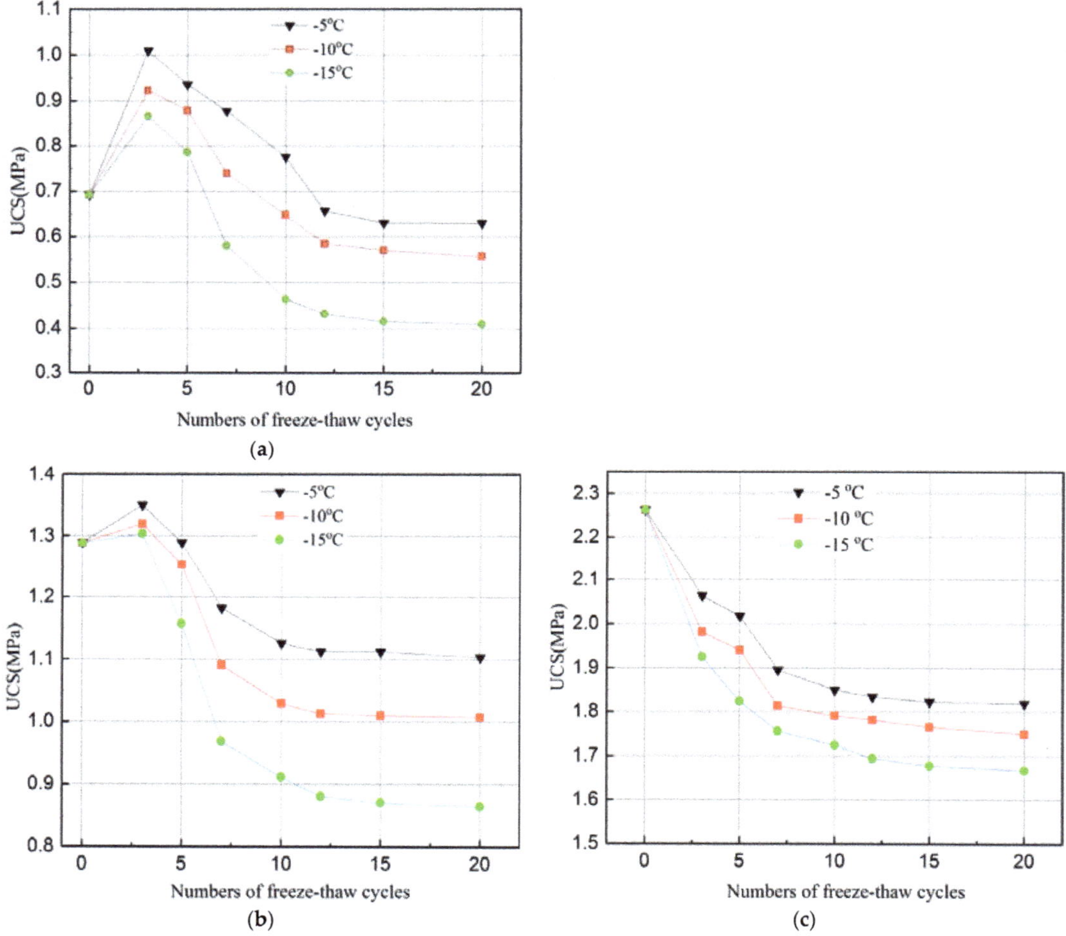

Figure 2. Effect of freeze–thaw cycles on UCS of tailings samples for different curing time and temperature. (**a**) After 3-day curing; (**b**) After 7-day curing; (**c**) After 28-day curing.

Figure 2a,b shows that the UCS of the 3- and 7-day samples increases in the first 3 cycles, and then decreases until the 12th cycle. It becomes flat during the 12–20 cycles. However, the UCS of the 28-day samples decreases in the first three cycles, and becomes stable during the 10–20 cycles, as can be seen in Figure 2c.

The hydration reaction of the samples with a short curing time (e.g., 3-days and 7-days) did not proceed completely since the free water converted to ice during freezing. Therefore, there were few hydration products but many internal pores in the samples. Although the volume expanded by about 9% due to water becoming ice, the pores in the samples provided sufficient space to accommodate the expansion. Thus, the swelling effect of the ice crystals on the samples is not significant. When the temperature rises, the ice melts, and the hydration reaction continues. The hydrated products, such as calcium silicate hydrate (C-S-H) gels and ettringite crystals, continue to fill the internal pores, reducing the porosity of the samples, and as such increasing the strength. This is why in the first three cycles of freeze and thaw, the UCS of the samples increased. At this stage, the effect of hydration reaction on the strength is greater than the damage of freeze–thaw cycles.

However, with the accumulation of damage by the freeze–thaw cycles, the UCS of the samples started to decrease when it reached a peak value. This can also be because more hydration reactions were completed. When the samples were cured for a longer time, e.g., 28-days, sufficient hydration products were produced which filled up the internal pores and overlapped on each other. Such samples had a relatively dense matrix with little room to accommodate the expansion of the ice crystal when the samples were frozen. Thus, any expansion of the ice crystal could destroy the dense matrix of the samples and hence reduced the UCS of the samples. This can be seen in Figure 2 where the curves flattened after 12 cycles of 3 day samples, after 10 cycles of 7- and 28-day samples, the UCS stabilized upon reaching a certain value. The main reason for such variations is that the internal structure of the samples was destroyed by the freeze–thaw cycles in the initial stage. Thus, as the number of freeze–thaw cycles increased, a new stable state was reached.

Figure 2 indicates that the temperature affects the UCS of the samples differently for different curing times. The UCS values of the samples frozen at $-10\ ^\circ$C were higher than that at $-15\ ^\circ$C but lower than that at $-5\ ^\circ$C for all curing time and freeze–thaw cycles. This phenomenon indicates that during the freeze–thaw cycles, the lower the temperature, the more destructive to the samples. Figure 2 also shows that the degree of temperature effect on samples was different with longer cured samples (i.e., 28-days). The reason for this is that when the samples were transferred from the curing box to the freezer, the freezing temperature was higher and the exchange rate between the samples was lower, and the the time required for the temperature to fall below the freezing point was higher. A hydration reaction is an exothermic process, a longer curing time means more hydration products during one freeze–thaw cycle and hence, less of an effect from temperature. Figure 2 further shows that, for the same curing time, the higher the freezing temperature, the greater the value of the UCS under the same freeze–thaw cycles. This is because when the temperature is below the freezing point, some of the crystal water is not converted to ice completely [46]. Being closer to the freezing point, means there is more free water in the hydration reaction which offsets the damage conducted by the freeze–thaw cycle to some extent. However, since the hydration reaction of the sample in 28-days almost consumed the raw materials, the influence of freezing temperature on the hydration reaction was reduced.

Figure 2 suggests that the UCS reduced less for higher freezing temperature, e.g., $-5\ ^\circ$C, than that for lower ones, this can be explained as follows. In general, there are two kinds of pore structures, which are closed and connected pores. When the freezing temperature is low, the water in the closed pores with rigid constraints of pore walls freezes rapidly, and the ice crystal pressure increases. When the ice crystal pressure exceeds the strength of the pore wall, the freeze–thaw damage occurs in the samples. For connected pores, the capillary force in the smaller pores will reduce the freezing point of water so that the water in the non-capillary pores freezes first. The small pores connected to it will be isolated after freezing since the water between the pores has no time to seep. As a result, the closed pore is formed with the same damage mechanism as described above. When the freezing temperature is high, the water freezing process is a quasi-static process. For closed pores, the ice crystal pressure increases simultaneously with the elastic deformation of the pore wall. The ice crystal pressure is converted into elastic deformation energy and stored in the pore wall. Therefore, the freeze–thaw damage generated in the closed pores is small. For the connected pores, an unfrozen water film is formed between the water and the pore wall during the freezing process. While freezing, the water in the pores penetrates through the unfrozen water film to other pores, which reduces the ice crystal pressure. Therefore, the freeze–thaw damage generated in the connected pores is small.

Compared with the UCS of the unfrozen samples in Figure 2, the UCS of samples with 3-, 7- and 28-day curing reduced by about 9–40%, 14–32%, and 19–26%, respectively, under the same testing conditions. This is because a short curing time results in fewer internal hydration products in the samples. In general, an increase in curing time can produce more hydration products in the samples with less water in the pores. However, more water in the pores of the samples promotes the destructive effect of ice crystals during the freezing and thawing. The damage by freeze–thaw cycles is irreversible and will eventually damage the samples with shorter curing time than those with longer curing time.

3.2. Crystalline Phases and Amount of Hydration Products of NCM

In the past, the research on freeze–thaw cycles mostly focused on rock and soil. These materials have no internal reactions, such as hydration reaction, during the freeze–thaw cycles, and their properties are relatively stable. However, for tailings samples, the hydration product is an important factor affecting the mechanical properties after freeze–thaw cycles. Therefore, it is important to analyze the change of the type and amount of hydration products in the samples before and after freeze–thaw cycles. The XRD test was used to identify the types of hydration products that were cured for 7 and 28-days after experiencing 20 freeze–thaw cycles at $-10\ ^{\circ}C$, with results shown in Figures 3 and 4.

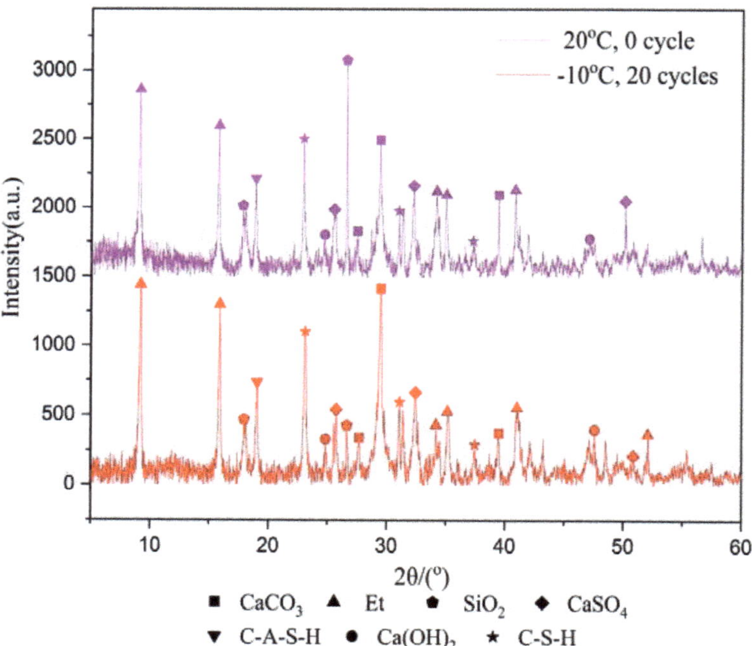

Figure 3. XRD image of 7-day samples.

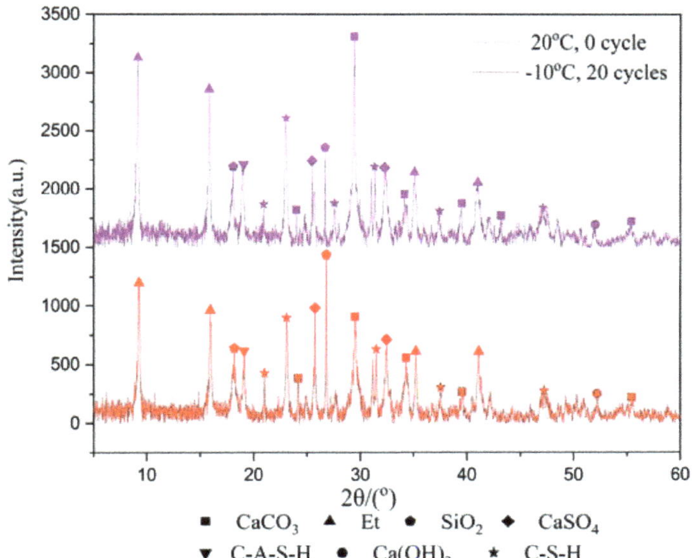

Figure 4. XRD image of 28-day samples.

Figures 3 and 4 show the products of calcium carbonate, ettringite, calcium hydroxide, hydrated calcium sulfoaluminate, calcium silicate hydrate, calcium sulfate and silica. Though the diffraction intensities of the substances are not the same at the same angle, the species phase does not change in general. The reason is that the hydration reaction still occurs during the freeze–thaw cycles, but it does not generate other substances due to the intermittent progress of the hydration reaction.

The TG analysis test was also used to determine the amount of hydration products of the samples cured for 7 and 28-days with 0 and 20 freeze–thaw cycles at the freezing temperature of $-5\ °C$, $-10\ °C$ and $-15\ °C$. The test data are plotted in Figures 5 and 6. The solid line in the figures indicates the weight loss of the cementitious material with the increase of temperature, and the corresponding dotted line indicates its first differential, that is, the rate of mass loss, which corresponds to different types of materials at different heating temperatures. The value of the heating temperature valley is different for the amount of substance.

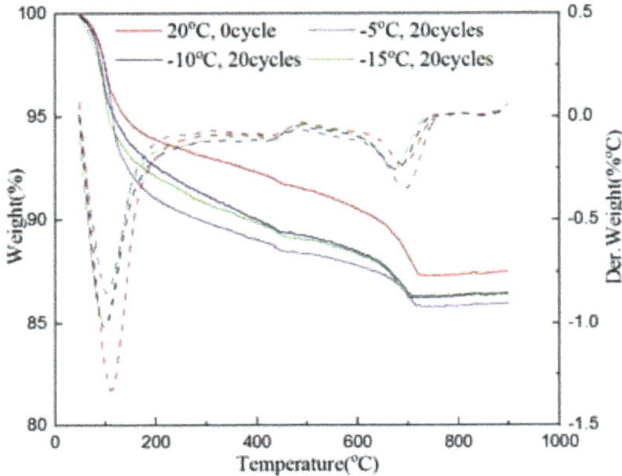

Figure 5. Hydration product of samples at different freezing temperatures after 7-day curing.

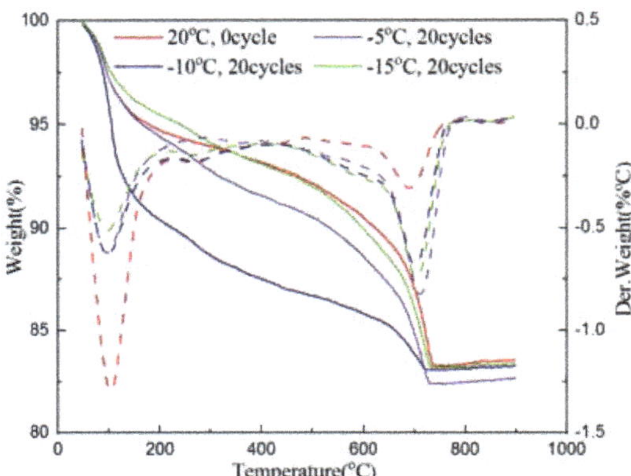

Figure 6. Hydration product of samples at different freezing temperatures after 28-day curing.

As can be seen from Figures 5 and 6, the troughs of the dotted lines in the temperature range of 50 °C to 800 °C represent the reduction of combined water, C-S-H gel, ettringite, calcium hydroxide and calcium carbonate. It can be seen from Figure 5, when the 7-day samples are subjected to unfrozen and 20 cycles of freeze–thaw at different freezing temperatures between 50 °C and 105 °C, the mass loss of the sample without freeze–thaw (i.e., 0 cycle) is large, while the samples experiencing 20 freeze–thaw cycles are relatively small. This shows that unfrozen samples contain more water, while the samples that experienced freeze–thaw cycles have less water due to the progress of the hydration reaction. It can also be seen that the sample with the freezing temperature of −5 °C has the most mass loss, indicating that the hydration reaction is more adequate at this temperature than that at the freezing temperature of −10 °C and −15 °C. The mass loss between 450 °C and 500 °C indicates the presence of calcium hydroxide. Though it is not clearly indicated, the curve of the unfrozen sample is relatively high. This proves the existence of hydration reactions during the freeze–thaw process again. The mass loss between 650–750 °C is due to the decomposition of calcium carbonate. The sample with 20 freeze–thaw cycles at the freezing temperature of −5 °C has the most mass loss, indicating that the hydration products produced after freezing and thawing at this temperature are the greatest. The reason is that the hydration reaction was still continuing during the freeze–thaw cycles, and the higher the freezing temperature is, the greater the degree of hydration reaction is.

Figure 6 shows that, at temperatures between 50 °C and 105 °C, the mass loss of the unfrozen sample is still the largest. However, compared with the 7-days sample, the difference in quality loss is not obvious. The reason is that the 7-day samples are rich in raw materials for hydration reaction, and the hydration reaction continues during the freezing and thawing process; while the raw material for hydration reaction in the 28-day samples has basically been consumed, and there is no difference before and after freeze–thaw cycles. Even between 450 °C and 500 °C, the decomposition of calcium hydroxide in the sample has not been seen. It can also be seen that the mass loss caused by the decomposition of calcium carbonate in the unfrozen samples between 650–750 °C is the smallest; while at the freezing temperature of −5 °C, the 20th freeze–thawed sample has the largest mass loss, but the difference is very small. In addition, the difference in the quality of the 28-day samples from the final remaining material is not as great as that of 7-day samples.

3.3. Surface Morphology Destroyed by Freeze–Thaw Cycles of the Samples

From SEM images presented in Figures 7–9, ettringite crystal, C-S-H gels, capillary pores and freeze–thaw cycles damaged pores can be studied. Only the samples frozen at −10 °C are presented in this section.

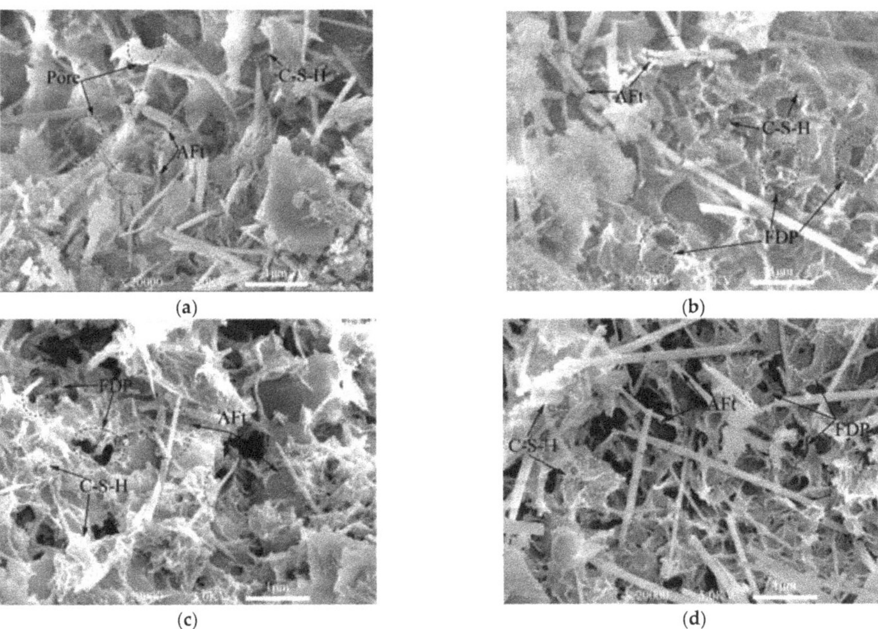

Figure 7. SEM images of 3-day samples under different freeze–thaw cycles. (**a**) Unfrozen; (**b**) After 5 freeze–thaw cycles; (**c**) After 10 freeze–thaw cycles; (**d**) After 20 freeze–thaw cycles.

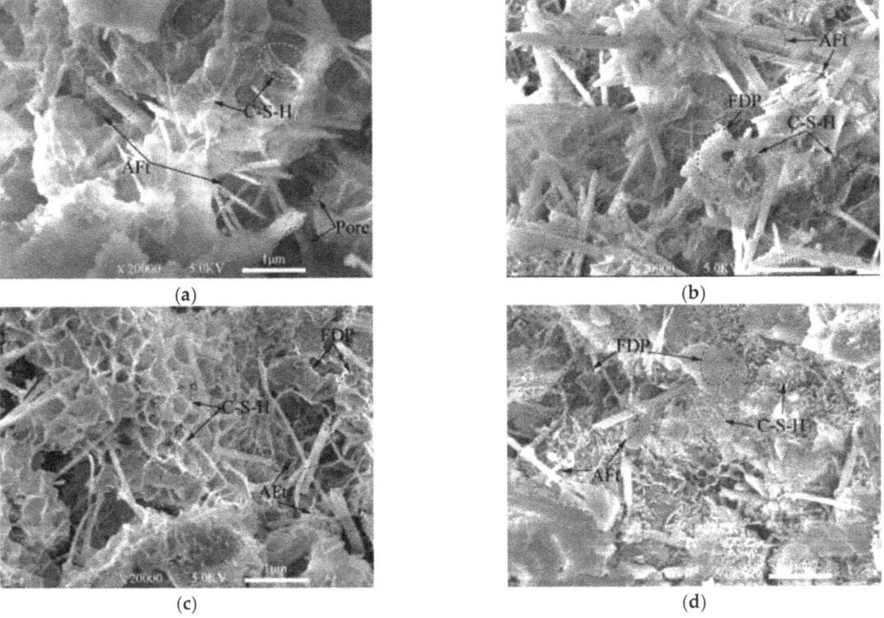

Figure 8. SEM images of 7-day samples under different freeze–thaw cycles. (**a**) Unfrozen; (**b**) After 5 freeze–thaw cycles; (**c**) After 10 freeze–thaw cycles; (**d**) After 20 freeze–thaw cycles.

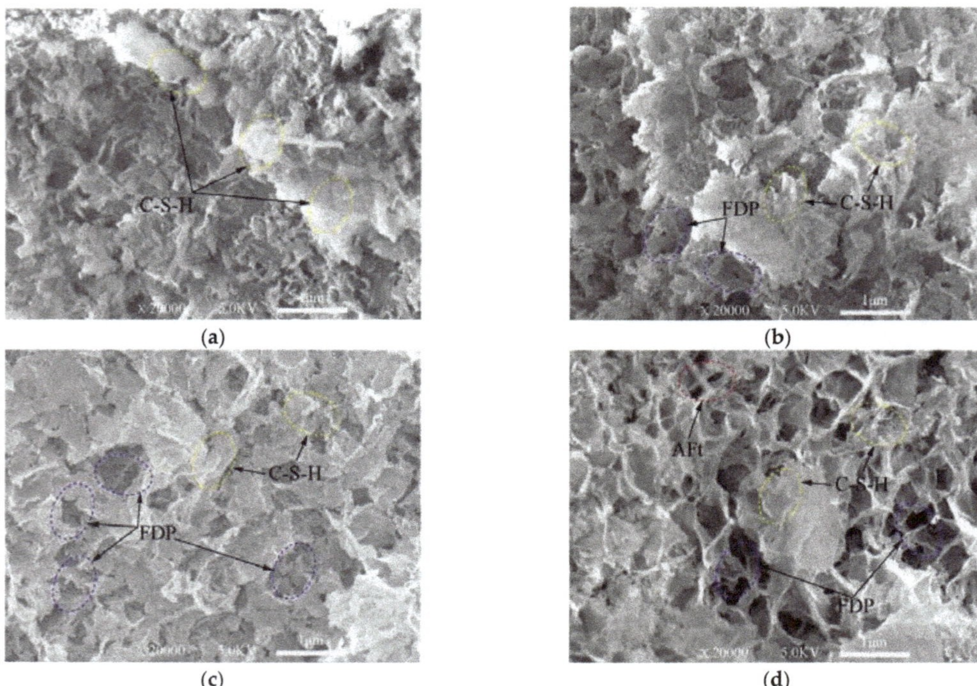

Figure 9. SEM images of 28-day samples under different freeze–thaw cycles. (**a**) Unfrozen; (**b**) After 5 freeze–thaw cycles; (**c**) After 10 freeze–thaw cycles; (**d**) After 20 freeze–thaw cycles.

Note: Aft (Ettringite crystal); C-S-H (calcium silicate hydrate); FDP: freeze–thaw cycles damaged pores.

The XRD test (Figures 3 and 4) shows the hydration products of the cementitious material elements, such as Ca, Al, O, Si, S, etc. According to the knowledge of mineralogy crystal morphology analysis [47], ettringite is needle-like and calcium silicate hydrate is gel-like. These morphological features are very obvious in the SEM test images. These topographic features are very obvious in the SEM test images of Figures 7–9. Figures 7a, 8a and 9a show SEM images of the 3-, 7- and 28-day samples, respectively. It can be seen that as the curing time increases, more and more C-S-H gels are produced and ettringites are surrounded by C-S-H gels. Obviously, only C-S-H gels can be seen on the surface for the 28-day samples. The dense internal structure of the hydration products makes the UCS values of the 28-day samples higher than those 3- and 7-day samples.

Figure 7 shows the SEM images of the samples cured for three days and then subjected to 0, 5, 10, 20 freeze–thaw cycles. Plenty of needle-like ettringites are observed in the matrix of the cement-based material in Figure 7a. Broken short columnar ettringite crystals due to the freeze–thaw damage can be observed in the pores in Figure 7b. In Figure 7c,d, needle-like crystals and C-S-H gels can be seen around the pores. However, it is obvious that there are a lot of pores on the surface of the gels, which are caused by the expansion of water converting to ice during the freeze–thaw cycles. Voids in the sample in Figure 7d increase compared with those in Figure 7b,c, indicating that the deterioration extent of the samples increases as the number of freeze–thaw cycles increases. It can also be seen from Figure 7b–d that, as the freeze–thaw cycles increased, proper and clear C–S–H gels in the samples still exist, even though they are damaged by the freeze–thaw cycles.

Figures 8 and 9 show the SEM images of the samples subjected to 0, 5, 10, 20 freeze–thaw cycles after cured for 7-days and 28-days. These images demonstrate that the hardened samples are perhaps more able to withstand freeze–thaw cycles. Highly saturated fresh samples may be severely damaged by a few cycles. The saturated water in the pores can freeze during cooling process. This difference will become clearer when comparing Figure 7 with Figure 9. There are more pores damaged by freeze–thaw cycles in the 3-day samples.

Figures 7d, 8d and 9d show that the defects are mostly micro-voids after experiencing 20 freeze–thaw cycles. However, in most cases, the degradation by freeze–thaw cycles is characterized by the gradual formation of microcracks in the samples. In the freezing process, water is redistributed throughout the mix by moving to the colder areas. If freezing is rapid, water has little chance to move towards the colder areas, thus creating a nearly uniform distribution of ice crystals. However, these crystals can still damage the immature cement-based material and weaken the bonds between the cement-based materials.

From the analysis above, the samples with short curing time continues to hydrate in the freeze–thaw cycles. As the number of freeze–thaw cycles increase, the amount of hydration products increases. There are many pores inside the samples, which slows down the damage caused by the expansion of the ice crystal. Thus, the influence of the hydration reaction on the samples at this stage is dominant. However, since the amount of final hydration product is only related to the cementitious material, slurry concentration and lime-sand ratio (although affected by temperature), the degree of hydration reaction is gradually smaller but maintained as the freeze–thaw cycle progresses. At a certain curing time, the effect of the freeze–thaw cycles on the samples plays a major role, which is manifested microscopically as the degree of deterioration increases with the number of freeze–thaw cycles. Macroscopically, the UCS of the samples is lower when the degree of microscopical damage is high.

3.4. Pore Size Distribution of the Samples

The MIP tests were carried out to study the effect of the freeze–thaw cycles on the pore size distribution of the tailings. In this study, the samples with 20 freeze–thaw cycles after cured for 7- and 28-days are chosen. The normalized volume from the MIP tests versus the pore diameter of the samples is plotted in Figures 10 and 11. The normalized volume curves show the variations of the mercury volume with the different pore sizes. From Equation (1), the intrusion pressure is inversely related to the pore diameter. In the intrusion process, the total mercury intrusion volume increases with the increase of intrusion pressure.

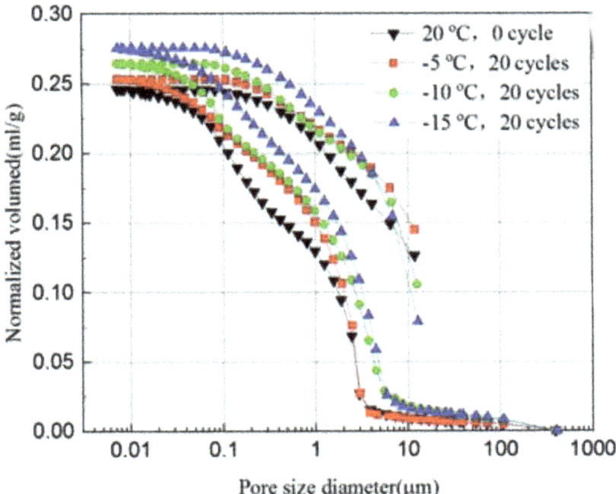

Figure 10. Intrusion and extrusion curves of 7-day samples at different cooling temperature.

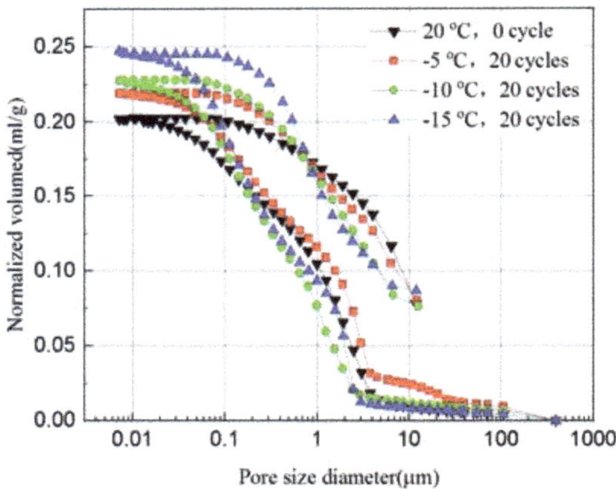

Figure 11. Intrusion and extrusion curves of 28-day samples at different cooling temperature.

It can be seen from Figures 10 and 11 that during the initial increase of external pressure, whether the samples were cured for 7-days or 28-days, the curves show similar trends. The intrusive volume of mercury increases slowly with the decrease of pore size, and the mercury mainly fills the gaps with larger diameters at this stage. For 7-day samples, the mercury mainly fills pores larger than 8.6 μm, but 2.8 μm for 28-day samples. It appears that tailing particles are redistributed to adapt to external pressure change during this process. The cumulative intrusive volume of mercury in the 7-day samples increases rapidly when the pore size is less than 2.5 μm. The cumulative intrusive volume of mercury in the 28-day samples increases rapidly when the pore size is less than 1.9 μm. This phenomenon indicates that there is a bottleneck effect during the invasion phase, and the external pressure forces the tailing particles to redistribute. After the mercury intrusion volume passes the bottleneck period, it will increase to a large value under a small pressure. This bottleneck period is the pressure required for the mercury liquid to destroy the pore wall. It can be seen that the starting point of the rapid increase in the volume of invaded mercury at different curing ages is different, and the starting points of the samples that were unfrozen and experienced 20 freeze–thaw cycles in the same curing time are similar. This shows that the curing time has a great impact on the development of pores, and the damage of freeze–thaw cycles to the pores may have a great impact at the micropore stage. However, even when the external pressure reaches the maximum value set by the instrument, it is still difficult for mercury to enter the smallest pores and closed pores. Thus, the mercury intrusion curve eventually tends to be moderate. The exit curve of mercury cannot return to the starting point because some mercury remains in the narrow pores, resulting in the exit of mercury less than the intrusion.

Comparing Figure 11 with Figure 10, the volume of mercury intrusion is reduced by 17.5%, 13.8%, 13.6%, 11.1% for samples after 20 freeze–thaw cycles at −5 °C, −10 °C, and −15 °C, respectively. The main reason for the decrease of mercury intrusion volume is that, as the curing time increases, the hydration products gradually accumulate and fill in the pores between the solid particles to make the internal structure more compact, which is supported from the UCS tests and the SEM tests. The hydration reaction of the 28-day samples has been completed, but the 7-day samples still have hydration reactions in the process of freeze–thaw cycles. The higher the freezing temperature are, the shorter the time needed to rise above zero, leading to longer hydration reaction time, and more hydration products in one freeze–thaw cycle. Likewise, the higher the freezing temperature is, the higher the reduction is in mercury intrusion volume. In Figure 10, for 7-day samples, the pore volume after 20 freeze–thaw cycles at freezing temperatures of −5 °C, −10 °C, and −15 °C is 3.1%, 7.6%, and 12.3% higher than that of unfrozen samples, respectively. In Figure 11, for 28-day samples, the pore volume after 20 freeze–thaw cycles at freezing temperatures of $-5\ ^{\circ}\text{C}$, $-10\ ^{\circ}\text{C}$, and $-15\ ^{\circ}\text{C}$ is 8.1%, 12.5%, and 21.1% higher than that of unfrozen samples, respectively. This observation is consistent with the UCS results of Figure 2c and it indicates that the porosity has a significant influence on the UCS of samples. It proves that the freeze–thaw cycles assist the pore development of the samples. The lower the temperature is, the more pore development there is. The increase in the pore volume of the 28-day samples is larger than that of the 7-day samples. This is because the hydration reaction of the 28-day samples is completed, while for the 7-day samples hydration continues, which slows down the development of pore volume to some extent.

Figures 12 and 13 present the log-differential mercury volume curves of samples at different cooling temperatures after 20 freeze–thaw cycles for both 7-day and 28-day curing times.

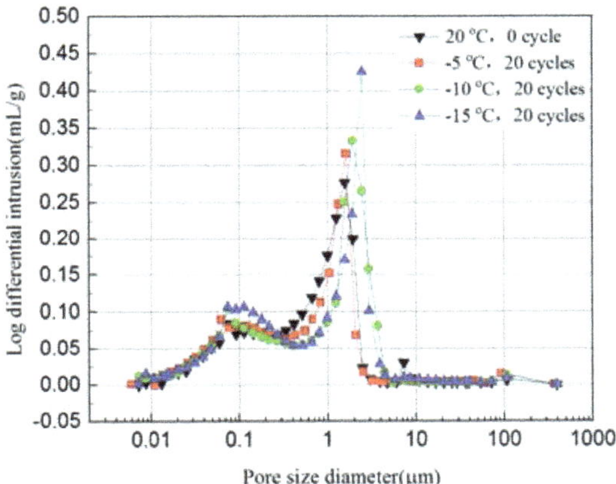

Figure 12. Log-differential pore volume curves of 7-day samples at different cooling temperature.

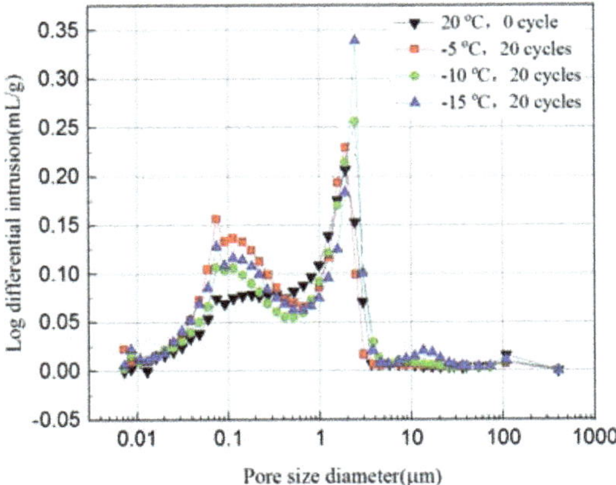

Figure 13. Log-differential pore volume curves of 28-day samples at different cooling temperature.

The variation of the log-differential mercury volume is the pore size distribution of the samples, where the abscissa values of the peak points are called the most probable pore sizes. It shows that the pores with the same sizes appear most within the samples. For these samples, the log-differential pore volume curves are mainly in the pore diameter ranging from 0.8 μm to 3.8 μm. In Figure 12, the most probable pore diameters of the 7-day samples frozen at $-5\ ^{\circ}\text{C}$, $-10\ ^{\circ}\text{C}$, $-15\ ^{\circ}\text{C}$ are 1.93 μm, 2.47 μm and 2.48 μm, respectively, after 20 freeze–thaw cycles, and increase by 0.7%, 21.9%, 29.1%, respectively, compared to the unfrozen samples. Figure 13 shows the most probable pore diameters of the 28-day samples frozen at $-5\ ^\circ C$, $-10\ ^\circ C$, $-15^\circ C$ are 1.62 μm, 1.91 μm and 2.47 μm, respectively, after 20 freeze–thaw cycles, increasing by 3.3%, 28.8% and 39.2%, respectively, compared to the unfrozen samples. The results show that whether the samples are cured for 7-days or 28-days, after freezing and thawing cycles, the most probable pore size is increasing, and the lower the freezing temperature is, the more probable the pore size increases, which indicates that the freezing temperature has a greater effect on samples. It also suggests that the freezing temperature has a significant effect on the micropores of the samples, which in turn affects the UCS of the samples.

4. Conclusions

In this paper, a comprehensive experiment was developed to investigate the impact of freeze–thaw cycles on the mechanical strength of the tailings reinforced with cement-based materials to mitigate environmental hazards and achieve sustainable production. Based on the test results and their analysis, the following conclusions can be reached:

(1) Freeze–thaw has a positive effect on UCS of tailings samples in the first three cycles for short curing times of 3- and 7-days but has a negative effect on the UCS for a normal curing time of 28-days under all freeze–thaw cycles. The frozen temperature has slight effect on UCS reduction for short curing time but has little effect for normal curing time.

(2) The larger the number of freeze–thaw cycles are, the more damage there is to the surface morphology and the matrix of the tailings, and the more severe the surface morphology damage is, the lower the UCS of the samples is.

(3) The freeze–thaw cycles have no effect on phases of the hydration products. The higher the freezing temperature is, the greater the amount of hydration products. Fuller hydration would result in a higher UCS of the samples.

(4) The mercury intrusion pressure is inversely related to the pore diameter of the samples. The lower the freezing temperature is the more mercury ingresses, and the most probable pore sizes increase after the freeze–thaw cycles, which in turn reduces the UCS of the samples.

From the results, we know freeze–thaw cycles have a significant impact on cemented tailings. In the application of real TSF embankment, when the ambient temperature drops to a certain value, we can consider taking thermal insulation measures for the TSF or increasing the amount of cementitious material to promote its hydration reaction. The freeze–thaw cycle creates cracks in the cemented tailings, and the TSF embankment is in danger of failure, we can reinforce the TSF embankment to make it stable and durable.

Temperature may not be the only factor for the change in strength. Next, we are going to study the amount of cementitious material and confining pressure on the stability of the cemented tailings in a low temperature environment, to find out the correlation among them; to determine under what conditions the amount of cementitious material should be increased, and under which circumstances the tailings dam should be strengthened, or if both should be carried out at the same time.

Author Contributions: P.D. and Y.H. conceived and designed the experiments; P.D., X.Z. and D.H. performed the experiments; S.C. and D.H. collected and analyzed the data; Y.H. contributed materials/analysis tools; P.D. and Y.H. wrote the paper. C.L. Writing-review & editing. Conceptualization: P.D. and Y.H.; Data curation, D.H.; Formal analysis, X.Z.; Funding acquisition, Y.H.; Investigation, X.Z. and S.C.; Methodology, P.D. and Y.H.; Resources, Y.H.; Software, D.H.; Supervision, Y.H.; Validation, D.H.; Visualization, S.C.; Writing—original draft, P.D.; Writing—review & editing, P.D. and C.L. All authors have read and agreed to the published version of the manuscript.

Funding: This work is supported by the National Natural Science Foundation of China (no.51674263) and the Fundamental Research Funds for the Central Universities (2011YZ02).

Data Availability Statement: The data presented in this study are available on request from the corresponding author.

Conflicts of Interest: The authors declare no conflict of interest.

References

1. David, J.W. Lessons from Tailings Dam Failures—Where to Go from Here? *Minerals* **2021**, *11*, 853. [CrossRef]
2. Liu, R.; Huang, F.; Du, R.; Zhao, C.; Li, Y.; Yu, H. Recycling and utilisation of industrial solid waste: An explorative study on gold deposit tailings of ductile shear zone type in China. *Waste Manag. Res.* **2015**, *33*, 570–577. [CrossRef]
3. Zhang, J.; Liu, J. The Statistics and Causes of Dam Break and Leakage in Chinese Tailings Pond. *Chin. Molybdenum Ind.* **2019**, *4*, 10–14. [CrossRef]
4. Yu, M.; Kong, X.; Huang, J.; Liu, J.; Li, J. Status of disposal of tailings as a solid waste and suggestions in China. *Ind. Miner. Processing* **2022**, *1*, 34–38, (In Chinese with English Abstract). [CrossRef]
5. Bolaños-Benítez, V.; Van Hullebusch, E.; Lens, P.L.; Quantin, C.; Van De Vossenberg, J.; Subramanian, S.; Sivry, Y. (Bio)leaching Behavior of Chromite Tailings. *Minerals* **2018**, *8*, 261. [CrossRef]
6. Anning, C.; Wang, J.; Chen, P.; Batmunkh, I.; Lyu, X. Determination and detoxification of cyanide in gold mine tailings: A review. *Waste Manag. Res.* **2019**, *37*, 1117–1126. [CrossRef] [PubMed]
7. Park, I.; Tabelin, C.B.; Jeon, S.; Li, X.; Seno, K.; Ito, M.; Hiroyoshi, N. A review of recent strategies for acid mine drainage prevention and mine tailings recycling. *Chemosphere* **2019**, *5*, 588–600. [CrossRef] [PubMed]
8. Sun, Q.; Wei, X.; Li, T.; Zhang, L. Strengthening Behavior of Cemented Paste Backfill Using Alkali-Activated Slag Binders and Bottom Ash Based on the Response Surface Method. *Materials* **2020**, *13*, 855. [CrossRef] [PubMed]
9. Qin, X.B.; Liu, L.; Wang, P.; Wang, M.; Xin, J. Microscopic Parameter Extraction and Corresponding Strength Prediction of Cemented Paste Backfill at Different Curing Times. *Adv. Civ. Eng.* **2018**, *4*, 2837571. [CrossRef]
10. Nasir, O.; Fall, M. Shear behavior of cemented pastefill–rock interfaces. *Eng. Geol.* **2008**, *101*, 146–153. [CrossRef]
11. Zhang, Q.L.; Chen, Q.S.; Wang, X.M. Cemented Backfilling Technology of Paste-Like Based on Aeolian Sand and Tailings. *Minerals* **2016**, *6*, 132. [CrossRef]
12. Opiso, E.M.; Tabelin, C.B.; Maestre, C.V.; Aseniero, J.P.J.; Park, I.; Villacorte-Tabelin, M. Synthesis and characterization of coal fly ash and palm oil fuel ash modified artisanal and small-scale gold mine (ASGM) tailings based geopolymer using sugar mill lime sludge as Ca-based activator. *Heliyon* **2021**, *7*, e06654. [CrossRef] [PubMed]
13. Hou, Y.B.; Tang, J.; Wei, S.X. Study on Tailings Consolidation Emissions Technology. *Chin. J. Met. Mine* **2011**, *40*, 59–62, (In Chinese with English Abstract).
14. Michael, P.; Arnel, B.; Aileen, O.; Bernardo-Arugay, I.; Resabal, V.J.; Villacorte-Tabelin, M.; Dalona, I.M.; Opiso, E.; Alloro, R.; Alonzo, D.; et al. Systems Approach toward a Greener Eco-efficient Mineral Extraction and Sustainable Land Use Management in the Philippines. *Chem. Eng. Trans.* **2021**, *88*, 1171–1176. [CrossRef]
15. Silwamba, M.; Ito, M.; Hiroyoshi, N.; Tabelin, C.B.; Fukushima, T.; Park, I.; Jeon, S.; Igarashi, T.; Sato, T.; Nyambe, I.; et al. Detoxification of lead-bearing zinc plant leach residues from Kabwe, Zambia by coupled extraction-cementation method. *J. Environ. Chem. Eng.* **2020**, *8*, 104197. [CrossRef]
16. Zhou, Z.; Ma, W.; Zhang, S.; Mu, Y.; Li, G. Effect of freeze-thaw cycles in mechanical behaviors of frozen loess. *Cold Reg. Sci. Technol.* **2018**, *146*, 9–18. [CrossRef]
17. Qiao, Y.F.; Sun, W.; Jiang, J. Damage process of concrete subjected to coupling fatigue load and freeze/thaw cycles. *Constr. Build. Mater.* **2015**, *93*, 806–811. [CrossRef]
18. Polat, R. The effect of antifreeze additives on fresh concrete subjected to freezing and thawing cycles. *Cold Reg. Sci. Technol.* **2016**, *127*, 10–17. [CrossRef]
19. Polat, R.; Demirboga, R.; Karakoc, M.B.; Türkmen, I. The influence of light weight aggregate on the physical-mechanical properties of concrete exposed to freeze–thaw cycles. *Cold Reg. Sci. Technol.* **2010**, *60*, 51–56. [CrossRef]
20. Cao, D.F.; Fu, L.Z.; Yang, Z.W. Experimental study on tensile properties of concrete after freeze-thaw cycles. *J. Build. Mater.* **2012**, *15*, 42–52. [CrossRef]
21. Qureshi, A.; Bussière, B.; Benzaazoua, M.; Lessard, F.; Boulanger-Martel, V. Geochemical Assessment of Desulphurized Tailings as Cover Material in Cold Climates. *Minerals* **2021**, *11*, 280. [CrossRef]
22. Hanjari, K.Z.; Utgenannt, P.; Lundgren, K. Experimental study of the material and bond properties of frost-damaged concrete. *Cem. Concr. Res.* **2011**, *41*, 244–254. [CrossRef]
23. Hanjari, K.; PerKettil, K.L. Modelling the structural behaviour of frost-damaged reinforced concrete structures. *Struct. Infrastruct. Eng.* **2013**, *9*, 416–431. [CrossRef]
24. Tang, Y.Q.; Li, J.; Wan, P.; Yang, P. Resilient and plastic strain behavior of freezing-thawing mucky clay under subway loading in Shanghai. *Nat. Hazards* **2014**, *72*, 771–787. [CrossRef]
25. Xie, S.B.; Qu, J.J.; Xu, X.T.; Pang, Y. Interactions between freeze–thaw actions, wind erosion desertification, and permafrost in the Qinghai–Tibet Plateau. *Nat. Hazards* **2017**, *85*, 829–850. [CrossRef]
26. Liu, L.; Ye, G.; Schlangen, E.; Chen, H.; Qian, Z.; Sun, W.; van Breugel, K. Modeling of the internal damage of saturated cement paste due to ice crystallization pressure during freezing. *Cem. Concr. Comp.* **2011**, *33*, 562–571. [CrossRef]
27. Liu, L.; Shen, D.; Chen, H.; Sun, W.; Qian, Z.; Zhao, H.; Jiang, J. Analysis of damage development in cement paste due to ice nucleation at different temperatures. *Cem. Concr. Comp.* **2014**, *53*, 1–9. [CrossRef]

28. Liu, L.; Wu, S.; Chen, H.; Haitao, Z. Numerical investigation of the effects of freezing on micro-internal damage and macro-mechanical properties of cement pastes. *Cold Reg. Sci. Technol.* **2014**, *106*, 141–152. [CrossRef]
29. Tang, S.; Yao, Y.; Andrade, C.; Li, Z. Recent durability studies on concrete structure. *Cem. Concr. Res.* **2015**, *78*, 143–154. [CrossRef]
30. Wang, Z.; Zeng, Q.; Wang, L.; Li, K.; Xu, S.; Yao, Y. Characterizing frost damages of concrete with flatbed scanner. *Constr. Build. Mater.* **2016**, *102*, 872–883. [CrossRef]
31. Ma, Q.; Ma, D.; Yao, Z. Influence of freeze-thaw cycles on dynamic compressive strength and energy distribution of soft rock specimen. *Cold Reg. Sci. Technol.* **2018**, *153*, 10–17. [CrossRef]
32. Liu, B.; Jiang, J.; Shen, S. Effects of curing methods of concrete after steam curing on mechanical strength and permeability. *Constr. Build. Mater.* **2020**, *256*, 119441. [CrossRef]
33. Shin, M.; Park, D.; Seo, Y. Response of subsea pipelines to anchor impacts considering pipe–soil–rock interactions. *Int. J. Impact Eng.* **2020**, *143*, 103590. [CrossRef]
34. Kalonji-Kabambi, A.; Bussière, B.; Demers, I. Hydrogeochemical Behavior of Reclaimed Highly Reactive Tailings, Part 2: Laboratory and Field Results of Covers Made with Mine Waste Materials. *Minerals* **2020**, *10*, 589. [CrossRef]
35. Pokharel, M.; Fall, M. Combined influence of sulphate and temperature on the saturated hydraulic conductivity of hardened cemented paste backfill. *Cem. Concr. Comp.* **2013**, *38*, 21–28. [CrossRef]
36. Fall, M.; Benzaazoua, M. Modeling the effect of sulphate on strength development of paste backfill and binder mixture optimization. *Cem. Concr. Res.* **2005**, *35*, 301–314. [CrossRef]
37. Hou, Y.; Ding, P.; Han, D.; Zhang, X.; Cao, S. Study on the Preparation and Hydration Properties of a New Cementitious Material for Tailings Discharge. *Processes* **2019**, *7*, 47. [CrossRef]
38. Li, W.; Fall, M. Sulphate effect on the early age strength and self-desiccation of cemented paste backfill. *Constr. Build. Mater.* **2016**, *106*, 296–304. [CrossRef]
39. Luo, T.; Zhang, C.; Sun, C.; Zheng, X.; Ji, Y.; Yuan, X. Experimental Investigation on the Freeze–Thaw Resistance of Steel Fibers Reinforced Rubber Concrete. *Materials* **2020**, *13*, 1260. [CrossRef]
40. Qiu, J.P.; Yang, L.; Sun, X.G.; Xing, J.; Li, S. Strength Characteristics and Failure Mechanism of Cemented Super-Fine Unclassified Tailings Backfill. *Minerals* **2017**, *7*, 58. [CrossRef]
41. You, Z.M.; Lai, Y.M.; Zhang, M.Y.; Liu, E. Quantitative analysis for the effect of microstructure on the mechanical strength of frozen silty clay with different contents of sodium sulfate. *Environ. Earth Sci.* **2017**, *4*, 143. [CrossRef]
42. Wu, S.Y.; Yang, J.; Yang, R.C.; Zhu, J.; Liu, S.; Wang, C. Investigation of microscopic air void structure of anti-freezing asphalt pavement with X-ray CT and MIP. *Constr. Build. Mater.* **2018**, *178*, 473–483. [CrossRef]
43. Lee, J.K.; Shang, J.Q. Micropore Structure of Cement-Stabilized Gold Mine Tailings. *Minerals* **2018**, *8*, 96. [CrossRef]
44. Zhang, Z.l; Cui, Z.D. Effects of freezing-thawing and cyclic loading on pore size distribution of silty clay by mercury intrusion porosimetry. *Cold Reg. Sci. Technol.* **2018**, *145*, 185–196. [CrossRef]
45. Cui, Z.D.; Tang, Y.Q. Microstructures of different soil layers caused by the high-rise building group in Shanghai. *Environ. Earth Sci.* **2011**, *63*, 109–119. [CrossRef]
46. Guo, B.; Zhou, Y.; Zhu, J.F.; Liu, W.; Wang, F.; Wang, L.; Jiang, L. An estimation method of soil freeze-thaw erosion in the Qinghai-Tibet Plateau. *Nat. Hazards.* **2015**, *78*, 1843–1857. [CrossRef]
47. Zhang, Y.; Zhang, S.; Ni, W.; Yan, Q.; Gao, W.; Li, Y. Immobilisation of high-arsenic-containing tailings by using metallurgical slag-cementing materials. *Chemosphere* **2019**, *223*, 117–123. [CrossRef] [PubMed]

 *sustainability*

Article

Characteristics of Overburden and Ground Failure in Mining of Shallow Buried Thick Coal Seams under Thick Aeolian Sand

Guangchun Liu [1,2]**, Youfeng Zou** [1]**, Wenzhi Zhang** [1,*] **and Junjie Chen** [1]

[1] School of Surveying and Land Information Engineering, Henan Polytechnic University, Jiaozuo 454000, China; liugch168@home.hpu.edu.cn (G.L.); zouyf@hpu.edu.cn (Y.Z.); chenjj@hpu.edu.cn (J.C.)
[2] School of Resources and Civil Engineering, Liaoning Institue of Science and Technoloy, Benxi 117004, China
* Correspondence: zhangwenzhi@hpu.edu.cn

Abstract: Mining can lead to overburden failure and ground damage, which are more severe in mining shallow buried thick coal seams (SBTCS) under thick aeolian sand (TAS). We attempted to discover characteristics of mining in this particular geological condition through theoretical derivation and numerical simulation, and field monitoring. Theoretical methods, combined with numerical simulation and field monitoring methods, reveal the essence of the development and distribution of surface cracks caused by mining SBTCS and depth to thickness ratio (DTR) to be 13.43, less than 15. The findings show that, when mining SBTCS, the overburden breaks down periodically, the initial collapse distance is greater than the collapse step, approximately 55 m on average, and the collapse step is approximately 45 m, on average, in the Daliuta Coal Mine. The collapsed blocks are stacked into goaf and form "masonry beams", and many cracks and pores are generated between the blocks. The weak stress of the aeolian sand layer causes the movement angle in the aeolian sand layer to be smaller than that in the bedrock, and leads to much sheer, tension and compression failure on the ground, and the main forms of cracks are compression uplift, tensile cracking, shear step.

Keywords: thick aeolian sand; shallow buried thick seam; overburden failure; ground damage; numerical simulation

Citation: Liu, G.; Zou, Y.; Zhang, W.; Chen, J. Characteristics of Overburden and Ground Failure in Mining of Shallow Buried Thick Coal Seams under Thick Aeolian Sand. *Sustainability* **2022**, *14*, 4028. https://doi.org/10.3390/su14074028

Academic Editors: Longjun Dong, Yanlin Zhao and Wenxue Chen

Received: 12 February 2022
Accepted: 23 March 2022
Published: 29 March 2022

Publisher's Note: MDPI stays neutral with regard to jurisdictional claims in published maps and institutional affiliations.

1. Introduction

After a coal seam is mined out from underground, the rock stratum above the goaf is affected by mining, resulting in collapses, cracks, and surface subsidence, water level decline, ecological environment deterioration, and so on [1–3]. When mining shallow buried thick coal seams (SBTCS), these phenomena and the damage are more serious and faster [4]. The main areas of coal exploitation in China have been transferred from east to west. Northwest China is enriched in coal with shallow buried, thick coal seams and simple overburden geological conditions [5]. The Daliuta Coalfield, in northern Shanxi Province, China [6], is located under a thick aeolian sand (TAS) layer of Quaternary sediment, mainly aeolian sand [7]. This is a typical coal seam occurrence condition with SBTCS. There are numerous problems associated with mining SBTCS, especially under TAS layers. After the coal seam is mined, a roof of coal collapse can cause grave damage to the overburden and ground covered by the aeolian sand [8–10]. Ground subsidence and cracks are another critical issue [11–13]. Many experts and scholars have conducted research on this problem of mining SBTCS and have achieved rich results. FAN De-yuan and others integrated physical and mechanical tests, theoretical analysis, similar material tests, numerical calculations, on-site industrial tests and other methods [14,15]. They systematically studied the deformation mechanism of the surrounding rock in mining under TAS and thick coal seam (TCS). They established a mechanical analysis model for the surrounding rock. The change in the regular pattern of displacement, stress, energy and other parameters were analyzed [16,17]. These studies did not consider thin, overburden and shallow burials. A numerical simulation method has be used to study the design of a retaining coal pillar width in mining SBTCS to ensure mining safety and improve coal yield [18–21]. Behrooz Ghabraie and others established a stress model and calculated and analyzed the stress related problems of mining TAS and TCS. According to the data of surface settlement monitoring and the geomechanical parameters of geological boreholes, the relevant parameters of the numerical model are corrected and adjusted. Numerical simulation and engineering analysis have established a mechanical theoretical basis for SBTCS [22–24]. Through theoretical derivation and numerical simulation analysis, Booth and others studied the surface deformation caused by overburden failure in mining SBTCS under TAS and established a prediction model for subsidence [25,26]. The existing prediction methods for mining subsidence cannot correctly explain deforming ground undulations. The research focuses on surface subsidence. Li Meng and others aimed at the complex conditions of ground undulations. Through similar material tests and numerical simulation analysis, a new prediction method for overburden movement was proposed based on key stratum theory and the rock Mohr–Coulomb failure criterion [15,27,28]. According to the occurrence characteristics of SBTCS, Park and others systematically studied the development and evolution regulations of overburden failure in shallow buried coal seams by using a similar 3D material physical test method. Ground subsidence of the SBTCS is mainly caused by aeolian sand layer settlement and bedrock breakage. The length and advancing distance of the workface directly affect the displacement of the surface and overburden [10].

However, most of the above research is mainly aimed at SBTCS with burial depths greater than 150 m, and depth (burial depth of coal seam) to thickness (thickness of coal seam) ratios (DTR) are more than 15 ($H/h > 15$). Some of them focus on TCS or TAC separately. Some scholars have used similar methods of test analysis, a small error can be amplified dozens or even hundreds of times because of the similarity ratio. Others have established corresponding mechanics based on the movement characteristics of the overburden. This method is based on theoretical calculation, and the essence of overburden failure can be obtained. Using the numerical simulation method based on theoretical calculations, the characteristics and regular pattern of overburden failure can be obtained [29]. In this study, the buried depth of SBTCS is approximately $H = 90$ m, and thickness of coal is $h = 6.7$ m (DTR = 13.43, less than 15) [30]. The thickness of aeolian sand is 38.5 m, accounting for approximately 42.6% of the buried depth, which is located in the Daliuta mine. According to the theory of the key stratum, the mechanical model of a key stratum collapse was established, and the initial failure distance and periodic failure distance of the key stratum were calculated. The model was verified by numerical simulation. This paper discusses the characteristics of overburden failure with mining SBTCS under TAS. It explains the ground damage characteristics and cracks development.

2. Materials and Methods

Academician Qian Minggao and others proposed the key stratum theory [31,32]. Due to the difference in thickness and properties of each rock layer in the overburden, its control and support of various overburdened rock strata are significantly different. The thickness and elastic modulus of the key stratum are large, and the strength is stronger than that of the other strata. The subsidence of key stratum is synchronous and coordinated in all or part of the upper overburden. Before the key stratum breaks, a slab or beam is formed to bear the weight of the upper rock. When it fails, it can continue to support the upper overburden with a "cantilever beam" as shown in Figure 1.

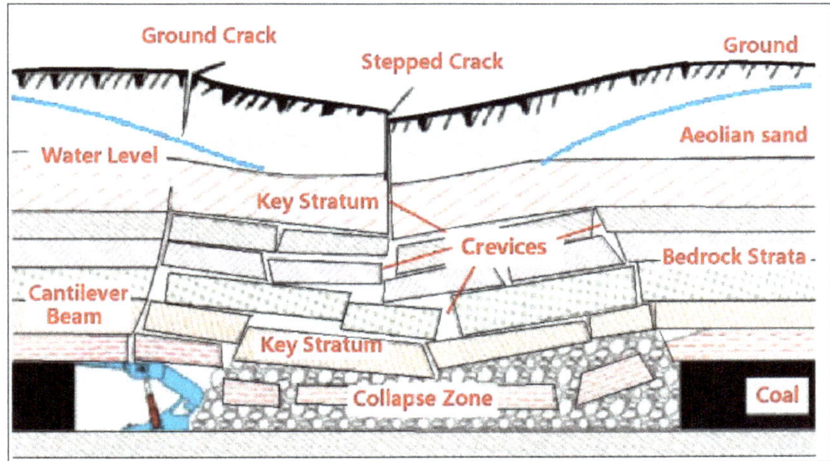

Figure 1. Schematic diagram of ground damage and of key stratum in overburden failure.

The overburden is simplified into a calculation model to analyze the failure principle of the key stratum. Assuming the bedrock layer of the overburden has n ($n > 2$) layers, the upper part of the bedrock is an aeolian sand layer with weak bearing capacity, and its load is q. According to the key stratum theory, the first key stratum controls and bears all the overburden and key stratum above, and it simultaneously coordinates deformation [33].

The following formula holds:

$$\frac{M_1}{E_1 I_1} = \frac{M_2}{E_2 I_2} = \frac{M_3}{E_3 I_3} = \cdots = \frac{M_i}{E_i I_i} \tag{1}$$

In the formula:
M_i is the bending moment of the i-th rock layer;
E_i is the elastic modulus of the i-th rock layer;
I_i is the inertia moment of the i-th rock layer.

According to the initial in situ stress conditions and the theory of key stratum, the first key stratum of overburden $1 \sim m$ ($m < n$) layers can be regarded as a composite rock beam structure, and its bending moments are:

$$M_a = M_1 + M_2 + M_3 + \cdots + M_m = \sum_{i=1}^{m} M_i \tag{2}$$

Similarly, the second key stratum of the overburden $m + 1 \sim n$ ($m < n$) layers can be regarded as a composite rock beam structure, and its bending moments are:

$$M_b = M_{m+1} + M_{m+2} + M_{m+3} + \cdots + M_n = \sum_{i=m+1}^{n} M_i \tag{3}$$

After transforming Formula (1), the relationship between the first key stratum and others is established, and then brought into Formula (2):

$$M_1 = \frac{E_1 I_1 M_a}{\sum_{i=1}^{m} E_i I_i} \tag{4}$$

Similarly, the second key stratum of the overburden $m + 1 \sim n$ layers can be expressed as:

$$M_{m+1} = \frac{E_{m+1} I_{m+1} M_b}{\sum_{i=m+1}^{n} E_i I_i} \tag{5}$$

When the mining distance is small, the stress of the first and second key stratum does not reached the destruction level, and the key stratum does not collapse. When the workface continues to advance and reaches the limit collapse distance, the first and second key stratum collapse simultaneously. With the workface advancing, the key stratum will collapse periodically. The key stratum block stacks into the collapse zone to form a "masonry beam". There are numerous large pores and crevices between the blocks.

Calculating the bearing capacity of the key stratum is usually carried out by using the recursive method. First, suppose that, when the first key stratum only controls itself, the load it bears is q_1; then, it can be calculated by Formula (6):

$$q_1 = \gamma_1 h_1 \tag{6}$$

In the formula, h_1 represents the gravity density and thickness of the rock layer.

Second, when the first key stratum can bear the first and second layers, the load to be borne is q_2, and then it can be calculated by the Formula (7):

$$q_2 = \frac{E_1 h_1^3 (\gamma_1 h_1 + \gamma_2 h_2)}{E_1 h_1^3 + E_2 h_2^3} \tag{7}$$

Next, recursively, the bearing capacity q_m of the m-th stratum can be obtained [34], which can be calculated by Formula (8):

$$q_m = \frac{E_m h_m^3 \left(\sum_{i=m}^{n} \gamma_i h_i + q\right)}{\sum_{i=m}^{n} E_i h_i^3} \tag{8}$$

In the formula:

h_m represents the thickness of the combined key stratum m;

E_m represents the elastic modulus of the combined key stratum m;

h_i is the i-th rock layer thickness.

According to the characteristics of the rock beam in the key layer theory, the coordinates are established as shown in Figure 2. To simplify the calculation, the broken distance of the key stratum is calculated using the fixed bracket mechanism model. The center of the rock beam and the left boundary of the rock beam are the origin, the positive direction of the X axis is horizontal to the right, and the positive direction of the Y axis is vertically upward. The length of the rock beam is L, the thickness is h, and the uniform load q of the loose layer is applied to the upper part of the rock beam.

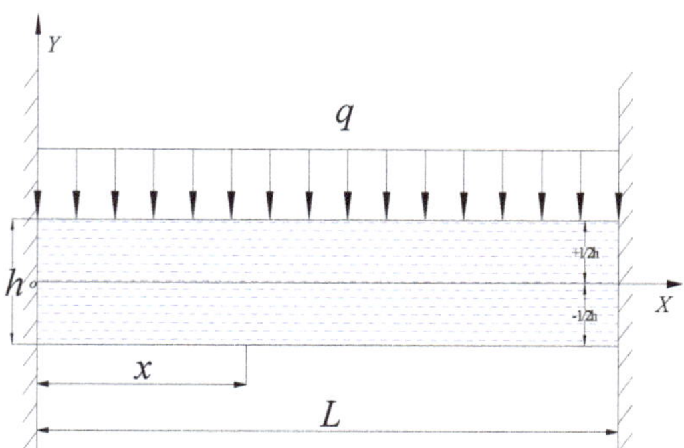

Figure 2. Schematic diagram of the key stratum fixed beam model.

The bending moment of any cross-section inside the rock beam of the key stratum can be expressed as [35]:

$$M_x = \frac{q}{12}(6Lx - 6x^2 - L^2) \tag{9}$$

In the formula:

M_x is the bending moment of any cross-section x in rock beam, Pa·m^2;

L is the span of the rock beam, m;

q is the rock beam load of the key stratum, Pa;

h is the thickness of the rock beam, m.

Therefore, the maximum bending moment at both ends of the beam is:

$$M_{\max} = \frac{qL^2}{12} \tag{10}$$

The central bending moment of the beam is:

$$M_{\max} = \frac{qL^2}{24} \tag{11}$$

According to mechanics, the failure of this rock beam is tensile failure. Assuming that the thickness of the rock beam is h, the normal stress σ of the beam upper and lower boundary y ($y = \pm\frac{1}{2}h$) is:

$$\sigma = \frac{12M_y}{h^3} = \pm\frac{q}{2h^2}\left(6Lx - 6x^2 - L^2\right) \tag{12}$$

The maximum stress at both ends of the beam is:

$$\sigma_{\max} = \frac{qL^2}{2h^2} \tag{13}$$

From Formula (13), the initial limit breaking distance of the i-th key stratum is:

$$L_i = h_i\sqrt{\frac{2\sigma_i}{q_i}} \tag{14}$$

In the formula:

L_i is the initial limit breaking distance of the i-th key stratum;

σ_i is the maximum tensile strength stress of the i-th key stratum;

h_i is the thickness of the i-th key stratum;

q_i is the load of the i-th key stratum above.

As the workface continues to advance, the overburden roof breaks and falls for the first time. The workface will break and collapse periodically. The period collapse step distance L_p is calculated by the Formula (15):

$$L_p = 2h_i\sqrt{\frac{\sigma_i}{3q_i}} \tag{15}$$

3. Results

3.1. Engineering Background

The study area selected is the Daliuta coal mine, which belongs to the mining area under the jurisdiction of Shendong Coal Industry Group. It is located on the Ulan Mulun River in the town of Daliuta, Shenmu County, Yulin City, Shanxi Province, China, which borders Ordos City in the Inner Mongolia Autonomous Region. This mine is one of China's most supersized and high yield mines, with an annual output of more than 21 million tons. Figure 3 shows its geographic location. A northwesterly wind prevails in spring and winter in the study area, with a speed of approximately 3.3 m/s. The annual average precipitation is less than 400 mm, and the annual variation is considerable. Sixty percent of the annual precipitation is mainly concentrated in July–September. This study area is located in a temperate semiarid desert plateau continental monsoon climate [36]. In the study area, there is a TAS layer covering the ground.

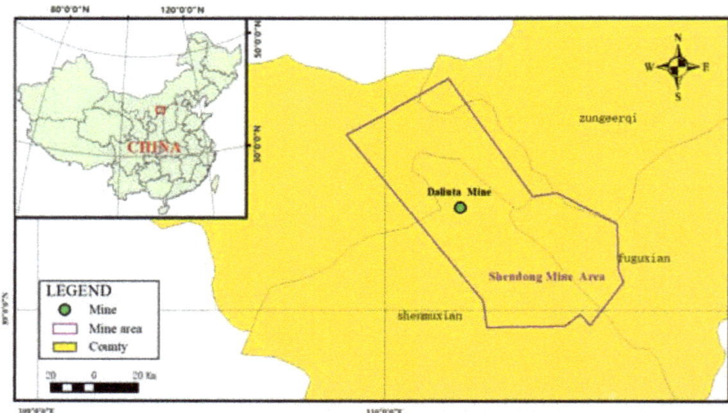

Figure 3. Geographical location of the study area.

During the advancement of the workface, there were many dense cracks on the surface of the Daliuta mining area, showing discontinuity and some regularity. Various types of surface cracks appear on the surface, which appear in the range of 10–100 m on the surface in front of the working face. These cracks can be subdivided into tensile cracks, compression cracks and stepped cracks. Different directions of stress can be analyzed through the crack shape on the surface. According to the data, the surface damage characteristics of the Daliuta mining area are as follows: tensile cracks appear on the surface in front of the workface, with a width of 0.3–0.5 m. The step crack drop is approximately 0.4–0.9 m, the maximum depth is approximately 3.5 m, and the crack spacing is only 2–3 m. The distribution density and width of compression cracks are slightly smaller than those of tensile cracks. Compression cracks with a length of 30–50 m appear at intervals of 2–3 m, and the uplift height is approximately 0.6 m.

Workface 1203 in the Daliuta mine was taken as the research object. The ground elevation of this workface is from 1083.4 m to 1229.3 m, and the coal seam floor elevation is from 981.3 m to 1032.8 m. The burial depth of the coal seam is 82–247 m, with an inclination angle of 1–3°, and the average burial depth of the coal seam is 90.3 m. The mining average thickness of the coal seam is designed to be 6.7 m. The surface above the workface is mostly covered by Quaternary loose sediment and mainly includes aeolian sand, with a thickness of 38.5 m. A bedrock layer with a thickness of approximately 51.8 m is mainly composed of siltstone mixed with sandy mudstone, coarse grained sandstone and fine grained sandstone. The coal floor is a fine grained sandstone layer with thickness of more than 3.0 m. The mining size of the workface is designed to be 301.3 m wide, and the advancing distance is 4268.8 m long. It adopts a fully mechanized and fully caving mining method with a large mining height and long working face. This high intensity mining method will lead to severe ground damage, such as subsidence, cracks, and even collapse pits.

3.2. Numerical Simulation Model Establishment

To more clearly study the principle of surface damage and rock stratum failure, a numerical simulation method was used to simulate and analyze the study area. The actual cause of surface cracks and damage is the surface subsidence caused by the mining process. However, the surface subsidence is caused by the settlement and damage of bedrock. The failure essence of the overburden lies in the failure of key stratum. The mechanical nature of the key stratum failure is the direct expression and essence of the surface and overburden failure. The numerical simulation method is a mechanical analysis method that can reflect the essential regular pattern of things, which is well recognized and practiced. Therefore, we use the numerical simulation method to analyze the overburden and surface damage caused by the mining of SBTCS and find the essential regular pattern of overburden damage.

FLAC3D software was selected to calculate the numerical model based on the geological conditions of the 1203 workface in the Daliuta Coal Mine. Tecplot software was used for drawing processing in the later stage. With the Mohr–Coulomb mechanics constitutive model, a numerical calculation 3D model was established for mining SBTCS under TAS [37,38]. The models in the X, Y and Z directions are 1000 m, 500 m and 100 m, respectively, with a total of 5,000,000 zones and six groups, as shown in Figure 4. Ensuring the mining size is as close to the real size as possible and reducing the influence of edge effects, 100 m coal pillars are reserved around the workface without mining. Due to a large number of nodes in the model, a high performance calculation is needed. The data solution takes a long time to reach the final mechanical equilibrium state.

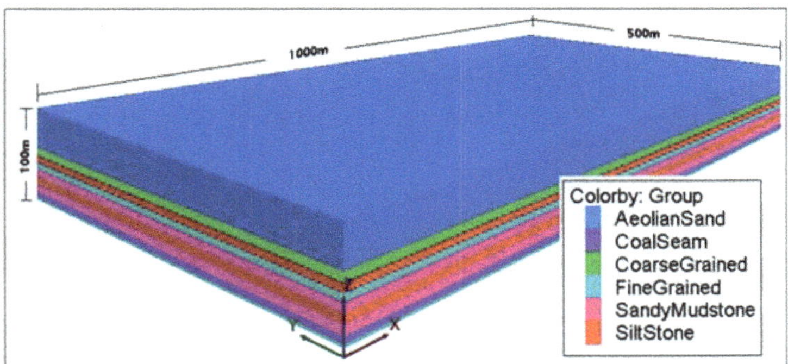

Figure 4. Schematic diagram of the 3D model of the numerical simulation based on FLAC3D.

To simulate the original stress of the underground rock stratum well, the model's boundary conditions must be limited to ensure that the model more truly reflects the actual geological conditions. The periphery of the model was fixed in the X and Y directions, to ensure that the boundary of the simulated stratum was not deformed in the horizontal direction. The bottom was completely fixed in the vertical Z direction, and the top was a free boundary, as shown in Figure 5. Mechanical parameters are the most important in the process of numerical simulation [39]. For the simulation of initial in situ stress, in FLAC, the initial stress is usually solved by setting the mechanical parameters and then solving to equilibrium. The original parameters are assigned according to the mechanical properties of the strata of the borehole. Under the action of gravity only, the initial in situ stress is solved to equilibrium (unbalance rate 10^{-6}). The selection and adjustment of the constitutive model and parameters is an essential link in numerical simulation. The selection of the constitutive model is related to the mechanical properties of the rock stratum, and the number and value of the corresponding parameters need to be selected for different constitutive models. The main mechanical parameters in the model came from the geological conditions of the Daliuta 1203 workface for calculating the stress in the numerical simulation. The physical and mechanical parameters of the rock stratum in the numerical simulation are shown in Table 1.

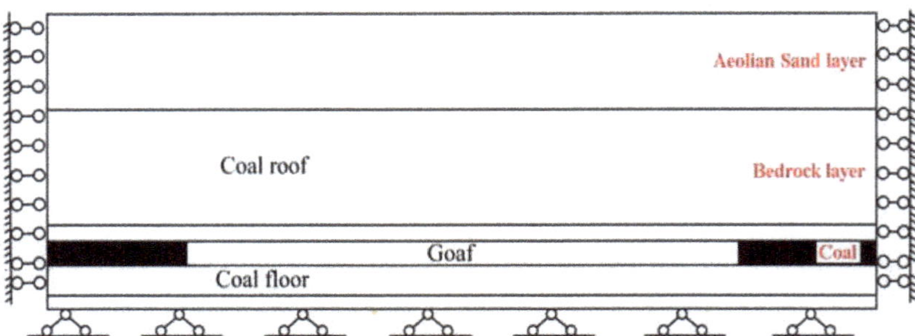

Figure 5. Boundary conditions of the 3D model for numerical simulation.

Table 1. Rock mechanics parameters in numerical simulation.

Lithology	h/m	E/Gpa	ρ	γ/KN/m3	R_m/Mpa	c/Mpa	φ/°
Aeolian Sand	38.5	12	0.3	15.8	0.0048	0.0016	27
Coarse Grained Sandstone	9.0	35	0.25	21	2.3	3.2	54
Siltstone	8.0	35	0.25	25	2.1	2.74	39
Fine Grained Sandstone	7.2	32	0.25	24.6	2.8	5.72	49
Sandy Mudstone	9.7	23	0.28	24.6	3.5	2.8	29
Siltstone	8.1	35	0.25	25	2.3	2.74	39
Sandy Mudstone	9.8	23	0.28	24.6	3.5	2.8	29
Coal	6.7	15	0.35	22.5	0.6	0.8	38
Fine Grained Sandstone	3.0	32	0.28	24.6	2.8	5.72	49

According to the key stratum theory, determining the key stratum position requires multiple and cyclic calculations. In the calculation process, if the collapse distance of the $i-1$ th stratum is smaller than the i th stratum, the load of the emphi -1 th stratum should be added to the i th stratum. Recalculate until the fracture distance of the $i+1$ th stratum is smaller than that of the i th stratum. The i th stratum is the key stratum. Using Formulas (14) and (15), the initial breaking and periodic fracture distances are calculated. According to the rock mechanics parameters in Table 1, we can calculate that the first key stratum is sandy mudstone, the immediate roof of coal. The initial failure distance is 52.85 m, and the periodic fracture distance is 43.12 m. Similarly, we can obtain that the second key stratum is sandy mudstone, which is between fine grained sandstone strata and siltstone strata. In the second key stratum, the initial failure distance is 56.37 m, and the periodic fracture distance is 46.03 m.

3.3. Analysis of Ground Damage and Field Measured Results

The simulation of the workface propulsion distance at 100 m, 200 m, 300 m, 400 m, 500 m, 600 m, 700 m and 800 m is extracted. According to different advancing distances, the settlement and stress changes in the surface and overburden are recorded as cloud maps. The cloud maps of the numerical simulation, such as the form of and change in ground damage, the settlement and stress distribution of the overburden, and the shape and field of plastic failure, are analyzed to obtain the failure characteristics of the overburden and surface movement and deformation.

For calibration and validation of the model, to obtain surface subsidence change data of SBTCS, the 46 monitoring points, where the distance between monitoring points is set to 20–25 m, with 21.7 m on average, were set up on the surface above the ground and monitored the settlement. Surface monitoring was carried out using traditional leveling methods. Two vertical observation lines were arranged along the ground to obtain the surface subsidence under the condition of different working face advancing distances, and then the subsidence caused by mining of SBTCS was analyzed. When the workface advanced by approximately 600 m, we selected 21 points for trend analysis between measured points and simulated calculated points. The maximum settlement of the surface was −3.898 m. The ground settlement, which was calculated by FLAC 3D, reached the maximum value of −3.78544 m (the data measured in the field and calculated by FLAC3D are shown in Figure 6). The resulting error ratio is about 2.89%. Through a comparison, it was found that the surface settlement value of the numerical simulation was very close to the field measured. The surface subsidence trend line is also in good agreement. This shows that the numerical simulation method is an effective method to calculate and predict the surface settlement caused by coal mining by establishing a reasonable 3D model, selecting an appropriate constitutive model and adjusting the corresponding mechanical parameters.

Figure 6. Comparison of calculated and measured values of ground subsidence.

Through numerical simulation, the shape of the surface subsidence basin is very close to the actual field. The subsidence at the edge of the basin is small, and there is more subsidence in the middle, which has a flat bottom. According to Figure 7, the settlement of the bedrock is large, and the settlement of the aeolian sand layer is small. There are large pores and cracks in the aeolian sand layer and the bedrock layer. The main reason is that aeolian sand has weak mechanical properties and good dispersion, and easily forms pores. The mechanical properties of the overburden are hard, brittle and easily broken, and collapsed. The settlement cloud map of the bedrock above the goaf shows that the settlement in the center is large and that along the edge it is smaller. The simulation calculation of the overburden settlement shows that the overburden failure caused by mining SBTCS under TAS has distinctive characteristics. The bedrock in the overburden generally has large settlements and has a large scale overall movement phenomenon, and the maximum subsidence is more than 6.5 m. The aeolian sand layer presents uniform settlement, the maximum subsidence is 3.5–4.0 m in the central basin. The movement and deformation in the central subsidence basin show uniform settlement and little change. With increasing mining distance, the surface subsidence gradually increases. When the working advance is approximately 300 m, it is close to full-mining under these geological conditions. Then, with the advance of mining distance, the ground subsidence basin appears to a flat bottom [40].

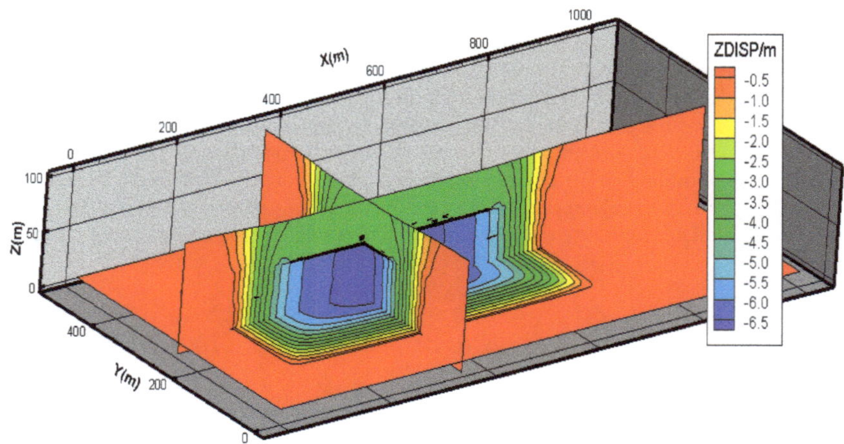

Figure 7. Cloud map of Deformation characteristics of overburden settlement.

As shown in Figure 8, many cracks are generated on the ground surface before and on both sides of the workface affected by mining. The surface has distinct damage. The cracks in front of the workface probably range from 20 m to 240 m, and the distance of crack is approximately 10–80 m.

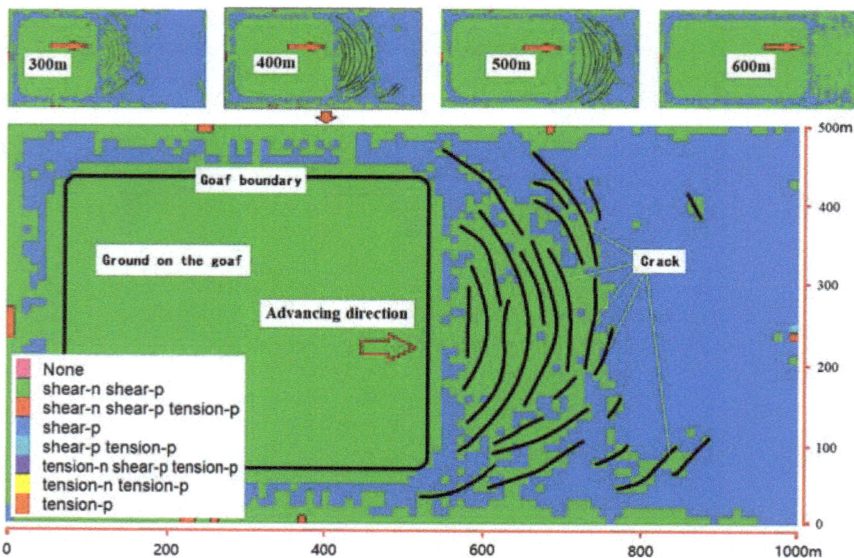

Figure 8. Deformation characteristics of ground.

3.4. Overburden Failure Analysis

With the continuous advancement of the working face, the roof of the coal seam collapses, and the collapsed rock fragments are superimposed on the goaf to form a masonry beam structure, the surface begins to subside, and the overburden under the aeolian sand begins to crack or even fracture. Finally, the original stress balance of the overburden, which is affected by mining, is broken.

A stress concentration phenomenon occurs in the overburden above the edge of the goaf that are affected by mining, as shown in Figure 9. After the collapse of the first key stratum, the second key stratum plays a major supporting role. The maximum vertical stress is more than 9.0 MPa and the direction is downward, and its support point moves upward, and such disasters caused by high stress often cause great damage to mining engineering activities [3,41,42]. An interlayer with weak stress appears between the aeolian sand layer and the bedrock layer. The reason for this is maybe that, after the mechanical state of bedrock layers and aeolian sand layer changes caused by mining, cracks and separation layers are generated between them.

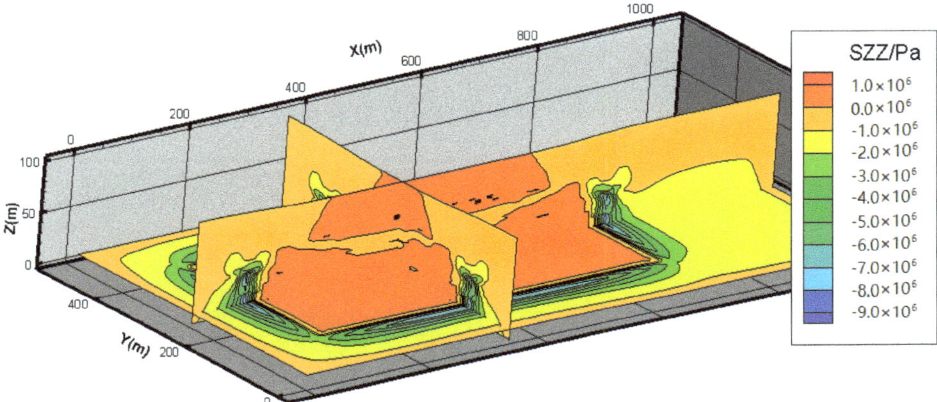

Figure 9. Cloud map of the overburden stresses.

As shown in Figure 10, shear failure and tensile failure show a layered structure in the horizontal direction and alternately appear in the vertical direction. The upper aeolian sand is mainly sheared, and the lower of aeolian sand is mainly tensile failure. The movement angle is different in the aeolian sand layer and the bedrock layer. The movement angle of the aeolian sand layer is smaller, while the movement angle of the bedrock layer is larger [43]. The movement angle in the TAS layer is smaller than that in the bedrock, which enlarges the range of the subsidence basin and expands the scope of ground damage.

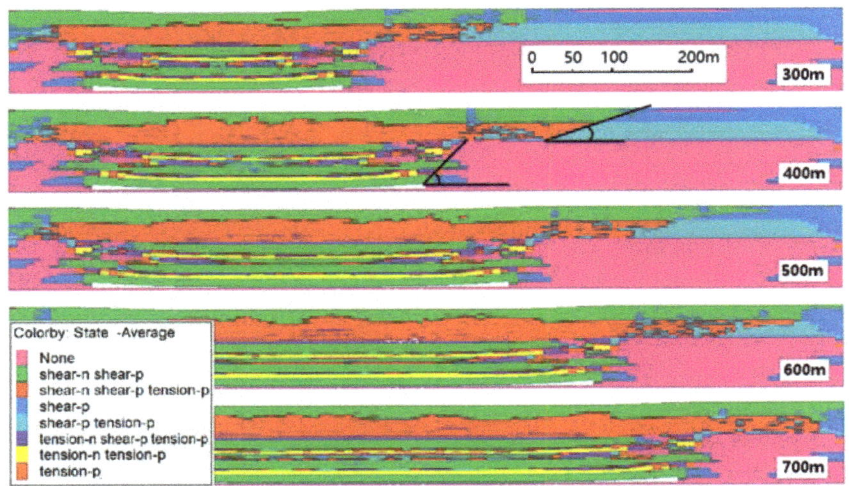

Figure 10. Plastic deformation of overburden at different schedule.

Overburden collapse can be divided into two stages, initial collapse and periodic collapse, as show in Figure 11. The initial fracture distance is usually larger than the periodic fracture distance. After the initial collapse, with the advance of the working face, the overburden forms a cantilever structure. When the maximum bearing capacity is reached, periodic collapse occurs. Then, the overburden directly collapses to the goaf, the height of the collapse zone is approximately 24 m. The collapse block is 40–50 m long, and it is consistent with the theoretical calculation. the collapse block is gradually crushed and compacted. The block provides a bearing capacity to support the upper broken key stratum.

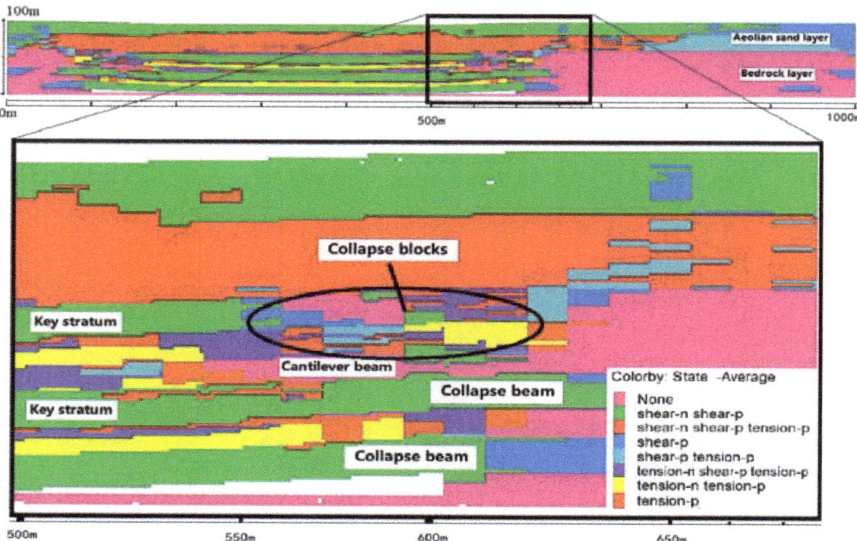

Figure 11. Local enlarged schematic diagram of key stratum collapse and cantilever beam formation process.

According to the cloud map of the numerical simulation, under this geological condition. The key stratum exists in the form of "cantilever beam" before collapse and "masonry beam" after collapse. There are a large number of fractures and fissures between blocks. With the continuous advancement of the working face, cracks in the overburden stratum continue to develop, and run through the whole bedrock layer, and there is no "bending zone" in the overburden. The cracks directly connect to the surface and form conduction with the surface.

4. Discussion

In this paper, we show that the overburden failure caused by the mining SBTCS, when the ratio of depth to thickness is less than $15(H/h = 13.43)$, can be divided into two stages: initial collapse and periodic collapse. According to the theoretical calculation, the initial collapse distance is greater than the collapse step, approximately 55 m on average, and the collapse step is approximately 45 m on average in the Daliuta Coal Mine.

The results of theoretical calculations are within the range of numerical simulation results. The movement angle in the loose thick aeolian sand layer is small, which expands the range of surface damage. A weak stress interlayer is generated between the overburden and the aeolian sand layer. This may be the cause of much sheer, tension and compression failure on the ground, as shown in Figure 12. This study provides a mechanical analysis basis for the surface damage caused by mining shallow and thick coal seams, and provides a basis for surface environmental treatment and ecological restoration.

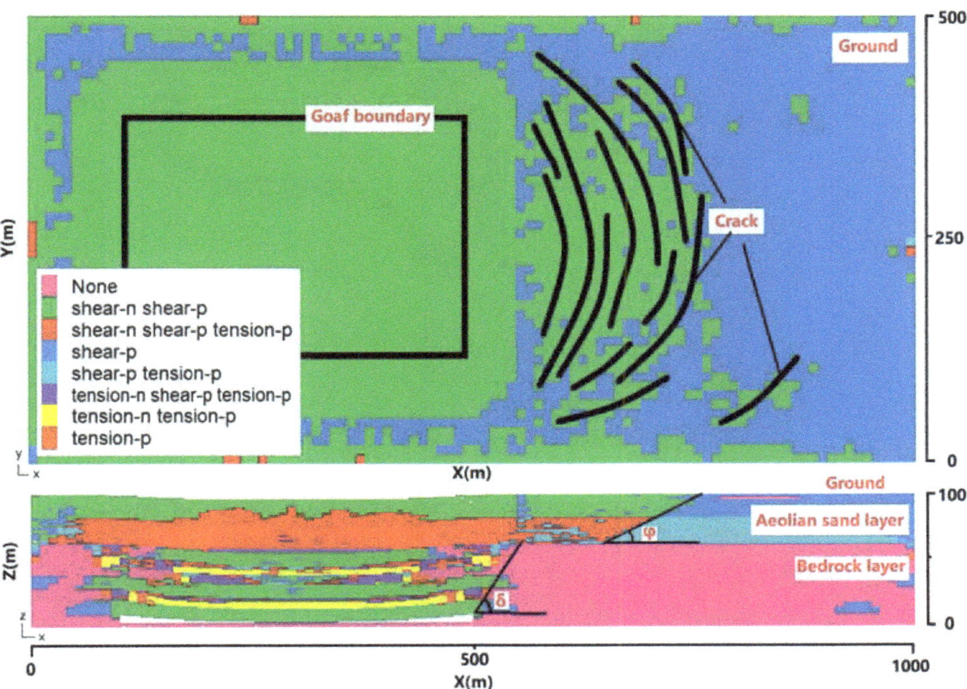

Figure 12. Comparison between overburden failure and surface failure.

The study of coal mine surface damage can improve our understanding of mine management and land reclamation, and be conducive to ecological mine restoration. The research on mining SBTCS has strong regional characteristics. In different geological conditions and rock stratum conditions, resulting overburden and surface damage also have different status [43–45]. Hou [34] and Huang [33] adopt the same calculation method as our study. Huang studies the failure mechanism of mining shallow coal seams and uses the result to control the ground pressure. Hou adopts the key stratum theory, puts forward the method of composite rock beams, and calculates the relevant parameters of rock beams. Although the stratigraphic conditions are quite different, the methods and ideas of theoretical analysis are correct. The study area and environment in Zou [4] is very close to our study. They obtained the conclusion that overburden failure presents periodic collapse and initial collapse, which is consistent with this study. However, we further analyzed the stress distribution characteristics of overburden and the variation regular pattern of movement angle in loose layers and bedrock layers.

The study area is located in an arid desert area in China. The influence of groundwater on overburden is not considered in this study. The study of the interaction between overburden and groundwater is a complex and long term process. Through the monitoring of surface water level, overburden pore water pressure and overburden stress change, the interaction relationship between overburden and water can be obtained, the coupling model of overburden failure and groundwater can be established, and the interaction relationship between overlying rock and groundwater can be obtained. The occurrence conditions and geological structure of coal seams in the study area are simple, belonging to horizontal coal seams with few faults. For complex geological conditions, such as coal seam dip angle and faults, further research will be carried out later in the research process. When there are dip angles and faults in overburden and coal seams, a variety of numerical simulation models with different dip angles and different fault forms can be constructed by simulation to study the characteristics of overburden and surface failure under complex conditions from the perspective of numerical simulation. According to mining subsidence theory and mechanical theory [44–46], the surface damage range can be calculated more accurately rather than qualitative analysis. When mining SBTCS, the DTR and aeolian sand thickness to bedrock thickness ratio (ASBR) should be considered. It reflects the relationship between coal seam thickness and buried depth, and the structural composition of bedrock. The change in DTR and ASTR have a key impact on the overburden and surface damage caused by mining SBTCS, which is also a subject worthy of in depth study in the future.

5. Conclusions

Based on the theory of key stratum and elastic mechanics, a mechanics model of key stratum failure was established in mining SBTCS under TAS. We calculated the breaking distance and periodic step distance of overlying strata. The key stratum were simplified as a fixed supporting beam structure to simplify the calculation formula. By calculation, we found that the initial failure distance is 52.81 m and the periodic breaking step distance is 43.12 m in the first key stratum. The initial failure distance is 56.37 m, and the periodic breaking step distance is 46.03 m in the second key stratum in Daliuta Coal Mine.

(1) According to the geological engineering conditions of the 1203 workface in the Daliuta Coal Mine, a numerical model was established by FLAC3D software. Compared with the field monitoring data of the surface, the numerical simulation results are very close to the measured values (the difference between the measured maximum settlement value and the simulated maximum settlement value is 0.11256 m (error accounts for 2.89%)). By correctly selecting the constitutive model and adjusting the appropriate parameters, the numerical simulation method is a method based on theoretical calculation, which can calculate and predict surface dependency caused by coal mining.

(2) The results show many cracks on the surface in front of the workface, which probably range from 240 m to 20 m. The distance between cracks is approximately from 10–80 m. The reason for the cracks on the ground is that the settlement of bedrock entirely exceeds that of the aeolian sand layer, and the shear stress of the aeolian sand layer by mining SBTCS.

(3) Although our research has achieved some results, we will further study and discuss the coupling effect of overburden and water, complex geological conditions and the influence of DTR on SBTCS.

Author Contributions: Conceptualization, Y.Z. and J.C.; methodology, Y.Z.; software, G.L.; validation, W.Z., G.L. and Y.Z.; formal analysis, G.L.; investigation, W.Z.; resources, W.Z.; data curation, G.L.; writing—original draft preparation, G.L.; writing—review and editing, G.L. and W.Z.; visualization, G.L.; supervision, W.Z.; project administration, W.Z.; funding acquisition, Y.Z. All authors have read and agreed to the published version of the manuscript.

Funding: This research was funded by the National Natural Science Foundation of China, grant number U1810203 and U21A20108, and This research was funded by Henan province science and technology tackling key project, China, grant number 212102310404.

Institutional Review Board Statement: Not applicable.

Informed Consent Statement: Not applicable.

Data Availability Statement: The data used to support the findings of this study are included within the article.

Conflicts of Interest: The authors declare no conflict of interest.

References

1. Dong, L.; Hu, Q.; Tong, X.; Liu, Y. Velocity-Free MS/AE Source Location Method for Three-Dimensional Hole-Containing Structures. *Engineering* **2020**, *6*, 827–834. [CrossRef]
2. Helm, P.R.; Davie, C.T.; Glendinning, S. Numerical modelling of shallow abandoned mine working subsidence affecting transport infrastructure. *Eng. Geol.* **2013**, *154*, 6–19. [CrossRef]

3. Dong, L.; Tong, X.; Ma, J. Quantitative Investigation of Tomographic Effects in Abnormal Regions of Complex Structures. *Engineering* **2020**, *7*, 1011–1022. [CrossRef]
4. Zou, Y. *Evolution Mechanism and Regulation of Surface Ecological Environment under High Intensity Mining*; Science Press: Beijing, China, 2019.
5. Yang, X.; Wen, G.; Dai, L.; Sun, H.; Li, X. Ground Subsidence and Surface Cracks Evolution from Shallow-Buried Close-Distance Multi-seam Mining: A Case Study in Bulianta Coal Mine. *Rock Mech. Rock Eng.* **2019**, *52*, 2835–2852. [CrossRef]
6. Ma, C.; Cheng, X.; Yang, Y.; Zhang, X.; Guo, Z.; Zou, Y. Investigation on Mining Subsidence Based on Multi-Temporal InSAR and Time-Series Analysis of the Small Baseline Subset—Case Study of Working Faces 22201-1/2 in Buertai Mine, Shendong Coalfield, China. *Remote Sens.* **2016**, *8*, 951. [CrossRef]
7. Li, Z.; Pang, Y.; Bao, Y.; Ma, Z. Research on Surface Failure Law of Working Faces in Large Mining Height and Shallow Buried Coal Seam. *Adv. Civ. Eng.* **2020**, *2020*, 8844249. [CrossRef]
8. Alejano, L.R.; Taboada, J.; García-Bastante, F.; Rodriguez, P. Multi-approach back-analysis of a roof bed collapse in a mining room excavated in stratified rock. *Int. J. Rock Mech. Min. Sci.* **2008**, *45*, 899–913. [CrossRef]
9. Kapusta, K.; Stańczyk, K.; Wiatowski, M.; Chećko, J. Environmental aspects of a field-scale underground coal gasification trial in a shallow coal seam at the Experimental Mine Barbara in Poland. *Fuel* **2013**, *113*, 196–208. [CrossRef]
10. Park, D.; Michalowski, R.L. Three-dimensional roof collapse analysis in circular tunnels in rock. *Int. J. Rock Mech. Min. Sci.* **2020**, *128*, 104275. [CrossRef]
11. Bell, F.G.; Stacey, T.R.; Genske, D.D. Mining subsidence and its effect on the environment: Some differing example. *Environ. Geol.* **2000**, *40*, 135–152. [CrossRef]
12. Kontogianni, V.; Pytharouli, S.; Stiros, S. Ground subsidence, Quaternary faults and vulnerability of utilities and transportation networks in Thessaly, Greece. *Environ. Geol.* **2007**, *52*, 1085–1095. [CrossRef]
13. Howladar, M.F.; Hasan, K. A study on the development of subsidence due to the extraction of 1203 slice with its associated factors around Barapukuria underground coal mining industrial area, Dinajpur, Bangladesh. *Environ. Earth Sci.* **2014**, *72*, 3699–3713. [CrossRef]
14. Fan, D.Y.; Liu, X.S.; Tan, Y.L.; Song, S.L.; Ning, J.G.; Ma, Q. Numerical simulation research on response characteristics of surrounding rock for deep super-large section chamber under dynamic and static combined loading condition. *J. Cent. South Univ.* **2020**, *27*, 3544–3566. [CrossRef]
15. Li, M.; Zhang, J.X.; Huang, Y.L.; Gao, R. Measurement and numerical analysis of influence of key stratum breakage on mine pressure in top-coal caving face with super great mining height. *J. Cent. South Univ.* **2017**, *24*, 1881–1888. [CrossRef]
16. Sepehri, M.; Apel, D.B.; Hall, R.A. Prediction of mining-induced surface subsidence and ground movements at a Canadian diamond mine using an elastoplastic finite element model. *Int. J. Rock Mech. Min. Sci.* **2017**, *100*, 73–82. [CrossRef]
17. Jawed, M.; Sinha, R.K. Design of rhombus coal pillars and support for Roadway Stability and mechanizing loading of face coal using SDLs in a steeply inclined thin coal seam—A technical feasibility study. *Arab. J. Geosci.* **2018**, *11*, 415. [CrossRef]
18. Unver, B.; Yasitli, N.E. Modelling of strata movement with a special reference to caving mechanism in thick seam coal mining. *Int. J. Coal Geol.* **2006**, *66*, 227–252. [CrossRef]
19. Zhu, H.Z.; Liu, P.; Tong, Z.Y. Numerical simulation research and application on protected layer pressure relief a ection under different coal pillar width. *Procedia Eng.* **2014**, *84*, 818–825.
20. Salmi, E.F.; Nazem, M.; Karakus, M. Numerical analysis of a large landslide induced by coal mining subsidence. *Eng. Geol.* **2017**, *217*, 141–152. [CrossRef]
21. Das, R.; Singh, P.K.; Kainthola, A.; Panthee, S.; Singh, T.N. Numerical analysis of surface subsidence in asymmetric parallel highway tunnels. *J. Rock Mech. Geotech. Eng.* **2017**, *9*, 170–179. [CrossRef]
22. Ghabraie, B.; Ren, G.; Zhang, X.; Smith, J. Physical modelling of subsidence from sequential extraction of partially overlapping longwall panels and study of substrata movement characteristics. *Int. J. Coal Geol.* **2015**, *140*, 71–83. [CrossRef]
23. Liu, W.; Pang, L.; Xu, B.; Sun, X. Study on overburden failure characteristics in deep thick loose seam and thick coal seam mining. *Geomat. Nat. Hazards Risk* **2020**, *11*, 632–653. [CrossRef]
24. Zhang, J.; Li, F.; Zeng, L.; Peng, J.; Li, J. Numerical simulation of the moisture migration of unsaturated clay embankments in southern China considering stress state. *Bull. Eng. Geol. Environ.* **2021**, *80*, 11–24. [CrossRef]
25. Booth, C.J. *The Effects of Longwall Coal Mining on Overlying Aquifers*; Special Publications; Geological Society: London, UK, 2002.
26. Dai, H.; Lian, X.; Liu, J.; Liu, Y.; Zhou, Y.; Deng, W.; Cai, Y. Model study of deformation induced by fully mechanized caving below a thick loess layer. *Int. J. Rock Mech. Min. Sci.* **2010**, *47*, 1027–1033. [CrossRef]
27. Sheorey, P.R.; Loui, J.P.; Singh, K.B.; Singh, S.K. Ground subsidence observations and a modifed infuence function method for complete subsidence prediction. *Int. J. Rock Mech. Min. Sci.* **2000**, *37*, 801–818. [CrossRef]
28. Ghabraie, B.; Ren, G.; Barbato, J.; Smith, J.V. A predictive methodology for multi-seam mining induced subsidence. *Int. J. Rock Mech. Min. Sci.* **2017**, *93*, 280–294. [CrossRef]
29. Wang, J.C.; Jiang, F.X.; Meng, X.J.; Wang, X.Y.; Zhu, S.T.; Feng, Y. Mechanism of Rock Burst Occurrence in Specially Thick Coal Seam with Rock Parting. *Rock Mech. Rock Eng.* **2016**, *49*, 1953–1965. [CrossRef]
30. Wang, T.; Ikarashi, K. Post Local Buckling Behavior and Ultimate Capacity of H–section Beam. *J. Asian Archit. Build. Eng.* **2011**, *10*, 429–436. [CrossRef]
31. Qian, M.; Miao, X.; Xu, J. Theoretical study on key stratum in strata control. *J. China Coal Soc.* **1996**, *21*, 12–23.

32. Xu, J.; Qian, M. Method for identifying the location of key overburden layers. *J. China Univ. Min. Technol.* **2000**, *29*, 463–467.
33. Huang, S. *Research on the Damage Mechanism and Control Methods of Cracks Caused by Mining in Shallow Coal Seams*; Xi'an University of Science and Technology: Xi'an, China, 2006; pp. 50–55.
34. Hou, Z. Application of combined key stratum theory and determination of its parameters. *Acta Coalae Sin.* **2001**, *26*, 611–615.
35. Li, H.; Yang, J.; Liu, M. Elastoplastic analysis of fixed beams at both ends under uniformly distributed load. *J. Hehai Univ. (Nat. Sci. Ed.)* **2005**, *4*, 447–451.
36. Liu, W.; Pang, L.; Liu, Y.; Du, Y. Characteristics Analysis of Roof Overburden Fracturein Thick Coal Seam in Deep Mining and Engineering Application of Super High Water Material in Back fill Mining. *Geotech. Geol. Eng.* **2019**, *37*, 2485–2494. [CrossRef]
37. Fujii, Y.; Ishijima, Y. Numerical simulation of microseismicity induced by deep longwall coal mining. *Min. Sci. Technol.* **1991**, *12*, 265–285. [CrossRef]
38. Eberhardt, E. The Hoek-Brown failure criterion. *Rock Mech. Rock Eng.* **2012**, *45*, 981–988. [CrossRef]
39. Swift, G. Relationship between joint movement and mining subsidence. *Bull. Eng. Geol. Environ.* **2014**, *73*, 163–176. [CrossRef]
40. Marschalko, M.; Yilmaz, I.; Lamich, D.; Drusa, M.; Kubečková, D.; Peňaz, T.; Burkotová, T.; Slivka, V.; Bednárik, M.; Krčmář, D.; et al. Unique documentation, analysis of origin and development of an undrained depression in a subsidence basin caused by underground coal mining (Kozinec, Czech Republic). *Environ. Earth Sci.* **2014**, *72*, 11–20. [CrossRef]
41. Alencar, A.; Galindo, R.; Melentijevic, S. Influence of the groundwater level on the bearing capacity of shallow foundations on the rock mass. *Bull. Eng. Geol. Environ.* **2021**, *80*, 6769–6779. [CrossRef]
42. Yan, W.; Chen, J.; Tan, Y.; Zhang, W.; Cai, L. Theoretical analysis of mining induced overburden subsidence boundary with the horizontal coal seam mining. *Adv. Civ. Eng.* **2021**, *2021*, 6657738. [CrossRef]
43. Unlu, T.; Akcin, H.; Yilmaz, O. An integrated approach for the prediction of subsidence for coal mining basins. *Eng. Geol.* **2013**, *166*, 186–203. [CrossRef]
44. Chen, C.; Hu, Z.; Wang, J.; Jia, J. Dynamic Surface Subsidence Characteristics due to Super-Large Working Face in Fragile-Ecological Mining Areas: A Case Study in Shendong Coalfield, China. *Adv. Civ. Eng.* **2019**, *2019*, 8658753. [CrossRef]
45. Sun, Y.; Zuo, J.; Karakus, M.; Wen, J. A Novel Method for Predicting Movement and Damage of Overburden Caused by Shallow Coal Mining. *Rock Mech. Rock Eng.* **2020**, *53*, 1545–1563. [CrossRef]
46. Yu, X.Y.; Wang, P.; Li, X.L. Analysis on characteristics of surface subsidence with Han Jia Wan coal mine. *Adv. Mater. Res.* **2012**, *524*, 520–527. [CrossRef]

Article

Effect of Chemical Corrosion and Axial Compression on the Dynamic Strength Degradation Characteristics of White Sandstone under Cyclic Impact

Jinchun Xue [1], Zhuyu Zhao [1], Longjun Dong [2,*], Jiefang Jin [3], Yingbin Zhang [1], Li Tan [1], Ruoyan Cai [1] and Yihan Zhang [2]

[1] School of Energy and Mechanical Engineering, Jiangxi University of Science and Technology, Nanchang 330013, China; xuejinchun@jxust.edu.cn (J.X.); zhuyuzhao@njfu.edu.cn (Z.Z.); zhangyingbin@jxust.edu.cn (Y.Z.); 2021028085400003@ecjtu.edu.cn (L.T.); exampie19951123@163.com (R.C.)

[2] School of Resources and Safety Engineering, Central South University, Changsha 410083, China; zyh6324@csu.edu.cn

[3] Jiangxi Provincial Key Laboratory of Environmental Geotechnology and Engineering Disaster Control, Jiangxi University of Science and Technology, Ganzhou 341000, China; jjf_chang@126.com

* Correspondence: lj.dong@csu.edu.cn

Abstract: Both chemical corrosion and axial compression impose critical influences on the internal microstructure of rock. Meanwhile, chemical corrosion can change a rock's mineral composition, which in turn affects the physical and mechanical properties of the rock. To investigate the dynamic strength characteristics of white sandstone under the coupling effect of axial load and chemical corrosion, a dynamic and static combined loading test device was adopted for performing cyclic impact tests on white sandstone immersed in chemical solution. The results show that with the increasing number of cycles under the same load, the peak strength of the rock presented a trend of 'strengthening first and then weakening'. The strength of rock resistance to impact failure reached its maximum when the solution of pH was 7 and axial pressure was 12.6 MPa. Under the same axial pressure, the effect of solution pH on the initial dynamic strength of white sandstone is a normal distribution. Acidic and alkaline environments are harmful to rocks during the initial impact, while neutral environments exert little effect and the pH of the solution influences the particle size of impact crushing particles. In addition, the chemical solution has a significant effect on the deterioration of rock strength during the process of initial impact, and the effect is inconspicuous in the later period.

Keywords: rock mechanics; cyclic impact; chemical corrosion; axial compression; strength degradation

Citation: Xue, J.; Zhao, Z.; Dong, L.; Jin, J.; Zhang, Y.; Tan, L.; Cai, R.; Zhang, Y. Effect of Chemical Corrosion and Axial Compression on the Dynamic Strength Degradation Characteristics of White Sandstone under Cyclic Impact. *Minerals* **2022**, *12*, 429. https://doi.org/10.3390/min12040429

Academic Editor: Gianvito Scaringi

Received: 3 March 2022
Accepted: 29 March 2022
Published: 31 March 2022

Publisher's Note: MDPI stays neutral with regard to jurisdictional claims in published maps and institutional affiliations.

1. Introduction

Certain large-scale infrastructure construction and mining projects are inseparable from geotechnical engineering. These include projects such as large-section tunnel excavation [1], resource exploitation [2,3], subgrade blasting [4], directional damming and interception projects [5,6]. The rock not only bears the static stress of structure during construction but also experiences the erosion of the chemical environment for a long time. In the process of blasting excavation, the stability of rock is inevitably influenced by frequent drilling and blasting vibrations [7]. The coupling of force and chemical corrosion has a critical influence on the stability of engineering rock mass in the long term [8].

Over the last decades, many scholars have performed a lot of research on investigating rock dynamic mechanical properties under various conditions. Li et al. [9] conducted a uniaxial cyclic impact compression test on granite through an improved split Hopkinson pressure bar (SHPB). The results show that the damage accumulation of the rock increased with the number of cyclic impacts. Gong et al. [10] carried out a one-dimensional experimental study on the dynamic characteristics of rock under combined dynamic and static loading, revealing that the impact strength is maximally strengthened with an axial compression ratio of 0.6~0.7. Zhou et al. [11] studied the dynamic mechanical behavior and failure characteristics of mudstone using the split Hopkinson pressure bar test device, and proposed a criterion for the dynamic strength of mudstone. Jin et al. [12] discussed the effects of different static loads on the failure pattern of rock subjected to cyclic impact and concluded that rock with an axial static load had extremity effects in the process of destruction. Ding et al. [13] investigated the failure process and the mechanical properties of limestone under different chemical solutions, finding that the strength of limestone decreased due to the chemical solutions. Xie et al. [14] conducted in situ stress restoration tests on cores of different burial depths and obtained rock behaviors under the action of in-situ stress, temperature, and pore pressure. Siddiqua et al. [15] explored the saturated mechanical behavior of light backfill and dense backfill, which clarified the strengthening effect of pore fluid chemistry on shear strength, stiffness, and yield behavior. Han et al. [16] compared microstructure, deformation characteristics, and the mechanical behavior of rock by chemical solutions with different pH, different concentrations, and different compositions, revealing that rock had a conversion trend of brittle to ductile after chemical corrosion. Xia et al. [17] conducted the whole process of compression failure of multi-crack limestone and limestone under different chemical solutions by adopting a self-developed micro loading instrument, which obtained the deformation characteristics of rock specimens, the mode of crack initiation, propagation and penetration as well as the different overlapping modes of rock bridges when they were damaged. Li et al. [18] studied the main components of calcareous cemented feldspar rock under different pH, and a rock damage model that could be applied to acidic solutions was put forward. Through scanning electron microscope (SEM) and X-ray diffraction (XRD) technology, Cui et al. [19] analyzed the mechanical and corrosion damage of the surface and mineral components of rock after the action of 0.01 mol/L NaOH solution with a pH of 12 and revealed internal changes in the rock under that solution.

Scholars both at home and abroad have conducted in-depth studies on the mechanical properties of rocks in static or quasi-static conditions [20–24]. Currently, studies on the mechanical properties of rock have mainly focused on a single factor, such as axial static load [25], impact load [26], and chemical corrosion [27]. However, there is less investigation on the rock cyclic impact failure and damage accumulation under the coupling of axial pressure and chemical corrosion. Based on split Hopkinson pressure bar (SHPB) test technology combined with one-dimensional stress wave theory, cyclic impact experiments were carried out on white sandstone immersed in chemical solutions with different pH and axial pressures, which explored fatigue damage mechanism and the characteristics of strength weakening of white sandstone under multi-factor coupling. The mechanism of strength and fatigue failure also provides a theoretical basis for the safe and effective construction management of blasting engineering under complex geological conditions.

2. Experimental Method

2.1. Rock Preparation

The rock material used in the present study was white sandstone that was taken from Kunming, Yunnan, China. These rock blocks are off-white, with good integrity and homogeneity as the research objects. Sandstone samples were subjected to professional elemental analysis and mineral identification, as shown in Figure 1. They were composed primarily of quartz, feldspar (e.g., potassium feldspar, sodium feldspar, and calcium feldspar), clay, calcite and a small number of hematite, quick lime and other minerals. The sandstone's quartz was identified as SiO_2. Feldspar was identified as a collection of $KAlSi_3O_8$, $NaAlSi_3O_8$ and $CaAl_2Si_3O_8$. Clay was identified as $xAl_3O_2 \cdot ySiO_2$. The main component of calcite is calcium carbonate. Furthermore, hematite and quick lime are Fe_2O_3 and CaO, respectively.

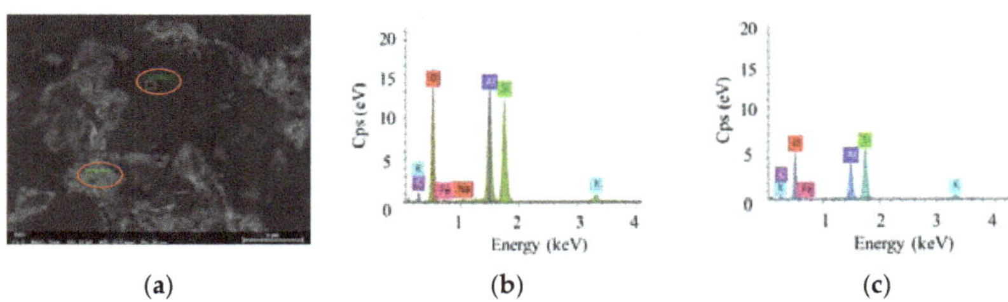

Figure 1. EDS analysis results. (**a**) EDS sampling point; (**b**) Element at point 1; (**c**) Element at point 2.

According to the size and accuracy requirements of the relevant guidelines of ISRM [28], the size of cylindrical white sandstone specimens is Φ50 mm $\times$ 50 mm, and that of unevenness and verticality are less than 0.05 mm by means of carefully grinding both ends of the rock specimens. The processed rock specimens are displayed in Figure 2. The rock specimens were dried and evacuated for 8 h, divided into eight groups, and sealed immersed in the chemical solutions presented in Table 1 for 240 days.

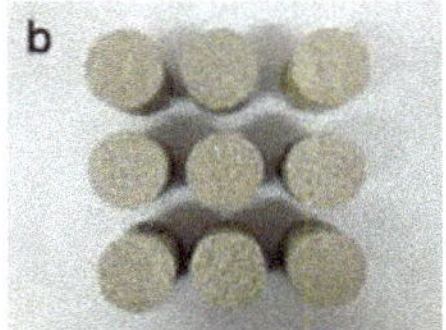

Figure 2. White sandstone specimens (**a–c**).

Table 1. Chemical solution.

Composition	Concentration (mol/L)	pH
NaCl	0.1	2, 7, 9, 12
Na_2SO_4	0.1	2, 7, 9, 12

2.2. Experimental Apparatus

In this study, based on the split Hopkinson pressure bar (SHPB) system established previously [29], an experiment system of SHPB is shown in Figure 3, using the laboratory dynamic cyclic impact compression experiment that was performed. The experiment facilities are mainly composed of a dynamic loading system (gas gun, launch cavity and shaped puncher), a test system (CS-1D strain instrument, DL-750 oscilloscope, and laser speedometer), a delivery system and an axial compression system. The delivery system mainly includes an incident bar, transmission bar, and buffer bar, the lengths of which are 2000 mm, 1500 mm, and 500 mm, respectively. All of the elastic bars were made of high-strength 40 Cr alloy steel, a wave velocity of 5400 m/s, a density of 7.81 g/cm^3, and a wave impedance of 4.2×10^7 MPa. The special-shaped punch is applied to eliminate PC oscillation and realize half-sine wave loading with constant strain rate loading [30–32].

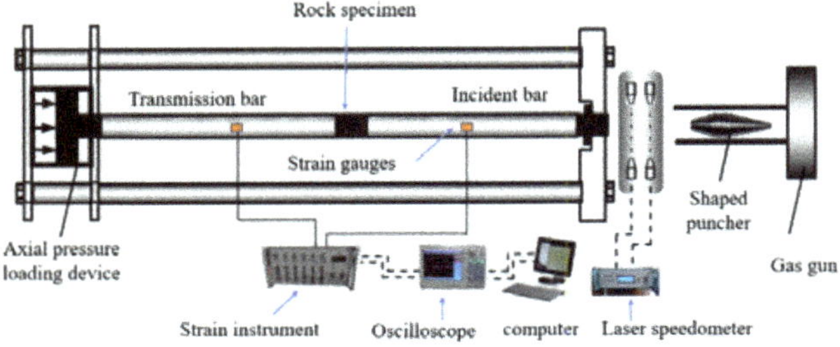

Figure 3. Split Hopkinson pressure bar (SHPB).

2.3. Experimental Method

Before conducting the cyclic impact test, a physico-mechanical parameters test was carried out on the white sandstone specimens by adopting the RMT-150 electro-hydraulic servo control material testing machine. The results of this test are shown in Table 2.

Table 2. Physico-mechanical parameters of white sandstone specimens.

Specimens	Density ($kg \cdot m^{-3}$)	Loading Rate ($MPa \cdot s^{-1}$)	Strain Rate (s^{-1})	Uniaxial Compression Strength (MPa)	Secant Modulus (GPa)	Poisson Coefficient
White sandstone	2321	0.23	1.00×10^{-5}	31.5	17.65	0.25

In order to ensure that cyclic impact test on the white sandstone with the chemical corrosion and axial pressure could be carried out successfully, the experimental method was designed as follows. Four solutions of pH at 2, 7, 9, and 12 were designed. According to the research conducted by Sun et al. [33], in the process of cyclic impact, rock specimens did not show the palpable end effect failure mode when there was no static load. Combined with the value of uniaxial compressive strength test, axial compression was set to four levels of uniaxial compressive strength of 20%, 40%, 65%, 85%, which were respectively 6.3 MPa, 12. 6 MPa, 20.5 MPa and 26.8 MPa. In addition, due to the existence of a threshold value in the cyclic impact load experiment [34], multiple pre-impact tests were required to determine the fixed air pressure in the high-pressure gas chamber of 0.8 MPa. The impact speed of the punch was 4.5 m/s.

The following conditions must be satisfied during the test process:

(1) The incident bar, transmission bar and buffer bar should be kept level.
(2) The magnitude of the axial load should be equal to the axial load value set in advance.
(3) To fix the impact load, it is necessary to strictly ensure that the pressure in the high-pressure air chamber is equal to the value set before each impact, and that the position of the punch in the launching chamber remains unchanged.
(4) The white sandstone specimens were sandwiched between the incident bar and the transmission bar, with petroleum jelly coated on both ends of the specimen to ensure good contact.

3. Typical Stress–Strain Curve during Cyclic Impact

Using a split Hopkinson pressure bar (SHPB), Jin et al. [35]. conducted a series of cyclic impact loading tests on sandstone under different static loading conditions, obtaining a typical dynamic stress–strain curve of sandstone during cyclic impact. The results of the experiments reveal that the sandstone's dynamic strength and the characteristics of deformation were affected by axial compression and the number of cycles. In this study, the cyclic impact test of white sandstone soaked in NaCl with pH of 9 and axial pressure of 6.3 MPa was taken as an example for analyzing the different stages of the test. Figure 4 presents an overlay chart of the typical waveform recorded during the cycle impact test. It can be observed from the figure that the amplitude of the incident wave has not changed significantly and that the superposition of waveform basically coincides, achieving the requirements of the constant-amplitude cyclic impact test. The reflected wave amplitude of the specimen increases and the transmitted wave amplitude decreases gradually with the increasing number of impacts, and the change of cyclic impact is obvious in the middle and later periods. This phenomenon occurs regardless of axial compression and pH. Since the size of the white sandstone specimen tested is small, the homogeneity hypothesis is introduced, satisfying the expression $\varepsilon_i(t) + \varepsilon_r(t) = \varepsilon_t(t)$ [36]. According to one-dimensional stress wave theory, the collected voltage signal is processed. Stress, strain, and average strain rate can be calculated using the following formula.

$$\sigma_s = \frac{A_e E_e}{2 A_s}(\varepsilon_I + \varepsilon_R + \varepsilon_T) \tag{1}$$

$$\varepsilon_S = -\frac{C_e}{L_s} \int_0^l (\varepsilon_I - \varepsilon_R - \varepsilon_T)\,dt \tag{2}$$

$$\dot{\varepsilon}_s = \frac{C_e}{L_s}(\varepsilon_I - \varepsilon_R - \varepsilon_T) \tag{3}$$

In these Equations, *As* and *Ls* are the cross-sectional area and length of the specimen, respectively; *Ae*, *Ee* and *Ce* represent elastic rod cross-sectional area, elastic modulus and longitudinal wave velocity, respectively; ε_I, ε_R and ε_T denote the incident, reflection and transmission wave signals, respectively.

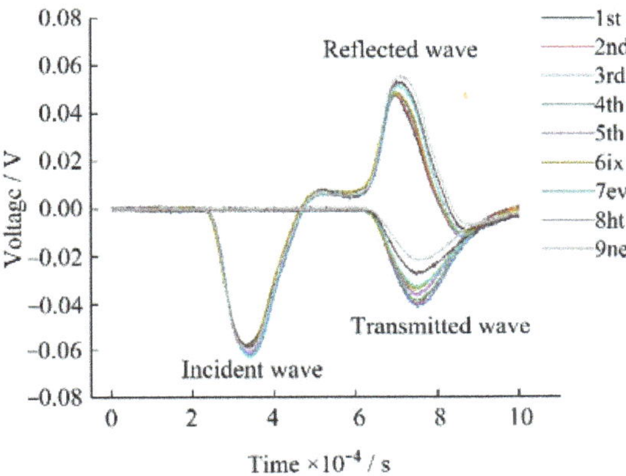

Figure 4. Typical stress waves during cyclic impact.

The typical stress–strain curves of the white sandstone under cyclic impact loading tests are shown in Figure 5. It can be observed from the figure that the stress–strain curve of the whole cyclic impact test can be divided into three types. The initial stress–strain curve (one to five impacts) is a semi-elliptic high flat curve. In addition, the elastic modulus, peak stress, and maximum strain have not changed significantly. The stress in the loading stage increases sharply with the increase of strain. The slope is steep, and the elastic modulus value is large. The maximum stress at this stage represents the maximum impact stress value that the specimen can withstand during the entire impact process, which is called the peak strength. However, the stress in the unloading stage drops sharply with the increasing strain.

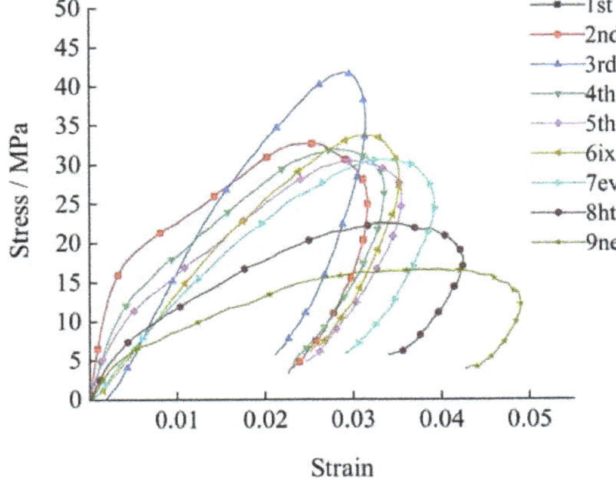

Figure 5. Typical stress-strain curve during cyclic impact.

In the mid-term (sixth to eighth impacts), there exists a gradient parabola. The changes in elastic modulus, peak stress, and maximum strain are more prominent than those in the initial stage. Besides, the change of stress with strain is larger than that of the initial stage in the loading section and unloading section. It shows that the micro-cracks in white sandstone are compacted and closed, resulting in small changes in modulus. Owing to the medium-term impact load, pore cracks begin to gradually expand and penetrate, and the bearing capacity suddenly decreases.

The later stage (the ninth impact) is a low flat parabolic curve, the peak stress is small, and the strain at the tail of the curve tends to decrease, which is consistent with the research results of the rock impact failure test under the one-dimensional dynamic and static combination determined by the literature [22]. This demonstrates that the white sandstone has not been completely destroyed and still possesses a certain load-bearing capacity, though it's capacity is weak. Therefore, it cannot be loaded under the same axial compression conditions.

4. Experimental Results and Discussion

This study lists the physical parameters and test data of a representative set of specimens, as listed in Table 3.

Table 3. Parameters of some specimens and the number of cyclic impacts.

σ_{as} (MPa)	Numbering	L (mm)	D (mm)	ρ (kg·cm^{-3})	pH	σ_{fd} (MPa)	σ_{md} (MPa)	σ_{cs} (MPa)	The Number of Cycle Impact
	w-062	47.03	49.08	2315	2	39.64	43.51	49.81	7
6.3	w-052	47.05	49.09	2329	7	43.12	43.20	49.50	9
	w-033	47.06	49.11	2312	9	42.68	42.68	48.98	8
	w-042	47.06	49.09	2322	12	42.03	49.43	55.73	7
	w-063	47.05	49.12	2327	2	37.16	41.42	54.02	5
12.6	w-053	47.06	49.09	2312	7	51.56	51.56	64.16	8
	w-034	47.05	49.09	2322	9	49.61	49.61	62.21	6
	w-043	47.07	49.10	2328	12	39.87	50.10	62.70	5
	w-064	47.06	49.14	2312	2	38.46	38.46	58.96	4
20.5	w-054	47.08	49.13	2329	7	43.11	43.11	63.61	5
	w-037	47.05	49.10	2329	9	40.72	40.72	61.22	5
	w-044	47.05	49.04	2320	12	40.75	42.83	63.33	2
	w-065	47.06	49.12	2308	2	27.26	31.92	58.72	2
26.8	w-055	47.06	49.05	2329	7	32.01	32.01	58.81	3
	w-036	47.06	49.06	2318	9	28.82	30.51	57.31	3
	w-045	47.07	49.02	2317	12	25.71	29.98	56.78	2

Note: σ_{as} is the axial pressure, ρ is density, σ_{fd} is initial impact stress peak, σ_{md} is the maximum stress peak of the cyclic impact, and σ_{cs} is the combined static–dynamic strength.

4.1. The Trend of Cyclic Impact Number

Figure 6 shows the relationship between the pH of the solution, the axial pressure, and the number of cyclic shocks. The number of cyclic impacts decreases with the increase of the axial pressure, and the number of impacts that can be endured under the axial pressure of 6.3 MPa is the most. As shown in the analysis, when the axial pressure is within a certain lower limit range, with the same load repeatedly repeating the impact, the original void inside the specimens will be closed first, and the ability and frequency of the sample to resist impact will be accordingly enhanced [37]. However, when the axial pressure is within a certain higher limit range, the secondary microcracks grow into pore cracks that penetrate the entire cross-section. During this process, the damage expands and the ability of the sample to resist external damage decreases sharply. In the middle and later stages of the impact stages, the whole rock specimen is finally damaged by tensile shear. Furthermore, this is similar to the dynamic characteristics of the cumulative damage evolution during excavation of underground jointed rock under repeated seismic load that is mentioned in Ref. [38]. With the pH of the solution, the number of shocks is normally distributed. The pH of the solution deviates from neutral, which indicates that the lower the number of shocks, the more obvious the corrosion effect of the solution.

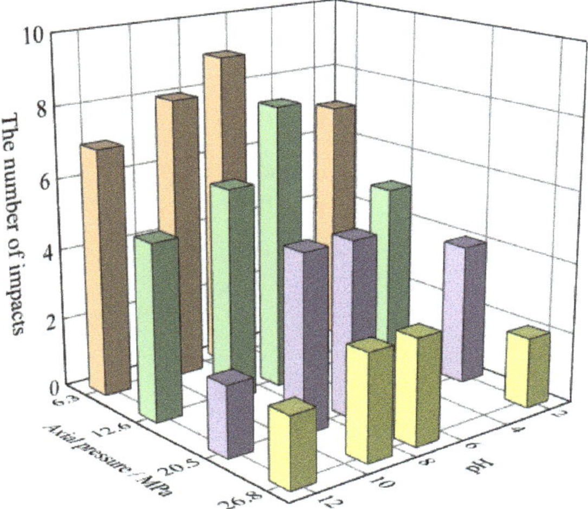

Figure 6. The trend of cyclic impact number.

4.2. Relationship between Initial Peak Stress and Axial Pressure

The peak strength of the rock represents the maximum stress caused by the impact resistance of the white sandstone specimens during impact. The strength of the specimen gradually weakens when the number of cyclic impacts increases. Although the overall peak deterioration law is similar, there also exist noticeable differences. As shown in Figure 7, with the increasing number of cycles under the axial pressure of 6.3 MPa and 12.6 MPa, the peak strength of the rock shows a trend of 'strengthening first and then weakening', which was different from the axial pressure of 20.5 MPa and 26.8 MPa. There exists no strength reinforcement at high pre-axial pressure.

The peak stresses of the initial impact under the axial pressure of 12.6 MPa are 42.03 MPa, 51.56 MPa, 49.61 MPa and 47.01 MPa, which are greater than the other three axial pressure conditions. When the number of impacts increases, the magnitude of peak stress degradation in the middle and later stages of this axial compression is more rapid than that of the other three axial compressions. This phenomenon occurs under any pH. This indicates that the dynamic strength of white sandstone at the initial stage is strengthened by 40% axial compression during cyclic impact. However, the axial compression at the middle and later stages will also aggravate the accumulation of internal damage of white sandstone specimens during impact.

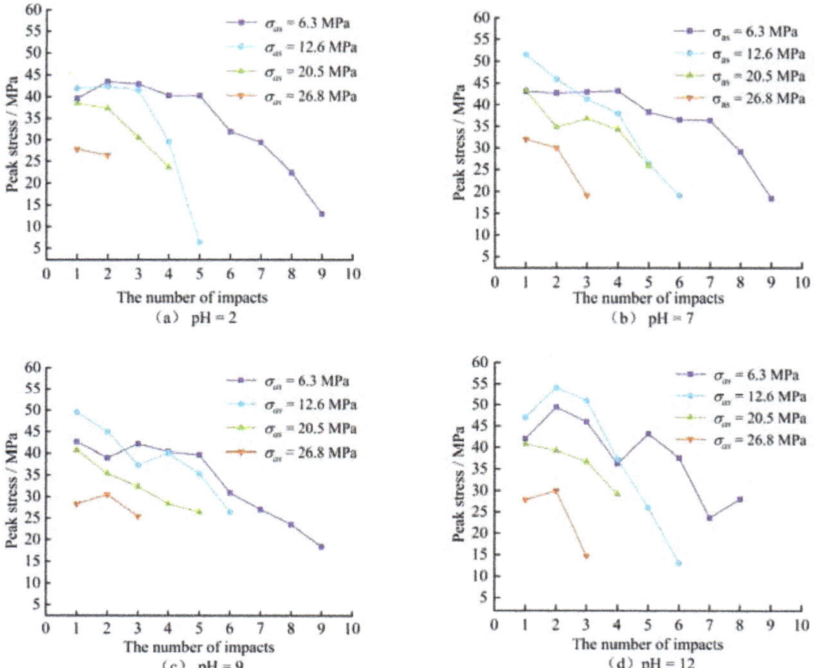

Figure 7. Relationship between peak stress and number of impacts. (**a**) pH = 2; (**b**) pH = 7; (**c**) pH = 9; (**d**) pH = 12.

4.3. Relationship between Peak Stress and Chemical Corrosion

The macroscopic manifestation of the water–rock reaction is the deterioration of the physical and mechanical properties of rock [39]. Figure 8 shows the relationship between the peak stress of the specimen and the chemical solution. Followed by the pH = 12 alkaline solution immersion test piece, the initial impact strength of the specimen is the lowest pH = 2 acid solution immersion. The maximum initial strength of the test piece is in a neutral solution at pH = 7. Complex physical and chemical reactions occur when rocks are immersed in a chemical solution. The physical interaction between water and rock mainly causes sandstone minerals to be dissolved in water. Changes in rock composition and structure are mainly due to the corrosion of minerals in chemical solutions. With the increase of acidity and alkalinity, the damage of rock in solution is mainly determined by chemical reactions. Mineral particles and the ion exchange reaction in the solution generate a new secondary mineral, changing the original mineral chemical composition, the generated secondary mineral composition and molecular weight. Density is very different from the original mineral. Inherent internal cementation way and structure characteristics change accordingly, eventually degrading sandstone in terms of its macro mechanical performance.

In acidic solution, alkaline cations on the surface of feldspar exchange with hydrogen ion in the solution, forming a weak acidic silicon-rich complex. Potassium, sodium, and calcium in feldspar detach from the rock mass skeleton and enter the solution [40].

$$KAlSi_3O_8 + 4H^+ + 4H_2O \rightarrow K^+ + Al^{3+} + 3H_4SiO_4 \tag{4}$$

$$NaAlSi_3O_8 + 4H^+ + 4H_2O \rightarrow Na^+ + Al^{3+} + 3H_4SiO_4 \tag{5}$$

$$CaAl_2Si_2O_8 + 2H^+ + H_2O \rightarrow Ca^{2+} + Al_2Si_2O_5(OH)_4 \tag{6}$$

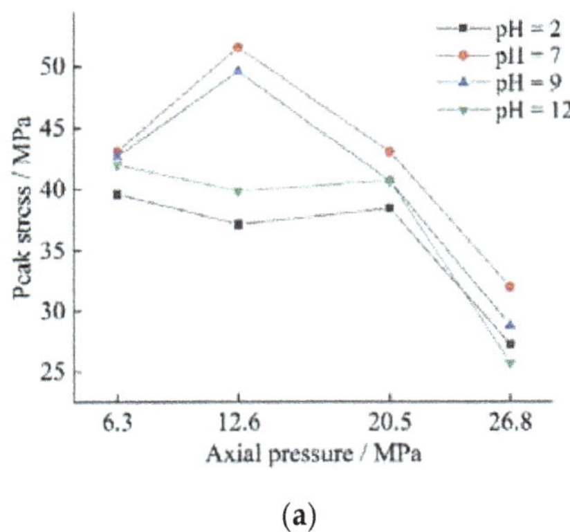 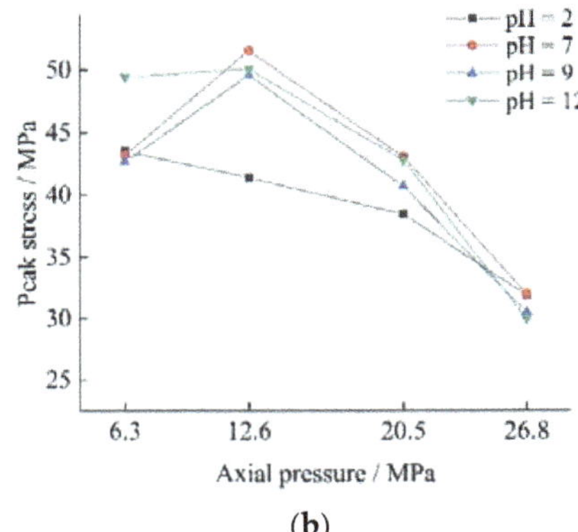

(**a**)

(**b**)

Figure 8. The trend of peak stress. (**a**) initial peak stress; (**b**) the maximum of peak stress.

In addition to the chemical reactions listed earlier, small amounts of hematite, and calcite in the mineral composition of sandstone also react with hydrogen ions in solution.

$$Fe_2O_3 + 6H^+ \rightarrow 2Fe^{3+} + 3H_2O \tag{7}$$

$$CaCO_3 + 2H^+ \rightarrow Ca^{2+} + H_2O + CO_2\uparrow \tag{8}$$

In the neutral solution, the solubility of sandstone components is low. Besides, it is difficult to react violently with the aqueous solution. Only a small amount of feldspar and quicklime participate in the chemical reaction and are stable.

In alkaline solution, hematite, calcite, quicklime, and other minerals in sandstone remain relatively stable. However, quartz and feldspar in sandstone are easy to react with hydroxide ion in alkaline solution.

$$SiO_2 + 2OH^- \rightarrow SiO_3^{2-} + H_2O \tag{9}$$

$$KAlSi_3O_8 + 6OH^- + 2H_2O \rightarrow K^+ + Al(OH)_4^- + 3H_2SiO_4^{2-} \tag{10}$$

$$NaAlSi_3O_8 + 6OH^- + 2H_2O \rightarrow Na^+ + Al(OH)_4^- + 3H_2SiO_4^{2-} \tag{11}$$

The failure mode of white sandstone is shown in Figure 9a–p under cyclic impact. White sandstone specimens mainly experience conjugate hyperbolic tensile shear failure, which was characterized by destructive body flaking around the hyperbola due to interfacial friction. Corrosion by acid and alkaline solutions of sandstone fragments' impact degree is more than with the neutral solution, with a smaller average size of fragments. In an acidic solution, the sandstone is completely disintegrated with a high proportion of fine particle size. Feldspar, clay, hematite, calcite and other minerals in sandstone are captured by solution, resulting in an increase in small particle size fragments. Structural cavities make rock structures become looser and more fragile. In an alkaline solution, the quartz particles on the surface of the specimen will be exfoliated by the solution, thereby increasing the number of fine particles. In neutral solutions, the solubility of mineral components is low, and the degree of fragmentation is mainly dependent on the cyclic shock load.

By testing the broken samples, it has been found that the solution corroded the sandstone and caused the internal cavity of the sandstone to produce structural defects. During cyclic impact, the rock exhibits a low-cycle fatigue damage and strength degradation. Repeated impact causes the rock particles to loosen and fall off. As the number of cycles increases, this phenomenon gradually penetrates into the interior of the sandstone. Meanwhile, the porosity of the internal structure allows the chemical solution to further penetrate into the interior of the sandstone. This interaction improves exposure between chemical solutions and mineral particles, thus allowing for a more thorough chemical reaction between water and rock. Therefore, the hydration solution and the cyclic shock influence the rate and degree of rock metamorphism while promoting each other in the process of rock metamorphism.

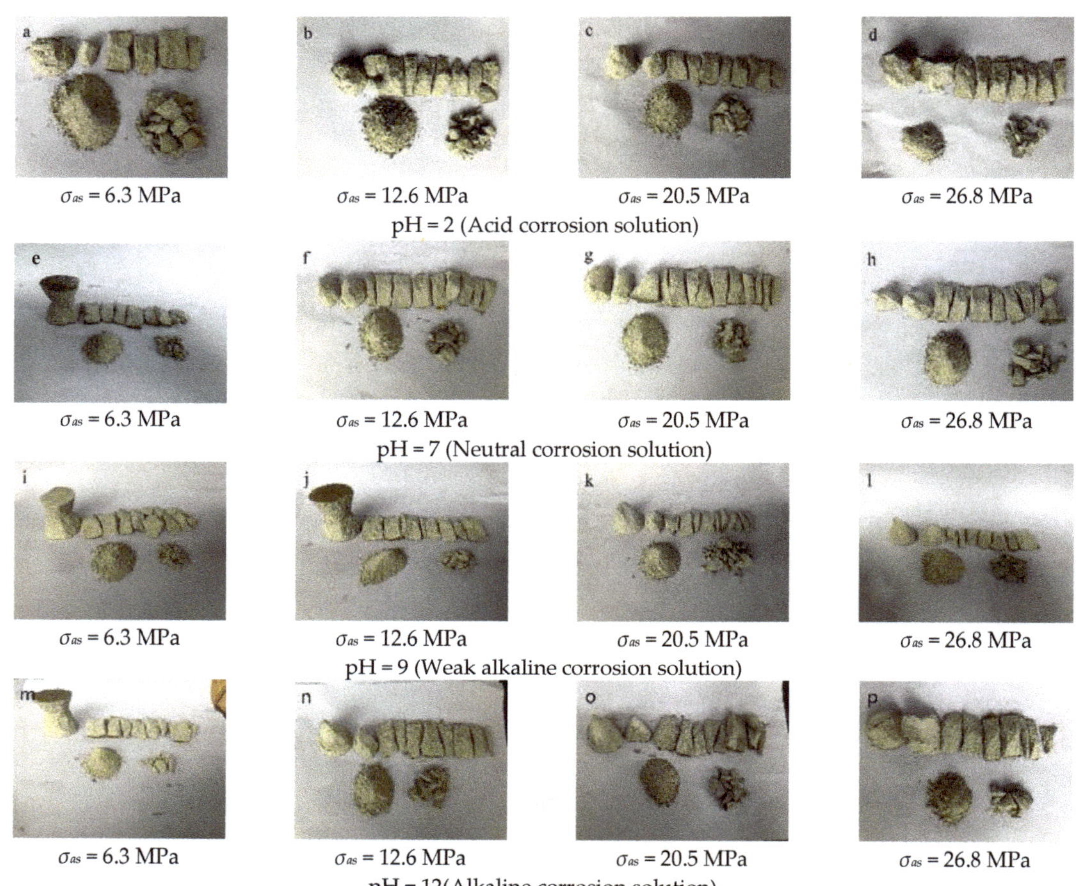

Figure 9. The destructed forms of white sandstone (**a–p**).

5. Conclusions

To conclude, this study introduces the cyclic impact test of white sandstone specimens under chemical corrosion and axial compression, as well as stress–strain and dynamic strength variation characteristics. Several vital conclusions are presented as follows:

(1) The stress–strain curve during the cyclic impact test of white sandstone can be divided into three types: an early semi-elliptic high flat curve, a mid-gradient parabola, and low flat parabolic curve.

(2) With the increase of the number of cycles under the same load, the peak strength of the rock shows a trend of 'strengthening first and then weakening'. The strength of stone resistance to impact failure reached its maximum at pH of 7 and axial pressure of 12.6 MPa.

(3) Under the same axial pressure, the effect of solution pH on the initial dynamic strength of white sandstone is a normal distribution. The acidic and alkaline environment damage is harmful to rocks, while neutral environment generates little effect. The

influence of solution pH on the dynamic strength of white sandstone is not apparent during the middle and late periods.

(4) Through carrying out a series of studies, the proper proportion of loading will make the micro-fractures in the white sandstone tend to be tight, and the impact strength of the white sandstone will be strengthened in the early stage. Overloading will aggravate the deterioration of the strength of the white sandstone under the impact and further intensify the damage in the middle and later stages. When the loading ratio exceeds a certain limit, the micro-cracks of the white sandstone will increase due to the increasing loading ratio. Furthermore, the structure of rock will be destroyed.

Author Contributions: Conceptualization, J.X. and L.D.; methodology, J.X.; software, Z.Z.; validation, L.D., Z.Z. and J.J.; formal analysis, Y.Z. (Yingbin Zhang); investigation, Y.Z. (Yihan Zhang); resources, L.T.; data curation, Y.Z. (Yingbin Zhang); writing—original draft preparation, J.X. and Z.Z.; writing—review and editing, J.X. and L.D.; visualization, Z.Z. and R.C.; supervision, L.D.; funding acquisition, J.X. All authors have read and agreed to the published version of the manuscript.

Funding: This research was supported financially by the National Natural Science Foundation of China (No. 51664016, No. 51664017) and the Jiangxi Provincial Department of Education (GJJ150694).

Data Availability Statement: The data used to support the findings of this study are available from the corresponding author upon request.

Conflicts of Interest: The authors declare no conflict of interest.

References

1. Sharifzadeh, M.; Kolivand, F.; Ghorbani, M.; Yasrobid, S. Design of sequential excavation method for large span urban tunnels in soft ground—Niayesh tunnel. *Tunn. Undergr. Space Technol.* **2013**, *35*, 178–188. [CrossRef]
2. Feng, J.; Wang, E.; Shen, R.; Chen, L.; Li, X.; Li, N. A source generation model for near-field seismic impact of coal fractures in stress concentration zones. *J. Geophys. Eng.* **2016**, *13*, 516–525. [CrossRef]
3. Guan, X.; Wang, X.; Zhu, Z.; Zhang, L.; Fu, H. Ground Vibration Test and Dynamic Response of Horseshoe-shaped Pipeline During Tunnel Blasting Excavation in Pebbly Sandy Soil. *Finite Elem. Anal. Des.* **2020**, *38*, 3725–3736. [CrossRef]
4. Dong, L.; Sun, D.; Shu, W.; Li, X. Exploration: Safe and clean mining on Earth and asteroids. *J. Clean. Prod.* **2020**, *257*, 120899. [CrossRef]
5. Bayraktar, A.; Sevim, B.; Altunişik, A.C. Finite element model updating effects on nonlinear seismic response of arch dam–reservoir–foundation systems. *Finite Elem. Anal. Des.* **2011**, *47*, 85–97. [CrossRef]
6. Dong, L.; Deng, S.; Wang, F. Some developments and new insights for environmental sustainability and disaster control of tailings dam. *J. Clean. Prod.* **2020**, *269*, 122270. [CrossRef]
7. Dong, L.-J.; Tang, Z.; Li, X.-B.; Chen, Y.-C.; Xue, J.-C. Discrimination of mining microseismic events and blasts using convolutional neural networks and original waveform. *J. Cent. South Univ.* **2020**, *27*, 3078–3089. [CrossRef]
8. Lu, Z.-D.; Chen, C.-X.; Feng, X.-T.; Zhang, Y.-L. Strength failure and crack coalescence behavior of sandstone containing single pre-cut fissure under coupled stress, fluid flow and changing chemical environment. *J. Cent. South Univ.* **2014**, *21*, 1176–1183. [CrossRef]
9. Li, X.; Li, C.; Cao, W.; Tao, M. Dynamic stress concentration and energy evolution of deep-buried tunnels under blasting loads. *Int. J. Rock Mech. Min. Sci.* **2018**, *104*, 131–146. [CrossRef]
10. Gong, F.Q.; Li, X.B.; Liu, X.L.; Zhao, J. Experimental study of dynamic characteristics of sandstone under one-dimensional coupled static and dynamic loads. *Chin. J. Rock Mech. Eng.* **2010**, *29*, 2076–2085.
11. Zhou, R.; Cheng, H.; Cai, H.; Wang, X.; Guo, L.; Huang, X. Dynamic Characteristics and Damage Constitutive Model of Mudstone under Impact Loading. *Materials* **2022**, *15*, 1128. [CrossRef] [PubMed]
12. Jin, J.F.; Li, X.B.; Wang, G.S.; Yin, Z.Q. Sandstone failure mode and mechanism under cyclic impact load. *J. Cent. South Univ. Nat. Sci. Ed.* **2012**, *43*, 254–262.
13. Ding, W.X.; Feng, X.T. Experimental study on mechanical effects of limestone under chemical corrosion. *J. Rock Mech. Geotech.* **2004**, *23*, 3571.
14. Xie, H.; Li, C.; He, Z.; Li, C.; Lu, Y.; Zhang, R.; Gao, M.; Gao, F. Experimental study on rock mechanical behavior retaining the in situ geological conditions at different depths. *Int. J. Rock Mech. Min. Sci.* **2021**, *138*, 104548. [CrossRef]
15. Siddiqua, S.; Siemens, G.; Blatz, J.; Man, A.; Lim, B.F. Influence of Pore Fluid Chemistry on the Mechanical Properties of Clay-Based Materials. *Geotech. Geol. Eng.* **2014**, *32*, 1029–1042. [CrossRef]
16. Han, T.L.; Shi, J.P.; Chen, Y.S.; Li, Z.H.; Ma, W.T. Experimental study on deterioration of mechanical properties and energy mechanism of calcareous sandstone under water chemistry. *J. Rock Mech. Geotech.* **2015**, *32*, 189–200.

17. Feng, X.-T.; Chen, S.; Li, S. Effects of water chemistry on microcracking and compressive strength of granite. *Int. J. Rock Mech. Min. Sci.* **2001**, *38*, 557–568. [CrossRef]

18. Li, N.; Zhu, Y.M.; Zhang, P.; Ge, X.R. Chemical damage model of calcareous cemented sandstone in acid environment. *Chin. J. Geotech. Eng.* **2003**, *25*, 395–399.

19. Cui, Q.; Feng, X.T.; Xue, Q.; Zhou, H.; Zhang, Z.H. Research on the mechanism of sandstone pore structure change under chemical corrosion. *Chin. J. Rock Mech. Eng.* **2008**, *27*, 134–141.

20. Mokhtarian, H.; Moomivand, H.; Moomivand, H. Effect of infill material of discontinuities on the failure criterion of rock under triaxial compressive stresses. *Theor. Appl. Fract. Mech.* **2020**, *108*, 102652. [CrossRef]

21. Karev, V.I.; Kilmov, D.M.; Kovalenko, Y.F.; Ustinov, K.B. Experimental Study of Rock Creep under True Triaxial Loading. *Mech. Solids* **2019**, *54*, 1151–1156. [CrossRef]

22. Wen, S.; Zhang, C.; Chang, Y.; Hu, P. Dynamic compression characteristics of layered rock mass of significant strength changes in adjacent layers. *J. Rock Mech. Geotech. Eng.* **2019**, *12*, 353–365. [CrossRef]

23. Kumar, J.; Rahaman, O. Lower Bound Limit Analysis Using Power Cone Programming for Solving Stability Problems in Rock Mechanics for Generalized Hoek–Brown Criterion. *Rock Mech. Rock Eng.* **2020**, *53*, 3237–3252. [CrossRef]

24. Yang, W.; Li, G.; Ranjith, P.; Fang, L. An experimental study of mechanical behavior of brittle rock-like specimens with multi-non-persistent joints under uniaxial compression and damage analysis. *Int. J. Damage Mech.* **2019**, *28*, 1490–1522. [CrossRef]

25. Kovrizhnykh, A.M.; Usol'Tseva, O.M.; Kovrizhnykh, S.A.; Tsoi, P.A.; Semenov, V.N. Investigation of Strength of Anisotropic Rocks under Axial Compression and Lateral Pressure. *J. Min. Sci.* **2017**, *53*, 831–836. [CrossRef]

26. Wu, H.; Dai, B.; Cheng, L.; Lu, R.; Zhao, G.; Liang, W. Experimental Study of Dynamic Mechanical Response and Energy Dissipation of Rock Having a Circular Opening Under Impact Loading. *Min. Met. Explor.* **2021**, *38*, 1111–1124. [CrossRef]

27. Zhou, M.; Li, J.; Luo, Z.; Sun, J.; Xu, F.; Jiang, Q.; Deng, H. Impact of water–rock interaction on the pore structures of red-bed soft rock. *Sci. Rep.* **2021**, *11*, 7398. [CrossRef]

28. Bieniawski, Z.T.H. Suggested methods for determining tensile strength of rock materials. *Int. J. Rock Mech. Min. Sci. Géoméch. Abstr.* **1978**, *15*, 99–103. [CrossRef]

29. Li, X.B.; Zhou, Z.L.; Ye, Z.Y.; Ma, C.D.; Zhao, F.J. Research on dynamic and static loading mechanical properties of rock. *J. Rock Mech. Eng.* **2008**, *27*, 96–104.

30. Zi, L.Z.; Xi, B.L.; Ai, H.L.; Yang, Z. Stress uniformity of split Hopkinson pressure bar under half-sine wave loads. *Int. J. Rock Mech. Min.* **2011**, *48*, 697–701. [CrossRef]

31. Zhang, Q.B.; Zhao, J. A Review of Dynamic Experimental Techniques and Mechanical Behaviour of Rock Materials. *Rock Mech. Rock Eng.* **2014**, *47*, 1411–1478. [CrossRef]

32. Cai, M. Practical Estimates of Tensile Strength and Hoek–Brown Strength Parameter m i of Brittle Rocks. *Rock Mech. Rock Eng.* **2010**, *43*, 167–184. [CrossRef]

33. Sun, B.; Ping, Y.; Zhu, Z.; Jiang, Z.; Wu, N. Experimental Study on the Dynamic Mechanical Properties of Large-Diameter Mortar and Concrete Subjected to Cyclic Impact. *Shock Vib.* **2020**, *2020*, 8861197. [CrossRef]

34. Gao, R.; Li, J. Equivalent constant-amplitude fatigue load method based on the energy equivalence principle. *Adv. Struct. Eng.* **2019**, *22*, 2892–2906. [CrossRef]

35. Jin, J.F.; Li, X.B.; Yin, Z.Q. Effects of axial pressure and cyclic impact times on dynamic mechanical properties of sandstone. *J. China Coal Soc.* **2012**, *37*, 923–930.

36. Wang, T.-T.; Shang, B. Three-Wave Mutual-Checking Method for Data Processing of SHPB Experiments of Concrete. *J. Mech.* **2014**, *30*, N5–N10. [CrossRef]

37. Michalowski, R.L.; Park, D. Stability assessment of slopes in rock governed by the Hoek-Brown strength criterion. *Int. J. Rock Mech. Min. Sci.* **2020**, *127*, 104217. [CrossRef]

38. Ma, M.; Brady, B. Analysis of the Dynamic Performance of an Underground Excavation in Jointed Rock under Repeated Seismic Loading. *Geotech. Geol. Eng.* **1999**, *17*, 1–20. [CrossRef]

39. Shang, D.; Zhao, Z.; Dou, Z.; Yang, Q. Shear behaviors of granite fractures immersed in chemical solutions. *Eng. Geol.* **2020**, *279*, 105869. [CrossRef]

40. Han, T.; Li, Z.; Shi, J.; Cao, X. Mechanical characteristics and freeze–thaw damage mechanisms of mode-I cracked sandstone from the Three Gorges Reservoir region under different chemical solutions. *Arab. J. Geosci.* **2021**, *14*, 944. [CrossRef]

Article

Random Forest Slurry Pressure Loss Model Based on Loop Experiment

Zengjia Wang [1], Yunpeng Kou [2,*], Zengbin Wang [3,*], Zaihai Wu [1] and Jiaren Guo [1]

[1] Backfilling Engineering Laboratory of Shandong Gold Group Co., Ltd., Laizhou 261441, China; wangzengjia1988@163.com (Z.W.); wuzaihai@sd-gold.com (Z.W.); guojiaren@sd-gold.com (J.G.)
[2] School of Civil and Resource Engineering, University of Science and Technology Beijing, Beijing 100083, China
[3] Qingdao Institute of Bioenergy and Bioprocess Technology, Chinese Academy of Sciences, Qingdao 266101, China
* Correspondence: kouyunpeng@126.com (Y.K.); wangzb@qibebt.ac.cn (Z.W.)

Abstract: A reasonable arrangement of filling pipelines can solve the problems of low line magnification, a high flow rate, large pipe pressure, etc., in deep well filling slurry transportation. The transportation pressure loss value of filling slurry is the main parameter for the layout design of filling pipelines. At present, pressure loss data are mainly obtained through the loop pipe experiment, which has problems such as a large amount of labor, high cost, low efficiency, and a limited amount of experimental data. In this paper, combined with a new generation of artificial intelligence technology, the random forest machine learning algorithm is used to analyze and model the experimental data of a loop pipe to predict the pressure loss of slurry transportation. The degree of precision reaches 0.9747, which meets the design accuracy requirements, and it can replace the loop pipe experiment to assist with the filling design.

Keywords: pipe transportation system test; pressure loss; random forest algorithm; filling-aided design

Citation: Wang, Z.; Kou, Y.; Wang, Z.; Wu, Z.; Guo, J. Random Forest Slurry Pressure Loss Model Based on Loop Experiment. *Minerals* **2022**, *12*, 447. https://doi.org/10.3390/min12040447

Academic Editors: Longjun Dong, Yanlin Zhao, Wenxue Chen and Yo-soon Choi

Received: 18 February 2022
Accepted: 31 March 2022
Published: 6 April 2022

Publisher's Note: MDPI stays neutral with regard to jurisdictional claims in published maps and institutional affiliations.

1. Introduction

The filling mining method is one of the most effective methods to ensure the safety of deep mining [1,2]. In the design of a deep well filling system, designing a reasonable arrangement of the underground filling pipeline is the main difficulty. The properties of tailings, the transportation conditions, and the pressure loss value of slurry transportation are different in different mines. At present, the theoretical calculation of the pressure loss of a high-concentration filling slurry is generally based on the Bingham rheological model, but there is a certain difference between the pressure loss value obtained when using slurry yield stress and plastic viscosity and the experimental value of the loop [3]. The main reasons for this are that the cross-sectional flow velocity is different during the transportation of high-concentration filling slurry, the flow velocity near the pipe wall is close to zero, the shear stress decreases with the increase in the shear rate, It is thixotropic and the flow curve is hysteretic [4,5]. For this reason, most designers need to master the pressure loss data of filling slurry while using the loop pipe experiment method. However, there are problems such as a large amount of labor and a long experimental period, and experimental variable parameters cannot fully simulate industrial filling pipelines.

In the filling and conveying theory, it is difficult to establish a transport model that can be used to calculate the pressure drop of the slurry by theoretical methods. With the development of artificial intelligence technology, methods to build predictive models based on existing data have gradually emerged. Abroad, Kumar et al. used the integral flow model to predict the pressure drop of multi-scale solid–liquid flow [4], but there is a problem in that the reverse analysis of the input limit parameters of numerical modeling and the generalization ability of curve fitting are poor. In China, Qi Chongchong of Central South University and others took the lead in proposing a "machine learning-assisted filling design" [6,7], And a variety of backfill system design prediction models have been established to promote the development of the traditional backfill field towards intelligence.

In this paper, the artificial intelligence random forest algorithm is used to analyze the experimental data of a loop pipe, establish the pressure loss prediction model of the filling slurry, and verify the feasibility of the proposed random forest model in the pressure loss prediction. This model can replace the traditional loop pipe experiment-aided filling piping system design.

2. Acquisition of Experimental Data of Loop Pipe

2.1. Construction of Loop Pipe Experiment System

The backfill engineering laboratory has an indoor loop test system. The test pipeline system is shown in Figure 1. P1–P2 is the pressure drop of the straight pipe section, P3–P4 is the pressure drop of the vertical section plus 90° elbow, and P5–P6 is the pipeline pressure drop of the inclined pipe section.

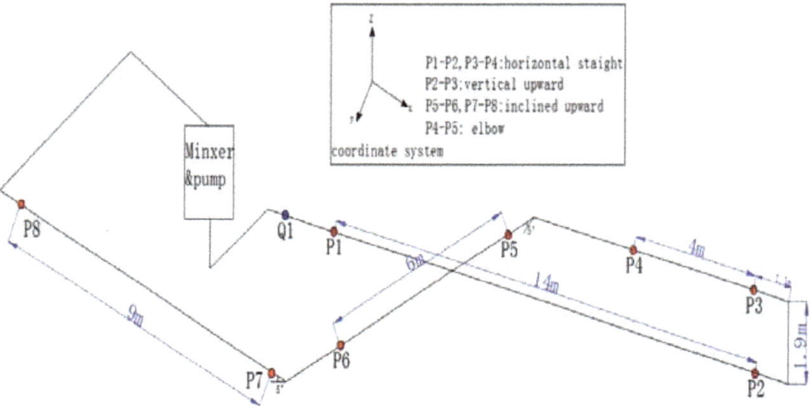

Figure 1. Schematic of the pipe transportation system.

The displacement of the loop pipe experiment system was 76 m^3/h, the outlet pressure was 1.5 Mpa, the inner diameter of the test pipe was 80 mm, and the pressure sensor adopted a flat model pressure transmitter with a range of 0–2 Mpa and an accuracy of ±0.25%. Pressure data were displayed and recorded using Wincc host computer software [8,9].

2.2. Relevant Parameter Experimental Data Acquisition

This loop pipe experiment was carried out with the tailings of Sanshandao Gold Mine; the mixing water was tap water, and the cementing material was the cementing material in use in the mine.

The particle size distribution of tailings was measured by a Malvern laser particle sizer, as shown in Table 1. When using XRD phase test analyzer, it was found that the main components of tailings were quartz and mica, non-toxic minerals, insoluble in water. The fluidity of the tailings backfill slurry and the strength of the backfill had little effect. The content of cementitious material + 80 microns was 9.5%, the initial setting time was 45 min, the final setting time was 8 h, and the specific surface area was 750 m^2/kg; thus, the material met the GB175-2007 "General Portland Cement" standard and could be used as a cementing material [10].

Table 1. Tailings grain composition.

Screen/Mesh	+100	−100~+200	−200~+320	−320~+400	−400
Full tailings proportion/%	30.07	14.36	12.06	1.06	42.5

The viscosity and yield stress of the slurry were tested with a BROOKFIELD RST rheometer in the United States. The test results are shown in Table 2. With an increase in slurry concentration, the plastic viscosity and yield stress also increase accordingly, so viscosity, yield stress, and slurry concentration are dependent variables and independent variables [11,12].

Table 2. Slurry rheological parameters.

Concentration	Lime–Sand Ratio	Plastic Viscosity (Pa·s)	Initial Yield Stress (Pa)
68%	0.25	0.159	40.549
70%	0.25	0.216	59.865
72%	0.25	0.322	95.158
74%	0.25	0.486	135.684

According to the above analysis, five independent variables, namely, tailings mass concentration, −400 mesh ratio, lime–sand ratio, flow rate, and pipeline structure, were selected for data modeling analysis. Under the premise of ensuring fluidity, the density test was configured with tailing concentrations of 68%, 70%, 72%, and 74%, with a lime–sand ratio of 1:4, 1:10, and 1:20, and the test trailer pump displacement was set to 7%, 13%, 18%, and 25% to record the pressure data of the horizontal section, the vertical section, and the slope section. This experiment obtained 144 sets of experimental data.

After the filling slurry was fully agitated, the loop conveying experiment was carried out. In this experiment, the filling pump is left in its preset speed and then enters the horizontal pipeline (P1–P2), passes through the vertical pipeline (P2–P3), passes through the horizontal pipeline (P3–P4), flows into the mixing tank through the inclined 5° slope pipeline (P5–P8), etc.

In order to eliminate the influence of the shear thinning of the slurry on the rheological behavior [13,14], each group of experiments lasted for half an hour. The stable pressure value was recorded when the conveying flow was stable, the pressure difference between the pressure monitoring points was calculated, and the pressure difference was divided by distance and converted to a pressure loss value over a distance of 1 km in Mpa/km. First, the loop transport experiment was carried out using the minimum concentration ratio; then, tailings and cementing materials were added to the stirring system; the ratio was changed; the concentration was increased; and the pump delivery flow was changed so as to change the independent variable parameters of the test slurry concentration, the ratio of tailings to cementitious material, the −400 mesh proportion, and the flow velocity [15]. The pressure loss values under different variable conditions were recorded.

3. Establishment and Analysis of Pressure Drop Prediction Model

The random forest algorithm was used to establish the relationship between the pressure drop in the slurry pipeline and its related variables. Random forest is an extended variant of bagging [16]. Classification tree is the theoretical basis of random forest. Random forest adopts the Bootstrap resampling method to build a decision tree model for each sample set, which can be used to solve classification and regression problems. It has a strong generalization ability and good noise immunity, and it has been successfully applied in many fields [17,18].

The random forest algorithm is not sensitive to multicollinearity and can reduce the impact of missing data and unbalanced data on the prediction results. The random forest algorithm is currently considered to be one of the optimal algorithms for nonlinear model prediction [19].

Building the Original Training Set

The random forest regression algorithm realizes the pressure loss prediction of the filling slurry pipeline. The random forest regression process is shown in Figure 2. The main steps are as follows:

(1) Use the sigmoid function to normalize the original data, re-extract b training sets from the data samples with Bootstrap, build a regression decision tree, and use the remaining samples as the test sample set.

(2) In the branching process, the variable smaller than the number of characteristic variables is randomly selected from all feature variables as an alternative branch, and the optimal branch is determined according to the principle of minimum node impurity.

(3) The regression decision tree uses top-down recursive branches, and the number of decision trees is the ntree value [20–22] as the growth termination condition.

(4) The decision trees produced by sampling are combined to form a regression model of random forest, and the mean of the predicted values of all decision trees is output as the prediction result.

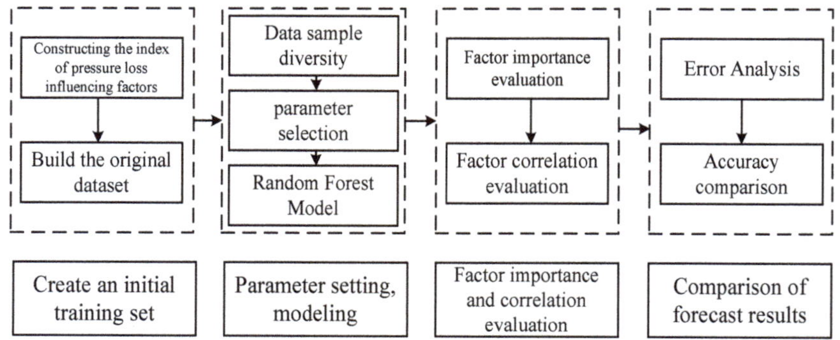

Figure 2. Random forest pressure drop prediction flowchart.

The slurry pressure drop data were obtained through the loop pipe experiment, and some data are shown in Table 3.

Table 3. Test sample data.

Serial Number	Influencing Factors					Evaluation Index
	Pipeline Angle°	Quality Concentration%	Lime–Sand Ratio	Flow Rate m/s	400 Mesh%	Pressure Loss Mpa/km
1	0	68%	0.25	1.32	42	1.177
2	0	70%	0.25	1.68	35	2.688
3	0	72%	0.25	2.2	38	4.111
4	0	74%	0.25	1.28	29	3.584
5	0	68%	0.1	1.36	37	0.987
⋮	⋮	⋮	⋮	⋮	⋮	⋮

For cross-validation, the main parameters of the random forest prediction model were the number of trees ntree = 400 and the number of variables of the random forest classification model mtry = 4. The program segment is shown in Figure 3.

```
def load_data(TrainStartPo, TrainEndPo, TestStartPo, TestEndPo, PredStartPo, PredEndPo, FeatureNum):
    workbook = xlrd.open_workbook(str('C:/Users/Desktop/1.xls'))
    sheet = workbook.sheet_by_name('Sheet1')
    train = [] # Training set
    test = []
    pred = [] # Prediction set
      for load_train in range(TrainStartPo-1, TrainEndPo):
          train.append(sheet.row_values(load_train))      # Load the test sample set
      for load_test in range(TestStartPo-1, TestEndPo):
          test.append(sheet.row_values(load_test))        # Load prediction sample set
      for load_pred in range(PredStartPo-1, PredEndPo):
          pred.append(sheet.row_values(load_pred))        #Transform sample set
    TrainSet = np.array(train)
    TestSet = np.array(test)
    PredSet = np.array(pred)              #Segmentation Features and Target Variables
    x1, y1 = TrainSet[:,:FeatureNum] , TrainSet[:,-1]
    x2, y2 = TestSet[:,:FeatureNum] , TestSet[:,-1]
    x3, y3 = PredSet[:,:FeatureNum] , PredSet[:,-1]
    return x1, y1 , x2, y2, x3, y3
```

Figure 3. Random forest program diagram (random classification of sample data).

4. Results and Discussion

4.1. Importance and Relevance Calculations

Random forest can calculate the importance of a single variable [23,24]. The number of data classifications for the classifier is M, and the results were compared with the correct classification and the random forest classifier. The number of errors of the statistical classifier is N, and the size of the data error can be expressed by Equation (1):

$$erroOB = \frac{N}{M} \tag{1}$$

According to the variable, x (importance score) can be expressed as:

$$score_i = \sum (erroOB_2 - erroOB_1)/ntree \tag{2}$$

where $erroOB_1$ is the out-of-bag data error of the decision tree for each lesson; $erroOB_2$ is the out-of-bag data error of feature x after adding noise interference [25–27]; and ntree is the number of decision trees. According to Formula (2), the importance of pressure drop influencing factors is scored; the sum of all importance scores is scaled to 1; and a stacked bar chart is drawn, which clearly reflects the importance scores of each factor as shown in Figure 4.

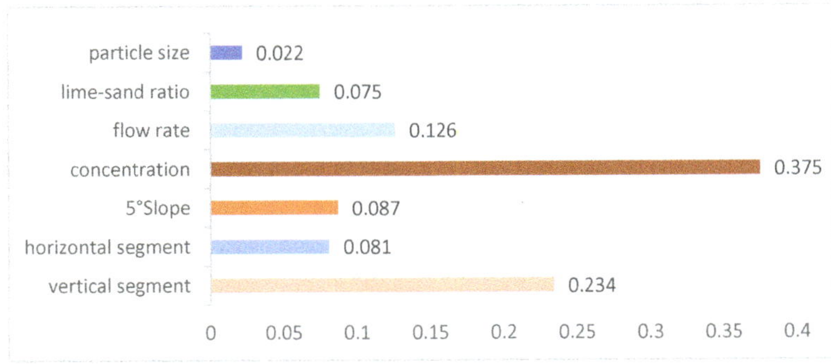

Figure 4. Test variable correlation score map. The abscissa is the weight, and the ordinate is the variable.

Through the analysis, it can be seen that the pipeline structure is the most important variable affecting the pressure drop, and the importance score was 0.402. The pressure drop predicted here is the total pressure drop, which includes the pressure loss pressure drop and the slurry static pressure drop. The importance scores of particle size, cementitious material-to-tailings mass ratio, flow rate, and concentration scores were 0.022, 0.075, 0.126, and 0.375, respectively. Slurry concentration is a secondary pressure drop factor to the pipeline structure, and relatively speaking, the proportion of −400 mesh particle size is the smallest factor.

Through the 5060 multi-function measuring instrument and the self-patented design of the filling pipeline pressure monitoring device, the long-term pressure data of the pipeline pressure during the full tailings filling of the Sanshandao Gold Mine were measured and analyzed. When calculating the pressure distribution of the filling pipeline, the pressure loss of the slurry in the complex area of the pipeline increased by 15% due to bending and joint wear, and the pressure loss of the vertical pipe section of the pipeline increased by 5%.

The correlation between two continuous variables was analyzed by the Pearson correlation coefficient [28,29], and the value range was [−1, 1]. The closer the absolute value of the sample correlation coefficient to 1, the higher the degree of correlation and the closer the relationship, and the correlation of the linear relationship between variables can be reflected by the correlation coefficient. The correlation function is presented as Formula (3):

$$\rho = \frac{\delta_{XY}}{\delta_X \delta_Y} \tag{3}$$

where δ_X and δ_Y are the standard deviation of random variables X and Y, respectively, and δ_{XY} is the covariance of X and Y.

The correlation table is used to express the correlation between different influencing factors and the pipeline pressure drop. The calculated correlation results are shown in Figure 5. Blue represents a positive correlation, red represents a negative correlation, and the area of the circle represents the strength of the correlation. It can be seen from the figure that the correlation between the concentration and the pipeline structure and pressure drop is the largest, and the importance score is the highest. The data come from the loop pipe experiment, and the test tailings are taken from the mine, which has guiding significance for the regulation of the mine filling slurry transportation. It is necessary to strengthen the management and control of these key factors during filling. When monitoring and regulating the abnormal state of the filling pipeline, strengthening the control weight of the factors provides a greater correlation [30].

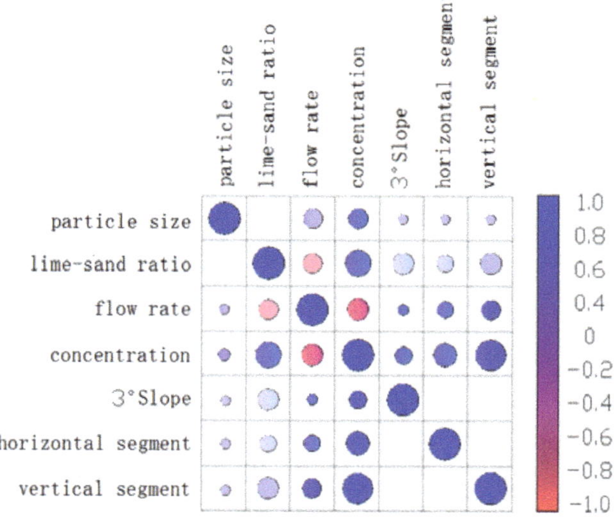

Figure 5. Correlation graph.

4.2. Pressure Drop Prediction Results

First, the original data were de-statically processed; the measured data were subtracted from the pressure change in the slurry due to the action of gravity; and the pressure P_i of the new data set was obtained by Formula (4), where the density of the test slurry was obtained by Formula (5) [31–33]:

$$P_i = \Delta P - \rho g h \tag{4}$$

$$\rho = \rho_0 \times 1 / \left(\frac{\rho_0 C_w \cdot (n\gamma_s + \gamma_c)}{\gamma_c \cdot \gamma_s (1+n)} + 1 - C_w \right) \tag{5}$$

where ΔP is the test pressure difference, Pa; ρ is the density of the test slurry, kg/m^3; g is the acceleration of gravity, m/s^2; is the vertical height difference, m; ρ_0 is the density of the water, and the industrial value is 1.02×10^3 kg/m^3; C_W is the slurry mass concentration, %; n is the lime–sand ratio; and γ_c is the true density of cementitious materials and the true density of tailings, kg/m^3.

A new data set was constructed, 121 sets of data were extracted from it to construct training samples of the random forest model, and the remaining 23 sets of data were test samples. Five eigenvalues were selected, and the Anaconda3 development environment was used to complete the model establishment using Python language [34,35]. First, the model was run using the samples during training, and the trained random forest model was used to perform regression fitting on the test sample set. The prediction result of the regression fitting on the test sample set is shown in Figure 6.

It can be seen in Figure 6 that the prediction accuracy of the pressure drop prediction model based on the random forest algorithm is high. The goodness of fit between the prediction and the actual value in the test set is 0.9747, and the mean square error is 0.0011; the value goodness of fit is 0.983. The value range of goodness of fit is [0, 1]. The larger the value, the higher the degree of fit [36,37], and the closer the mean square error MSE is to 0, indicating that the error between the predicted data and the original data is smaller. The predicted value of the model is very close to the measured value, which verifies the feasibility of the random forest model to predict the slurry pressure drop in the loop experiment [38,39].

4.3. Comprehensive Evaluation of Forecast Results

In order to further verify the accuracy of the random forest model algorithm in the application of slurry pressure drop loss prediction [40–42], the BP artificial neural network and the polynomial linear fitting method were used for comparison [43], and the goodness of fit R2 and the mean square error MSE were used to determine the prediction accuracy of the model. The obtained error comparison results are shown in Table 4.

Table 4. Prediction model error comparison table.

Model	Performance	
	R^2	MSE
Random forest algorithm	0.9747	0.0011
BP artificial neural network	0.9538	0.0512
Linear fit	0.9326	0.1862

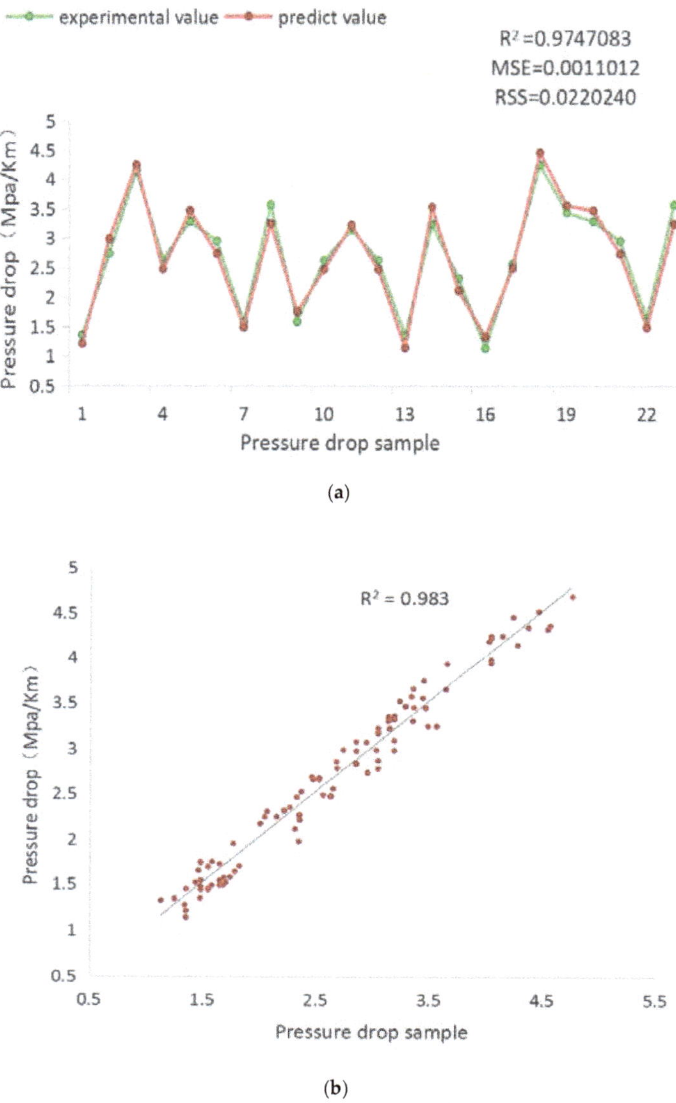

Figure 6. Model prediction: (**a**) results of fitting between the predicted values of the test set and the actual values; (**b**) results of fitting between the predicted values of the training set and the measured values.

5. Results

The following conclusions can be drawn from the analysis of the experimental data of the tailings loop and the established random forest prediction model at this stage:

(1) The biggest factor affecting the change in slurry pressure is the vertical pipeline structure, followed by the slurry concentration. The proportion of the tailings −400 mesh particle size has little effect on the slurry pressure drop.

(2) The correlation analysis of the data shows that the slurry pressure loss is positively correlated with the slurry concentration, ratio, and flow rate. The fluidity of the slurry is negatively correlated with the pressure drop, concentration, and the ratio of the slurry.

(3)   The random forest pressure loss model established based on the experimental data of the loop pipe has a high prediction accuracy. The goodness of fit between the experimental value and the predicted value on the test set and training set is 0.9747 and 0.983, The prediction accuracy is higher than BP neural network. Based on polynomial linear fitting, it can replace the complex loop experiment to carry out the intelligent aided design of filling systems.

(4)   The algorithm model can be used to predict the pressure of the filling pipeline by learning the pressure distribution data of the mine filling pipeline. To provide ideas for follow-up research, the algorithm can be used in combination with the automatic system to realize the judgment and early warning of the abnormal state of the filling pipeline by predicting the pipeline pressure, and the development of "smart back-filling" technology can be promoted.

Author Contributions: Conceptualization, Z.W. (Zengjia Wang) and Y.K.; methodology, Z.W. (Zengjia Wang) and Z.W. (Zengbin Wang); software, Z.W. (Zengjia Wang) and Z.W. (Zengbin Wang); validation, Z.W. (Zengbin Wang), Z.W. (Zaihai Wu) and J.G.; formal analysis, Y.K.; investigation, Z.W. (Zengjia Wang) and Z.W. (Zaihai Wu); resources, J.G.; data curation, Z.W. (Zengbin Wang); writing—original draft preparation, Z.W. (Zengbin Wang); Z.W. (Zengjia Wang) and Y.K. serve as directors of the program, and Obtained all necessary funds. All authors have read and agreed to the published version of the manuscript.

Funding: This research was supported by the National Key R&D Program of the 13th Five-Year Plan (2018YFC0604600) and the Shandong Provincial Major Science and Technology Innovation Project (2019SDZY05).

Conflicts of Interest: The authors declare no conflict of interest.

References

1. Cai, M.; Xue, D.; Ren, F. Current status and development strategy of metal mines. *J. Eng. Sci.* **2019**, *41*, 417–426.
2. Li, X.; Zhou, S.; Zhou, Y.; Min, C.; Cao, Z.; Du, J.; Luo, L.; Shi, Y. Durability Evaluation of Phosphogypsum-Based Cemented Backfill Through Drying-Wetting Cycles. *Minerals* **2019**, *9*, 321. [CrossRef]
3. Wang, Z.; Qi, Z.; Kou, Y.; Yang, J.; Song, Z.; Jia, H. Smart filling system enables new development of mines? *Min. Res. Dev.* **2022**, *1*, 156–161.
4. Kumar, U.; Mishra, R.; Singh, S.N.; Seshadri, V. Effect of particle gradation on flow characteristics of ash disposal pipelines. *Powder Technol.* **2003**, *132*, 39–51. [CrossRef]
5. Qi, C.; Yang, X.; Li, G.; Xu, Y.; Kiu, X.; Cao, X.; Huang, C.; Liu, E.; Qian, S.; Liu, X.; et al. Research status and perspectives of the application of artificial intelligence in mine backbackfilling. *J. China Coal Soc.* **2021**, *46*, 688–700.
6. Qi, C.; Fourie, A. Cemented paste backfill for mineral tailings management: Review and future perspectives. *Miner. Eng.* **2019**, *144*, 106025. [CrossRef]
7. Dong, L.; Tong, X.; Ma, J. Quantitative Investigation of Tomographic Effects in Abnormal Regions of Complex Structures. *Engineering* **2021**, *7*, 1011–1022. [CrossRef]
8. Wei, C.; Wang, X.; Zhang, Y.; Zhang, Q. Paste-like cemented backfilling technology and rheological characteristics analysis based on jigging sands. *J. Central South Univ.* **2017**, *24*, 155–167. [CrossRef]
9. Ouattara, D.; Yahia, A.; Mbonimpa, M.; Tikou, B. Effects of superplasticizer on rheological properties of cemented paste backfills. *Int. J. Miner. Process.* **2017**, *161*, 28–40. [CrossRef]
10. Zhou, Z. *Machine Learning*; Tsinghua University Press: Beijing, China, 2016; pp. 178–181.
11. Russell, S.; Norvig, P. *Artificial Intelligence: A Modern Approach*, 3rd ed.; Pearson: London, UK, 2011; Volume 175, pp. 935–937.
12. Zhang, J.; Guo, W.; Song, B.; Zhou, Y.; Zhang, Y. Performance Prediction of Asphalt Pavement Based on Random Forest. *J. Beijing Univ. Technol.* **2021**, *47*, 1256–1263.
13. Ji, W.; Liu, Y.; Cai, J.; Wang, B. Mineral pressure prediction method based on random fores. *Chin. J. Min. Rock Form. Control Eng.* **2021**, *3*, 71–81.
14. Yao, D.; Yang, J.; Zhan, X. Feature selection algorithm based on random forest. *J. Jilin Univ. Eng. Sci.* **2014**, *44*, 137–141.
15. Pearson, K. Notes on the History of Correlation. *Biometrika* **1920**, *13*, 1–16. [CrossRef]
16. Hu, Y.; Zhang, L.; Yuan, F.; Li, T.; Wu, X.; Deng, T. Research on concrete strength prediction based on random forest. *Constr. Technol.* **2020**, *49*, 89–94.
17. Zhang, Q.; Liu, W.; Wang, X.; Chen, Q. Optimal prediction model of backfill paste rheological parameters. *J. Cent. South Univ. Sci. Technol.* **2018**.

18. Wang, Z.; Wu, A.; Wang, H. Prediction on the InterfaceShear Strength of Backfill and Surrounding Rock Based on PSO-BPNN Algorithm. *Min. Res. Dev.* **2020**, *40*, 130–134.

19. Hojamberdiev, M.; Arifov, P.; Tadjiev, K. Processing of refractory materials using various magnesium sources derived from Zinelbulak talc-magnesite. *J. Miner. Metall. Mater. Eng. Ed.* **2011**, *18*, 10. [CrossRef]

20. Qi, C.; Fourie, A.; Chen, Q. Neural network and particle swarm optimization for predicting the unconfined compressive strength of cemented paste backfill. *Constr. Build. Mater.* **2018**, *159*, 473–478. [CrossRef]

21. Sun, W.; Wu, A.X.; Hou, K.P.; Yang, Y.; Liu, L.; Wen, Y.M. Experimental study on the microstructure evolution of mixed disposal paste in surface subsidence areas. *Minerals* **2016**, *6*, 43. [CrossRef]

22. Dong, L.; Zhou, Y.; Deng, S.; Wang, M.; Sun, D. Evaluation methods of man-machine-environment system for clean and safe production in phosphorus mines: A case study. *J. Central South Univ.* **2021**, *28*, 3856–3870. [CrossRef]

23. Wu, A.; Ruan, Z.; Bürger, R.; Yin, S.; Wang, J.; Wang, Y. Optimization of flocculation and settling parameters of tailings slurry by response surface methodology. *Miner. Eng.* **2020**, *156*, 106488. [CrossRef]

24. Wu, J.; Yin, Q.; Gao, Y.; Meng, B.; Jing, H. Particle size distribution of aggregates effects on mesoscopic structural evolution of cemented waste rock backfill. *Environ. Sci. Pollut. Res.* **2021**, *28*, 16589–16601. [CrossRef] [PubMed]

25. Cao, G.; Wei, Z.; Wang, W.; Zheng, B. Shearing resistance of tailing sand waste pollutants mixed with different contents of fly ash. *Environ. Sci. Pollut. Res.* **2020**, *27*, 8046–8057. [CrossRef] [PubMed]

26. Qiu, J.; Guo, Z.; Yang, L.; Jiang, H.; Zhao, Y. Effect of tailings fineness on flow, strength, ultrasonic and microstructure characteristics of cemented paste backfill. *Constr. Build. Mater.* **2020**, *263*, 120645. [CrossRef]

27. Gao, J.; Fourie, A. Spread is better: An investigation of the mini-slump test. *Miner. Eng.* **2015**, *71*, 120–132. [CrossRef]

28. Li, H.; Wu, A.; Cheng, H. Analysis of conical slump shape reconstructed from stereovision images for yield stress prediction. *Cem.Concr. Res.* **2021**, *150*, 106601. [CrossRef]

29. Qi, C.; Guo, L.; Wu, Y.; Zhang, Q.; Chen, Q. Stability Evaluation of Layered Backfill Considering Filling Interval, Backfill Strength and Creep Behavior. *Minerals* **2022**, *12*, 271. [CrossRef]

30. Wu, S.Z.; Wang, M.N.; Yu, L.; Liu, D.-G.; Huang, Q.-W. Model test study on stress characteristics of backfill to segment in TBM tunnel. *Rock Soil Mech.* **2018**, *39*, 3976–4009.

31. Benzaazoua, M.; Fall, M.; Belem, T. A contribution to understanding the hardening process of cemented pastefill. *Miner. Eng.* **2004**, *17*, 141–152. [CrossRef]

32. Li, S.; Zhang, R.; Feng, R.; Hu, B.; Wang, G.; Yu, H. Feasibility of Recycling Bayer Process Red Mud for the Safety Backfill Miningof Layered Soft Bauxite under Coal Seams. *Minerals* **2021**, *11*, 722. [CrossRef]

33. Qi, C.; Xu, X.; Chen, Q. Hydration reactivity difference between dicalcium silicate and tricalcium silicate revealed from tructural and Bader charge analysis. *Int. J. Miner. Met. Mater.* **2022**, *29*, 335–344. [CrossRef]

34. Le, Z.-H.; Yu, Q.-L.; Pu, J.-Y.; Cao, Y.-S.; Liu, K. A Numerical Model for the Compressive Behavior of Granular Backfill Based on Experimental Data and Application in Surface Subsidence. *Metals* **2022**, *12*, 202. [CrossRef]

35. Zhang, P.F.; Zhang, Y.B.; Zhao, T.B.; Tan, Y.L.; Yu, F.H. Experimental research on deformation characteristics of waste-rock materialin ynderground backfill mining. *Minerals* **2019**, *9*, 102. [CrossRef]

36. Li, M.; Li, A.L.; Zhang, J.X.; Huang, Y.L.; Li, J.M. Effects of particle sizes on compressive deformation and particle breakage of gangue used for coal mine goaf backfill. *Powder Technol.* **2020**, *360*, 493–502. [CrossRef]

37. Meng, G.H.; Zhang, J.X.; Li, M.; Zhu, C.L.; Zhang, Q. Prediction of compression and deformation behaviours of gangue backfill materials under multi-factor coupling effects for strata control and pollution reduction. *Environ. Sci. Pollut. Res. Int.* **2020**, *27*, 36528–36540. [CrossRef]

38. Wang, X.; Xie, J.; Xu, J.; Zhu, W.; Wang, L. Effects of Coal Mining Height and Width on Overburden Subsidence in Longwall Pier-Column Backfilling. *Appl. Sci.* **2021**, *11*, 3105. [CrossRef]

39. Wang, R.; Zeng, F.; Li, L. Applicability of Constitutive Models to Describing the Compressibility of Mining Backfill: A Comparative Study. *Processes* **2021**, *9*, 2139. [CrossRef]

40. Keita, A.M.T.; Jahanbakhshzadeh, A.; Li, L. Numerical analysis of the stability of arched sill mats made of cemented backfill. *Int. J. Rock Mech. Min. Sci.* **2021**, *140*, 104667. [CrossRef]

41. Cui, L.; Fall, M. A coupled thermo–hydro-mechanical–chemical model for underground cemented tailings backfill. *Tunn. Undergr. Space Technol.* **2015**, *50*, 396–414. [CrossRef]

42. Huan, C.; Zhang, S.; Zhao, X.; Li, S.; Zhang, B.; Zhao, Y.; Tao, P. Thermal Performance of Cemented Paste Backfill Body Considering Its Slurry Sedimentary Characteristics in Underground Backfill Stopes. *Energies* **2021**, *14*, 7400. [CrossRef]

43. Zhao, P.; Li, X.Z.; Zhang, Y.; Liu, K.; Lu, M.H. Stratified thermal response test measurement and analysis. *Energy Build.* **2020**, *215*, 109865. [CrossRef]

sensors

Article

Automatic Implementation Algorithm of Pressure Relief Drilling Depth Based on an Innovative Monitoring-While-Drilling Method

Zheng Wu, Wen-Long Zhang * and Chen Li

School of Energy and Mining Engineering, China University of Mining and Technology (Beijing),
Beijing 100083, China; bqt1900101036@student.cumtb.edu.cn (Z.W.); 13120008810@163.com (C.L.)
* Correspondence: wenlong0523@163.com

Abstract: An innovative monitoring-while-drilling method of pressure relief drilling was proposed in a previous study, and the periodic appearance of amplitude concentrated enlargement zone in vibration signals can represent the drilling depth. However, there is a lack of a high accuracy model to automatically identify the amplitude concentrated enlargement zone. So, in this study, a neural network model is put forward based on single-sensor and multi-sensor prediction results. The neural network model consists of one Deep Neural Network (DNN) and four Long Short-Term Memory (LSTM) networks. The accuracy is only 92.72% when only using single-sensor data for identification, while the proposed multiple neural network model could improve the accuracy to being greater than 97.00%. In addition, an optimization method was supplemented to eliminate some misjudgment due to data anomalies, which improved the final accuracy to the level of manual recognition. Finally, the research results solved the difficult problem of identifying the amplitude concentrated enlargement zone and provided the foundation for automatically identifying the drilling depth.

Keywords: vibration signals; neural network; drilling state identification algorithm; drilling depth; monitoring-while-drilling method

Citation: Wu, Z.; Zhang, W.-L.; Li, C. Automatic Implementation Algorithm of Pressure Relief Drilling Depth Based on an Innovative Monitoring-While-Drilling Method. *Sensors* **2022**, *22*, 3234. https://doi.org/10.3390/s22093234

Academic Editors: Longjun Dong, Yanlin Zhao and Wenxue Chen

Received: 22 March 2022
Accepted: 21 April 2022
Published: 22 April 2022

Publisher's Note: MDPI stays neutral with regard to jurisdictional claims in published maps and institutional affiliations.

1. Introduction

Vibration signals are often used in sensor monitoring [1–3], defect diagnosis [4–7] and engineering applications, such as gearboxes [8–10], aero-engines [11,12] and wind turbines [13]. Vibration signals in underground coal mining are often used to help the analysis of mining dynamic disasters, stress environments, drilling process and the state of surrounding rocks [14–19]. Microseismic (MS) monitoring, aimed at monitoring vibration signals, can detect the dynamic event of the surrounding rock and predict the rock burst disaster. Acoustic emission (AE) technology is widely used in the field of geotechnical engineering [20–22]. The AE signal contains spatial information about the complex structural distribution inside the material [23,24] and vast key information about the rock fracture evolution process [25]. The AE tomography can detect early internal damages, faults and abnormal regions with the distribution of velocity field in the drilling process [26,27]. A monitoring-while-drilling (MWD) system can employ the drilling signals to monitor the quality of borehole constructions, obtain the information of surrounding rocks and other useful information during the drilling process [28,29].

MS monitoring has gradually become the most conventional means of coal mine vibration signal analysis and is now widely used in China's underground coal mines [30–33]. The mature technology of MS monitoring has solved a series of problems for actual underground coal mining. Similarly, drilling construction is an indispensable process for coal mine production, such as providing support, pressure relief and other required drilling engineering. Pressure relief drilling (PRD) is often used to reduce stress concentrations and avoid dynamic hazard events, such as rock bursts [34,35]; signal analysis during the drilling process can provide guidance for the drilling quality analysis and construction process. In the previous studies, Zhou et al. [36] proposed a hybrid rock recognition approach that combined Gaussian process regression with clustering, and employed MWD data and the adjusted penetration rate to achieve automated rock recognition. Liu et al. [37] analyzed the relationship between the transverse, longitudinal and torsional vibration of the drill rod and the properties of the rock being drilled. Zhang et al. [38] studied the drilling amplitude signals collected by the MS equipment, which can determine the drilling difficulty areas, sticking drilling and vibration events. Pu et al. [39] investigated the performance of ten frequently used machine learning models for MS/blasting event recognition. Faradonbeh et al. [40] discussed the applicability of three data mining techniques along with five conventional criteria to predict the occurrence of rock bursts in a binary condition.

The study results of [38] (Figure 1) pointed out that the periodic appearance of amplitude clusters can represent the drilling depth, which was of great significance to the automatic statistics of drilling depth and the supervision of workload. However, there is a lack of a method that can automatically and accurately identify amplitude clusters in the entire process of borehole construction. There are misjudgments caused by human subjective consciousness as only relying on the manual identification of amplitude clusters, which is inefficient and cannot meet the requirements of the efficient and safe operation of the mine.

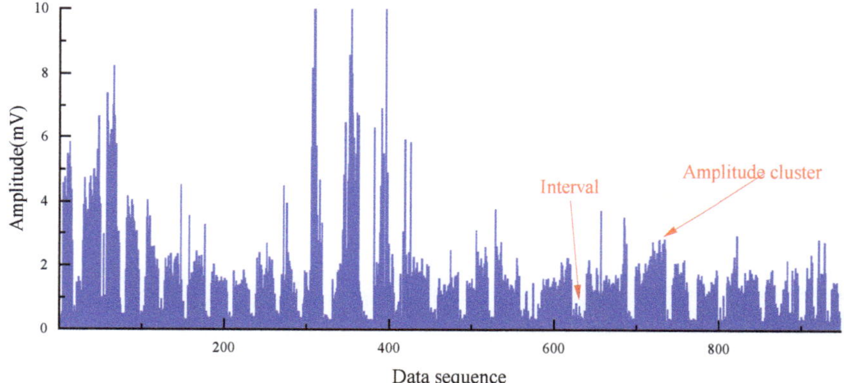

Figure 1. Amplitude clusters and intervals in amplitude data during drilling.

In this study, a deep learning method is used to analyze the drilling amplitude signals collected by MS monitoring equipment. A fusion neural network is obtained based on Deep Neural Network (DNN) [41–44] and Long Short-Term Memory (LSTM) [45–48] algorithms, which can automatically distinguish and determine drilling information, such as the amplitude clusters, drilling start point, termination point and drilling duration of each section in an efficient and accurate manner.

The proposed rig drilling status identification algorithm can efficiently and accurately identify and analyze drilling operations, such as pressure relief drilling, support drilling in the underground coal mine, and obtain the actual construction length, construction sequence, time spent on each process and other details of the construction operations, as well as additional information, such as drilling difficulty and abnormal vibration information. It has far-reaching significance to ensure the quality and safe operation of underground construction and obtaining the properties of the roadway surrounding rocks.

The main contributions are as follows:

(1) Divide all data into the training set, validation set and test set. The test set is not involved in the training and tuning of the neural network and is only used as the data for the final model effect evaluation to avoid the problem of information leakage that leads to the fake high identification accuracy of the neural network. The training and validation sets are divided by the stratified K-fold cross-validation method to find the optimal hyperparameters in the model training and tuning, which eliminates the influence of the imbalanced amount of data between the two categories on the model.

(2) An efficient, automatic and precise neural network model is proposed to identify the drilling status of drilling rigs by drilling amplitude signals, which can fuse the data from single and multiple sensors, and the identification results from different neural networks.

(3) An optimization method is presented, which is similar to "submerge" for two types of recognition anomalies caused by data in drilling state recognition by the neural network identification algorithm.

The remainder of this paper is organized as follows. In Section 2, we describe where and how we collected the drilling signal data and introduce basic concepts related to the neural network algorithm that is used in this paper. In Section 3, we demonstrate the composition and division of the dataset of the neural network in this paper, the preprocessing of the data and the structure of the proposed neural network recognition algorithm. In Section 4, we present and analyze the recognition results of the proposed algorithm, perform recognition error analysis, propose an error handling method and show the final recognition results. Finally, conclusions and future works are provided in Section 5.

2. Research Methods

2.1. Data Collection Method

MS monitoring equipment was selected to monitor the drilling process of PRD boreholes, which were located in three different underground coal mines in the Shandong Province, China. The PRD boreholes were drilled by a CMS1-6500/75 drill rig (shown in Figure 2a) with a length of 1 m per drill rod, and a new drill rod was added after each rod was drilled. The PRD boreholes were located at the side of the roadway, 1.5 m away from the roadway floor, with a drilling diameter of 150 mm and a design drilling depth of 30 m.

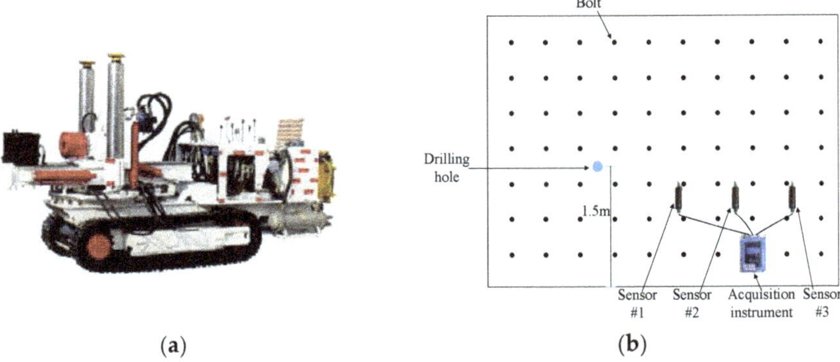

Figure 2. Drill rig and MS equipment layout. (**a**) CMS1-6500/75 drill rig; (**b**) Drill hole and MS equipment layout at roadway side.

The MS monitoring system was arranged on one side of the PRD borehole, and contained three sensors, which are arranged as shown in Figure 2b. The three sensors are at different distances with the same PRD drilling hole, and the amplitudes of the drilling signals collected for the same drilling hole are different. When the drilling position changes, it cannot be guaranteed that a certain sensor always collects the maximum, minimum or middle amplitudes. In order to eliminate the influence of the distance from the borehole, the average value of the vibration amplitude signals collected by the three sensors was calculated, so that the amplitude signals of Sensor #1, Sensor #2 and Sensor #3 would be arranged from the smallest to the largest.

2.2. Neural Network Algorithm

For the problem of identifying the drilling status of the drilling rig, the collected drilling signals were learned and trained with the help of the excellent judgment accuracy of the deep neural network. For the identification of the drilling state of the drilling rig, it can be simplified as a binary classification [49]: the signal points in the drilling state as 1 (positive sample) and the signal points in extension of the drill rod, the non-drilling state, as 0 (negative sample). The middle layers of the neural network use a Rectified Linear Unit (ReLU) as the activation function, and the last layer uses a sigmoid activation function to output a probability value in the range of 0 to 1. The ReLU function resets all negative values to zero, while the sigmoid function "compresses" any value to the interval [0, 1], and its output value can be regarded as a probability value; the expressions of the two activation functions are given in Equations (1) and (2). Therefore, the network uses a binary cross-entropy loss function to calculate the loss, as in Equation (3).

$$\text{ReLU}(x) = \left\{ \begin{array}{ll} x & if\ x > 0 \\ 0 & if\ x \leq 0 \end{array} \right. \tag{1}$$

$$S(x) = \frac{1}{1 + e^{-x}} \tag{2}$$

$$loss = -\frac{1}{N} \left[\sum_{i=1}^{N} y_i \cdot \log(p(y_i)) + (1 - y_i) \cdot \log(1 - p(y_i)) \right] \tag{3}$$

where N represents the number of samples; y_i represents the label value of sample i; and $p(y_i)$ represents the predicted probability value of the label value of sample i.

The practice of training and testing the model on all datasets is problematic, which can lead to the rapid over-fitting of the model on that dataset. Therefore, we divided the dataset into training set, verification set and test set in order to obtain a generalized model. The model was trained and learned on the training set, and the hyperparameters of the model were adjusted using the performance of the model on the validation set as a feedback signal. However, this causes information leakage when the model parameters are tuned multiple times on the validation set, and the model quickly over-fits on the validation set. So, we created a completely unused dataset, a test set, to evaluate the model, and the best parameters were determined by grid search [50] techniques. Then, the optimal parameters were used to re-train the model on all the training sets, and the effect of the model was finally evaluated on the test set.

Depending on the number of data points and the way the validation set is divided, it may result in a large variance in the validation scores, which makes it impossible to evaluate the model reliably. In this case, the best practice is to use the K-fold cross-validation [51] shown in Figure 3. This method divides the available data into K folds, instantiates K identical models, trains each model on K-1 folds and evaluates it on the remaining one. The validation score of the model is equal to the average of the K validation scores.

However, some classification problems may also show a large imbalance in the distribution of target classes, for example, there may be several times more negative samples than positive ones. In such cases, stratified sampling is used to ensure that the relative class frequencies are approximately the same in each training and validation fold. Stratified K-fold cross-validation [52] (Figure 4) is a variant of K-fold cross-validation, which returns stratified folds, with each fold containing roughly the same percentage of samples for each target category as the entire collection.

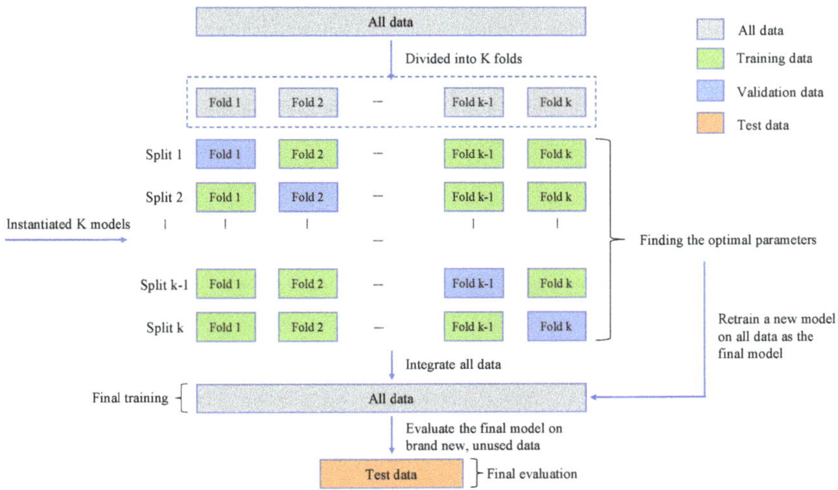

Figure 3. Schematic diagram of the K-fold cross-validation.

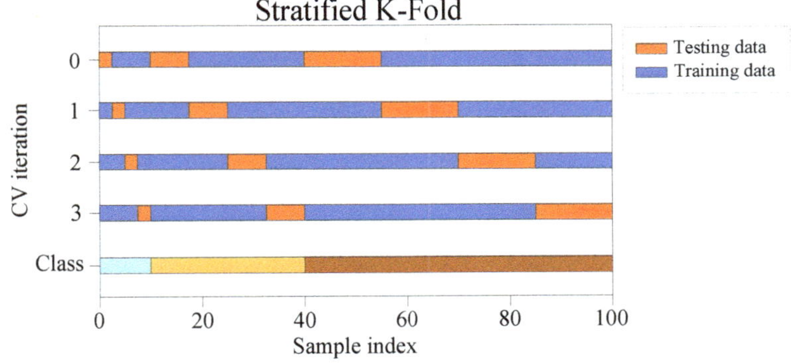

Figure 4. Schematic diagram of stratified K-fold cross-validation.

3. Design of Experiment

3.1. Composition of Experimental Data

We carried out drilling signal data acquisition in three coal mines (marked as mine A, B and C) in the Shandong Province, China. For mines A and B, the drilling signal data of one borehole was collected in each mine, marked as A1, B1. Additionally, the drilling signal data of five boreholes were collected in mine C, marked as C1~C5, with seven boreholes drilling rig amplitude data in total. In order to train and obtain a generalized neural network model, the amplitude data of one borehole A1 in mine A and two boreholes (C1, C3) in mine C were used as training data. The designed neural network was trained and verified by stratified K-fold cross-validation, and the data of one borehole B1 in B mine and three other holes (C2, C4, C5) in C mine are used as the test dataset to test the identification accuracy of the final model.

3.2. Pre-Processing of Experimental Data

The PRD drilling amplitude data from the training and validation sets were sorted, and the vibration signals within the drilling time were filtered according to the actual drilling time. The data points were manually labelled as 0 or 1 according to the time corresponding to the drilling state of the rig recorded during the field construction and the size of the drilling signal amplitude value (1 is the drilling state, 0 is the state of connecting the drill rod). The collected raw vibration signal data are used as raw data, as shown in Figure 5.

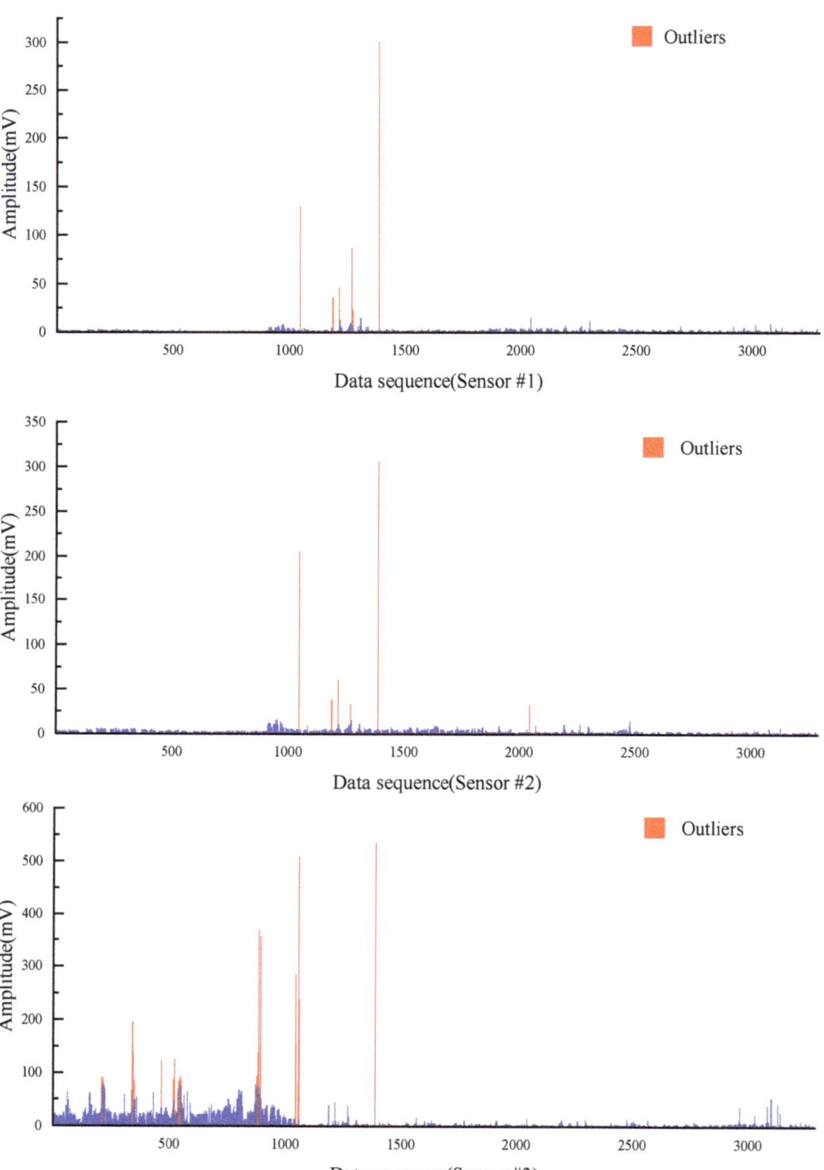

Figure 5. Original data of all 3 sensors.

It can be easily seen from the Figure 5 that the collected vibration signal data contain some points with abnormally large amplitude values, which seriously deviate from the range of other amplitude value distributions; these outliers are randomly present in two different drilling states and the locations of outliers collected by different sensors may be different from each other. If these outliers are retained as inputs to the model, they have a great disturbance and impact on the subsequent training of the model and the accuracy of the final model; thus, it is necessary to remove the outliers from the input data before the model training process to avoid the model learning the wrong information and to ensure that the training and accuracy of the model are not affected by the outliers. Therefore, the 3σ principle [53,54] was used to filter the raw data in order to eliminate the influence of outliers on the model, that is, the $(\mu - 3\sigma, \mu + 3\sigma)$ in each set of datasets is taken as the screening criterion for the outlier data. For the vibration signal data collected by each sensor, the abnormal value points exceeding 3σ are removed, which is shown in Figure 6.

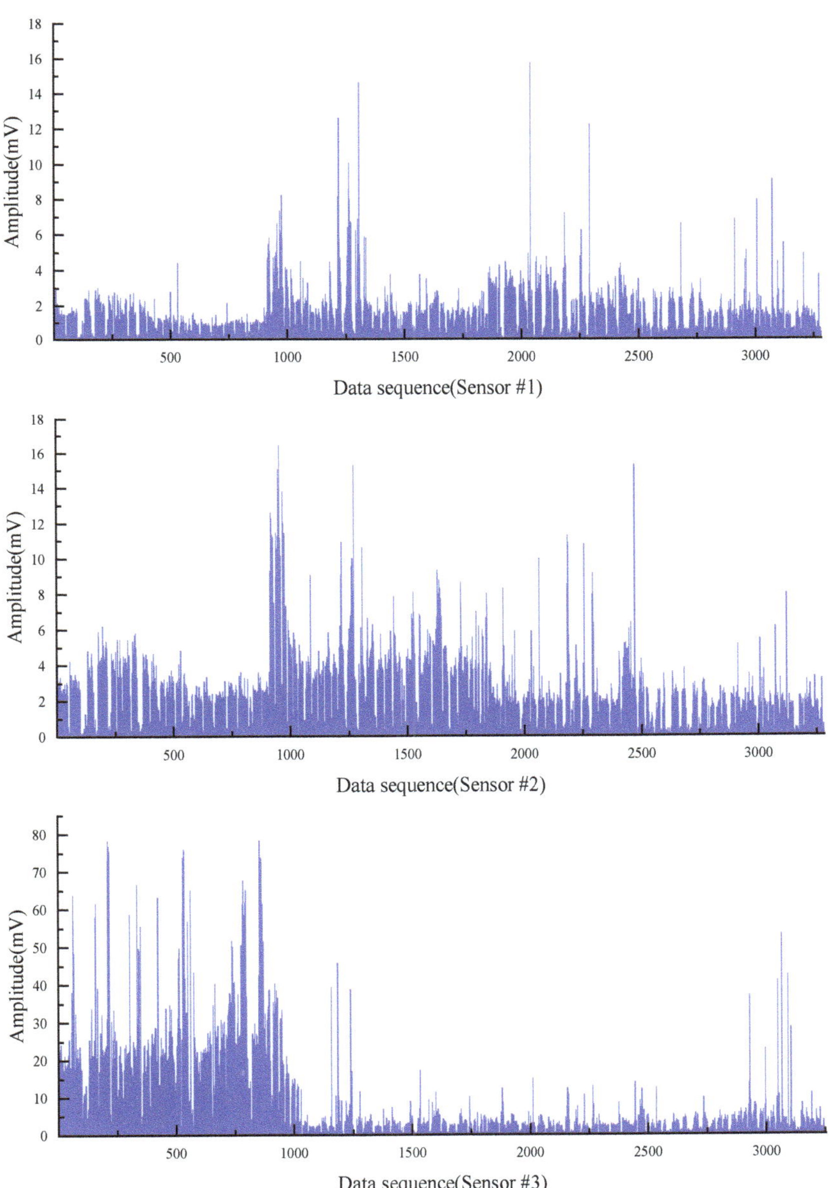

Figure 6. Amplitude signal data after removing the outliers.

The distribution of amplitude cluster and intervals can be seen clearly after removing the outliers, and the purpose of removing outliers and revealing the characteristics of the data was preliminarily achieved. To determine whether the two types of labels of the data are distinguishable in terms of the vibration amplitude index, the vibration signals data were represented according to the label classification as shown in Figure 7. For Sensor #1, the average amplitude of the drilling state is 1.85 and the average amplitude of the connecting state is 0.38. For Sensor #2, the average amplitude of the drilling state is 3.04 and the average amplitude of the connecting state is 0.47. For Sensor #3, the average amplitude of the drilling state is 9.87 and the average amplitude of the connecting state is 3.63. The vibration signals of the two states are well differentiated in terms of amplitude mean and maximum amplitude, and the difference features can be trained and learned by the designed neural network. Since the raw data are a combination of drilling amplitude values from three different boreholes in two different mines, there are some differences between the magnitude of amplitude values in different boreholes. In each individual borehole drilling data, the amplitude values of the two drilling states are still well separable. Therefore, the vibration amplitude can be used as a classification indicator to distinguish between the two drilling states.

Since the experimental data were divided into two categories and the number of data contained within the two categories was not equal, to make the model fully learn the characteristics of different types of data and improve its prediction accuracy, the stratified K-fold cross-validation method was used to avoid the model learning the characteristics only from one type of data, while the features of the other type are not sufficient learned. This study used stratified 10-fold stratified cross-validation. The training data were divided into training sets and validation sets in order to avoid information leakage caused by the model being adjusted directly on the test set during the learning process, which means that the entire training data were divided into 10 folds and the proportion of the two categories in each fold was approximately the same as the proportion in the total data set. The training was performed on 9 folds of the data each time, and the remaining 1 fold was used as the validation set to verify the model; the model effect was tested on a separate unused test set after the final model was trained.

3.3. Drilling State Identification Neural Network

The data, after the outlier removal and normalization process conducted in the previous section, were used as the input data of the neural network, and the neural network model was established using the LSTM and DNN methods. Three independent LSTM networks were built using single data from each of the three sensors as input data; one LSTM network and one DNN network were built using all the amplitude data collected by all three sensors as the input data. A total of five neural networks were established, and the architecture of the neural network is shown in Figure 8. Each neural network model was trained and validated separately, and the five neural networks jointly judged the drilling status of the rig using the respective vibration amplitudes and all vibration amplitudes collected by different sensors at the same moment. The entire process was constructed so that the amplitude data collected by the three sensors as a whole were input to the LSTM network and DNN network as input data, and the three different single sensor data were input to the LSTM networks #1, #2 and #3. The five sub-neural networks reveal the discrimination results of drilling state of drilling rig. For a certain moment, three drilling amplitude signals of the rig and five judgment results are obtained. Then, the three or more than three same drilling states were taken as the final drilling state result according to the majority rule.

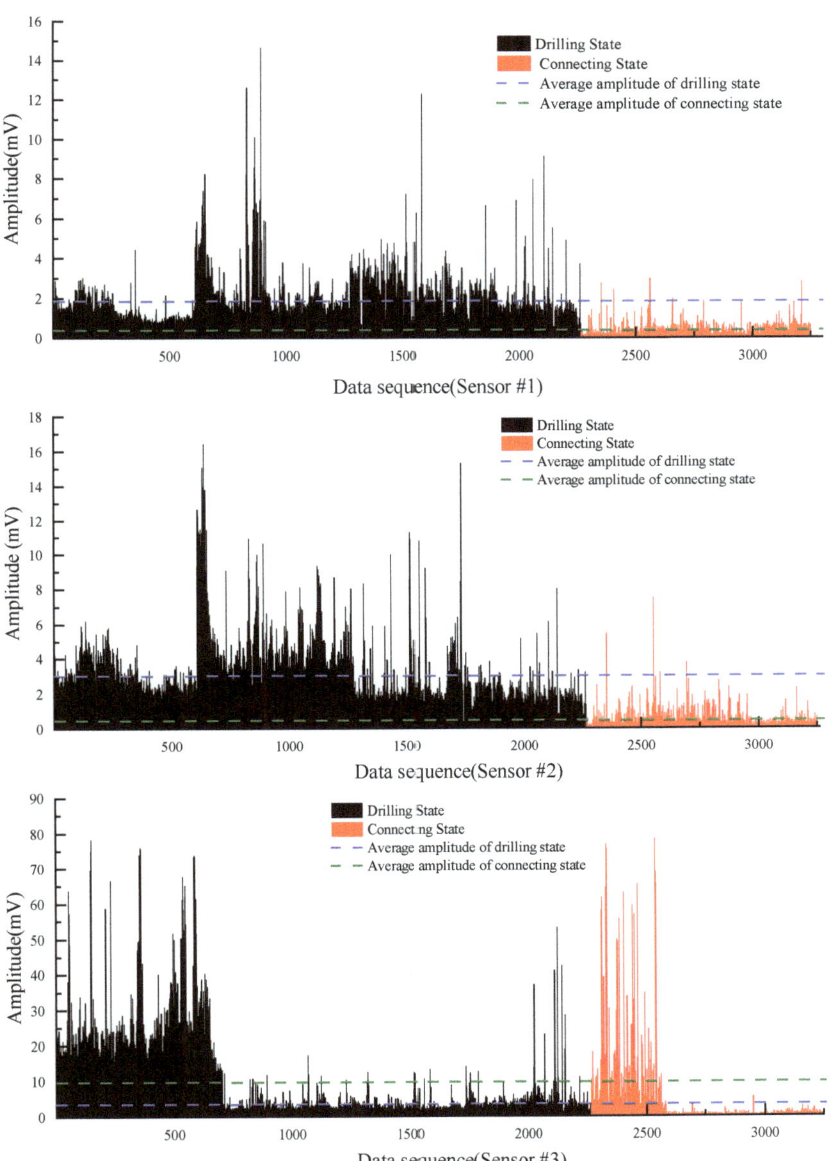

Figure 7. Separability of the two states.

The five sub-neural network models were trained and evaluated separately using the data in the training set as input data, and the feedback data (accuracy, Receiver Operating Characteristic and Area Under ROC Curve) obtained on the validation set were used to adjust and optimize the structure and parameters of the neural network (such as epochs, number of layers of deep neural network and number of neurons per layer). The training and adjustment were continued until the judgment performance of each neural network reached a good judgment accuracy rate. The final neural network model structure and parameters were determined after the comparison of the accuracy of model prediction results with different parameters.

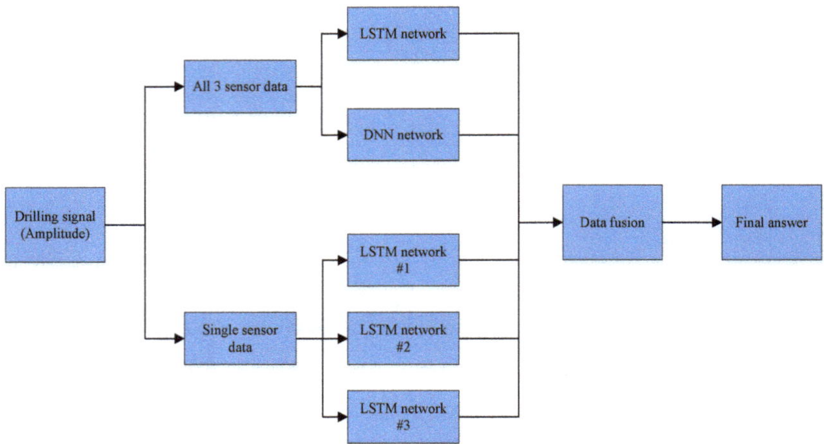

Figure 8. The architecture of the proposed identification neural network model.

After determining the final parameters, the training and validation sets were integrated into one training set. A new neural network was re-established according to the optimal neural network structure parameters and the training was restarted to ensure that each model could learn from the entire training set. That is, all the data were initially divided into training sets and verification sets to obtain the final neural network model, which was then applied to the data in the test set to finally evaluate the effect of the model.

4. Analysis and Discussion of the Experimental Results

4.1. Analysis of the Experimental Results

The four unused borehole data used as the test data were input into the final trained neural network model; the accuracy of the model is shown in Table 1 and the accuracy of each neural network in the model is shown in Table 2. The trained neural network models have a good identification accuracy, which are all over 97.00% and the average is 97.65%; their accuracy can effectively recognize the drilling state of the drilling signals collected in the field. The accuracy of the identification may not be satisfactory when only using single-sensor data for the identification, which is only about 92.72%, and the accuracy may become worse when encountering a more complex situation. Fusing the information of the recognition results of multiple sensors can effectively improve the identification accuracy of the final model and make the identification accuracy of the recognition model more robust.

Table 1. The accuracy of the neural network on the test data.

	Test Data #1	Test Data #2	Test Data #3	Test Data #4
Recognition accuracy	97.00%	98.47%	97.99%	97.15%

Table 2. The accuracy of each sub-neural network on the test data.

Network Type	Test Data #1	Test Data #2	Test Data #3	Test Data #4
LSTM #1	93.72%	92.72%	95.61%	97.72%
LSTM #2	95.15%	96.17%	97.37%	96.96%
LSTM #3	93.44%	92.15%	94.49%	96.20%
LSTM all	96.72%	93.10%	98.25%	96.96%
DNN	96.86%	98.47%	98.62%	97.15%

4.2. Error Analysis

Comparing the identification results of the final model with the ground true, it was found that there are some discriminative abnormal points in the result. Taking part of the data in the Test #1 borehole as an example, and the partial discriminant abnormal point data shown in Table 3, we can observe that the data collected by the sensors corresponding to these distinguishing abnormal points are usually very different from the data of the surrounding points, which can also be seen in other datasets. Therefore, it can be assumed that the appearance of these discriminative anomalies has little to do with the discriminative neural network, but there are anomalies in the collected data.

Table 3. Part of the outlier data points in the Test #1 dataset.

Data Sequence Number	Sensor #1 Amplitude	Sensor #2 Amplitude	Sensor #3 Amplitude	Drilling State Ground True	Drilling State Judged by the Network
. . .	. . .	. . .	. . .	. . .	. . .
17	2.707553	2.619514	6.796628	1	1
18	1.930003	2.793286	5.172647	1	1
19	1.963386	2.451638	5.28292	1	1
20	0.502925	0.59154	1.494293	1	0
21	1.324366	2.157405	4.256246	1	1
22	3.292366	3.701357	7.12655	1	1
23	3.088607	3.210542	7.949529	1	1
. . .	. . .	. . .	. . .	. . .	. . .
94	0.391242	0.756341	0.875585	0	0
95	0.225544	0.427893	0.178449	0	0
96	0.334536	0.450832	0.215356	0	0
97	1.083765	1.435536	2.665392	0	1
98	0.483574	0.586286	0.426804	0	0
99	0.425458	0.389063	0.18947	0	0
100	0.204848	0.269115	0.353566	0	0
. . .	. . .	. . .	. . .	. . .	. . .

The misjudgment of the identification results can be roughly divided into two categories. The first one is the "1110111" type of signals, that is, continuously or discontinuously sporadic signals in the continuous drilling state (1 state) are judged to be in the connecting state (0 state). In fact, the amplitude values of these points are usually very smaller, always one-half or one-third, than other points around them in a time series, so they can be regarded as anomalies. Combined with the actual situation on site, these points may be the sticking drill on-site. The other is the "0001000" type of signals, that is, sporadic signals mixed in the continuous connecting state (0 state) are judged as drilling state (1 state). Similarly, these points are abnormally different from other nearby points, usually 3 times or more higher than nearby points, and have a very short duration of 1 point with occasional cases lasting 2 points (8 s), so they can also be considered as anomalies. In the actual situation, these points may be the percussion made by the rig operator in order to lengthen the drill rod or faraway blasting, rock burst events or other strong vibration events that are collected by the sensor.

In order to eliminate the influence of the two types of anomalies mentioned above on the identification of drilling state, we propose an optimization method similar to "submerge". The idea is that, when the state of a point is different from the state of the two points around it and the state of those two points is the same, the state of the point is modified to be the same as theirs. In addition, sometimes there are two consecutive points that are abnormal points (00011000, 11100111), but, in the actual drilling construction, the drilling state cannot only last for 8 s and the extension of the drill rod only takes 8 s. So, there is also a need for a separate "submerge" process for such consecutive abnormal points.

Based on this fact, we wrote a program to perform the "submerge" process of the 0 state abnormal points for each group of identification results, perform the "submerge" process of the 1 state abnormal points for the obtained results and process the points with consecutive outliers. In this way, we optimized the identification results and further improved the final accuracy of the discriminant. The final identification results after the "submerge" smoothing process are, respectively, shown in Figures 9–12. The final identification case results of Test #1 are shown in detail, while the results of the other test groups are shown in abbreviated form. The final identification accuracy is almost the same as that of the manually labeled drilling state.

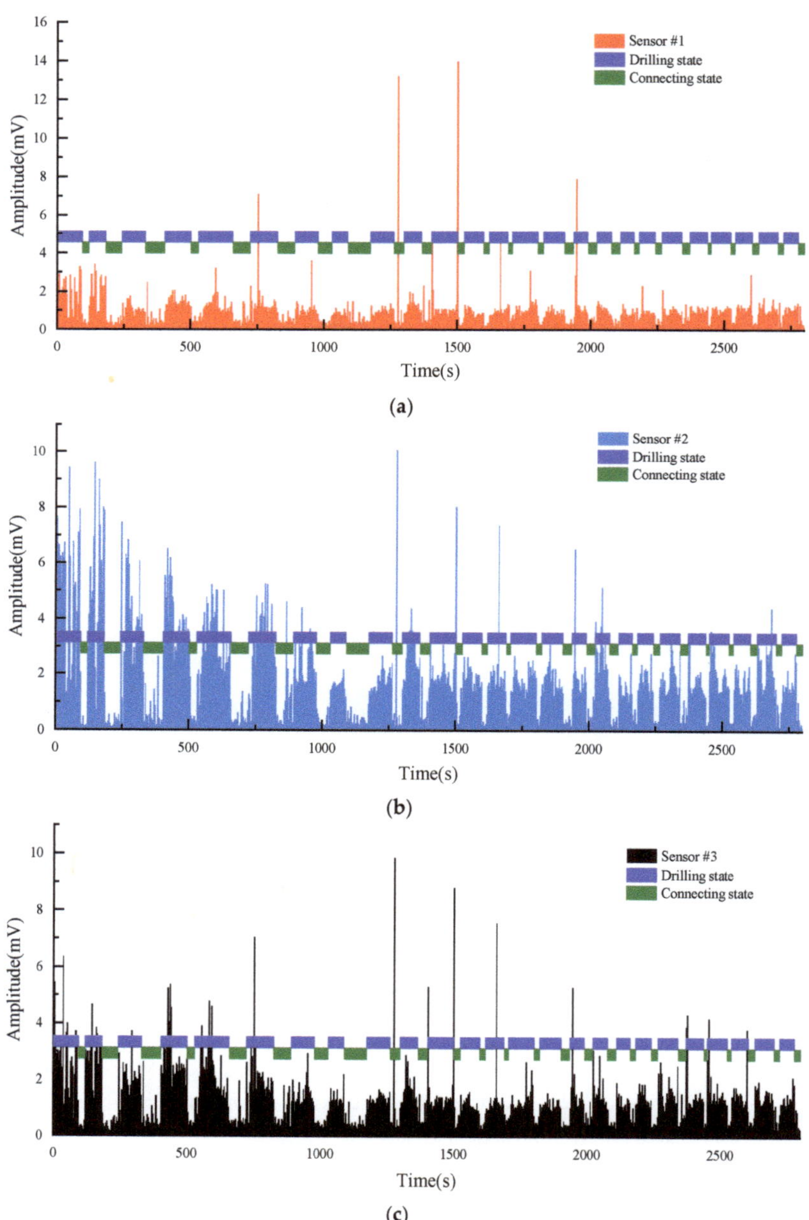

Figure 9. *Cont.*

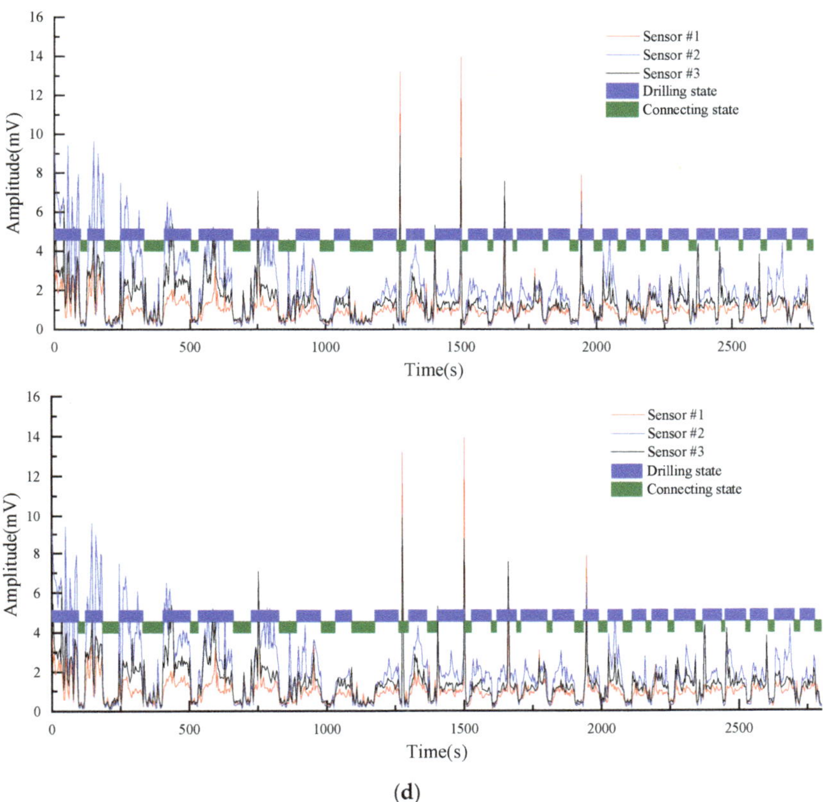

Figure 9. Test#1 drilling state final identification results. (**a**) Drilling state final identification result and vibration signal of Sensor #1. (**b**) Drilling state final identification result and vibration signal of Sensor #2. (**c**) Drilling state final identification result and vibration signal of Sensor #3. (**d**) Drilling state final identification result and vibration signal of all sensors.

As shown in Figure 9, the blue and green rectangles in each graph are the identification results of the drilling status judged by the neural network algorithm. The length of each rectangle on the time axis is the time consumed by that section of the construction. The length of each drilling state is the length of one drill rod. Therefore, the length of each blue rectangle in the graph shows the time consumed by each drill rod, and thus indirectly indicates the drilling difficulty at that depth. The green rectangle between two adjacent blue rectangles is the non-drilling state, such as drill rod connection.

Test #1 is the borehole drilling amplitude data collected in mine C. Figure 9a–c shows the amplitude of the vibration signals collected by each sensor, while Figure 9d shows the amplitude signals of the three sensors fused into one graph. Since the three sensors were installed at different distances from the borehole, from Figure 9, we can see that the amplitude of the vibration signals collected by each sensor at the same time are different, but the overall trend is the same. It can be clearly seen that the drilling state identified by the neural network algorithm matches the area with high amplitude values due to the borehole construction, and accordingly, the connecting state matches the area with low amplitude values.

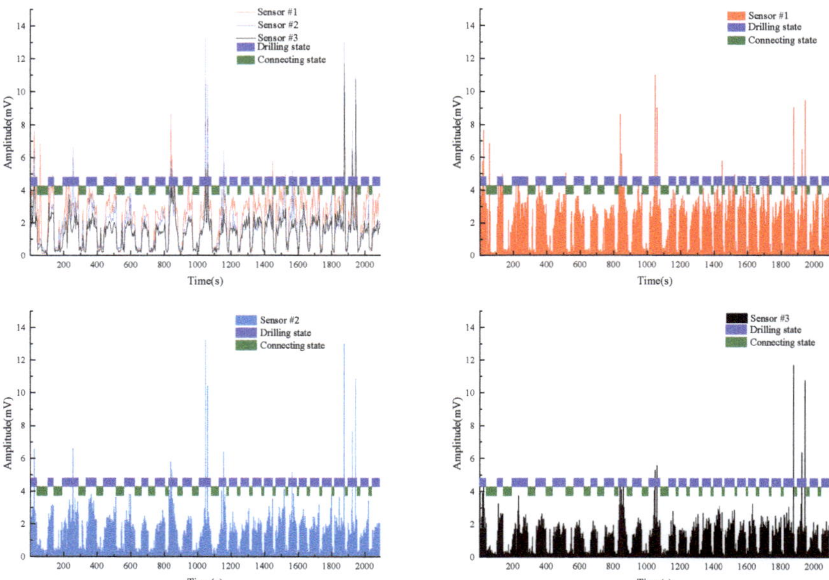

Figure 10. Test#2 drilling state final identification results.

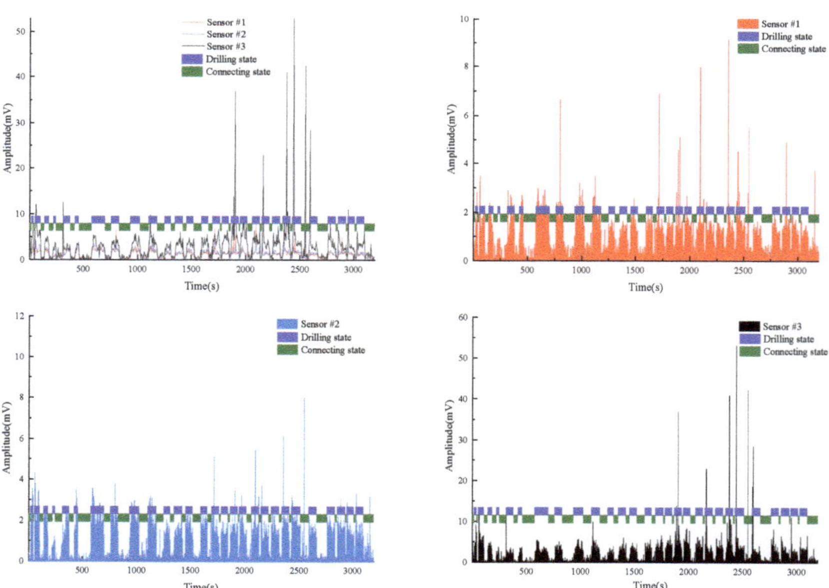

Figure 11. Test#3 drilling state final identification results.

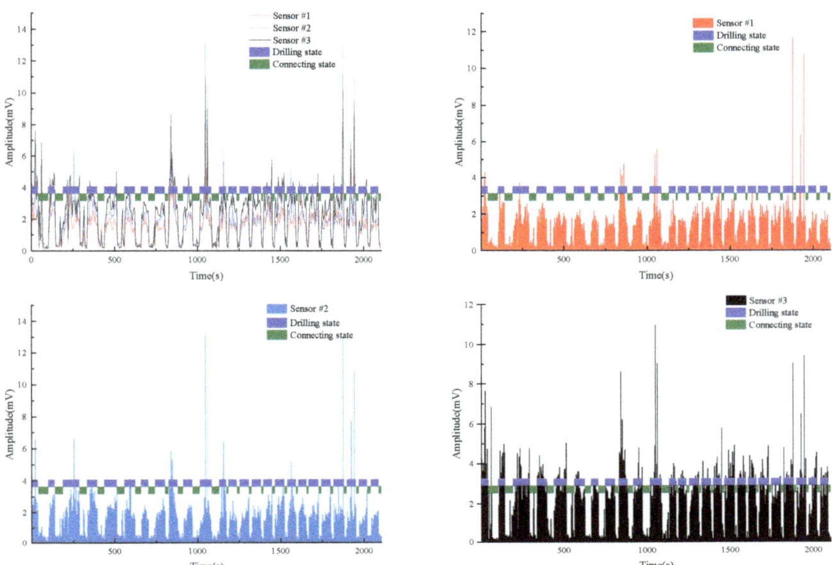

Figure 12. Test#4 drilling state final identification results.

Test #2 is the borehole drilling amplitude data collected in mine B. Unlike Test #1, the source of this dataset, mine B, was not trained in the model. The model only learned the vibration amplitude data collected in mines A and C, which means that the model is completely unaware about the information related to mine B. As the four datasets used as the test set were not used in the model training and tuning process, there is no information leakage problem. This set of vibration signal data, from a completely new mine, has greater significance for the evaluation of model performance. This dataset eliminates the problem of the model having better identification accuracy in other boreholes in the same mine due to the knowledge of the mine's geological conditions. The performance of the model on this dataset can better reflect the generalization ability of the model. As can be seen in Figure 10, the model has very good identification accuracy on this dataset, which comes from a brand new condition.

Test #3 and Test #4 are the other two borehole drilling amplitude data collected in mine C. It is also evident from Figures 11 and 12 that the model still has a good recognition accuracy on these two test sets.

By using the above method, the state of the drilled hole, the amplitude clusters, intervals and the starting and ending points of drilling or drill rod connecting can be automatically and efficiently recognized with a high recognition accuracy. The vibration signals of downhole drilling can be identified and monitored continuously and efficiently. It can obtain, in a timely and accurate manner, the information of the whole process of the drilling construction, such as the construction depth of borehole drilling construction, construction time of each section of drill rod and the difficulty of construction at different depths. It can also ensure that the underground PRD boreholes are constructed according to the designed depth and the construction quality is effectively monitored, which is of great significance to ensure mine safety and obtain rock mass information.

5. Conclusions and Future Work

5.1. Conclusions

A neural network model for drilling rig drilling status identification that fuses single-sensor and multi-sensor prediction results was proposed in this study. The vibration signals of the drilling status of a same borehole were collected using multiple sensors, and the information of the identification results of single and multiple sensors were fused. The method was tested and verified; the identification accuracy of all four test datasets were over 97.00%, and the final identification accuracy was almost the same as that of the manually labeled drilling state after using the "submerge" optimization method. The results show that this method makes up for the deficiency of large errors (up to 6.32%) in the identification results due to the use of a single-sensor data source, and effectively improves the identification accuracy. The study is of great engineering significance to effectively identify and judge the construction length of underground borehole drilling and monitor the drilling information in the entire process of construction.

We innovatively proposed a drilling rig with a drilling status identification neural network algorithm that uses single-sensor and multi-sensor data and fuses multiple sub-neural network identification results. Several LSTM and DNN sub-neural networks were constructed using different drilling amplitude signal data sources. A new optimization method was proposed for the misjudgment caused by data anomalies in the identification results. It is a drilling state identification results optimization method that considers both the amplitude data of the time point and its neighboring points in the time dimension.

The main conclusions are as follows:

(1) A high-accuracy neural network algorithm for the automatic identification of the drilling status of drilling rigs was proposed. The method uses single-sensor and multi-sensor data from the same borehole as input data and fuses the identification results from different types of sub-neural networks using different inputs, effectively improving the final identification accuracy. The identification accuracy of four test datasets of borehole amplitude data from two different mines were all above 97.00%.

(2) An optimization method was proposed to deal with two types of misjudgment in the identification results due to data anomalies, and the optimized identification results are almost the same as the drilling status marked manually according to the actual construction status on-site.

5.2. Future Work

The underground environment is complex and there are a large number of noise sources, such as coal cutting by shearer loaders, blasting of heading face and overlying rock breaking. The noise signals are inevitably collected by the amplitude sensors, and are not only meaningless for the judgment of the drilling state of the drilling rig, but also affect the accuracy of model judgment. In addition, dynamic events, such as rock burst and coal and gas outburst, are completely eliminated in this study, which is very meaningful for research. Therefore, the next step of this study is to achieve the accuracy of noise elimination and the discriminative analysis of other dynamic events signals.

Author Contributions: Conceptualization, Z.W. and W.-L.Z.; methodology, Z.W.; software, Z.W.; validation, W.-L.Z. and C.L.; formal analysis, Z.W. and W.-L.Z.; investigation, C.L.; resources, W.-L.Z.; data curation, Z.W.; writing—original draft preparation, Z.W.; writing—review and editing, W.-L.Z. and C.L. All authors have read and agreed to the published version of the manuscript.

Funding: This research was funded by National Natural Science Foundation of China, grant number 52004289.

Institutional Review Board Statement: Not applicable.

Informed Consent Statement: Not applicable.

Data Availability Statement: The data presented in this study are available upon request from the corresponding author.

Conflicts of Interest: The authors declare no conflict of interest.

References

1. Fan, X.; Liang, M.; Yeap, T. A joint time-invariant wavelet transform and kurtosis approach to the improvement of in-line oil debris sensor capability. *Smart Mater. Struct.* **2009**, *18*, 085010. [CrossRef]
2. Zhao, Y.; Zhang, L.; Liao, J.; Wang, W.; Liu, Q.; Tang, L. Experimental study of fracture toughness and subcritical crack growth of three rocks under different environments. *Int. J. Geomech.* **2020**, *20*, 04020128. [CrossRef]
3. Wang, K.; Liu, Z.; Liu, G.; Yi, L.; Yang, K.; Peng, S.; Chen, M. Vibration Sensor Approaches for the Monitoring of Sand Production in Bohai Bay. *Shock. Vib.* **2015**, *2015*, 591780. [CrossRef]
4. Zhao, Y.; Wang, Y.; Wang, W.; Tang, L.; Liu, Q.; Cheng, G. Modeling of rheological fracture behavior of rock cracks subjected to hydraulic pressure and far field stresses. *Theor. Appl. Fract. Mech.* **2019**, *101*, 59–66. [CrossRef]
5. Feng, K.; Jiang, Z.; He, W.; Qin, Q. Rolling element bearing fault detection based on optimal antisymmetric real Laplace wavelet. *Measurement* **2011**, *44*, 1582–1591. [CrossRef]
6. Sheen, Y.-T. An envelope detection method based on the first-vibration-mode of bearing vibration. *Measurement* **2008**, *41*, 797–809. [CrossRef]
7. Baydar, N.; Ball, A. Detection of Gear Failures Via Vibration and Acoustic Signals Using Wavelet Transform. *Mech. Syst. Signal Processing* **2003**, *17*, 787–804. [CrossRef]
8. Singh, A.; Parey, A. Gearbox fault diagnosis under non-stationary conditions with independent angular re-sampling technique applied to vibration and sound emission signals. *Appl. Acoust.* **2019**, *144*, 11–22. [CrossRef]
9. Li, C.; Liang, M. Time–frequency signal analysis for gearbox fault diagnosis using a generalized synchrosqueezing transform. *Mech. Syst. Signal Processing* **2012**, *26*, 205–217. [CrossRef]
10. Yu, X.; Feng, Z.; Liang, M. Analytical vibration signal model and signature analysis in resonance region for planetary gearbox fault diagnosis. *J. Sound Vib.* **2021**, *498*, 115962. [CrossRef]
11. Fei, C.W.; Bai, G.C. Wavelet correlation feature scale entropy and fuzzy support vector machine approach for aeroengine whole-body vibration fault diagnosis. *Shock. Vib.* **2013**, *20*, 341–349. [CrossRef]
12. Wang, S.; Chen, X.; Tong, C.; Zhao, Z. Matching Synchrosqueezing Wavelet Transform and Application to Aeroengine Vibration Monitoring. *IEEE Trans. Instrum. Meas.* **2017**, *66*, 360–372. [CrossRef]
13. Koukoura, S.; Carroll, J.; McDonald, A.; Weiss, S. Comparison of wind turbine gearbox vibration analysis algorithms based on feature extraction and classification. *Iet Renew. Power Gener.* **2019**, *13*, 2549–2557. [CrossRef]
14. Chen, J.; Yue, Z.Q. Weak zone characterization using full drilling analysis of rotary-percussive instrumented drilling. *Int. J. Rock Mech. Min. Sci.* **2016**, *89*, 227–234. [CrossRef]
15. Zhao, Y.; Zhang, L.; Wang, W.; Liu, Q.; Tang, L.; Cheng, G. Experimental study on shear behavior and a revised shear strength model for infilled rock joints. *Int. J. Geomech.* **2020**, *20*, 04020141. [CrossRef]
16. Chen, J.; Yue, Z.Q. Ground characterization using breaking-action-based zoning analysis of rotary-percussive instrumented drilling. *Int. J. Rock Mech. Min. Sci.* **2015**, *75*, 33–43. [CrossRef]
17. Qin, M.; Wang, K.; Pan, K.; Sun, T.; Liu, Z. Analysis of signal characteristics from rock drilling based on vibration and acoustic sensor approaches. *Appl. Acoust.* **2018**, *140*, 275–282. [CrossRef]
18. Zhao, Y.; Zhang, C.; Wang, Y.; Lin, H. Shear-related roughness classification and strength model of natural rock joint based on fuzzy comprehensive evaluation. *Int. J. Rock Mech. Min. Sci.* **2021**, *137*, 104550. [CrossRef]
19. Zhang, W.; Ma, N.; Ren, J.; Li, C. Peak particle velocity of vibration events in underground coal mine and their caused stress increment. *Measurement* **2021**, *169*, 108520. [CrossRef]
20. Cheng, T.; Zhang, Z.; Yi, Q.; Yin, B.; Yuan, H. Denoising method of rock acoustic emission signal based on improved VMD. *J. Min. Strat. Control. Eng.* **2022**, *4*, 023011.
21. Lan, S.; Song, D.; Li, Z.; Liu, Y. Experimental study on acoustic emission characteristics of fault slip process based on damage factor. *J. Min. Strat. Control. Eng.* **2021**, *3*, 033024.
22. Wu, Y.; Dong, Q.; He, J. The Effect of Chemical Corrosion on Mechanics and Failure Behaviour of Limestone Containing a Single Kinked Fissure. *Sensors* **2021**, *21*, 5641. [CrossRef] [PubMed]
23. Dong, L.; Tong, X.; Hu, Q.; Tao, Q. Empty region identification method and experimental verification for the two-dimensional complex structure. *Int. J. Rock Mech. Min. Sci.* **2021**, *147*, 104885. [CrossRef]
24. Suwansin, W.; Phasukkit, P. Deep Learning-Based Acoustic Emission Scheme for Nondestructive Localization of Cracks in Train Rails under a Load. *Sensors* **2021**, *21*, 272. [CrossRef] [PubMed]
25. Dong, L.; Zhang, Y.; Sun, D.; Chen, Y.; Tang, Z. Stage characteristics of acoustic emission and identification of unstable crack state for granite fractures. *Chin. J. Rock Mech. Eng.* **2022**, *41*, 120–131.
26. Dong, L.; Tong, X.; Ma, J. Quantitative Investigation of Tomographic Effects in Abnormal Regions of Complex Structures. *Engineering* **2021**, *7*, 1011–1022. [CrossRef]
27. Su, Y.; Dong, L.; Pei, Z. Non-Destructive Testing for Cavity Damages in Automated Machines Based on Acoustic Emission Tomography. *Sensors* **2022**, *22*, 2201. [CrossRef]
28. Hatherly, P.; Leung, R.; Scheding, S.; Robinson, D. Drill monitoring results reveal geological conditions in blasthole drilling. *Int. J. Rock Mech. Min. Sci.* **2015**, *78*, 144–154. [CrossRef]
29. Ghosh, R.; Schunnesson, H.; Gustafson, A. Monitoring of Drill System Behavior for Water-Powered In-The-Hole (ITH) Drilling. *Minerals* **2017**, *7*, 121. [CrossRef]

30. Ma, K.; Wang, S.-J.; Yuan, F.-Z.; Peng, Y.-L.; Jia, S.-M.; Gong, F. Study on Mechanism of Influence of Mining Speed on Roof Movement Based on Microseismic Monitoring. *Adv. Civ. Eng.* **2020**, *2020*, 8819824. [CrossRef]

31. Li, D.; Zhang, J. Rockburst Monitoring in Deep Coalmines with Protective Coal Panels Using Integrated Microseismic and Computed Tomography Methods. *Shock. Vib.* **2020**, *2020*, 8831351. [CrossRef]

32. Zhao, Y.; Wang, C.; Bi, J. Analysis of fractured rock permeability evolution under unloading conditions by the model of elastoplastic contact between rough surfaces. *Rock Mech. Rock Eng.* **2020**, *53*, 5795–5808. [CrossRef]

33. Jiang, R.; Dai, F.; Liu, Y.; Li, A. A novel method for automatic identification of rock fracture signals in microseismic monitoring. *Measurement* **2021**, *175*, 109129. [CrossRef]

34. Zhang, W.; Li, C.; Jin, J.; Qu, X.; Fan, S.; Xin, C. A new monitoring-while-drilling method of large diameter drilling in underground coal mine and their application. *Measurement* **2021**, *173*, 108840. [CrossRef]

35. Zhang, C.; Tu, S.; Chen, M.; Zhang, L. Pressure-relief and methane production performance of pressure relief gas extraction technology in the longwall mining. *J. Geophys. Eng.* **2017**, *14*, 77–89. [CrossRef]

36. Zhou, H.; Hatherly, P.; Monteiro, S.T.; Ramos, F.; Oppolzer, F.; Nettleton, E.; Scheding, S. Automatic rock recognition from drilling performance data. In Proceedings of the 2012 IEEE International Conference on Robotics and Automation, Saint Paul, MN, USA, 14–18 May 2012; pp. 3407–3412.

37. Liu, S.; Fu, M.; Jia, H.; Li, W.; Luo, Y. Numerical simulation and analysis of drill rods vibration during roof bolt hole drilling in underground mines. *Int. J. Min. Sci. Technol.* **2018**, *28*, 877–884. [CrossRef]

38. Zhang, W.; Li, C.; Ren, J.; Wu, Z. Measurement and application of vibration signals during pressure relief hole construction using microseismic system. *Measurement* **2020**, *158*, 107696. [CrossRef]

39. Pu, Y.; Apel, D.B.; Hall, R. Using machine learning approach for microseismic events recognition in underground excavations: Comparison of ten frequently-used models. *Eng. Geol.* **2020**, *268*, 105519. [CrossRef]

40. Shirani Faradonbeh, R.; Taheri, A. Long-term prediction of rockburst hazard in deep underground openings using three robust data mining techniques. *Eng. Comput.* **2018**, *35*, 659–675. [CrossRef]

41. Feng, S.; Zhou, H.Y.; Dong, H.B. Using deep neural network with small dataset to predict material defects. *Mater. Des.* **2019**, *162*, 300–310. [CrossRef]

42. Yang, Y.L.; Fu, P.Y.; He, Y.C. Bearing Fault Automatic Classification Based on Deep Learning. *Ieee Access* **2018**, *6*, 71540–71554. [CrossRef]

43. Li, Y.; Cheng, G.; Liu, C.; Chen, X.H. Study on planetary gear fault diagnosis based on variational mode decomposition and deep neural networks. *Measurement* **2018**, *130*, 94–104. [CrossRef]

44. Zhang, R.; Peng, Z.; Wu, L.F.; Yao, B.B.; Guan, Y. Fault Diagnosis from Raw Sensor Data Using Deep Neural Networks Considering Temporal Coherence. *Sensors* **2017**, *17*, 549. [CrossRef] [PubMed]

45. Yu, Y.; Si, X.S.; Hu, C.H.; Zhang, J.X. A Review of Recurrent Neural Networks: LSTM Cells and Network Architectures. *Neural Comput.* **2019**, *31*, 1235–1270. [CrossRef]

46. Shen, S.L.; Njock, P.G.A.; Zhou, A.N.; Lyu, H.M. Dynamic prediction of jet grouted column diameter in soft soil using Bi-LSTM deep learning. *Acta Geotech.* **2021**, *16*, 303–315. [CrossRef]

47. Yuan, X.H.; Chen, C.; Jiang, M.; Yuan, Y.B. Prediction interval of wind power using parameter optimized Beta distribution based LSTM model. *Appl. Soft Comput.* **2019**, *82*, 105550. [CrossRef]

48. Yang, Y.; Zeng, Q.L.; Yin, G.J.; Wan, L.R. Vibration Test of Single Coal Gangue Particle Directly Impacting the Metal Plate and the Study of Coal Gangue Recognition Based on Vibration Signal and Stacking Integration. *Ieee Access* **2019**, *7*, 106783–106804. [CrossRef]

49. Luo, J.-S.; Lo, D.C.-T. Binary Malware Image Classification Using Machine Learning with Local Binary Pattern. In Proceedings of the 2017 IEEE International Conference on Big Data (Big Data), Boston, MA, USA, 11–14 December 2017; IEEE: Piscataway, NJ, USA, 2017; pp. 4664–4667.

50. Huang, Q.; Mao, J.; Liu, Y. An Improved Grid Search Algorithm of SVR Parameters Optimization. In Proceedings of the 2012 IEEE 14th International Conference on Communication Technology, Chengdu, China, 9–11 October 2012; IEEE: Piscataway, NJ, USA, 2012; pp. 1022–1026.

51. Wong, T.-T.; Yeh, P.-Y. Reliable accuracy estimates from k-fold cross validation. *IEEE Trans. Knowl. Data Eng.* **2019**, *32*, 1586–1594. [CrossRef]

52. Al-Badarneh, A.; Al-Shawakfa, E.; Bani-Ismail, B.; Al-Rababah, K.; Shatnawi, S. The impact of indexing approaches on Arabic text classification. *J. Inf. Sci.* **2017**, *43*, 159–173. [CrossRef]

53. Li, L.; Wen, Z.; Wang, Z. Outlier detection and correction during the process of groundwater lever monitoring base on pauta criterion with self-learning and smooth processing. In *Theory, Methodology, Tools and Applications for Modeling and Simulation of Complex Systems*; Springer: Berlin/Heidelberg, Germany, 2016; pp. 497–503.

54. Shen, C.; Bao, X.J.; Tan, J.B.; Liu, S.T.; Liu, Z.J. Two noise-robust axial scanning multi-image phase retrieval algorithms based on Pauta criterion and smoothness constraint. *Opt. Express* **2017**, *25*, 16235–16249. [CrossRef] [PubMed]

 *sustainability*

Article

Study on Crack Classification Criterion and Failure Evaluation Index of Red Sandstone Based on Acoustic Emission Parameter Analysis

Jiashen Li [1,2], Shuailong Lian [1,2], Yansen Huang [1,2] and Chaolin Wang [1,2,*]

[1] College of Civil Engineering, Guizhou University, Guiyang 550025, China; gs.lijs19@gzu.edu.cn (J.L.); gs.sllian19@gzu.edu.cn (S.L.); yshuang_gu@126.com (Y.H.)
[2] Guizhou Provincial Key Laboratory of Rock and Soil Mechanics and Engineering Safety, Guiyang 550025, China
* Correspondence: clwang3@gzu.edu.cn

Abstract: The acoustic emission (AE) characteristics of rock during loading can reflect the law of crack propagation and evolution in the rock. In order to study the fracture mode in the process of rock fracture, the AE characteristics and crack types of red sandstone during fracture were investigated by conducting Brazilian indirect tensile tests (BITT), direct shear tests (DST), and uniaxial compression tests (UCT). The evolution law of AE event rate, RA and AF values, and the distribution law of RA–AF data of red sandstone samples in three test types were analyzed. Based on the kernel density estimation (KDE) function and the coupling AE parameters (RA–AF values) in DST and BITT, the relatively objective dividing line for classifying tensile and shear cracks was discussed, and the dividing line was applied to the analysis of fracture source evolution and the failure precursor of red sandstone. The results show that the dividing line for classifying tensile and shear cracks of red sandstone is AF = 93RA + 75. Under uniaxial compression loading, the fracture source of red sandstone is primarily shear source in the initial phase of loading and tensile source in the critical failure phase, and the number is far greater than shear source. K = AF/(93RA + 75) can be defined as the AE parameter index, and its coefficient of variation CV (k) can be used as the failure judgment index of red sandstone. When CV (k) < 1, it can be considered that red sandstone enters the instability failure phase.

Keywords: acoustic emission; sensor; parameter analysis; RA and AF; crack classification criterion

Citation: Li, J.; Lian, S.; Huang, Y.; Wang, C. Study on Crack Classification Criterion and Failure Evaluation Index of Red Sandstone Based on Acoustic Emission Parameter Analysis. *Sustainability* **2022**, *14*, 5143. https://doi.org/10.3390/su14095143

Academic Editors: Longjun Dong, Yanlin Zhao and Wenxue Chen

Received: 18 March 2022
Accepted: 21 April 2022
Published: 24 April 2022

Publisher's Note: MDPI stays neutral with regard to jurisdictional claims in published maps and institutional affiliations.

1. Introduction

Brittle rock is a complex geological medium, in which microcracks will occur under loading, and with the propagation and connection of microcracks on different scales, the rock will be damaged [1–4]. Furthermore, in a variety of rock engineering applications, red sandstone, as sedimentary rock, has been used widely [5–7]. Hence, studying the failure characteristics of red sandstone is of great significance for stability monitoring and disaster early warning in engineering projects [8–10]. When the rock is damaged, it will produce an acoustic emission (AE), which is essentially the elastic wave released in the process of crack generation and propagation [11–14]. In fact, the AE characteristics of rock during loading can reflect the law of crack propagation and evolution in the rock [15–19].

AE parameters can be divided into time domain parameters and frequency domain parameters, which are all extracted from the AE time domain waveform [20–23]. The time domain parameters of AE signals most widely used to reveal the rock failure mechanism are: rise time, duration, AE count, maximum amplitude, energy, and average frequency (AF) [24–28], as shown in Figure 1a. The frequency domain parameters of AE signal most widely used are peak frequency, frequency centroid, and partial power; peak frequency is the point corresponding to the maximum amplitude in the frequency spectrum, the frequency centroid, representing the center of mass of the AE signal, is calculated from a sum of magnitude time frequencies divided by a sum of magnitude, and partial power is calculated from the sum of the frequency spectrum within a specified range divided by the total power of all frequencies [29,30]. In addition, in AE field monitoring, in order to improve the accuracy of monitoring and disaster early warning, the location of the fracture source has gradually become an important development direction [31].

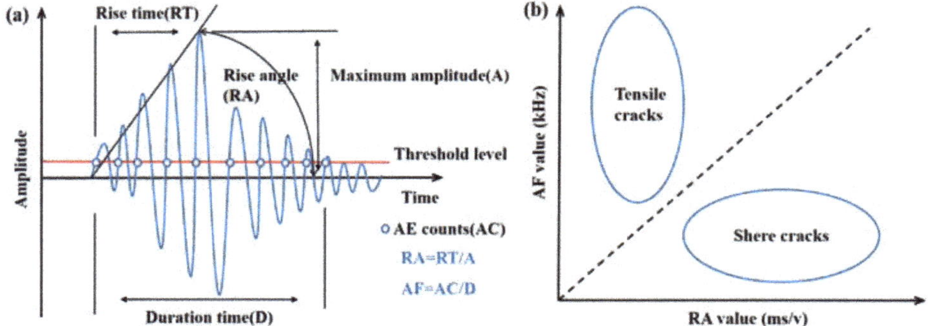

Figure 1. Typical AE waveform and micro fracture (crack) classification. (**a**) AE parameter in a hit; (**b**) crack classification based on RA/AF value.

As regards the cracking type of brittle materials, AE signals from tensile cracking and shear cracking have different characteristics. The source type can be classified by tracking the characteristics of an AE signal to improve the understanding of the rock cracking mode. The time domain parameters commonly used to classified cracking type are the rise angle (RA) and average frequency (AF) [32–34]. The RA value is defined as the ratio of rise time to amplitude, in ms/V; the AF value, the number of threshold crossings (i.e., counts) over the duration of the AE signal, is the ratio of counts to duration, in kHz [35–38]. Generally, tensile fracture corresponds to an AE signal with low RA value and high AF value, while shear fracture corresponds to an AE signal with high RA value and low AF value [13,39], as shown in Figure 1b. Based on the above conclusions, the damage mechanism and failure mode classification of brittle materials can be analyzed by the RA and AF values of the AE signals in previous studies. Based on the analysis of RA and AF values, some scholars have investigated the damage mechanism and failure mode of different materials, such as hollow slab specimens of calcite and marble, ice structures, rubber powder concrete, and other rock types, in several basic lab tests [40–43]. Meanwhile, many scholars have investigated the effects of strain rate, brittleness, bedding, microwave radiation, and crack on rock failure mode by analyzing RA and AF values [44–50]. In addition, Muñoz-Ibáñez et al. compared the advantages and disadvantages of semicircular bending (SCB) and pseudo compact stretching (PCT) by analyzing the RA and AF values [51]. In addition, frequency domain parameters are often used to classified rock fracture types. For example, in order to effectively classify source types, based on peak frequency or partial power, Zhang et al. proposed a new source classification criterion [30]. Li et al. studied the dominant frequency characteristics of the AE signal of marble based on direct tensile tests, and showed that the low-frequency waveform represents tensile cracking and the high-frequency waveform represents shear cracking [52].

However, when focusing on the cracking type classification of brittle materials based on RA and AF values, there is no clear standard for the boundary between RA and AF values. Niu et al. determined that the proportional relationship between RA and AF values is 2:1 in their evaluation for the classification of fracture modes of flawed red sandstone under uniaxial compression [53]. Wang et al. proposed that the proportional relationship between RA and AF values is 1:3.75 when investigating the influence of multi-stage cyclic loading on the classification of marble fracture mode [54,55]. Yao et al. proposed that the proportional relationship between RA and AF values is 1:2 when investigating the effect of moisture on the failure mode in coal [56]. Fan et al. proposed that the proportional relationship between RA and AF values is 50:1 when investigating the fracture behavior of fully graded concrete under three-point bending loading at different loading rates [57]. Moreover, the intercept was introduced when some scholars investigated crack classification criteria based on RA and AF values. Das et al. proposed that the optimal dividing line for classifying the fracture type of strain hardening cementitious composite (SHCC) specimens is AF = 26.9841RA − 268.6918 [58]. Du et al. proposed that the optimal dividing line for classifying the fracture type of marble is AF = 400RA + 50, when studying the AE characteristics of marble [26]. In addition, when analyzing the precursory characteristics of rock instability, Dong et al. proposed that the ratio of RA to AF is 1:200, and found that the anisotropic characteristics of AE event rate can effectively reveal the failure precursors of rock mass, and determine the direction of principal stress [59].

It can be seen from the above analysis that the rock cracking type classification based on RA and AF values is mainly derived from the empirical relationship between RA and AF values, which is uncertain and empirical; that is, the boundary between shear cracking and tensile cracking has not been determined. Therefore, it is particularly important to determine the relatively objective boundary between tensile and shear cracking. Nowadays, many scholars use the cluster analysis method to determine the optimal dividing line [58,60,61], and many scholars use the kernel density estimation (KDE) function, a non-parametric density estimation method [53–55]. In addition, some scholars investigated the RA–AF characteristics of rock under direct tensile failure modes by conducting a direct tensile test, to determine the optimal dividing line [25,52]. However, most of the above research methods focused on a single loading condition, and the single application of a mathematical analysis method could not fully reflect the fracture type of rock. In addition, the reliability of the dividing line verified in the above research is still low. In this paper, the AE characteristics of red sandstone during the fracture process were investigated by conducting BITT, DST, and UCT. At the same time, based on the KDE function, the AE data collected in BITT and DST are coupled to discuss the dividing line for classifying red sandstone cracking type, and the reliability of the dividing line is verified. In addition, the corresponding precursory characteristic parameters of rock failure are proposed based on the determined dividing line.

2. Materials and Experimental Methods

2.1. Specimen Preparation

The origin of the light brown sandstone blocks used in the experimental testing is the northwestern Sichuan Basin, China. The P-wave and average density of sandstone blocks are measured as 3270 m/s and 2390 kg/m^3, respectively. All specimens used in the lab tests came from the same rock block and were cut in the same direction to avoid specimen dispersion. In this testing, nine specimens were prepared in three sizes. The side of the cube specimen used in DST was 100 mm, the size of disc specimen used in BITT was Φ50 × H25 mm, and the size of the cylindrical specimens used in UCT was Φ50 × H100 mm. The geometry and dimensions of the specimens in the three test types are shown in Figure 2. The accuracy of each specimen is within the range specified by ISRM.

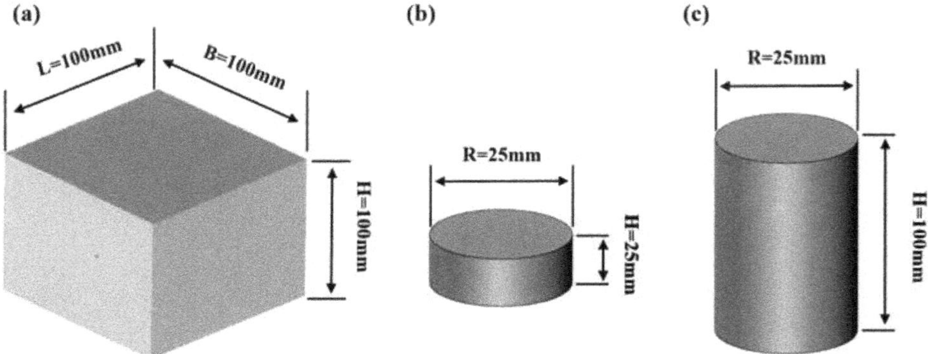

Figure 2. The geometry and dimensions of the specimens in the three types. (**a**) DST; (**b**) BITT; (**c**) UCT.

2.2. Experimental Equipment and Setup

(1) Loading equipment

The designed DST was conducted on the WDAJ-600 rock shear testing machine. The maximum loading value of the WDAJ-600 test machine in the vertical and horizontal directions is 600 kN, and its loading accuracy is ±0.5%. In addition, the designed BITT and UCT were conducted in the DSZ-1000 stress–strain controlled testing system. The maximum loading value of the DSZ-1000 test machine is 1000 kN, and its loading accuracy is also ±0.5%.

In this study, in order to ensure that the shape of the specimen will not be damaged by the indenter of the testing machine when specimen failure occurs, the displacement control mode was selected as the loading mode. In DST and UCT, the loading rate was 0.1 mm/min. In BITT, since the length of the specimen along the loading direction was half of those in the DST and UCT, the displacement rate was 0.05 mm/min. The normal stress of DST was 1 MPa. Before the formal loading, the force of 0.5 kN was applied to the specimens in DST and BITT, and the force of 1 kN was applied to the specimens in UCT, which ensured that the specimens were in full contact with the loading device, so as to further eliminate the noise generated during the contact between the specimens and the loading device in the formal test.

(2) AE monitoring system

The PCI-2 AE monitoring system was used to collect the AE signals during the deformation and damage of the specimen, and its manufacturer was the American Acoustics Company. The preamplifier gain of the AE monitoring system was set to 40 dB, which was used to increase the anti-interference ability of weak signals. The threshold was also set to 40 dB; that is, AE signals whose amplitude exceeded 40 dB were recorded. The sampling length of the single waveform and sampling rate was set to 2 k and 5 MSPS, respectively. The sensor used was a PICO sensor with a resonant frequency of 150 kHz. The operating frequency range was set to 20–400 kHz.

When installing the sensor on the surface of the sample, firstly, a layer of coupling agent was applied on the sensor, to ensure that there was no gap between the sensor and the rock surface; secondly, four sensors coated with coupling agent were placed on the surface of the rock sample using black insulating tape. Black insulating tape has good elasticity, so it can stabilize the sensor and also protect it from being crushed. The above measures ensured that the collected acoustic emission signal was not distorted.

Before the formal loading, a "pencil lead breaking test" was conducted to check whether all channels were connected normally, so as to further ensure that the AE signals can be collected effectively. The pencil lead fracture is a practical pulse simulation source.

Pencil lead fractures are used to simulate the acoustic emission signal generated by concrete deformation and fracture.

All testing systems used are shown in Figure 3. The types of tests, the distribution of AE sensors and damaged rock specimens are shown in Figure 4.

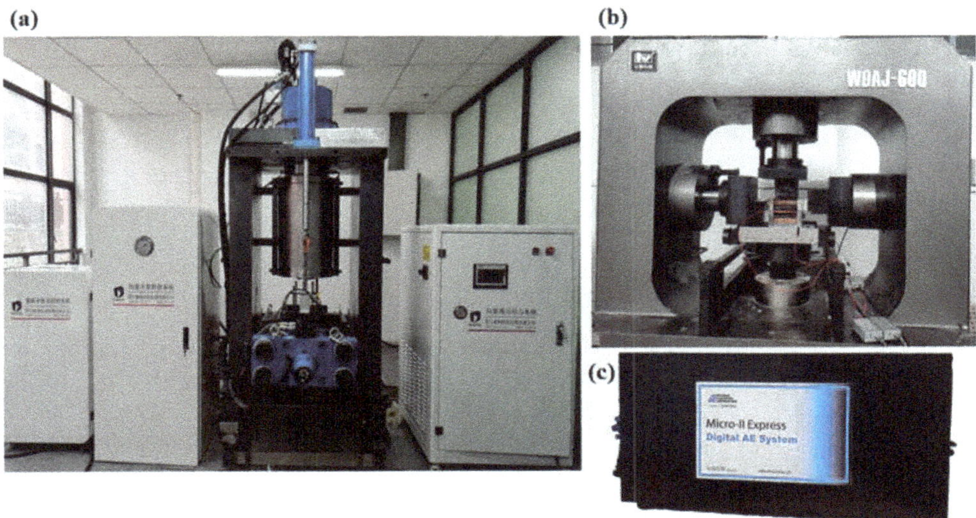

Figure 3. Testing systems used in this study: (**a**) DSZ-1000 stress–strain controlled testing system; (**b**) WDAJ-600 rock shear testing machine; (**c**) PCI-2 AE monitoring.

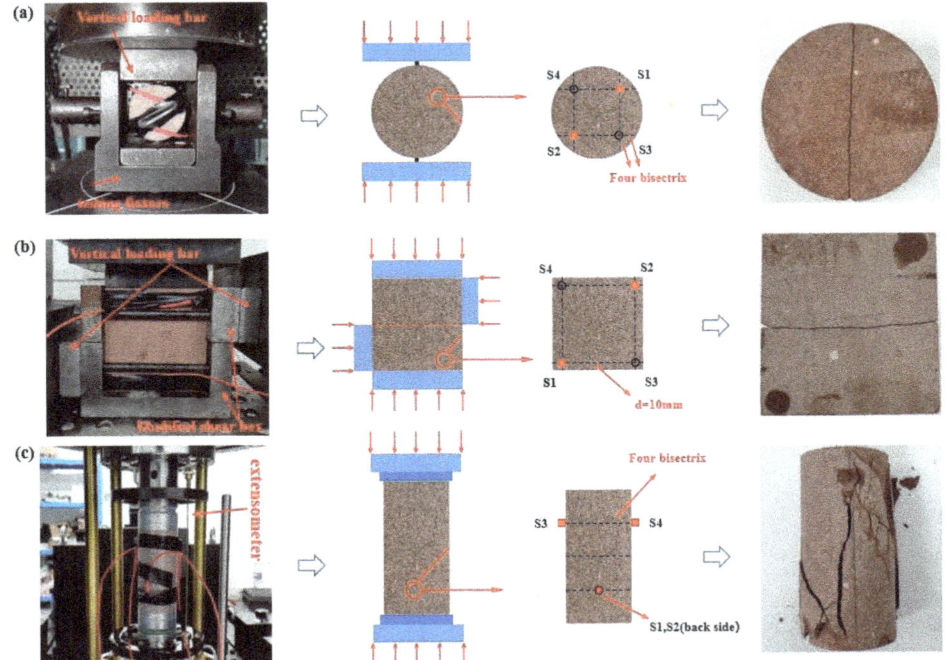

Figure 4. Types of test, the distribution of AE sensors and damaged rock specimens. (**a**) BITT; (**b**) DST; (**c**) UCT.

Three identical rock samples were set up in each test type to eliminate the effect of discreteness on the test results. The stress–displacement curve or stress–strain curve of the specimens in each of the three test types is shown in Figure 5. The strain data were collected by an extensometer, so the stress–strain curve could be drawn in UCT, while in DST and BITT, only stress displacement curves could be drawn. In addition, the number and strength of all specimens are shown in Table 1, in which σ_t, σ_s and σ_c denote the tensile strength (Mpa), the shear strength (Mpa), and the uniaxial compressive strength (Mpa), respectively.

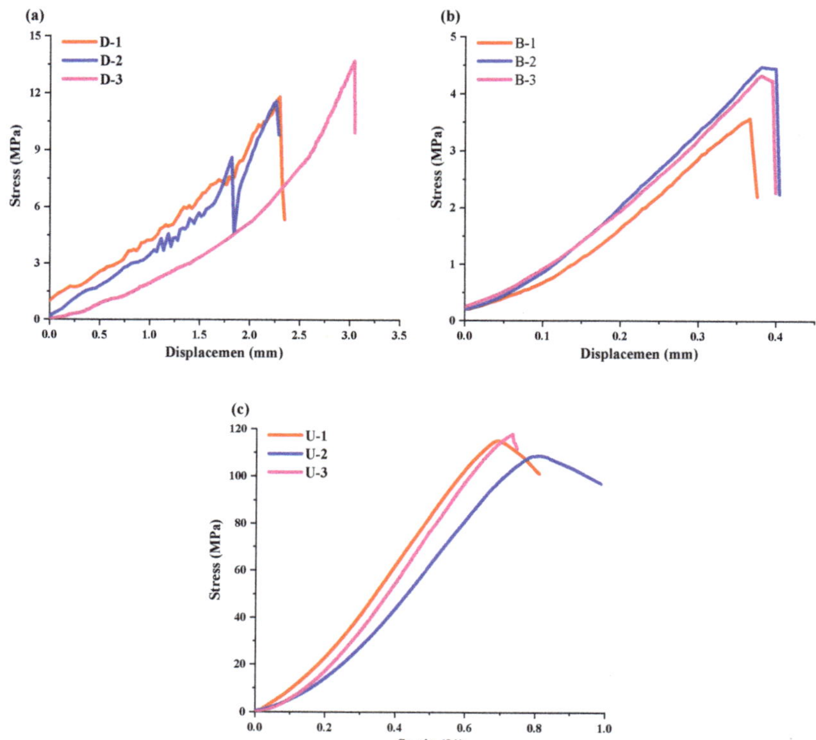

Figure 5. The stress–displacement curve or stress–strain curve of the specimen in three test types. (**a**) DST; (**b**) BITT; (**c**) UCT.

Table 1. Basic physical and mechanical parameters of red sandstone.

Number	Type of Test	Loading Rate (mm/min)	$\sigma_t/\sigma_s/\sigma_c$ (MPa)
B-1			7.06
B-2	BITT	0.05	8.50
B-3			8.80
D-1			12.48
D-2	DST	0.1	11.79
D-3			13.68
U-1			115.27
U-2	UCT	0.1	108.79
U-3			118.10

3. AE Data Processing Methods

The AE event rate is represented by event interval function F(τ), which can reveal the crack propagation of rock specimens, from microcracking to macrocracking. Based on the RA and AF values, the fracture mode of the red sandstone in the loading process was analyzed, and the classification method of the red sandstone fracture mode was determined. In addition, the KDE function was adopted to visualize the RA–AF data density maps. In this chapter, we focus on the three adopted AE data processing methods.

3.1. Inter-Event Time Function F(τ) Theory

The basic connotation of the inter-event time function F(τ) is the average occurrence frequency of N AE events that move continuously, and τ represents the time interval of N AE events [62]. The derivation process is as follows [16]:

$$\Delta t_i = t_i - t_{i-1}, \ i = 2, 3, \ldots \tag{1}$$

where t_i is the time of the i-th AE event, and t_{i-1} is the time of the previous AE event.

$$\tau_i = \frac{t_{N+i-1} - t_{i-1}}{N}, \ i = 2, 3, \ldots \tag{2}$$

$$\tau_1 = \frac{t_N - t_1}{N} \tag{3}$$

$$F(\tau_i) = \tau_i^{-1}, \ i = 1, 2, \ldots \tag{4}$$

When calculating F(τ) in this paper, the N-value of all samples is taken as 50. Furthermore, it is worth noting that F(τ) for N $-$ 1 acoustic emission events cannot be defined, but this does not affect the accuracy of the overall test results.

3.2. RA and AF Values Method

Since the unit of amplitude is dB, it is necessary to convert the unit dB into voltage unit V, and the conversion formula is shown in Formula (5) [37]. RA and AF values are calculated according to the basic parameters of the AE signal, and the calculation method is shown in Formulas (6) and (7) [32].

$$B(\text{mV}) = 10^{\frac{A(dB)}{20} - 1} \tag{5}$$

$$RA\ value = \frac{Rise\ time}{Maximum\ amplitude} \tag{6}$$

$$AF\ value = \frac{Count}{Duration\ time} \tag{7}$$

3.3. Kernel Density Estimation (KDE) Method

The KDE method is widely used in data analysis. In this study, the KDE method was utilized to identify and visualize high-density regions of RA and AF values. The basic principles of the KDE approach are as follows [63].

The basic idea of the KDE method is that each point in the estimated data contributes an "atom" of probability density to the estimate, and $p(z)$ is used to represent the true density estimate of multivariate data. The formula is as follows:

$$\hat{p}(z) = \frac{1}{nh} \sum_{i=1}^{n} K\left(\frac{z - Z_i}{h}\right) \tag{8}$$

where n is the total number of estimated samples, h is the smoothing parameter controlling the atomic width, and Z_i represents the i-th data point. $K(x)$ is a kernel function. In this paper, the multivariate Gaussian function is taken as the kernel function, and its formula is as follows:

$$K(\underline{x}) = \frac{1}{(2\pi)^{\frac{d}{2}}} exp\left(-\frac{1}{2}\|x\|^2\right) \tag{9}$$

where d denotes the dimension of data space.

The probability density function values for all sample points can be simply obtained by establishing an estimate $p(z)$.

The accuracy of the estimate is affected by the value of h and the size of the sample, so the value of h must be reasonably determined. Since the least squares cross-validation method has a large advantage in reducing the squared error between the density estimate and the actual density, this method has been selected to determine the value of h in this paper. The calculation formula of square error is as follows:

$$J[\hat{p}] = \int [p(\underline{x}) - \hat{p}(\underline{x})]^2 d\underline{x} \tag{10}$$

When the real density is Gaussian, the optimal smoothing parameters can be determined by the following formula:

$$h^* = Bn^{1/(d+4)} \tag{11}$$

where parameter B is related to parameter d, and its expression of multivariate distribution is shown as:

$$B = \begin{cases} 1 \ when \ d = 2 \\ (\frac{4}{d+2})^{\frac{1}{d+4}} \ otherwise \end{cases} \tag{12}$$

4. Experimental Results

The AE data of different samples have good consistency in the BITT, DST, and UCT. Therefore, the data of the specimen with the greatest strength were selected for analysis, such as B-3, D-3, and U-3. Simultaneously, the characteristic of the AE event rate in the whole process of the test was analyzed to reveal the AE activity in the process of sample damage and fracture by the time function f between events F(τ). In this section, the evolution law of the AE event rate, the RA and AF values, and the distribution law of RA–AF data in three test types were analyzed.

4.1. AE Event Rate Monitoring

Figure 6 shows the evolution characteristics of the AE event rate and stress with time for the red sandstone samples in the three types of tests. In Figure 6, the vertical coordinate includes stress (black), the number of accumulative AE events (blue), and its corresponding AE event rate (F(τ), pink); the horizontal coordinate is the time of each test type. The whole process of rock damage and fracture is divided into phase-1 and phase-2, according to the evolution law of the AE event rate with time. In phase-1, the AE event rate showed a relatively steady state, and the cumulative AE events grew slowly, which indicates that the crack initiation and expansion activities in rock are moderate. Therefore, phase-1 is called the gentle growth period of AE events (microcrack generation phase). Further, phase-2 is called the sharp growth period of AE events (macrocrack generation phase). In phase-2, the AE event rate increases gradually with a state of high and low fluctuation, and the cumulative AE events increase sharply, which indicates that the crack propagation activity in the critical failure stage in the rock is intense, and macroscopic cracks gradually form, showing a failure trend. The critical time point between adjacent phases is considered to be T_t, and F_t is the corresponding load at T_t. F_t is expressed as $F_t = kF_p$, where F_p is the peak load and k is the ratio of F_t to F_p. In addition, the proportion of AE events in phase-1 and phase-2 to the total number of events in each test type is shown in Figure 7.

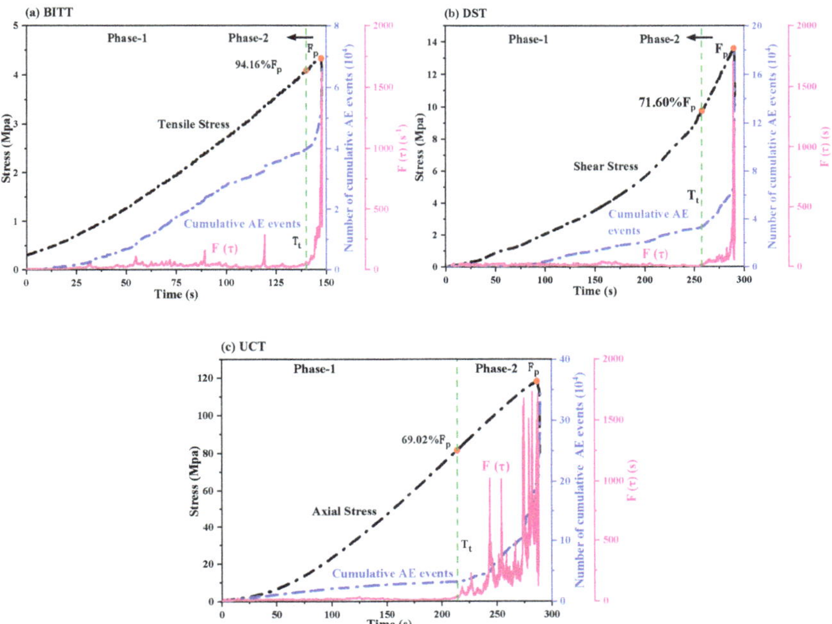

Figure 6. Variations of stress, cumulative AE events and corresponding AE event rate with time. (**a**) BITT; (**b**) DST; (**c**) UCT.

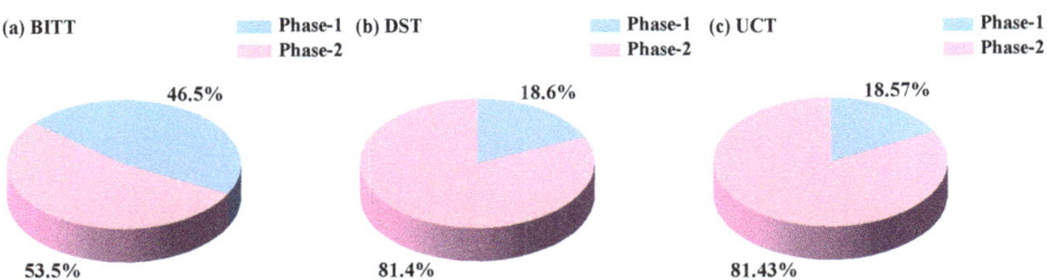

Figure 7. The proportion of AE events in phase-1 and phase-2 to the total number of events. (**a**) BITT; (**b**) DST; (**c**) UCT.

In BITT, the number of AE events in phase-1 and phase-2 accounts for 46.48% and 53.52% of the total events, respectively. In DST, the number of AE events in phase-1 and phase-2 accounts for 18.57% and 81.43% of the total events, respectively. In UCT, the number of AE events in phase-1 and phase-2 accounts for 17.33% and 82.67% of the total events, respectively. On the whole, nearly half of the AE events in BITT occurred before loading F_t, indicating that the rock has been damaged to a certain extent in the microcrack generation phase. However, in DST and UCT, the number of AE events is less before loading F_t, which indicates that the rock damage mainly occurs in the macrocrack generation phase. In addition, the F_t in BITT, DST, and UCT are 94.16% F_p, 69.02% F_p and 71.81% F_p, respectively. It can be seen that the macrocrack generation phase in BITT is significantly shorter than that in DST and UCT. The above results are due to the different fracture modes of rocks under different loading conditions. In BITT, tensile fracture mainly occurs; in DST, shear fracture mainly occurs; and in UCT, tensile and shear fracture always occur together.

4.2. Evolution of RA and AF Values

In this section, only the data collected by one channel were analyzed to avoid the overlapping of signals collected by different channels. The data collected by sensor-1 were selected to analyze, and the moving average of these parameters was calculated from 50 AE events.

The evolution law of the RA and AF values of red sandstone in the three test types with time is shown in Figure 8. The RA and AF values show ups and downs before T_t; however, they show obvious trends after T_t. When the loading time is in the interval between T_t and the final failure time (T_f), there is an obvious downward trend in RA value and an obvious upward trend in AF value. In addition, when rock failure occurs, the RA value will suddenly decrease, and the AF value will suddenly increase in the three test types.

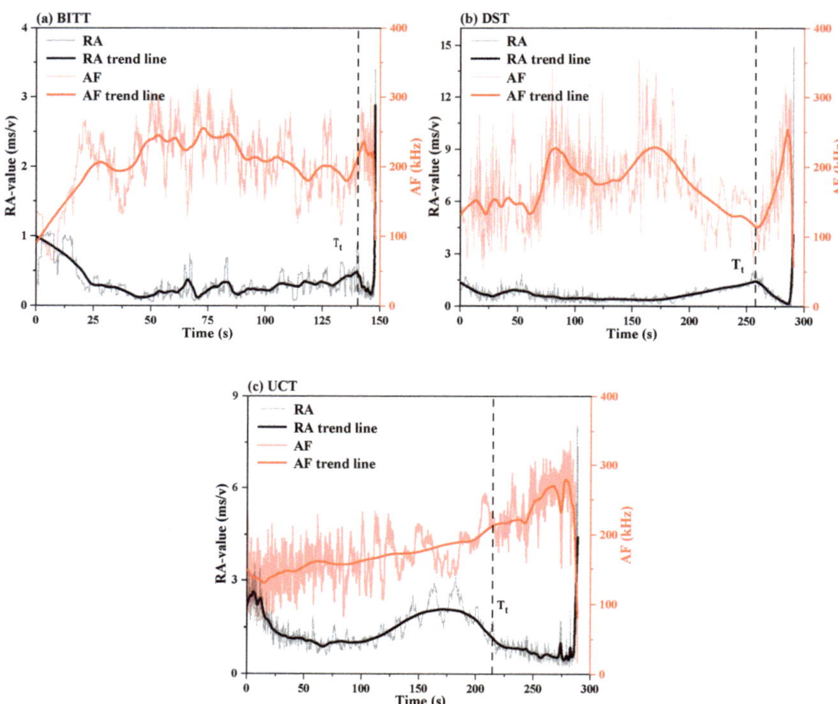

Figure 8. Temporal change of RA and AF values in different test types. (**a**) BITT; (**b**) DST; (**c**) UCT.

4.3. The Kernel Density Distribution of RA–AF Values

The RA–AF distribution can qualitatively describe the variation trend of shear and tensile fracture in the tested specimens. The density maps of RA–AF data in the three types of tests are shown in Figures 9–11, in which the density of data is lower in the red region and higher in the purple and blue regions. A change in color from red to purple indicates that the data distribution has changed from sparse to dense. With the color changing from red to purple, the distribution of data changes from sparse to dense. The purple and blue areas are called the main data distribution areas, which are marked with the black dot–dash square frame. Furthermore, a manual straight line of 45, passing through two points (0, 0) and (30, 1100) with a slope of 36.67, is drawn as the reference line to distinguish the RA–AF distribution. It can be seen from Section 4.2 that RA and AF values vary greatly in different time periods. Therefore, T_t is selected as the dividing point to divide the loading process of each test into two phases, in which the RA–AF distribution is compared. Here, t represents the actual time, and T_f represents the moment when the sample failure occurs.

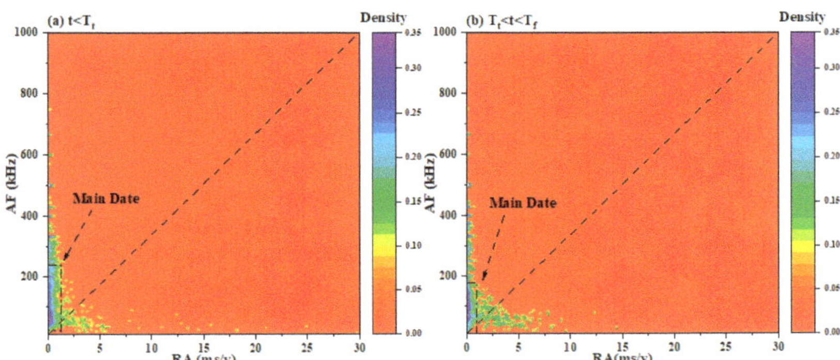

Figure 9. RA–AF data density maps in BITT. (**a**) t < T_t; (**b**) T_t < t < T_f.

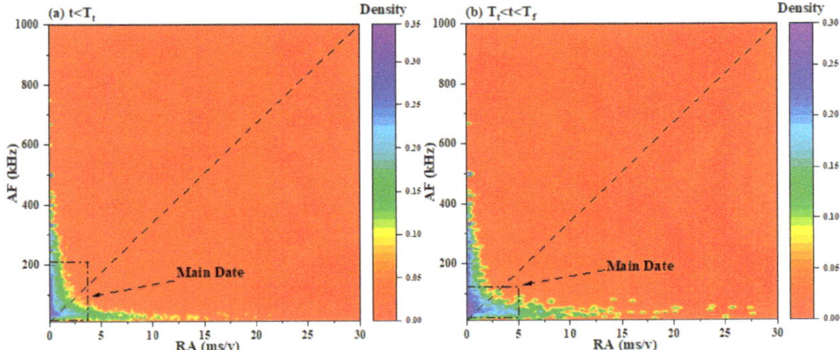

Figure 10. RA–AF data density maps in DST. (**a**) t < T_t; (**b**) T_t < t < T_f.

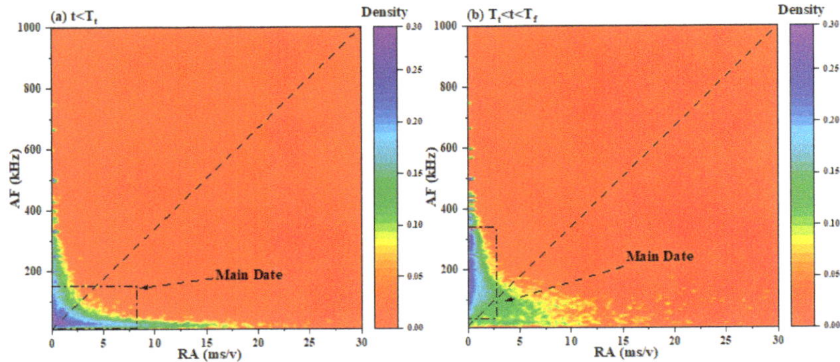

Figure 11. RA–AF data density maps in UCT. (**a**) t < T_t; (**b**) T_t < t < T_f.

Figure 9 shows that the data are mainly distributed above the reference line in BITT, and the RA–AF distribution in phase-1 and phase-2 is similar. From the main data, we can see that the main distribution range of AF values is 0–240 KHz, while the distribution of RA values is in a smaller range (0–1 ms/v). Only a small amount of data is distributed below the reference line, indicating that the shear characteristic signal is dominant in BITT. Figure 10 shows that the main data above the reference line are still dominant in DST, while the main data below the reference line also increases significantly. There are differences in RA–AF distribution between phase-1 and phase-2. From the main data, we can see that the range in AF values in the two phases is 0–150 kHz, and the range in RA value in phase-1 is 0–3 ms/V, while the range in RA values in phase-2 is significantly increased, to 0–5 ms/V. This suggests that the shear characteristic signals increase clearly with the increase in loading in DST. Figure 11 shows that the distribution of RA–AF data in the two phases of UCT is different. In phase-1, the main data are evenly distributed on the upper and lower sides of the reference line, and their distribution range is the rectangular area bounded by the RA value range of 0–8 ms/V and the AF value range of 0–180 kHz. In phase-2, the data are mainly distributed above the reference line. The range of AF values is 0–2.5 ms/V, and the range of RA values is 30–340 kHz. This suggests that shear and tensile characteristic signals are generated simultaneously in the microcrack generation phase in UCT, while tensile characteristic signals are mainly generated in the macrocrack generation phase.

4.4. Classification of Tensile and Shear Cracks

It can be seen from Section 4.3 that the RA–AF distribution in BITT and DST mainly presents obvious tensile characteristic signals and shear characteristic signals, respectively. Therefore, the RA–AF data in BITT and DST were selected for analysis to determine the dividing line of RA–AF distribution between tensile and shear cracks in red sandstone samples. The method used to determine the dividing line as to plot the RA–AF data in BITT and DST on the same scatter diagram, and then find a straight line so that the data proportion under the straight line in BITT is basically the same as that on the straight line in DST [26]. An enlargement of the main data in Figures 10b and 11b is shown in Figure 12, in which it can be seen that the straight line (AF = 75 kHz) is a reference line. There are obvious differences between the RA–AF distributions above and below the reference line, regardless of whether we are using BITT or DST. Hence, the value of 75 kHz has been selected as the intercept of the dividing line. Next, the dividing line was determined by constantly changing the slope. When the slope of the dividing line reached the value of 93, both the data proportion in BITT below the line and that in DST above the line were 39.9%. Therefore, the straight line, AF = 93RA + 75, was determined as the dividing line between the tensile crack and shear crack in the RA–AF scatter plots, as shown in Figure 13.

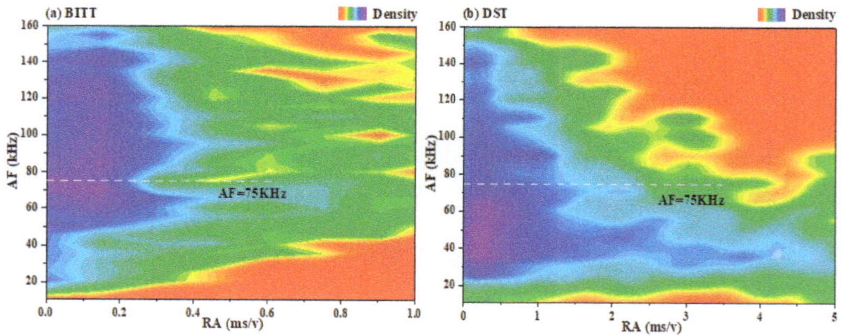

Figure 12. Enlarged figure of the main data in BITT and DST. (**a**) BITT; (**b**) DST.

As shown in Figure 13, we have determined that the dividing line between tensile fracture and shear fracture in the RA–AF scatter plots of red sandstone is AF = 93RA + 75. The RA–AF data distributed above the dividing line represent the data generated by tensile fracture, and the RA–AF data distributed below the dividing line represent the data generated by shear fracture. Therefore, the AE monitoring technology can classify between tensile fracture and shear fracture in the whole process of rock loading. When a fracture occurs, the RA–AF value of the collected AE signal is distributed above the dividing line (AF > 93RA + 75), which can be considered as a tensile fracture. On the contrary, if the RA–AF value is distributed below the division line (AF < 93RA + 75), the fracture can be considered as a shear fracture.

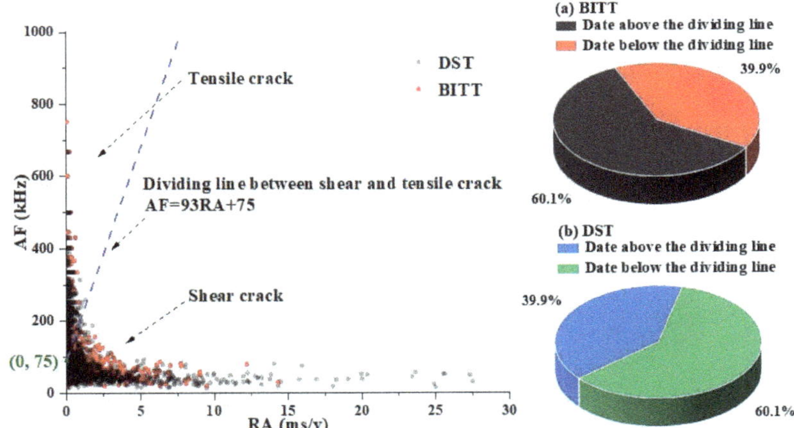

Figure 13. Determinations of the dividing line between tensile and shear cracks. (**a**) BITT; (**b**) DST.

4.5. Statistics of Tension and Shear Fracture, and Analysis of Failure Mechanism

According to the dividing line determined above, the microcracks of red sandstone under uniaxial compression test loading are statistically analyzed. The statistics of the tensile and shear cracks of the U-1, U-2 and U-3 specimens are shown in Figure 14. The total numbers of AE events in the U-1, U-2 and U-3 specimens are 38,783, 51,200 and 45,264, respectively. It can be seen in Figure 14 that tensile cracks account for a large proportion (more than 67%) in the U-1 specimen, while shear cracks account for a large proportion (about 60%) in the U-2 and U-3 specimens, which indicates that the fracture mode of rock under uniaxial compression loading is complex. In addition, by observing the failure modes of the U-1, U-2 and U-3 specimens (Figure 15), it can be found that the U-1 specimen with more tensile cracks is broken, while the U-2 and U-3 specimens with more shear cracks are more complete. The macroscopic cracks on the surface of each specimen are mainly shear cracks, accompanied by a certain number of tensile cracks. These results correspond to the results reflected by AE parameters, but they still need to be further analyzed from the perspective of the rock failure mechanism.

Griffith crack exists in rock material at the initial state. Griffith cracks in rocks can be formed by pores, voids, soft or hard nodules or particles, particle boundaries, etc. [64–66]. When the shear stress on the fracture surface exceeds the shear strength, shear fracture will occur in the rock, which is the main fracture mode.

When the extension direction of the crack in the rock is approximately parallel to the direction of the compressive stress, the tensile stress will concentrate at the tip of the crack under the compressive stress [67]. If the tensile stress concentration is large and reaches the tensile strength of the material, the crack will begin to expand. In this case, the pressure will continue to increase, and the crack will expand rapidly, which may eventually lead to a macrofracture.

If the crack extension direction in the rock intersects with the compressive stress direction at a small angle, we take out an "isolator" containing an inclined crack AB from the rock sample along the axis, as shown in Figure 16. With the gradual increase in axial pressure, shear slip will occur along the crack's surface in the rock specimen. In the above case, normal stress N and friction force F occur on the shear slip surface, and the combined stress along the axial direction is less than the axial stress in the rock specimen; otherwise there will be no slip (shear) fracture. Therefore, there must be shear stress on the BG surface to balance the axial stress. In addition, the shear slip action will also produce a tensile stress perpendicular to the axial direction, which increases with the increase in the slip surface. Obviously, in the compression process, without confining pressure or with very small confining pressure, a tensile fracture along the axial direction will occur with the increase in the shear slip surface. After a tensile fracture surface appears, the axial shear and the tensile stress of the rock specimens below it (along the slip surface) will be reduced to zero. Then, with the continuous expansion of the shear surface, tensile fractures will occur one by one [68,69]. Therefore, when there are many Griffith cracks in the rock that are approximately parallel to the direction of the compressive stress, or within a small angle, shear fracture and tensile fracture occur simultaneously in the rock.

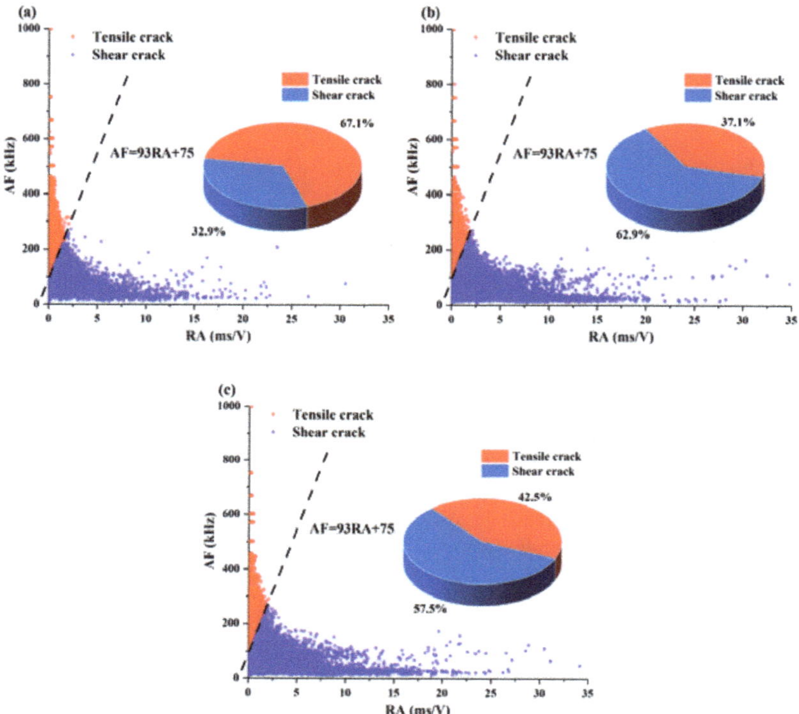

Figure 14. The statistics of tensile and shear cracks of the U-1, U-2 and U-3 specimens. (**a**) U-1; (**b**) U-2; (**c**) U-3.

Figure 15. The failure modes of the U-1, U-2 and U-3 specimens. (**a**) U-1; (**b**) U-2; (**c**) U-3.

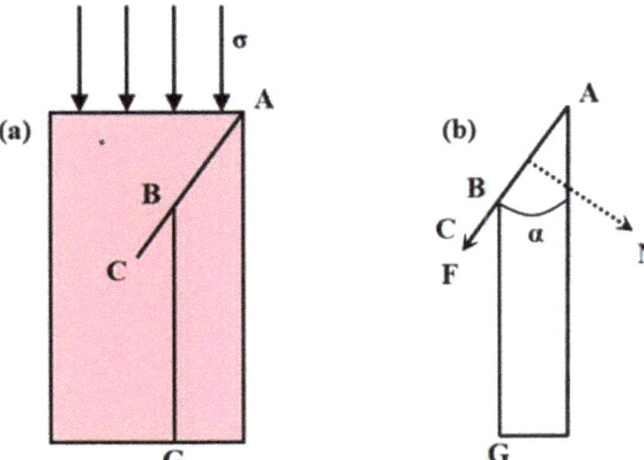

Figure 16. Shear and tensile failure mechanism of rock specimen. (**a**) Isolator; (**b**) stress diagram.

Obviously, the above rock failure mechanism corresponds to the rock failure characteristics reflected by the AE parameters. Therefore, the proposed dividing line of RA–AF scatter plots can be used for classifying tensile and shear fractures to determine the fracture mode of red sandstone.

5. Discussion

Having determined the dividing line (AF = 93RA + 75) between shear and tensile cracks, we focus on the evolution law of tension and the shear sources of sandstone in UCT on the basis of RA and AF values. In addition, combined with the statistical index analysis of the b-value, the failure precursor index of red sandstone is here discussed based on the dividing line. In this section, the analysis is based on the data of U-1, U-2 and U-3 specimens.

5.1. Evolution Characteristics of Tensile and Shear Sources in UCT

When the signal generated by a fracture is received by multiple sensors (greater than or equal to four) at the same time, a positioning signal, the fracture source, will be formed. Because it is difficult for a fracture signal to be collected by multiple sensors at the same time, the location source data in this section are much fewer than the number of acoustic emission events in the above section.

The two-dimensional spatial distribution of tensile and shear AE sources in each sample is shown in Figure 17, in which the distribution of the AE source of red sandstone is relatively scattered under the condition of uniaxial compression, which is generally consistent with the position of the macro failure surface of the specimen. The total numbers of sources of U-1, U-2, and U-3 specimens are 1062, 1001, and 1388, respectively, most of which are tensile sources. The percentages of tensile sources of the U-1, U-2, and U-3 specimens in the total number of sources are 84.7%, 77.4%, and 71.4%, respectively, as shown in Figure 18. Compared with Figure 14, it is found that although shear cracks account for a large proportion of microcracks in U-2 and U-3, the fracture sources are mainly tensile sources, indicating that tensile fracture more easily forms a positioning signal.

Figure 19 shows the cumulative number of tensile and shear AE sources corresponding to axial strain throughout the whole process for three specimens, including the total stress–strain curve. According to the deformation characteristics of specimens, the approximate straight line section of the stress–strain curve can be defined as the phase from the elastic deformation to the stable development of microcracks (II), and the phase before elastic deformation to the stable development of microcracks (II) can be defined as the micropore compaction phase (I). In addition, after the elastic deformation reaches the stable development phase of microcracks (II), before peak stress occurs, the unstable development phase of microcracks emerges (III). The slopes of the approximate straight lines of the stress–strain curves for the U-1, U-2, and U-3 specimens are 210.89, 188.31, and 214.78, respectively.

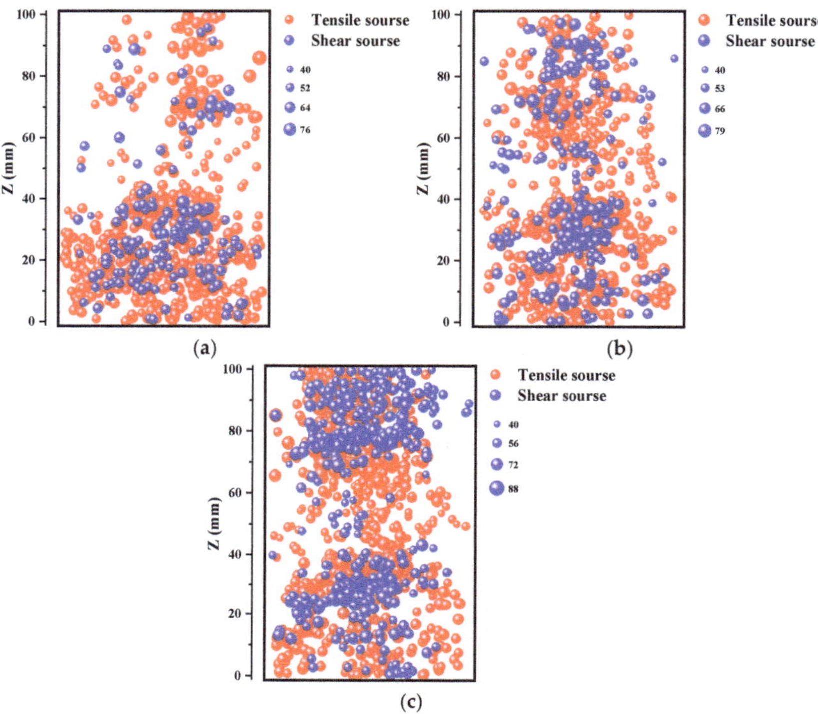

Figure 17. Spatial distribution of AE sources: (**a**) U-1; (**b**) U-2; (**c**) U-3. **Note:** The size of the scatter diagram indicates the source amplitude.

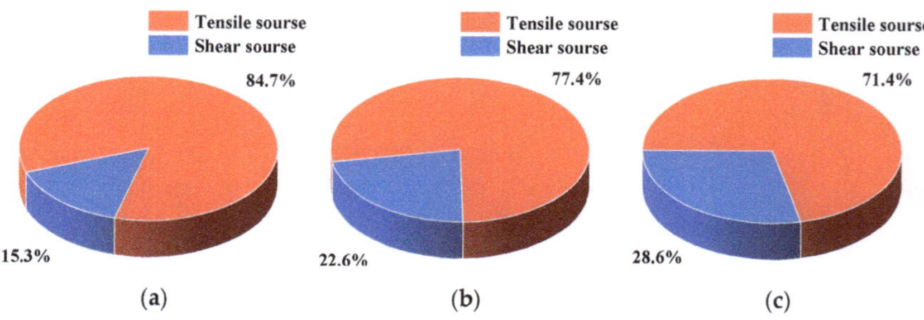

Figure 18. Percentage of different types of AE sources: (**a**) U-1; (**b**) U-2; (**c**) U-3.

A fascinating phenomenon was found, whereby shear sources always grow prior to tensile sources in the initial phase I. This phenomenon depends on the initial micro pore and microcrack state of each specimen, to a certain extent. With the increase in loading stress, there is a rapid growth point in the number of tensile sources. The stress corresponding to the rapid growth point of the tensile source of the U-1, U-2, and U-3 specimens is $83.09\%\sigma_t$, $57.22\%\sigma_t$ and $70.25\%\sigma_t$, respectively (σ_t is the peak stress). The rapid growth point of the tensile source of the U-1 specimen is at the critical point of phase II and phase III, and the rapid growth points of the tensile sources of the U-2 and U-3 samples are all at phase II. However, in phase I, the shear source increases rapidly; in phase II, the growth of the shear source is slow, and its cumulative curve is roughly "horizontal". The rapid growth point of the shear source appears in phase III, which is close to rock failure. The stress corresponding to the rapid growth points of the shear sources of the U-1, U-2, and U-3 specimens is 99.49%σ_t, 95.46%σt, and 95.93%σ_t, respectively. Owing to the strength of the U-2 specimen being lower than that of the U-1 and U-3 specimens, the rapid growth point of the U-2 specimen occurs earlier, and especially the rapid growth point of the tensile source.

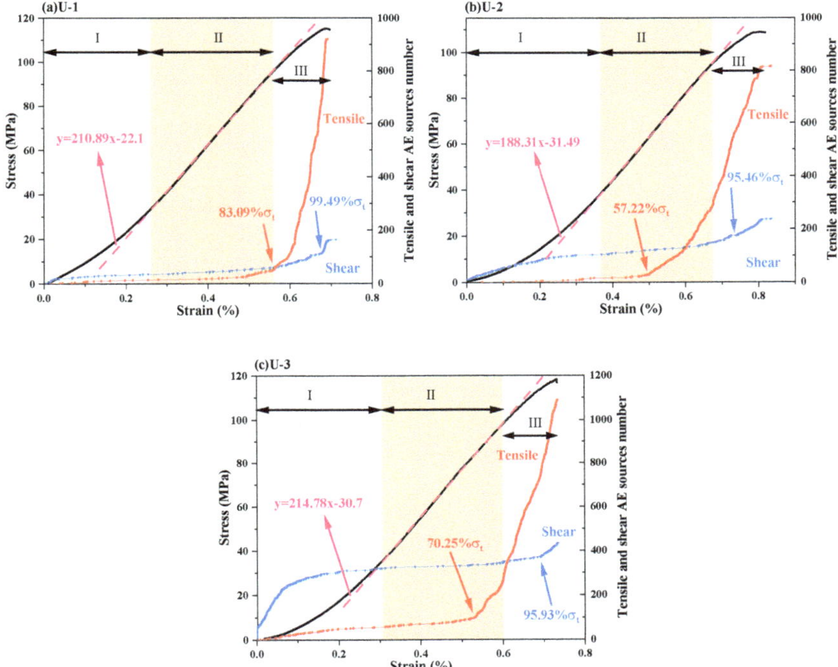

Figure 19. Curves of AE source numbers with axial strain: (**a**) U-1; (**b**) U-2; (**c**) U-3.

It can be seen from the above analysis that under uniaxial compression loading, the fracture source of red sandstone is primarily the shear source in the initial phase of loading and the tensile source in the critical failure phase, and the number is far greater than that of the shear source.

5.2. Failure Precursor Index of Rock Based on k Value

There will be certain contingencies in the parameters obtained in AE monitoring. The instantaneous growth of a certain index may be unable to objectively reflect the intensification of fracture in the rock. Therefore, in the AE monitoring based on parameter analysis, the AE parameters are usually statistically processed to further obtain the corresponding statistical indicators.

Among the statistical indicators based on amplitude, the most commonly used is the b-value statistic. The b-value originates from the Gutenberg Richter (G-R) relationship in seismology; that is, the logarithm of cumulative times (N), greater than magnitude (M), is linear with magnitude (M) [70], as shown in Formula (13).

$$\log N = a - bM \tag{13}$$

where a and b are constants. In the analysis of AE parameters, the amplitude can usually be divided by 20 to represent the AE magnitude M, i.e., M = A/20. In the calculation of the b-value, the unit of A is dB.

The B-value is mainly used to measure the relative number of small-magnitude fracture events and large-magnitude fracture events in rocks under compression, which can represent the scale of magnitude distribution of AE events. Therefore, it is widely used to analyze and forecast the precursors of rock fracture [71]. When the b-value is larger, it indicates that small- and medium-scale fracture events account for a large proportion; otherwise, it indicates that large-scale fracture events are dominant. In laboratory test, the corresponding b-value is about 1 ($\pm$0.5), when the rock mass fails.

In addition, in previous studies, the dividing line between shear and tensile crack was in the form of y = kx. In the analysis of the fracture mechanism, the slope of the dividing line, k = AF/RA, was used as an index to classify the shear fracture and tensile fracture. However, the intercept of the dividing line proposed in this paper is not zero, so the k-value, k = AF/(93RA + 75), is selected as the AE parameter index to estimate the damage degree of the red sandstone specimens. When the number of signals with large k-value increases, it indicates that the proportion of tensile fracture in the specimen increases, and the damage intensifies. For k-value, the instantaneous accidental value is also not enough to explain the intensification of fracture. Therefore, it is difficult to estimate the severity of specimen failure when only using the absolute k-value of one or several AE signals. In this study, the coefficient of variation (CV) is selected as the statistical index, and the dispersion of k-value distribution is used to describe the fracture of specimens. As a normalized measure indicator, CV is defined as the ratio of the standard deviation to the average.

Figure 20 shows the statistical results of the b-value and the CV (k) of the U-1, U-2, and U-3 specimens, taking the strain as the independent variable. In the process of calculation, the AE data of each specimen were equally divided into 100 segments, and then the corresponding b-value and CV (k) of each segment were calculated. The sample sizes of the U-1, U-2, and U-3 specimens are 240, 199, and 269 respectively. In Figure 20, the phase partition from Figure 19 is used again. Since the data in the pink dotted rectangle in the figure are too dense, the data in this area are enlarged. Furthermore, two kinds of reference lines are set in the figure. One is a horizontal dotted line, which corresponds to b-value = 1 and CV (k) = 1; the other is the vertical dotted line, which corresponds to the rapid growth point of the tensile source, as described in Section 5.1.

It can be seen from Figure 20 that when the strain of all specimens reaches the corresponding rapid growth point of the tensile source, the b-values decrease significantly, and when the b-values decrease to less than 1, the specimen will enter phase-III. In phase-III, except for individual points, the other b-values are less than 1, and when the specimen is close to complete failure, the b-values decrease significantly again. Additionally, when the strain of all specimens approaches the rapid growth point of the corresponding tensile source, CV (k) shows a downward trend, and the first CV (k) behind the vertical reference line is close to 1. In phase-III, the CV (k) of all specimens is stable at less than 1. Differently from the b-value, when the specimen is close to complete failure, the CV (k) increases significantly and will exceed 1.

In the early loading phase of all specimens, the b-values were basically distributed between 0.7 and 1, and then increased to 1–1.5, which indicates that the micropores of the specimens were gradually compacted at the initial phase of loading, and then small fractures dominated in the specimens. However, until the specimens approached the failure phase and were in the failure phase, the b-value began to decrease sharply. The b-value of the U-2 specimen suddenly decreased in the early phase of loading, but then returned to a higher level, which indicates that some large-scale fractures appeared in this phase, but did not continue to develop, and returned to the fracture development mode dominated by small-scale fractures.

In the early loading phase of all specimens, the span of CV (k), basically distributed between 1 and 3.4, was large, which means that the dispersion of the k-value was large, indicating that there were great differences, with tensile cracks and shear cracks occurring together, in the crack types of the specimens in the early loading phase. However, when the strain reached the corresponding rapid growth point of the tensile source, especially after the specimen entered phase-III, the CV (k) was relatively stable, which means that the dispersion of the k-value was small, indicating that tensile cracks were mainly produced. Meanwhile, in phase-III, the CV (k) also increased locally on the basis of relative stability, especially when the specimens were close to complete failure, which increased significantly. This shows that a small number of shear cracks were also generated in the specimen in phase-III, and when the specimen was in complete failure, a large number of shear cracks were generated.

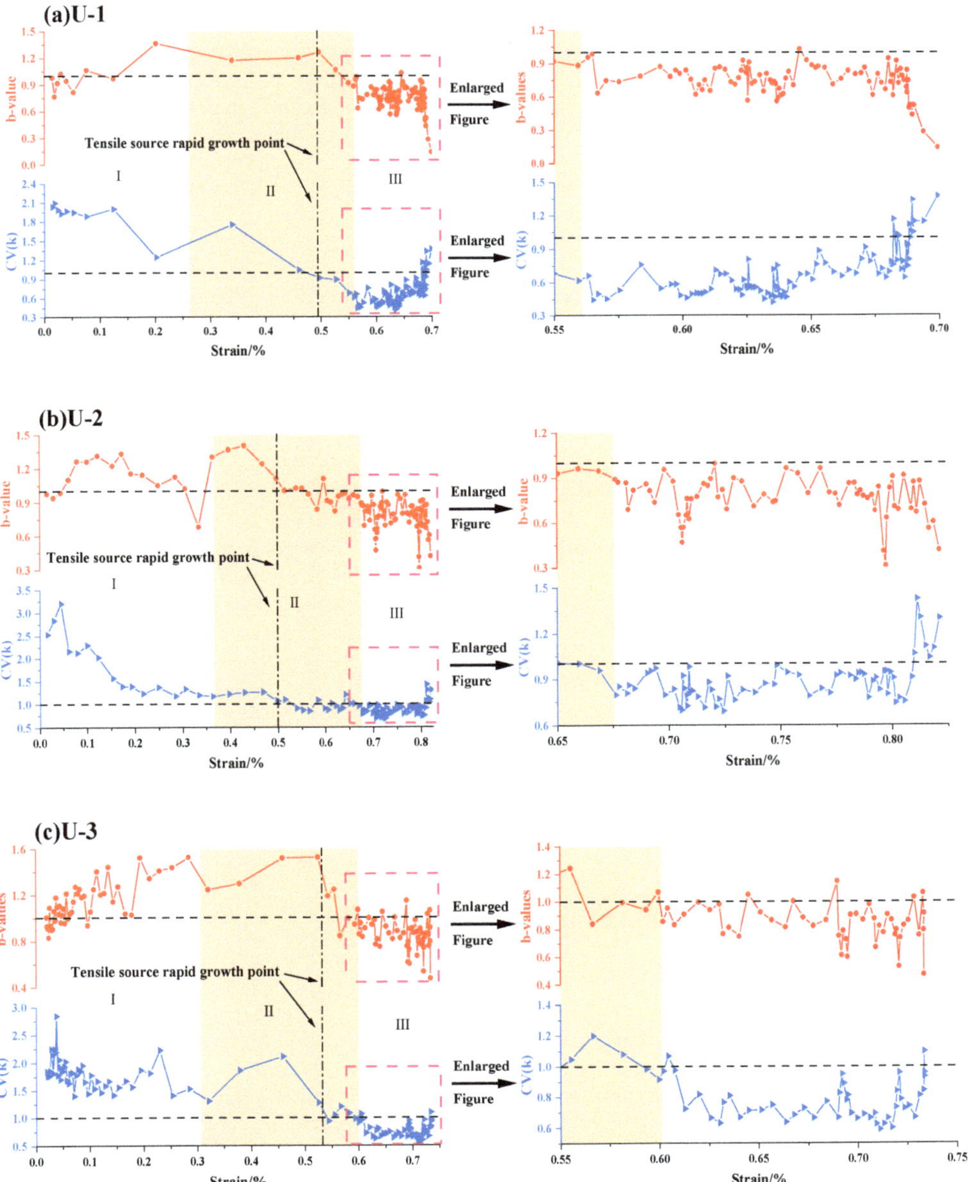

Figure 20. Comparison of CV(k) and b-values among different specimens. (**a**) U-1; (**b**) U-2; (**c**) U-3.

From the above results, it can be found that the combination of b-value and CV (k) can reveal the failure process of a specimen. In the early (stable) failure phase, the b-value was mainly distributed between 1 and 1.5, while the CV (k) value's distribution span was large, ranging from 1 to 3.4; in the instability failure phase, both the b-value and CV (k) were less stable than 1. Only when the final failure occurred did CV (k) increase significantly, and come to exceed 1. Therefore, 1.0 can also be used as the recommended judgment value of the CV (k) index.

In a practical sense, when the CV (k) is greater than 1, the dispersion of the k-value exceeds 100%. It can be seen from the above conclusions that in the unstable failure phase, tensile fracture is the main fracture type, and the corresponding k-values are large. With the aggravation of the fracture, when the specimen is in complete failure, the proportion of shear fracture signal increases, and more signals with small k-value appear.

The dispersion of k-values exceeding 100% indicates that the number of AE signals with small k-value increased significantly, and the proportion of shear fracture signals increased significantly. Therefore, taking CV (k) as the fracture failure judgment value of red sandstone has practical significance.

6. Conclusions

The AE characteristics of red sandstone in BITT, DST, and UCT were analyzed using AE monitoring technology. The variation law of the RA and AF values was analyzed to study the fracture mode and propagation characteristics of cracks in red sandstone. Based on the kernel density estimation (KDE) function and coupling the AE parameters (RA–AF values) in DST and BITT, the classification method of red sandstone fractures was determined. The reliability of this method was verified by the results of uniaxial compression tests. The conclusions are as follows:

(1) AE event rate can reflect the transformation of rock samples from microcracks to macrocracks. The macrocrack generation phase in UCT was the longest, that in DST was the second longest, and that in BITT was the shortest;

(2) The KDE method can effectively identify and visualize the high-density areas of RA and AF values. In the failure mode dominated by tensile fracture, the RA value was low and the AF value was high. On the contrary, in the failure mode dominated by shear fracture, the RA value was high and the AF value was low. When rock failure occurred, the RA and AF values both developed in opposite directions;

(3) It was determined that the dividing line for classifying tensile and shear cracks in the RA and AF value data is AF = 93RA + 75. The reliability of the dividing line has been verified by analyzing the failure mode and fracture mechanism of the sample;

(4) Under uniaxial compression loading, the fracture source of red sandstone was mainly the shear source in the initial phase of loading, and the tensile source in the critical failure phase, and the number of the latter was far greater than that of the shear source;

(5) $K = AF/(93RA + 75)$ was proposed as an AE parameter index to reflect the internal fracture of the red sandstone specimen. Further, the corresponding reference judgment value CV (k) = 1 was proposed. It can be considered that the test sample entered the instability failure phase when CV (k) < 1.

Author Contributions: Conceptualization, J.L. and C.W.; data curation, J.L.; formal analysis, Y.H.; funding acquisition, C.W. and S.L.; investigation, J.L.; methodology, C.W.; project administration, Y.H.; resources, C.W.; software, S.L.; supervision, C.W. and S.L.; validation, S.L. and C.W.; visualization, Y.H.; writing (original draft), J.L.; writing (review and editing), J.L. and S.L. All authors have read and agreed to the published version of the manuscript.

Funding: This research was funded by National Natural Science Foundation of China (No. 52004072, No. 52064006, and No. 52164001), the Guizhou Provincial Science and Technology Foundation (No. [2020]4Y044, No. [2021]292, [2021]4023, and [2021]N404), and the Guizhou Provincial Graduate Research Foundation (YJSCXJH[2020]088).

Institutional Review Board Statement: Not applicable.

Informed Consent Statement: Informed consent was obtained from all subjects involved in the study.

Data Availability Statement: Some or all data, models, or codes that support the findings of this study are available from the corresponding author (fgan@gzu.edu.cn.) upon reasonable request.

Conflicts of Interest: The authors declare no conflict of interest.

Abbreviations

AE	Acoustic emission
BITT	Brazilian indirect tensile test
DST	Direct shear test
UCT	Uniaxial compression test
KDE	Kernel density estimation
RA	RA = rise time/amplitude
AF	Average frequency
σ_s	Shear strength
σ_c	Uniaxial compressive strength
$F(\tau)$	The inter-event time function/AE event rate
Tt	Time at the beginning of drastic increase in AE events
Ft	Load at the beginning of drastic increase in AE events
F_p	Peak load during the test
CV	The coefficient of variation

References

1. Bi, J.; Zhou, X.P.; Qian, Q.H. The 3D numerical simulation for the propagation process of multiple pre-existing flaws in Rock-Like materials subjected to biaxial compressive loads. *Rock Mech. Rock Eng.* **2016**, *49*, 1611–1627. [CrossRef]
2. Zhao, Y.; Zhang, C.; Wang, Y.; Lin, H. Shear-related roughness classification and strength model of natural rock joint based on fuzzy comprehensive evaluation. *Int. J. Rock Mech. Min.* **2021**, *137*, 104550. [CrossRef]
3. Zhao, Y.; Zhang, Y.; Yang, H.; Liu, Q.; Tian, G. Experimental study on relationship between fracture propagation and pumping parameters under constant pressure injection conditions. *Fuel* **2022**, *307*, 121789. [CrossRef]
4. Tang, L.; Zhao, Y.; Liao, J.; Liu, Q. Creep experimental study of rocks containing weak interlayer under multilevel loading and unloading cycles. *Front. Earth Sci.* **2020**, *480*, 519461. [CrossRef]
5. Zhao, Y.; He, P.; Zhang, Y.; Wang, C. A new criterion for a Toughness-Dominated hydraulic fracture crossing a natural frictional interface. *Rock Mech. Rock Eng.* **2019**, *52*, 2617–2629. [CrossRef]
6. Zhao, Y.; Bi, J.; Wang, C.; Liu, P. Effect of unloading rate on the mechanical behavior and fracture characteristics of sandstones under complex triaxial stress conditions. *Rock Mech. Rock Eng.* **2021**, *54*, 4851–4866. [CrossRef]
7. Zhao, Y.; Liu, Q.; Zhang, C.; Liao, J.; Lin, H.; Wang, Y. Coupled seepage-damage effect in fractured rock masses: Model development and a case study. *Int. J. Rock Mech. Min.* **2021**, *144*, 104822. [CrossRef]
8. Zhao, Y.; Wang, Y.; Wang, W.; Tang, L.; Liu, Q.; Cheng, G. Modeling of rheological fracture behavior of rock cracks subjected to hydraulic pressure and far field stresses. *Theor. Appl. Fract. Mech.* **2019**, *101*, 59–66. [CrossRef]
9. Zhao, Y.; Zhang, L.; Liao, J.; Wang, W.; Liu, Q.; Tang, L. Experimental study of fracture toughness and subcritical crack growth of three rocks under different environments. *Int. J. Geomech.* **2020**, *20*, 04020128. [CrossRef]
10. Zhao, Y.; Wang, C.L.; Bi, J. Analysis of fractured rock permeability evolution under unloading conditions by the model of elastoplastic contact between rough surfaces. *Rock Mech. Rock Eng.* **2020**, *53*, 5795–5808. [CrossRef]
11. Zhang, S.; Wu, S.; Chu, C.; Guo, P.; Zhang, G. Acoustic emission associated with Self-Sustaining failure in Low-Porosity sandstone under uniaxial compression. *Rock Mech. Rock Eng.* **2019**, *52*, 2067–2085. [CrossRef]
12. Liang, Z.; Xue, R.; Xu, N.; Li, W. Characterizing rockbursts and analysis on frequency-spectrum evolutionary law of rockburst precursor based on microseismic monitoring. *Tunn. Undergr. Space Tech.* **2020**, *105*, 103564. [CrossRef]
13. Aggelis, D.G. Classification of cracking mode in concrete by acoustic emission parameters. *Mech. Res. Commun.* **2011**, *38*, 153–157. [CrossRef]
14. Aggelis, D.G.; Mpalaskas, A.C.; Matikas, T.E. Acoustic signature of different fracture modes in marble and cementitious materials under flexural load. *Mech. Res. Commun.* **2013**, *47*, 39–43. [CrossRef]
15. Zhao, Y.; Ding, D.; Bi, J.; Wang, C.; Liu, P. Experimental study on mechanical properties of precast cracked concrete under different cooling methods. *Constr. Build. Mater.* **2021**, *301*, 124141. [CrossRef]
16. Zhang, J.; Zhou, X. AE event rate characteristics of flawed granite: From damage stress to ultimate failure. *Geophys. J. Int.* **2020**, *222*, 795–814. [CrossRef]
17. Jiang, R.; Dai, F.; Liu, Y.; Li, A.; Feng, P. Frequency characteristics of acoustic emissions induced by crack propagation in rock tensile fracture. *Rock Mech. Rock Eng.* **2021**, *54*, 2053–2065. [CrossRef]

18. Zhou, X.; Zhang, J.; Qian, Q.; Niu, Y. Experimental investigation of progressive cracking processes in granite under uniaxial loading using digital imaging and AE techniques. *J. Struct. Geol.* **2019**, *126*, 129–145. [CrossRef]
19. Fu, X.; Ban, Y.; Xie, Q.; Abdullah, R.A.; Duan, J. Time delay mechanism of the kaiser effect in sandstone under uniaxial compressive stress conditions. *Rock Mech. Rock Eng.* **2021**, *54*, 1091–1108. [CrossRef]
20. Zhao, Y.; Zheng, K.; Wang, C.; Bi, J.; Zhang, H. Investigation on model-I fracture toughness of sandstone with the structure of typical bedding inclination angles subjected to three-point bending. *Theor. Appl. Fract. Mec.* **2022**, *119*, 103327. [CrossRef]
21. Yang, H.; Lin, H.; Chen, Y.; Wang, Y.; Zhao, Y.; Yong, W.; Gao, F. Influence of wing crack propagation on the failure process and strength of fractured specimens. *Bull. Eng. Geol. Environ.* **2022**, *81*, 71. [CrossRef]
22. Liao, J.; Zhao, Y.; Tang, L.; Liu, Q.; Sarmadivaleh, M. Experimental Studies on Cracking and Local Strain Behaviors of Rock-Like Materials with a Single Hole before and after Reinforcement under Biaxial Compression. *Geofluids* **2021**, 8812006. [CrossRef]
23. Zhao, Y.; Wang, Y.; Tang, L. The compressive-shear fracture strength of rock containing water based on Druker-Prager failure criterion. *Arab. J. Geosci.* **2019**, *12*, 452. [CrossRef]
24. Lacidogna, G.; Piana, G.; Accornero, F.; Carpinteri, A. Multi-technique damage monitoring of concrete beams: Acoustic Emission, Digital Image Correlation, Dynamic Identification. *Constr. Build. Mater.* **2020**, *242*, 118114. [CrossRef]
25. Zhang, Z.; Deng, J. A new method for determining the crack classification criterion in acoustic emission parameter analysis. *Int. J. Rock Mech. Min.* **2020**, *130*, 104323. [CrossRef]
26. Du, K.; Li, X.; Tao, M.; Wang, S. Experimental study on acoustic emission (AE) characteristics and crack classification during rock fracture in several basic lab tests. *Int. J. Rock Mech. Min.* **2020**, *133*, 104411. [CrossRef]
27. Zhao, Y.; Zhang, L.; Wang, W.; Tang, J.; Lin, H.; Wan, W. Transient pulse test and morphological analysis of single rock fractures. *Int. J. Rock Mech. Min.* **2017**, *91*, 139–154. [CrossRef]
28. Zhao, Y.; Liu, Q.; Liao, J.; Wang, Y.; Tang, L. Theoretical and numerical models of rock wing crack subjected to hydraulic pressure and far-field stresses. *Arab. J. Geosci.* **2020**, *13*, 926. [CrossRef]
29. Chai, M.; Hou, X.; Zhang, Z.; Duan, Q. Identification and prediction of fatigue crack growth under different stress ratios using acoustic emission data. *Int. J. Fatigue* **2022**, *160*, 106860. [CrossRef]
30. Zhang, F.; Yang, Y.; Fennis, S.A.A.M.; Hendriks, M.A.N. Developing a new acoustic emission source classification criterion for concrete structures based on signal parameters. *Constr. Build. Mater.* **2022**, *318*, 126163. [CrossRef]
31. Dong, L.; Tong, X.; Ma, J. Quantitative investigation of tomographic effects in abnormal regions of complex structures. *Engineering* **2021**, *7*, 1011–1022. [CrossRef]
32. Gan, Y.; Wu, S.; Ren, Y.; Zhang, G. Evaluation indexes of granite splitting failure based on RA and AF of AE parameters. *Rock Soil Mech.* **2020**, *41*, 2324–2332.
33. Zhao, Y.; Zhang, L.; Wang, W.; Pu, C.; Wan, W.; Tang, J. Cracking and stress–strain behavior of Rock-Like material containing two flaws under uniaxial compression. *Rock Mech. Rock Eng.* **2016**, *49*, 2665–2687. [CrossRef]
34. Tang, Y.; Lin, H.; Wang, Y.; Zhao, Y. Rock slope stability analysis considering the effect of locked section. *Bull. Eng. Geol. Environ.* **2021**, *80*, 7241–7251. [CrossRef]
35. Ohno, K.; Ohtsu, M. Crack classification in concrete based on acoustic emission. *Constr. Build. Mater.* **2010**, *24*, 2339–2346. [CrossRef]
36. *JCMS-III B5706*; Monitoring Method for Active Cracks in Concrete by Acoustic Emission. Federation of Construction Materials Industries: Tokyo, Japan, 2003.
37. Ohtsu, M. Rilem Technical Committee (Ohtsu M. Recommendation of RILEM TC 212-ACD: Acoustic emission and related NDE techniques for crack detection and damage evaluation in concrete. *Mater. Struct.* **2010**, *43*, 1187–1189.
38. Zhao, Y.; Liao, J.; Wang, Y.; Liu, Q.; Lin, H.; Chang, L. Crack coalescence patterns and local strain behaviors near flaw tip for rock-like material containing two flaws subjected to biaxial compression. *Arab. J. Geosci.* **2020**, *13*, 1251. [CrossRef]
39. Wang, H.; Liu, D.; Cui, Z.; Cheng, C.; Jian, Z. Investigation of the fracture modes of red sandstone using XFEM and acoustic emissions. *Theor. Appl. Fract. Mech.* **2016**, *85*, 283–293. [CrossRef]
40. Lotidis, M.A.; Nomikos, P.P. Acoustic emission location analysis and microcracks' nature determination of uniaxially compressed calcitic marble hollow plates. *Geomech. Geophys. Geo-Energy Geo-Resour.* **2021**, *7*, 38. [CrossRef]
41. Li, D.; Du, F. Monitoring and evaluating the failure behavior of ice structure using the acoustic emission technique. *Cold Reg. Sci. Technol.* **2016**, *129*, 51–59. [CrossRef]
42. Han, Q.; Yang, G.; Xu, J.; Fu, Z.; Lacidogna, G.; Carpinteri, A. Acoustic emission data analyses based on crumb rubber concrete beam bending tests. *Eng. Fract. Mech.* **2019**, *210*, 189–202. [CrossRef]
43. Triantis, D. Acoustic emission monitoring of marble specimens under uniaxial compression. Precursor phenomena in the near-failure phase. *Proced. Struct. Integr.* **2018**, *10*, 11–17. [CrossRef]
44. Xie, Q.; Li, S.; Liu, X.; Gong, F.; Li, X. Effect of loading rate on fracture behaviors of shale under mode I loading. *J. Cent. South Univ.* **2020**, *27*, 3118–3132. [CrossRef]
45. Wang, Y.; Peng, K.; Shang, X.; Li, L.; Liu, Z.; Wu, Y.; Long, K. Experimental and numerical simulation study of crack coalescence modes and microcrack propagation law of fissured sandstone under uniaxial compression. *Theor. Appl. Fract. Mech.* **2021**, *115*, 103060. [CrossRef]
46. Liu, X.; Liu, Z.; Li, X.; Gong, F.; Du, K. Experimental study on the effect of strain rate on rock acoustic emission characteristics. *Int. J. Rock Mech. Min.* **2020**, *133*, 104420. [CrossRef]

47. Zhang, H.; Wang, Z.; Song, Z.; Zhang, Y.; Wang, T.; Zhao, W. Acoustic emission characteristics of different brittle rocks and its application in brittleness evaluation. *Geomech. Geophys. Geo-Energy Geo-Resour.* **2021**, *7*, 48. [CrossRef]
48. Meng, Y.; Jing, H.; Liu, X.; Yin, Q.; Wei, X. Experimental and numerical investigation on the effects of bedding plane properties on the mechanical and acoustic emission characteristics of sandy mudstone. *Eng. Fract. Mech.* **2021**, *245*, 107582. [CrossRef]
49. Ge, Z.; Sun, Q. Acoustic emission characteristics of gabbro after microwave heating. *Int. J. Rock Mech. Min.* **2021**, *138*, 104616. [CrossRef]
50. Chang, X.; Zhang, X.; Dang, F.; Zhang, B.; Chang, F. Failure behavior of sandstone specimens containing a single flaw under true triaxial compression. *Rock Mech. Rock Eng.* **2022**. [CrossRef]
51. Muñoz-Ibáñez, A.; Delgado-Martín, J.; Herbón-Penabad, M.; Alvarellos-Iglesias, J. Acoustic emission monitoring of mode I fracture toughness tests on sandstone rocks. *J. Petrol. Sci. Eng.* **2021**, *205*, 108906. [CrossRef]
52. Li, L.R.; Deng, J.H.; Zheng, L.; Liu, J.F. Dominant frequency characteristics of acoustic emissions in white marble during direct tensile tests. *Rock Mech. Rock Eng.* **2017**, *50*, 1337–1346. [CrossRef]
53. Niu, Y.; Zhou, X.; Berto, F. Evaluation of fracture mode classification in flawed red sandstone under uniaxial compression. *Theor. Appl. Fract. Mec.* **2020**, *107*, 102528. [CrossRef]
54. Wang, Y.; Zhang, B.; Gao, S.H.; Li, C.H. Investigation on the effect of freeze-thaw on fracture mode classification in marble subjected to multi-level cyclic loads. *Theor. Appl. Fract. Mech.* **2021**, *111*, 102847. [CrossRef]
55. Wang, Y.; Meng, H.; Long, D. Experimental investigation of fatigue crack propagation in interbedded marble under multilevel cyclic uniaxial compressive loads. *Fatigue Fract. Eng. Mater. Struct.* **2021**, *44*, 933–951. [CrossRef]
56. Yao, Q.; Chen, T.; Tang, C.; Sedighi, M.; Wang, S.; Huang, Q. Influence of moisture on crack propagation in coal and its failure modes. *Eng. Geol.* **2019**, *258*, 105156. [CrossRef]
57. Xiangqian, F.; Shengtao, L.; Xudong, C.; Saisai, L.; Yuzhu, G. Fracture behaviour analysis of the full-graded concrete based on digital image correlation and acoustic emission technique. *Fatigue Fract. Eng. Mater. Struct.* **2020**, *43*, 1274–1289. [CrossRef]
58. Das, A.K.; Suthar, D.; Leung, C.K.Y. Machine learning based crack mode classification from unlabeled acoustic emission waveform features. *Cem. Concr. Res.* **2019**, *121*, 42–57. [CrossRef]
59. Dong, L.; Chen, Y.; Sun, D.; Zhang, Y. Implications for rock instability precursors and principal stress direction from rock acoustic experiments. *Int. J. Min. Sci. Technol.* **2021**, *31*, 789–798. [CrossRef]
60. Prem, P.R.; Murthy, A.R. Acoustic emission monitoring of reinforced concrete beams subjected to four-point-bending. *Appl. Acoust.* **2017**, *117*, 28–38. [CrossRef]
61. Sagar, R.V.; Srivastava, J.; Singh, R.K. A probabilistic analysis of acoustic emission events and associated energy release during formation of shear and tensile cracks in cementitious materials under uniaxial compression. *J. Build. Eng.* **2018**, *20*, 647–662. [CrossRef]
62. Triantis, D.; Kourkoulis, S.K. An alternative approach for representing the data provided by the acoustic emission technique. *Rock Mech. Rock Eng.* **2018**, *51*, 2433–2438. [CrossRef]
63. Rippengill, S.; Worden, K.; Holford, K.M.; Pullin, R. Automatic classification of acoustic emission patterns. *Strain* **2003**, *39*, 31–41. [CrossRef]
64. Wawersik, W.R. *Detailed Analysis of Rock Failure in Laboratory Compression Tests*; University of Minnesota: Minneapolis, MN, USA, 1968.
65. Hoek, E. Brittle fracture of rock. *Rock Mech. Eng. Pract.* **1968**, *130*, 9–124.
66. Jaeger, J.C. *"Brittle Fracture of Rocks." The 8th US Symposium on Rock Mechanics (USRMS)*; OnePetro: Richardson, TX, USA, 1966.
67. Li, M.; Qu, Y.; Qin, S.; Yan, X.; Sun, Q.; Ma, P. Criteria of swelling potential and analysis on failure mechanism for mud rock. *J. Eng. Geol.* **2009**, *17*, 633–637.
68. Chen, J.; Li, S. Fracture failure mechanism of rock. *Adv. Earth Sci.* **2004**, *19*, 275–278.
69. Zhao, Y.; Wang, C.; Ning, L.; Zhao, H.; Bi, J. Pore and fracture development in coal under stress conditions based on nuclear magnetic resonance and fractal theory. *Fuel* **2022**, *309*, 122112. [CrossRef]
70. Sagasta, F.; Zitto, M.E.; Piotrkowski, R.; Benavent-Climent, A.; Suarez, E.; Gallego, A. Acoustic emission energy b-value for local damage evaluation in reinforced concrete structures subjected to seismic loadings. *Mech. Syst. Signal Process.* **2018**, *102*, 262–277. [CrossRef]
71. Guo, P.; Wu, S.; Zhang, G.; Chu, C. Effects of thermally-induced cracks on acoustic emission characteristics of granite under tensile conditions. *Int. J. Rock Mech. Min.* **2021**, *144*, 104820. [CrossRef]

Article

Modelling Coal Dust Explosibility of Khyber Pakhtunkhwa Coal Using Random Forest Algorithm

Amad Ullah Khan [1], Saad Salman [2], Khan Muhammad [2,3,*] and Mudassar Habib [1]

[1] Department of Chemical Engineering, University of Engineering and Technology, Peshawar 25000, Pakistan; amadullah@uetpeshawar.edu.pk (A.U.K.); muddasarhabib@uetpeshawar.edu.pk (M.H.)
[2] National Centre of Artificial Intelligence, Intelligent Information Processing Laboratory, University of Engineering and Technology, Peshawar 25000, Pakistan; saad_salman@uetpeshawar.edu.pk
[3] Department of Mining Engineering, University of Engineering and Technology, Peshawar 25000, Pakistan
* Correspondence: khan.m@uetpeshawar.edu.pk; Tel.: +92-91-9216796-8 (ext. 3031)

Abstract: Coal dust explosion constitutes a significant hazard in underground coal mines, coal power plants and other industries utilising coal as fuel. Knowledge of the explosion mechanism and the factors causing coal explosions is essential to investigate for the identification of the controlling factors for preventing coal dust explosions and improving safety conditions. However, the underlying mechanism involved in coal dust explosions is rarely studied under Artificial Intelligence (AI) based modelling. Coal from three different regions of Khyber Pakhtunkhwa, Pakistan, was tested for explosibility in 1.2 L Hartmann apparatus under various particle sizes and dust concentrations. First, a random forest algorithm was used to model the relationship between inputs (coal dust particle size, coal concentration and gross calorific value (GCV)), outputs (maximum pressure (P_{max}) and the deflagration index (K_{st})). The model reported an R^2 value of 0.75 and 0.89 for P_{max} and K_{st}. To further understand the impact of each feature causing explosibility, the random forest AI model was further analysed for sensitivity analysis by SHAP (Shapley Additive exPlanations). The study revealed that the most critical parameter affecting the explosibility of coal dust were particle size > GCV > concentration for P_{max} and GCV > Particle size > Concentration for Kst. Mutual interaction SHAP plots of two variables at a time revealed that with <200 gm/L concentration, −73 μm size and a high GCV coal was the most explosive at a high concentration (>400 gm/L), explosibility is relatively lower irrespective of GCV and particle sizes.

Keywords: coal dust explosibility; random forest; SHAP

Citation: Khan, A.U.; Salman, S.; Muhammad, K.; Habib, M. Modelling Coal Dust Explosibility of Khyber Pakhtunkhwa Coal Using Random Forest Algorithm. *Energies* **2022**, *15*, 3169. https://doi.org/10.3390/en15093169

Academic Editors: Longjun Dong, Yanlin Zhao and Wenxue Chen

Received: 25 March 2022
Accepted: 24 April 2022
Published: 26 April 2022

Publisher's Note: MDPI stays neutral with regard to jurisdictional claims in published maps and institutional affiliations.

1. Introduction

The explosion is "an event that once initiated, grows rapidly and initially unbounded" [1]. Therefore, the need for coal dust explosion investigation is a factor for safety in the chemical process industries and its storage for potential energy management [2]. Furthermore, coal dust explosion in a confined environment (coal mine and chemical process industries) results in the production of high pressure due to heating and the expansion of air and gases produced, which leads to destruction and human loss.

Therefore, understanding coal dust explosions is significant to finding the governing factors to mitigate them for increasing safety in industrial working environments. It has been reported that coal dust explosibility is affected by particles size, amount of fines [3], ignition temperature [4,5], air quantity [6] and concentration of coal [7–9] in an explosive environment [10]. The P_{max} (in MPa or bar) indicates the maximum destructive pressure released from a coal dust explosion, and the deflagration index (K_{st} in MPa or bar-m/s) is reported as the measure of explosibility [11]. The strength of explosibility is represented by the K_{st} values from no explosion (0) to weak (0–200), intense (200–300) and powerful (> 300) [12].

Sensors are available to monitor dust concentration and size [13], supported by communication through cloud computing [14]. With the emerging fields of IoT and real-time prediction, e.g., water in-rush [15], tailing dam stability [16], coal fire in sealed off regions [17], and stress monitoring in underground mines [18], available sensors for measuring coal quality, size and dust concentration could be used to predict explosibility as an early warning to prevent explosions. However, apart from other causes such as lack of awareness and safety hazard violations, many coal mining accidents are caused by a lack of calibration of sensors and the non-availability of coal dust prediction systems [19].

In the past few years coal dust explosibility has been studied extensively [5–15]. Coal dust emanates from heavy cutting machines (e.g., longwall mining), crushers and during loading on conveyor belts [20]. The particles may remain suspended where air ventilation velocity is high and later settle on the surface after some time in the accessways and mine entries [21]. The water spray lets the dust settle down, which is powdered with an inert material to reduce its explosibility [22]. Recent trends follow the use of AI to model flame propagation [23] of settled coal dust in galleries. Computational fluid dynamics (CFD) based models have been used to model flame propagation using airflow and coal dust measures [24–27]. Multilinear regression models have been used to model moisture vs. coal explosibility [3]. A generalised model for understanding how different types of explosible coal dust are affected by the coal characteristics and other coal parameters remains a challenge. Rarely has explosibility been modelled using an AI algorithm for investigating various aspects of coal explosibility. Particle dispersion and turbulence [24,28] are vital factors governing dust explosibility, which is dependent on particle concentration, size and shape [26–29]. Therefore, measuring coal dust characteristics during the air suspension phase can enable early monitoring and warning, and it can be an indirect measure to estimate the inert material required after dust suppression. This work is carried out to address the modelling part for its possible subsequent use in these environments in connection with IoT sensors to predict explosibility before time.

The phenomenon of explosibility was modelled using data collected from three different regions of Khyber Pakhtunkhwa (KP) and tested in a 1.2 L Hartman equipment. Using the fractional factorial design, the required number of tests have been conducted to generate data for modelling explosibility by an AI algorithm. A random forest regression algorithm was used to model the effect of coal properties on the response, i.e., the maximum pressure (P_{max}) and the deflagration index (K_{st}). A game theory-based method, Shapley Additive exPlanations (SHAP) [30], explains how the variation in coal dust properties affects the response. Furthermore, sensitivity analysis is performed to quantify this effect and to identify the safe limits for each parameter to mitigate coal explosibility in the KP coal mines.

Literature Review

Many researchers have investigated coal dust explosibility to measure the main factors influencing explosibility [2,4–11,31–34]. Moradi et al. [35] investigated the effect of sizes of coal particles from different mines on the burning rate of coal using a 2-litre closed chamber. The coal dust concentration, pressure and initial temperature were constant at 10000 g/m^3, 1.5 bar and 25 °C, respectively. The response of varying particle sizes was recorded, keeping all the parameters stable. The maximum pressure rate and the explosibility index reported an inverse relationship with the particle size, i.e., 44µm and 37 µm had a higher burning velocity than other dimensions [35]. Another study was conducted on coal dust to measure the explosion severity and the ignition sensitivity of different ranks of coal. Lower ranks of coal are reported to be easily ignited with severe explosion due to the highly volatile content and pyrolytic property of coal [35]. Cao et al. [33] used experimental and numerical analysis to understand the explosion severity of coal dust. The simulations showed the behaviour of coal dust particles after the explosion. These results were consistent with the experimental observations; hence, the simulations can be reliably used to model coal dust explosion. Cashdollar [36] used a US Bureau of Mines (USBM) 20-L laboratory chamber to measure the effect of coal dust explosibility. The 20-litre chamber data agree relatively well with those from full-scale experimental mine tests.

Particle size, volatile and oxygen [37] contents are almost equally important in governing the strength of coal dust explosibility. Tan et al. [31] used a pipe apparatus to analyse the effect of change in dust particle size, concentration and a mixture of methane-coal dust on the explosion pressure. Both the particle size and the concentrations varied at five different levels. A high explosibility index K_{st} and the maximum pressure P_{max} were recorded for nano-sized particles compared to micro-particles. Similarly, a 38 L explosive chamber was used for testing coal dust explosibility ignition at different concentrations [32]. A 5 KJ Sobbe igniter ignited coal dust to test if the coal was under or over-fueled at different concentrations. It was observed that coal dust concentrations below 100 g/m^3 and 200 g/m^3 failed to deflagrate because of insufficient fuel.

Furthermore, higher dust concentrations above 1200 g/m^3 and 1400 g/m^3 significantly affected the maximum pressure as less oxygen was available to detonate the coal dust sample. When conducting a coal dust explosibility experiment, the range of each parameter plays a vital role in governing the response to the dust explosion. The parameters studied previously with the respective level of variation are reported in Table 1.

Table 1. Ranges of different coal dust parameters in previous research projects.

Sr. No.	Concentration (g/m^3)	Particle Size μm	Volatile Matter (%)	References
1.	213–1282	<38, 38–74, 74–212	17.73	[7]
2.	400,500	5, 11, 33, 95, 145 and 190		[38]
3.	250, 500, 1000, 1500, 2000, 2500, 3000		32.7–44.4	[39]
4.		104-200	Qualitative	[40]
5.	400	20,38,75,250,350		[9]
6.	60, 125, 250, and 500		14.03–40.97	[41]
7.	100 to 600	53, 75	6–38.5	[42]
8.	150, 250, 500	0–56, 56–71 and 71–90		[43]
11.	60, 80, 100, 120, 150, 200, 250, 300, 350, 400, 450 and 500	38,53,73,104,147,300	19.69–27.18	[4]

2. Materials and Methods

2.1. Samples Collection

Three different coal type samples from two regions (Figure 1), i.e., Cherat and Darra localities of KP province in Pakistan, were collected and analysed for the proximate analysis reported in Table 1. After the study, the samples were ground to prepare samples of −43 μm, −53 μm, −75 μm, −120 and +125 μm sizes. The coal particle sizes were tested at concentrations of 50, 100 to 800 gm/L in the 1.2 L Hartmann apparatus. Various particle sizes of coal dust from three different coal types on different concentrations were tested for explosion in the 1.2 L Hartman apparatus and modelled using an RF-based model. This study would help to identify the response parameters, and the sensitivity analysis will help to control coal dust explosions.

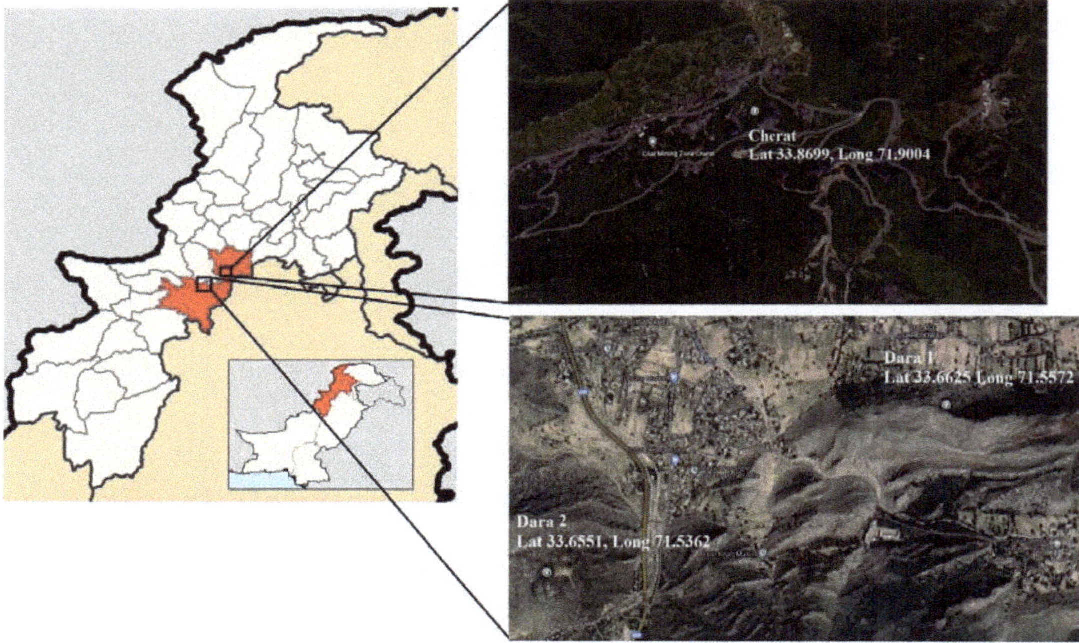

Figure 1. The regions from which coal samples were collected.

2.2. Experimental Setup

Cerchar first created the Hartman apparatus in the 1970s [44], consisting of a glass tube with a dispersion cup. The powder was dispersed through an air blast and whirled up by the pressurised air. The test substance was then subjected to a flame to determine if it could be ignited by the ignition source [44,45]. The concentration of the dust–air mixture in the Hartmann tube could be varied incrementally between 100 g/m^3 and 4000 g/m^3. Dust could be classified as a dust explosion hazard by this screening if ignition occurred with one of the two ignition types (spark ignition or filament ignition). A complete exclusion of the dust explosion hazard was not possible when screening in the Hartmann tube. Suppose the result of the spark ignition and the filament ignition was negative. In that case, the dust explosion hazard must be examined in the 20 L apparatus in order to make a reliable statement on the dust explosion hazard as stipulated in the standard DIN EN ISO 80079-20-2.

A total of 8 samples, each of 1.5 kg, were collected from the respective sites Dara 1, Dara 2, and Cherat localities. The representative samples from each site were collected, including subsamples using the coning and the quartering method. The samples were crushed in a disc crusher in the Department of Geology, Universiti Teknologi Malaysia (UTM) Johor Bahru. Explosibility tests were conducted in the Explosibility Lab of the Energy Engineering Department UTM Malaysia. The chemical analysis of samples was conducted in the Pakistan Council of Scientific and Industrial Research (PCSIR) Lab Peshawar, Pakistan.

A schematic view of the 1.2 L Hartmann stainless steel tube is presented in Figure 2. An ignition electrode is located on the vertically symmetric axis of the Hartmann tube, approximately 6 inches from the bottom. The minimum ignition energy was the minimum spark energy that could ignite dust and maintain combustion. The ignition source used for the tests is a continuous spark generated by a high voltage transformer between two standardised electrodes placed near the bottom of the cylindrical 1.2 L Hartmann tube. The energy content of the spark corresponds to the equivalent energy of about 10 Joules of a discharge (temporary) spark. The known concentration and particle size coal sample was placed on an umbellate diffuser arranged in the lower part of the device and used for dust diffusion into a 1.2 L Hartmann tube.

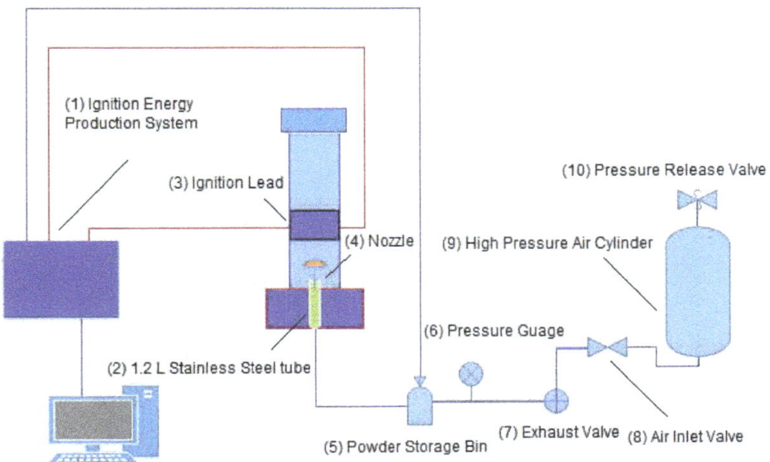

Figure 2. Schematic diagram of the experimental setup for coal dust explosion experiments.

2.3. Random Forest Algorithm

Leo Breiman [46] first proposed the random forest (RF) algorithm as a supervised ensemble learning approach to handle classification and regression tasks. An ensemble learner combines the output from many predictors (trees in this case), referred to as a forest [47], to learn complex relationships. The fundamental unit predictor, i.e., each tree, is obtained by deriving nodes and branches from bootstrapping, using roughly 63% of the original data for training [48]. At the same time, the remaining samples are termed 'out-of-bag' (OOB) samples [49]. Each node in the tree represents the splitting of this training data on an input variable homogeneously, starting from the root node at the top, branching down through subsequent splitting and nodes selection, downward to the leaf (terminal) node representing an output variable. Thus, each node on any tree represents data splitting on a variable recursively down the tree until the terminal node or other stopping criterion is reached. A node variable, i.e., an input variable and a respective cutoff is selected among the 'p' out of 'm' input variables (p < m) and possible cutoff values. A critical trait of RF variable selection during node formation and branching is that the 'p' predictors considered are a subset of 'm' predictors in the bootstrap data, thus resulting in uncorrelated outputs from a collection of B trees termed a forest. Therefore, uncorrelated outputs from each tree in a forest reduces the RF model's variance. For a chosen B possible bootstrapped training data sets, resulting in $b = 1, 2 \ldots B$ trees, a bth bootstrapped training set provide the output $fb(x)$, and finally the average output from all trees report the final estimate

$$fbag\,(x) \;=\; \frac{1}{B}\sum fb(x)$$

Tree building in the random forest regression model commences from the starting node and moves downward using the following steps:

- Divide the predictor space, i.e., the set of all features into J distinct non-overlapping regions $R_1, R_2 \ldots R_j$ (cut values)
- For every observation that falls into a region, make the exact prediction, i.e., the mean of all the response values which fall into that region
- Select the cut value which has the minimum residual sum of squares (RSS)

$$\mathrm{RSS} = \sum_{j=1}^{J} \sum_{i \in R_j} \left(y_i - \hat{y}_{R_j} \right)^2$$

- The same process is repeated at each node until any of the stopping criteria is met. Stopping criteria for a regression tree depth can be any of the following:
- Minimum observation at internal node
  - Minimum number of sample observations required for a further split at a node is not met
- Minimum observation at a leaf node
  - Minimum number of sample observations needed is not found at each node after splitting
- Maximum depth of the tree
  - Maximum layers of the tree are reached

When running a machine learning model on any dataset, the variables that govern the model's structure are called hyperparameters. After setting the base model, the hyperparameters need to be tuned to improve the machine learning model's performance. For example, in a random forest regression model, the four primary hyperparameters are as follows:

1. "n_estimators": The number of estimators refers to the number of decision trees built by the random forest regression model before taking the maximum average of predictions. A higher number of trees improve the performance at the cost of computational expense.
2. The "max_depth" maximum depth hyperparameter is the depth of each decision tree in a random forest model. A very high value of the maximum depth hyperparameter leads to overfitting the model.
3. "min_samples_split" is the minimum number of data points placed in a node before splitting the node.
4. "min_samples_leaf" is the minimum number of data points allowed in a leaf node.

2.4. Sensitivity Analysis of the Model

Interpretation of most machine learning models, such as ensemble methods or deep networks, is often complex and commonly referred to as a "black box" [30,50]. In recent times Explainable AI (XAI) algorithms have seen a rising trend, where XAI algorithms like (SHAP, Partial Dependence Plots (PDP), Accumulated Local Effects (ALE), etc.) are used to explain the predictions by AI algorithms [51,52]. SHAP [30] is a novel approach that unveils the learned complexity of these machine learning prediction models. It is a valuable tool for exploring the response of individual variables to output variables as it breaks down the predictions into individual feature impacts [52]. The SHAP feature importance chart reports the importance of each input variable in descending order affecting the output in absolute terms. The feature importance value of an input feature is based on the mean absolute magnitude of the SHAP values over all instances. A summary plot of SHAP values explains the output variable's sensitivity to the concerned input variable. The summary plot can describe the cause-and-effect relationship through high or low values (represented by red to blue colours) of the input feature and the respective SHAP value (of the output) on the horizontal axis. A positive (SHAP value) on the horizontal axis for high values (red) of the input variable refers to a direct relationship while an inverse relationship occurs if the input feature has low (blue) values. In contrast, input features with high values (red) reporting negative (SHAP value) on the horizontal axis would refer to an inverse relationship of the input with the output (response) variable, while a low (blue) value of the input variable would indicate a direct relationship with the output. Jittered points placed densely on the graph represent the same SHAP value reported in numerous instances.

SHAP dependence scatter plots show the effect a feature has on the predictions made by the model. Each point in the scatter plot is a single prediction from the dataset. The *x*-axis represents the value of the feature, and the SHAP value on the vertical axis illustrates the effect of a feature's value on the model's output. The colour corresponds to a second feature that interacts with the first feature of concern.

Furthermore, to analyse the outcome of the RF model for the change in each respective input feature, the model was also fed with different values of the feature understudy, holding the mean of all other input features constant.

3. Modelling Explosibility of Coal from Khyber Pakhtunkhwa Province of Pakistan Using Random Forest Regression Model

3.1. Data Collection

The design of experiments is necessary to generate the number of experimental runs with the minimum essential experimentation [53]. The fractional factorial design [54] was used to carry out the experimental runs to investigate these variables' effect on explosibility. Following the design of experiments, different levels of the two input variables, i.e., particle size and concentration, are presented in Table 2. The fractional factorial design suggested 84 runs as an optimal representation of the extent of the full factorial.

Table 2. Coal Types with Proximate Analysis Results.

Coal Type	Total Moisture wt% (Air Dried Basis)	Ash wt%	Volatile Matter wt%	Fixed Carbon wt%	Sulphur wt%	GCV kcal/kg
Cherat	6	26	15.45	49.7	2.81	4901
Dara 1	6.4	17.50	12.26	60.51	3.32	6318.2
Dara 2	7.03	18.57	10.28	61.07	3.02	6120.2

Various concentrations of 0.1, 0.2, 0.3, 0.4, 0.5, 0.6, 0.7, 0.8 and 0.9 g of coal sample were obtained after passing from sieves of the required sizes through disc crusher. Samples were fed to the 1.2 L Hartmann tube and sealed by tightening the screws at the bottom of the tube. Next, the air inlet unit forced compressed air at 15 psi pressure into the air storage tank. Then, a magnetic valve was opened to disperse the coal particles as dust into the tube, followed by the final ignition of the electrode. The pressure sensor transmitted the amplitude of explosion at millisecond intervals for registering values to the LabVIEW software in the personnel computer (PC) connected to the system. The recorded data was used to draw the graphs of pressure change against time to find (dp/dt), for reporting the explosibility (K_{st}). The sensor did not report the dust ignition if the flame propagated less than 60 mm from the spark position; the powder would be considered explosible if the dust fired or exploded during the tests. If no dust fired or exploded in three series of tests for any concentration, the powder was considered not explosible under the test conditions.

The raw data is read from a .csv file with >75,000 readings recorded at milliseconds (ms) interval and plotted against the corresponding pressure (in bar) to determine the maximum slope line (dP/dt) max, i.e., 'maximum rate of pressure rise' (Figure 3). A tangent is fitted to this point and extended to find $dP = (P_{max} - 0)$ and dt, i.e., the difference between times @ P_{max} and intercept at $P = 0$. Finally, the K_{st} values were determined for each test according to the equation given below:

$$K_{st} = \left(\frac{dP}{dt}\right)_{max} V^{\frac{1}{3}}$$

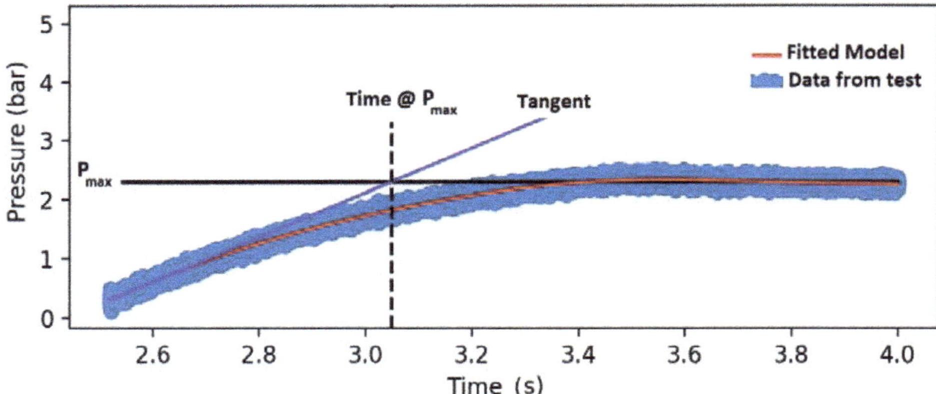

Figure 3. Pressure vs. time and determination of $(dP/dt)_{max}$ for a single explosibility test of 1 gm sample (from Cherat region Khyber Pakhtunkhwa Pakistan) tested in 1.2 L Hartman apparatus with 95 µm size and 500 mg/L concentration.

3.2. Data Preparation

A total of 84 tests were conducted, out of which 81 reported meaningful results. Two samples that reported no explosibility values were removed; one outlier that reported a very high value of K_{st} was also removed. The variables were normalised using the Min–Max scalar to adjust various scales of variables to a standard between 0–1. Principal component analysis (PCA) was applied to sample features reported in Tables 2 and 3, including the outputs P_{max} and K_{st}. Based on selected Eigen loadings (Table 4) of the most critical components defined by Eigenvalues (Table 4), GCV, particle size and concentration were selected. At the same time, redundant features were ignored during further processing. GCV can be considered as an indirect representation of the coal type. Table 5 shows the descriptive statistics of selected features for all 81 samples.

Table 3. Variation of Coal dust concentration and particle size.

Coal Dust Concentration (mg/L)	50	100	200	300	400	500	600	700	800
Size (µm) < (−sign) or >(+sign)	−43	−53	−73	−120	+125				

Table 4. PCA Eigen Vectors and Loadings.

	PC 1	PC 2	PC 3	PC 4	PC 4	PC 5	PC 6	PC 7	PC 8
Eigen Values	182,759.4	63,250.0	519.6	2.7	0.6	0.004	0	0	0
Conc.	−0.0	0.99	0.0	0.0	−0.0	−0.0	0	0	0
size	0.0	−0.0	0.99	0.01	0.01	−0.0	0	0	0
T. Moist	−0.0	0	−0.0	0.2	0.04	−0.0	0.2	0.77	−0.56
Ash	0.0	0	0.0	−0.02	−0.0	0.0	0.6	−0.6	−0.6
V. Matter	0.0	0	0.0	−0.7	−0.2	0.01	−0.5	0.0	−0.5
F. Carbon	−0.0	0	−0.0	0.6	0.14	−0.01	−0.6	−0.3	−0.4
GCV	−0.99	−0.0	0.03	−0.007	−0.001	0.0	0.007	−0.0	−0.0
P_{max}	0.0	−0.0	0.01	0.22	−0.97	−0.14	0	0	0
K_{st}	0.0	−0.0	0.003	0.05	−0.13	0.99	0	0	0

Table 5. Data Descriptive Statistics.

	Concentration mg/L	Particle Size (Micron)	GCV kcal/kg	P_{max} bar	K_{st} bar·m/s
Count	81	81	81	81	81
Mean	405.55	−78.66	6072.73	2.058	0.422
Std dev	251.49	33.42	427.30	0.877	0.165
Min	50	−43	4901	0.15	0.023
Max	800	+125	6318.2	3.41	0.661

Coal dust concentration, particle size and GCV were used as the independent variables and P_{max}, K_{st} as the dependent variables for AI-based modelling. Previously, it was reported that volatile matter is one of the most critical parameters that influence coal dust's explosibility characteristics [34,36]. However, in this case, the range and the magnitude of volatile matter were low (10.28% to 15.45%); therefore, GCV represented a particular coal type. In contrast, the volatile matter content in prior research was 11.19 to 42.26, reporting a more significant effect on the explosibility beyond 20% [34].

The random forest (RF) regression model was applied using the Python Scikit-learn package [55]. The model was trained on 75% of the training data, while the remaining 25% of data was used to test the model accuracy using the train test split method of model selection in sklearn library. To find the best set of hyperparameters of the RF algorithm, the parameters were varied using the grid search CV method [56], as shown in Table 6. A total of 1200 combinations were investigated during hyperparameter tuning.

Table 6. Hyperparameters ranges tried.

S. No	Description	Range	Number of Values
1	"min_samples_split"	2, 3, 4	3
2	"min_samples_leaf"	1, 2, 3, 4	4
3	"n_estimators"	10, 20, 30, 50, 100	5
4	"max_depth"	2, 5, 6, 10	4

$(3 \times 4 \times 5 \times 4) \times 5$ fold cross-validation = 1200 combinations

The 5-fold cross-validation split the data (80:20 train test split ratio) 5 times for each case of the chosen hyperparameters within the grid search and performed the training and testing each time. This hyperparameter tuning process was repeated three times to investigate the effect of random initiation on the hyperparameters. Following the best-chosen hyperparameters based on the grid search, the data was randomly split using the train_test_split method into train:test datasets based on a 70:30 split to train and to test the model. Similarly, the testing accuracy was reported, and the model was further investigated for sensitivity analysis. The most widely used regression evaluation metrics, i.e., coefficient of determination (R^2) [57] and root mean square error (RMSE) [57], were used to evaluate the performance of the RF regression model.

4. Results

Table 7 reports the optimised value for each hyperparameter during the tuning of the model. A Pearson R of 0.86 and 0.94 were observed for the P_{max} and K_{st}, respectively, with the corresponding R^2 of 0.75 and 0.89 for the best combination of hyperparameters (Table 7).

Table 7. Optimised hyperparameters after the Grid search method.

Parameter Name	Description	Value
n_estimators	The number of trees in RF	50
max_depth	The maximum depth of the tree	2
min_samples_leaf	The minimum number of samples required to be at a leaf node	2
min_samples_split	The minimum number of samples required to split an internal node	3

Figure 4 shows the actual vs. the predicted values for both variables. The corresponding root mean square Error (RMSE) were reported as 0.14 and 0.07 for the P_{max} and K_{st} during the testing phase.

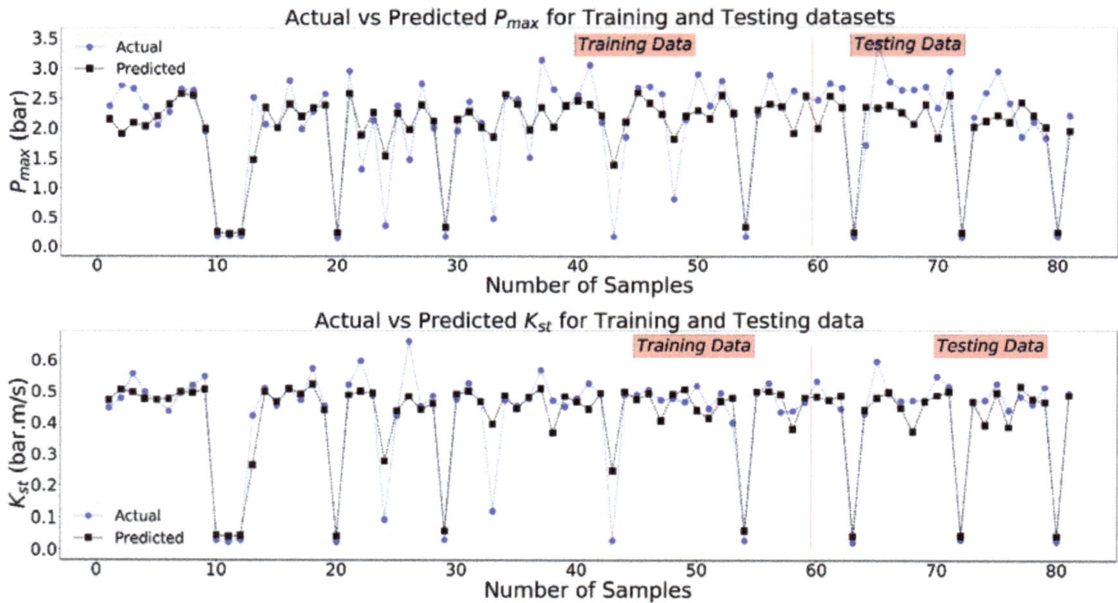

Figure 4. Actual vs. Predicted P_{max} (**top**) and K_{st} (**bottom**) Values.

The magnitude of importance of different input variables was visualised using the SHAP feature importance chart given in Figure 5a,b. The most critical parameter was particle size and GCV for predicting P_{max}, while for K_{st} particle size was the most crucial variable. Observing the SHAP summary plots in Figure 6, concentration indicated an inverse relationship (negative correlation) with P_{max} and K_{st}. Smaller particle sizes reported lower P_{max} and K_{st}, while a mix of different particle sizes caused higher P_{max} and K_{st}, which required further investigation. Similarly, higher GCV coal caused lower or higher P_{max}/K_{st}, while lower GCV coal was positively correlated with P_{max} and K_{st}. Such nonlinear relationships ought to be explored well by the SHAP interaction summary plot given in Figure 7 and the interaction dependence plots in Figure 8 that report the effect of two features on the predicted outcome of a model. It explores if the relationship between the target and the variables is linear, monotonic, or more complex.

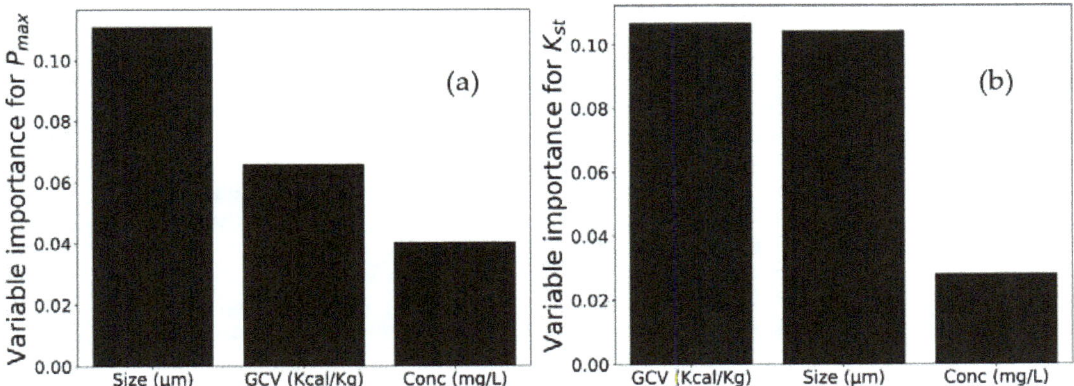

Figure 5. SHAP feature importance chart for (a) P_{max} and (b) K_{st}.

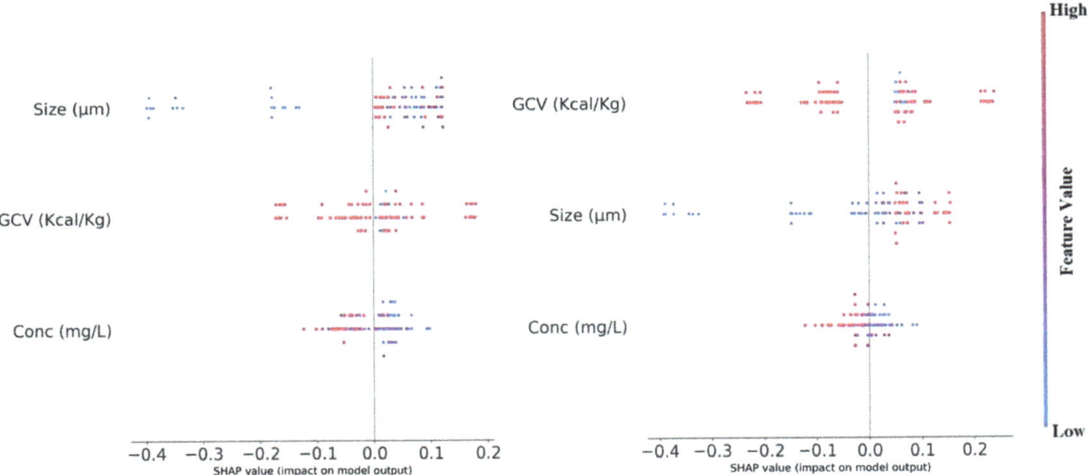

Figure 6. SHAP Summary Plot for P_{max} (**left**) and K_{st} (**right**).

The clearest (and most significant) interaction effect occurs between the size and GCV, indicated by the greatest horizontal spread followed by concentration and size (Figure 7). Conversely, GCV and concentration had a minor interaction. Therefore, to explore further, dependence interaction curves are reported in Figure 8.

Generally, explosibility (K_{st}) decreased with increasing concentration, as seen in Figure 8a. At smaller concentrations, i.e., <200 mg/L, explosibility was higher. Fine sizes, i.e., <-73, had a high positive contribution to K_{st} compared to larger particle sizes (>-73 as red dots), showing a low positive contribution to K_{st} (SHAP values are close to 0). At greater concentration (>500 gm/L), particle sizes <-73 have a higher negative impact on K_{st} than greater particle sizes. Between 300–400 mg/L concentration, the positive impact of K_{st} exists at a lower strength (lower positive SHAP values), as a threshold concentration at which the high impact of coarse particle sizes starts becoming evident and then subsides beyond 500 gm/L concentration. Figure 8c also reports the general decreasing trend of K_{st} with the increase in concentration. Figure 8c,d show that at smaller concentrations (<200 gm/L), higher GCV coal increased explosibility (K_{st}) compared to lower GCV coal which had a minimal effect on explosibility. At 400 gm/L concentration, the overall impact is positive on K_{st}. However, beyond a 400 gm/L concentration, high GCV coal negatively impacted explosibility (K_{st}), while low GCV coal had nearly no impact on K_{st}. Furthermore, Figure 8d shows that lower GCV (Cherat coal) did not affect explosibility, whereas medium GCV (Darra 2) coal reported higher K_{st}, while lower K_{st} was reported by Darra 2 coal (GCV = 6138). Larger particle sizes increased explosibility (Figure 8e) for even lower GCV Cherat coal. In comparison, medium GCV Darra 2 coal had a minimum effect on explosibility for larger sizes, while fine sizes caused higher explosibility with high positive SHAP values (Figure 8f). Moreover, the higher GCV Darra 1 coal reported the least overall explosibility and a relatively lower explosibility for smaller sizes than larger.

In the case of dependence interaction plots for P_{max}, a decrease in P_{max} is seen with increasing concentration (Figure 9a), similar to Figure 8a. The horizontally opposite Figures complimented each other during interpretation. For example, the value for size is read from Figure 9b, while explaining the concentration interaction with size in Figure 9a. Similarly, interpreting Figure 9b concerning the importance of concentration to size, the concentration values can be read from Figure 9a.

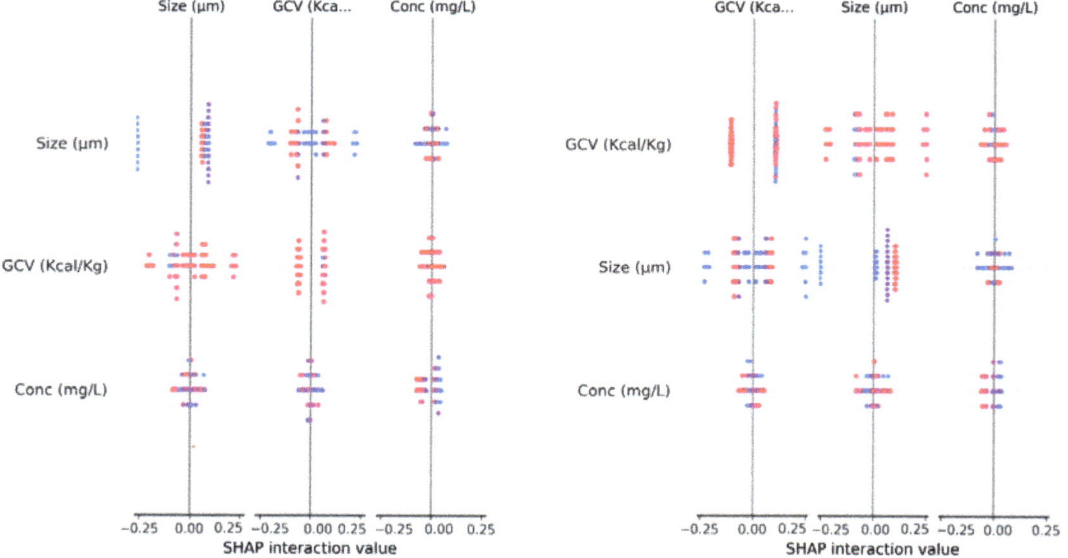

Figure 7. SHAP interaction summary plot of variables for P_{max} (**left**) and K_{st} (**right**).

A nonlinear relationship is shown between concentration vs. size for P_{max}, where at a low concentration (<−200 mg/L), smaller sizes report a highly positive effect on P_{max} (Figure 9a); −53 µm size being the dominating cause of high P_{max}, as seen in Figure 9b. Beyond 500 gm/L concentration, negative SHAP values indicate a higher concentration inverse effect P_{max} irrespective of particle sizes. However, between 400–500 mg/L, larger particle sizes have a higher positive impact on P_{max}, and finer sizes have a lower impact (SHAP values are close to 0). The −43 micron sizes exhibit very low SHAP values, indicating they highly reduce explosibility, as shown in Figure 9b.

The particle size showed an increasing trend for P_{max} beyond −53 µm size in Figure 9b due to the nonlinear concentration vs. size interaction explained in Figure 9a. Figure 9c also showed a decreasing trend of P_{max} with an increase in concentration, as in Figure 9a. In Figure 9c, at higher concentrations, >500 mg/L, a drop is observed in P_{max} for both low and high GCV coal. Figure 9e shows that P_{max} reported an increase as the size increased. Cherat coal (lower GCV) reported a slight positive correlation with P_{max} (Figure 9f), while Darra 2 coal (6120 kcal/kg GCV) had a greater positive P_{max} for smaller particle sizes compared to those that were larger (Figure 9f). The impact of Darra 1 coal (6318 kcal/kg GCV) was highly negative (inverse) among the coal types. A similar effect was seen for smaller sizes of this coal type than larger sizes.

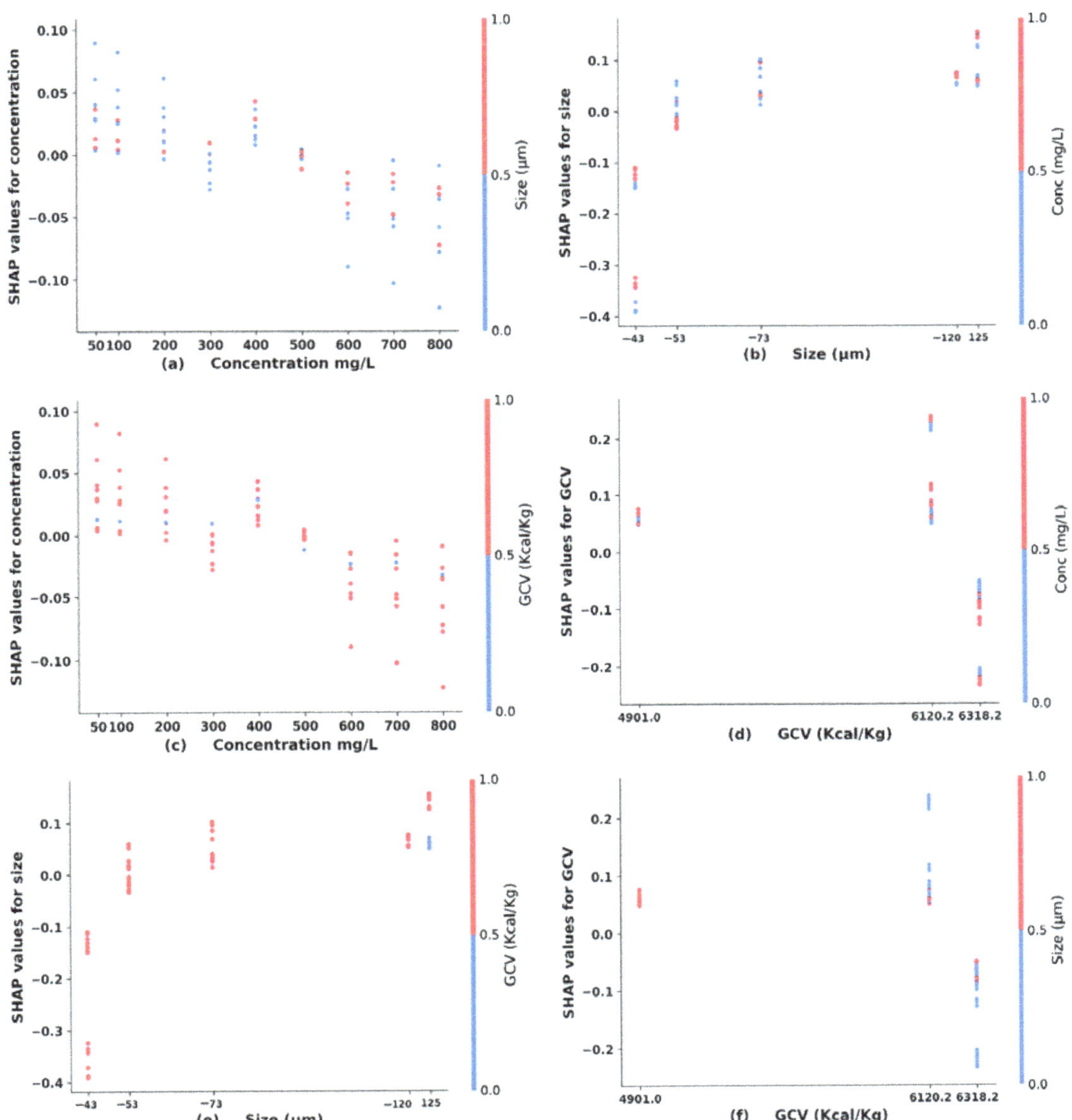

Figure 8. Interaction dependence plots of (**a**) concentration vs. size, (**b**) size vs. concentration, (**c**) concentration vs. GCV, (**d**) GCV vs. concentration, (**e**) size vs. GCV and (**f**) GCV vs. size for K_{st} from SHAP values for explaining the RF model.

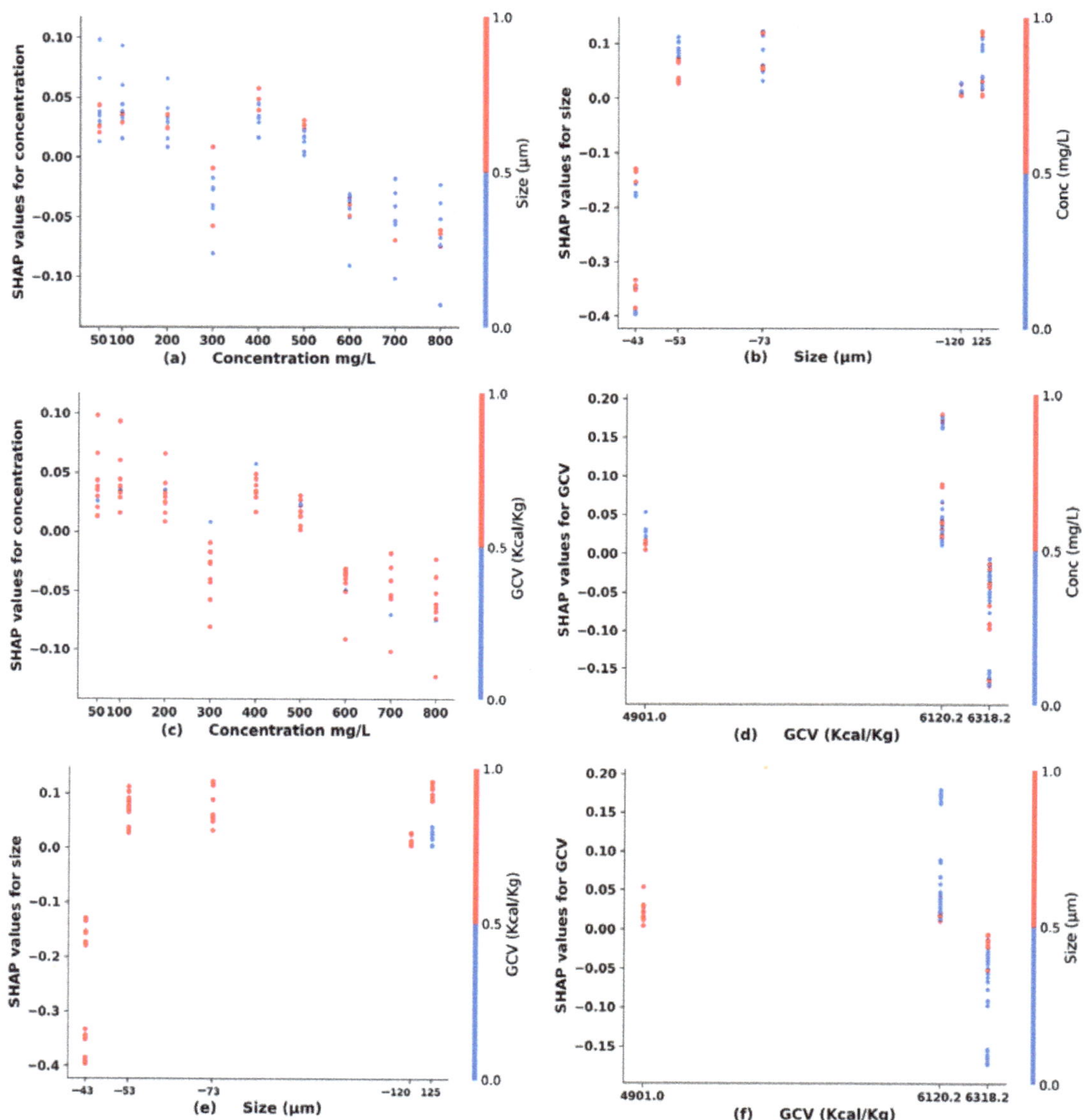

Figure 9. Interaction dependence plots of (**a**) concentration vs. size, (**b**) size vs. concentration, (**c**) concentration vs. GCV, (**d**) GCV vs. concentration, (**e**) size vs. GCV and (**f**) GCV vs. size for P_{max} from SHAP values for explaining the RF model.

To further explore the sensitivity of output variables, one of the input variables (particle size, concentration and GCV) was varied while keeping other input variables constant at the mean values given in Table 5. The results were plotted and reported in Figure 10. Figure 10a,b showed that the P_{max} and K_{st} decreased with increased concentration, as reported in Figure 8a,b. Figure 10c,d indicated that P_{max} and K_{st} increased with particle sizes and dropped beyond −53 microns and −120 microns, respectively. P_{max} and K_{st} are the least for the highest GCV as in Figure 10e,f. These results give similar results to the

SHAP dependence plots, but the SHAP dependence plots are more detailed, highlighting the interaction between two variables.

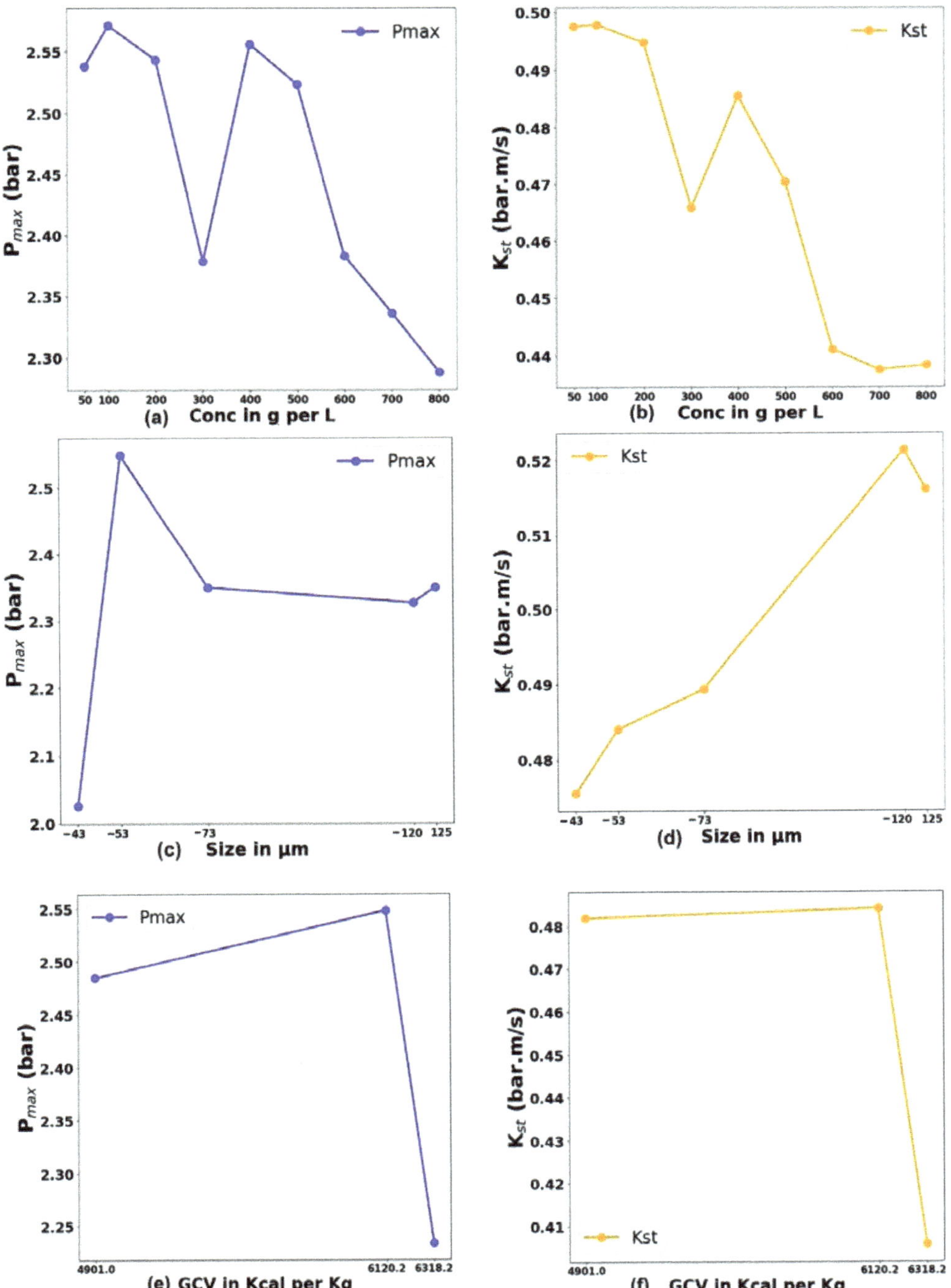

Figure 10. Sensitivity Analysis Graphs for each input variable.

5. Discussion

Machine learning models are considered black-box models that do not explain cause-and-effect relationships. However, SHAP gave some interesting insights into the cause-and-effect relationship between inputs (size, concentration and coal type represented by GCV and outputs (explosibility (K_{st}) and maximum pressure (P_{max})). It is evident from Figure 8a,b,e,f and Figure 9a,b,e,f that different particle sizes of coal dust have a significant effect on both P_{max} and K_{st}. Smaller particle sizes cause smooth and robust deflagration due to greater intermolecular forces [49] and a higher contribution to volatilisation [41]. The model clearly shows this effect at lower concentrations <300 gm/L. Larger particles are relatively dispersed in the form of flakes with edges and corners [31]. However, larger particle sizes exhibited stronger explosibility (K_{st}) and maximum pressure (P_{max}) at medium concentrations (400 gm/L). Moreover, because the particle size ranges -43, -53, -73, and -120 µm were cumulative, there was higher size dispersity as the particle size increased.

Moreover, coal dust lifted by impact airflow has dispersion and sedimentation processes. Therefore, the dispersion and the sedimentation rates of coal dust with different particle sizes are also different, which may have influenced the suspension state of coal powder. In addition, the coal powder was not added with any anti-agglomerating substance; thus, agglomeration may occur in the explosion, thereby influencing the combustion performance of coal dust. Additionally, the +125 µm particle sizes that exclude sizes smaller than this range also report high P_{max} and K_{st} in Figures 8b and 9b. Such an effect may occur if the participation of coal dust in the explosion belonged to a gas–solid reaction; hence, the reaction process and the mechanism may be more complicated.

In this study, higher GCV (6120 and 6318) and small sizes significantly affected both explosibility and the maximum pressure. Dara 1 coal (GCV: 6318) reduces the P_{max} and K_{st} at smaller sizes than a larger size, and Dara 2 coal (GCV: 6120) increases both P_{max} and K_{st} at smaller sizes compared to those that are larger. Additionally, the lower GCV (4901) Cherat coal had minimal effect on P_{max} and K_{st}.

The concentration, in general, has an inverse relationship with both P_{max} and K_{st}, i.e., lower concentration coal has high explosibility and vice versa. This may be because there is sufficient oxygen for combustion reaction at low/medium concentrations or medium carbon content represented by lower or medium GCV, aiding in faster heat transfer [58].

Without oxygen at higher concentrations, the densely concentrated dust particles cannot get enough oxygen molecules to deflate completely. Furthermore, the leftover unreacted molecules absorb the heat, leading to low explosibility [58]. Therefore, a lower explosibility zone is mainly associated with high concentration (>500 mg/L). These results indicate that the oxygen fuel is deficient in higher concentrations (>500 gm/L), causing lower P_{max} and K_{st}; therefore, the 1.2 L Hartmann is limited generally for conducting tests at higher concentrations. It is also observed that smaller particle sizes at lower concentrations are more explosible, i.e., positively contribute to K_{st} and P_{max}, but this effect is reversed at higher concentrations, i.e., smaller particle sizes negatively impacted P_{max} and K_{st}. Additionally, the larger particle sizes showed a minimal impact on explosibility with rising concentrations as they have low SHAP values.

Sensor deployment, supported by the proposed AI model, may be implemented in zones of high air velocities [19], where dust is in suspension with ample oxygen supply. This safety monitoring system may be deployed with other dust suppression/control measures that must also be in place to suppress dust. At 400 to 800 gm/L, the experiments were more dominating due to the lower oxy-fuel ratio; therefore, further experiments at these ranges must be conducted in a 20 L or 1 m^3 chamber to explore explosibility characteristics at these concentrations. An IoT sensor system + AI-based explosibility model may be analysed using coal dust concentration, dust particle size, and coal type parameter as a monitoring and warning system to further this research. The quantity of coal dust concentration and suppression can also be linked to estimating the amount of inert material required to reduce the explosive properties of coal to a safe level.

6. Conclusions

The objective of the work was to address the knowledge gap by AI-based modelling to investigate the effect of coal concentration, coal type and particle size on explosibility K_{st} and P_{max}. Coal from three different localities was tested in 84 experiments to measure the maximum pressure P_{max} and the explosibility index K_{st} in the 1.2 L Hartman apparatus for various concentrations and particle sizes. K_{st} were determined by fitting a tangent to each curve of the experimental run. Samples were recorded for GCV, dust concentration and particle size as input variables to predict P_{max} and K_{st}. The random forest algorithm was applied to the input data for modelling the outputs, reporting R^2 scores of 0.75 and 0.89 for P_{max} and K_{st}, respectively. The Shapley Additive exPlanations (SHAP) algorithm explained the behaviour/prediction of the random forest model, identifying the essential input variable with a sensitive limit. A SHAP based summary plot and interaction dependence curves were plotted to get an insight into the cause-and-effect relationship learned by the model. Additionally, the model's response to the change in each feature was derived through sensitivity analysis, where each feature was varied at different levels. At the same time, all the other components were held constant at the mean values of each respective feature.

The following are the essential conclusions:

- The coal dust samples of the KP region have low volatile matter (10–16%); hence GCV was more representative of coal type as reported by PCA results.
- Initially, SHAP plots reported the parameters influencing the coal dust explosibility in descending order were particle size > GCV > concentration for P_{max} and GCV > particle size > Concentration for K_{st}.
- SHAP plots reliably interpreted the random forest explosibility model explaining the complex inter variable phenomenon in greater detail.
- A SHAP interaction plot revealed that the concentration of coal dust particles has an inverse relationship with P_{max} and K_{st}. A lower concentration results in a higher P_{max} and K_{st}, which is a consequence of sufficient oxygen available to deflagrate all the coal dust particles. High GCV coal is utilised to its maximum at lower concentrations, causing higher explosibility.
- At higher concentrations, there is not enough oxygen for the complete reaction of the coal dust particles. The excess molecules absorb the heat of coal dust explosion, resulting in an overall drop in maximum pressure and explosibility.
- At concentration <200 gm/L, lower particle size (−73 μm) and high GCV coal have the highest explosibility, and more significant particle sizes have no impact.
- 400 gm/L is a threshold concentration as a high impact of fine particle sizes exists, and explosibility is also positively related to this concentration.
- At greater concentrations (>500 gm/L), particle sizes <−73 μm negatively impact K_{st} and P_{max} compared to more significant particle sizes for even lower GCV coal.
- The increase in concentration beyond 400 gm/L decreased explosibility, high GCV coal caused a negative impact on explosibility.
- Concentration vs. size interaction plot showed that lower concentration (<200 gm/L) and fine size (<−73 μm) reported higher P_{max}. While at >500 gm/L concentration, P_{max} decreased irrespective of particle sizes.
- At concentrations between 400–500 mg/L, larger particle sizes positively correlate with P_{max} compared to finer sizes.
- Cherat coal (lower GCV) reported a slight positive correlation with P_{max}, while Darra 2 coal (6120 kcal/kg GCV) had a greater positive P_{max} for smaller particle sizes than those that were larger.
- The impact of Darra 1 coal (6318 kcal/kg GCV) was highly negative (inverse) among the coal types; a similar effect was observed for smaller sizes of this coal type than larger sizes.
- The model is valid for access airway vicinities with high air velocities, where dust is suspended at a lower concentration with ample oxygen supply.

- Further explosibility experiments must be conducted at higher concentrations (400 to 800 gm/L) in a 20 L or 1 m^3 chamber to overcome the lower oxy-fuel ratio within the Hartmann apparatus. These will benefit modelling explosibility in regions of high dust concentrations, e.g., underground mines with medium/lower air velocity.
- An IoT sensor system may be developed by deploying the AI-based explosibility model to use coal dust concentration, particle size, and coal type parameters as a monitoring and warning system.

Author Contributions: Conceptualisation, A.U.K., K.M. and M.H.; data curation, A.U.K. and K.M.; formal analysis, A.U.K., S.S. and K.M.; investigation, A.U.K., S.S., K.M. and M.H.; methodology, A.U.K. and K.M.; software, S.S. and K.M.; supervision, M.H.; validation, S.S. and K.M.; writing—original draft, S.S. and K.M.; writing—review & editing, A.U.K., S.S., K.M. and M.H. All authors have read and agreed to the published version of the manuscript.

Funding: Higher Education Commission of Pakistan: National Centre of Artificial Intelligence.

Institutional Review Board Statement: Not applicable.

Informed Consent Statement: Not applicable.

Data Availability Statement: Links to publicly archived datasets analysed or generated during the study: "Not applicable".

Acknowledgments: Rafiziana Md. Kasmani, Chemical Engineering Universiti Teknologi Malaysia is acknowledged for her kind support and supervision to allow the author to conduct experiments in this research at UTM labs.

Conflicts of Interest: The authors declare no conflict of interest.

References

1. Keller, J.O.; Gresho, M.; Harris, A.; Tchouvelev, A.V. What is an explosion? *Int. J. Hydrog. Energy* **2014**, *39*, 20426–20433. [CrossRef]
2. Rice, G.S. *Explosibility of Coal Dust*; USGS Bulletin, Govt. Printing Office: Washington, WA, USA, 1910; Volume 425.
3. Moradi, H.; Sereshki, F.; Ataei, M.; Nazari, M. Evaluation of the effect of the moisture content of coal dust on the prediction of the coal dust explosion index. *Rud. Geol. Naft. Zb.* **2020**, *35*, 37–47. [CrossRef]
4. Mittal, M. Limiting oxygen concentration for coal dusts for explosion hazard analysis and safety. *J. Loss Prev. Process Ind.* **2013**, *26*, 1106–1112. [CrossRef]
5. Gou, X.; Wu, J.; Liu, L.; Wang, E.; Zhou, J.; Liu, J.; Cen, K. Study on factors influencing pulverized coal ignition time. *Adv. Mater. Res.* **2013**, *614–615*, 120–125. [CrossRef]
6. Gokhale, A. Effect of Particle-Size and Air Flow Rates on the Ignition Temperature of Sub Bituminous Coal. *Int. J. Mod. Trends Eng. Res.* **2016**, *3*, 89–97. [CrossRef]
7. Azam, S.; Mishra, D.P. Effects of particle size, dust concentration and dust-dispersion-air pressure on rock dust inertant requirement for coal dust explosion suppression in underground coal mines. *Process Saf. Environ. Prot.* **2019**, *126*, 35–43. [CrossRef]
8. Liu, Q.; Bai, C.; Li, X.; Jiang, L.; Dai, W. Coal dust/air explosions in a large-scale tube. *Fuel* **2010**, *89*, 329–335. [CrossRef]
9. Harris, M.L.; Sapko, M.J.; Zlochower, I.A.; Perera, I.E.; Weiss, E.S. Particle size and surface area effects on explosibility using a 20-L chamber. *J. Loss Prev. Process Ind.* **2015**, *37*, 33–38. [CrossRef]
10. Rybak, W.; Moroń, W.; Ferens, W. Dust ignition characteristics of different coal ranks, biomass and solid waste. *Fuel* **2019**, *237*, 606–618. [CrossRef]
11. Reyes, O.J.; Patel, S.J.; Mannan, M.S. Quantitative structure property relationship studies for predicting dust explosibility characteristics (Kst, Pmax) of organic chemical dusts. *Ind. Eng. Chem. Res.* **2011**, *50*, 2373–2379. [CrossRef]
12. Kazmi, M.Z. Experimental Study of Polyethylene and Sulfur Dust Explosion Characteristics. Master's Thesis, Texas A&M University, College Station, TX, USA, 2018.
13. Lebecki, K.; Małachowski, M.; Sołtysiak, T. Continuous dust monitoring in headings in underground coal mines. *J. Sustain. Min.* **2016**, *15*, 125–132. [CrossRef]
14. Zou, Z.; Zheng, L.; Huang, J.; Li, J.; Wang, J. Design and Implementation of Coal Mine Dust Monitoring System Based on Cloud Platform. In Proceedings of the CSAE '18: 2nd International Conference on Computer Science and Application Engineering, Hohhot, China, 22–24 October 2018; Volume 31, pp. 1–8.
15. Ma, K.; Sun, X.Y.; Tang, C.A.; Wang, S.J.; Yuan, F.Z.; Peng, Y.L.; Liu, K. An early warning method for water inrush in Dongjiahe coal mine based on microseismic moment tensor. *J. Cent. South Univ.* **2020**, *27*, 3133–3148. [CrossRef]

16. Dong, L.; Shu, W.; Sun, D.; Li, X.; Zhang, L. Pre-Alarm System Based on Real-Time Monitoring and Numerical Simulation Using Internet of Things and Cloud Computing for Tailings Dam in Mines. *IEEE Access* **2017**, *5*, 21080–21089. [CrossRef]

17. Kumari, K.; Dey, P.; Kumar, C.; Pandit, D.; Mishra, S.S.; Kisku, V.; Chaulya, S.K.; Ray, S.K.; Prasad, G.M. UMAP and LSTM based fire status and explosibility prediction for sealed-off area in underground coal mine. *Process Saf. Environ. Prot.* **2021**, *146*, 837–852. [CrossRef]

18. Dong, L.J.; Tang, Z.; Li, X.B.; Chen, Y.C.; Xue, J.C. Discrimination of mining microseismic events and blasts using convolutional neural networks and original waveform. *J. Cent. South Univ.* **2020**, *27*, 3078–3089. [CrossRef]

19. Li, X.; Cao, Z.; Xu, Y. Characteristics and trends of coal mine safety development. *Energy Sources Part A Recovery Util. Environ. Eff.* **2020**, 1–19. [CrossRef]

20. Brodny, J.; Tutak, M. Exposure to harmful dusts on fully powered longwall coal mines in Poland. *Int. J. Environ. Res. Public Health* **2018**, *15*, 1846. [CrossRef]

21. Shahan, M.R.; Seaman, C.E.; Beck, T.W.; Colinet, J.F.; Mischler, S.E. Characterization of airborne float coal dust emitted during continuous mining, longwall mining and belt transport. *Min. Eng.* **2017**, *69*, 61–66. [CrossRef]

22. Zlochower, I.A.; Sapko, M.J.; Perera, I.E.; Brown, C.B.; Harris, M.L.; Rayyan, N.S. Influence of specific surface area on coal dust explosibility using the 20-L chamber. *J. Loss Prev. Process Ind.* **2018**, *54*, 103–109. [CrossRef]

23. Liu, T.; Cai, Z.; Wang, N.; Jia, R.; Tian, W. Prediction Method of Coal Dust Explosion Flame Propagation Characteristics Based on Principal Component Analysis and BP Neural Network. *Math. Probl. Eng.* **2022**, *2022*, 1–8. [CrossRef]

24. Di Benedetto, A.; Russo, P.; Sanchirico, R.; Di Sarli, V. CFD Simulations of Turbulent Fluid Flow and Dust Dispersion in the 20 Liter Explosion Vessel. *AIChE J.* **2013**, *59*, 2485–2496. [CrossRef]

25. Di Sarli, V.; Russo, P.; Sanchirico, R.; Di Benedetto, A. CFD simulations of the effect of dust diameter on the dispersion in the 20 l bomb. *Chem. Eng. Trans.* **2013**, *31*, 727–732. [CrossRef]

26. Di Sarli, V.; Danzi, E.; Marmo, L.; Sanchirico, R.; Di Benedetto, A. CFD simulation of turbulent flow field, feeding and dispersion of non-spherical dust particles in the standard 20 L sphere. *J. Loss Prev. Process Ind.* **2019**, *62*, 103983. [CrossRef]

27. Di Sarli, V.; Russo, P.; Sanchirico, R.; Di Benedetto, A. CFD simulations of dust dispersion in the 20 L vessel: Effect of nominal dust concentration. *J. Loss Prev. Process Ind.* **2014**, *27*, 8–12. [CrossRef]

28. Yueze, L.; Akhtar, S.; Sasmito, A.P.; Kurnia, J.C. Prediction of air flow, methane, and coal dust dispersion in a room and pillar mining face. *Int. J. Min. Sci. Technol.* **2017**, *27*, 657–662. [CrossRef]

29. Sanchirico, R.; Russo, P.; Di Sarli, V.; Di Benedetto, A. On the explosion and flammability behavior of mixtures of combustible dusts. *Process Saf. Environ. Prot.* **2015**, *94*, 410–419. [CrossRef]

30. Lundberg, S.M.; Lee, S.I. A unified approach to interpreting model predictions. *Adv. Neural Inf. Process. Syst.* **2017**, *30*, 4766–4775.

31. Tan, B.; Liu, H.; Xu, B.; Wang, T. Comparative study of the explosion pressure characteristics of micro- and nano-sized coal dust and methane–coal dust mixtures in a pipe. *Int. J. Coal Sci. Technol.* **2020**, *7*, 68–78. [CrossRef]

32. Eades, R.; Perry, K. Evaluation of a 38 L Explosive Chamber for Testing Coal Dust Explosibility. *J. Combust.* **2019**, *2019*, 5810173. [CrossRef]

33. Cao, W.; Qin, Q.; Cao, W.; Lan, Y.; Chen, T.; Xu, S.; Cao, X. Experimental and numerical studies on the explosion severities of coal dust/air mixtures in a 20-L spherical vessel. *Powder Technol.* **2017**, *310*, 17–23. [CrossRef]

34. Wang, J.; Meng, X.; Zhang, Y.; Chen, H.; Liu, B. Experimental Study on the Ignition Sensitivity and Explosion Severity of Different Ranks of Coal Dust. *Shock Vib.* **2019**, *2019*, 2763907. [CrossRef]

35. Moradi, H.; Sereshki, F.; Ataei, M.; Nazari, M. Experimental study of the effects of coal dust particle size on laminar burning velocity in a mixture of coal dust and methane. *Min.-Geol.-Pet. Eng. Bull.* **2019**, *34*, 47–56. [CrossRef]

36. Cashdollar, K.L. Coal dust explosibility. *J. Loss Prev. Process Ind.* **1996**, *9*, 65–76. [CrossRef]

37. Song, T.; Zhang, J.; Wang, G.; Wang, H.; Xu, R. Influencing factors of the explosion characteristics of modified coal used for blast furnace injection. *Powder Technol.* **2019**, *353*, 171–177. [CrossRef]

38. Liu, S.H.; Cheng, Y.F.; Meng, X.R.; Ma, H.H.; Song, S.X.; Liu, W.J.; Shen, Z.W. Influence of particle size polydispersity on coal dust explosibility. *J. Loss Prev. Process Ind.* **2018**, *56*, 444–450. [CrossRef]

39. Moroń, W.; Ferens, W.; Czajka, K.M. Explosion of different ranks coal dust in oxy-fuel atmosphere. *Fuel Process. Technol.* **2016**, *148*, 388–394. [CrossRef]

40. Sapko, M.J.; Cashdollar, K.L.; Green, G.M. Coal dust particle size survey of US mines. *J. Loss Prev. Process Ind.* **2007**, *20*, 616–620. [CrossRef]

41. Li, Q.; Wang, K.; Zheng, Y.; Ruan, M.; Mei, X.; Lin, B. Experimental research of particle size and size dispersity on the explosibility characteristics of coal dust. *Powder Technol.* **2016**, *292*, 290–297. [CrossRef]

42. Man, C.K.; Gibbins, J.R. Factors affecting coal particle ignition under oxyfuel combustion atmospheres. *Fuel* **2011**, *90*, 294–304. [CrossRef]

43. Kuracina, R.; Szabová, Z. The Maximum Explosion Pressure of Lignite in Dependence on Particle Size. *Res. Pap. Fac. Mater. Sci. Technol. Slovak Univ. Technol.* **2019**, *27*, 57–64. [CrossRef]

44. Janes, A.; Chaineaux, J.; Carson, D.; Le Lore, P.A. MIKE 3 versus HARTMANN apparatus: Comparison of measured minimum ignition energy (MIE). *J. Hazard. Mater.* **2008**, *152*, 32–39. [CrossRef] [PubMed]

45. Danzi, E.; Franchini, F.; Dufaud, O.; Pietraccini, M.; Marmo, L. Investigation of the fluid dynamic of the modified hartmanntube equipment by high-speed video processing. *Chem. Eng. Trans.* **2021**, *86*, 367–372. [CrossRef]

46. Breiman, L. Random forests. *Mach. Learn.* **2001**, *45*, 5–32. [CrossRef]
47. Efron, B. Bootstrap Methods: Another Look at the Jackknife. In *Breakthroughs in Statistics*; Springer Series in Statistics; Kotz, S., Johnson, N.L., Eds.; Springer: New York, NY, USA, 1992; pp. 569–593.
48. Dasgupta, A.; Sun, Y.V.; König, I.R.; Bailey-Wilson, J.E.; Malley, J.D. Brief review of regression-based and machine learning methods in genetic epidemiology: The Genetic Analysis Workshop 17 experience. *Genet. Epidemiol.* **2011**, *35*, 5–11. [CrossRef] [PubMed]
49. Bu, Y.; Li, C.; Amyotte, P.; Yuan, W.; Yuan, C.; Li, G. Moderation of Al dust explosions by micro- and nano-sized Al_2O_3 powder. *J. Hazard. Mater.* **2020**, *381*, 120968. [CrossRef]
50. Bataineh, M.; Steenhard, D.; Singh, H. Feature Impact for Prediction Explanation. In Proceedings of the 19th Industrial Conference on Data Mining, New York, NY, USA, 17–21 July 2019; pp. 160–167.
51. Zou, J.; Petrosian, O. Explainable AI: Using shapley value to explain complex anomaly detection ML-based systems. *Front. Artif. Intell. Appl.* **2020**, *332*, 152–164. [CrossRef]
52. Wieland, R.; Lakes, T.; Nendel, C. Using SHAP to interpret XGBoost predictions of grassland degradation in Xilingol, China. *Geosci. Model Dev. Discuss.* **2020**, 1–28. [CrossRef]
53. Bauser, G.M.; Sauer, K.S. *Extrusion*, 2nd ed.; ASM International: Materials Park, OH, USA, 2006; Volume 200, ISBN 9780871708373.
54. Aoki, S.; Takemura, A. Design and analysis of fractional factorial experiments from the viewpoint of computational algebraic statistics. *J. Stat. Theory Pract.* **2012**, *6*, 147–161. [CrossRef]
55. Pedregosa, F.; Varoquaux, G.; Gramfort, A.; Michel, V.; Thirion, B.; Grisel, O.; Blondel, M.; Prettenhofer, P.; Weiss, R.; Dubourg, V.; et al. Scikit-learn: Machine learning in Python. *J. Mach. Learn. Res.* **2011**, *12*, 2825–2830.
56. Liashchynskyi, P.; Liashchynskyi, P. Grid Search, Random Search, Genetic Algorithm: A Big Comparison for NAS. *arXiv* **2019**, arXiv:1912.06059.
57. Emmert-streib, F.; Dehmer, M. Evaluation of Regression Models: Model Assessment, Model Selection and Generalization Error. *Mach. Learn. Knowl. Extr.* **2019**, *1*, 521–551. [CrossRef]
58. Su, J.; Cheng, Y.; Song, S.; Ma, H.; Wang, W.; Wang, Y.; Zhang, S. Explosion Characteristics and Influential Factors of Coal dust/sodium Chlorate Mixture on Basis of an Explosion Accident in China. *Combust. Sci. Technol.* **2021**, *193*, 1313–1325. [CrossRef]

Article

Low-Cost Electromagnetic Split-Ring Resonator Sensor System for the Petroleum Industry

Alejandro Rivera-Lavado [1,2], Alejandro García-Lampérez [1], María-Estrella Jara-Galán [3], Emilio Gallo-Valverde [3], Paula Sanz [4] and Daniel Segovia-Vargas [1,*]

[1]   Signal Theory and Communications Department, Carlos III University of Madrid, 28903 Madrid, Spain; arivera@ing.uc3m.es (A.R.-L.); alamperez@tsc.uc3m.es (A.G.-L.)
[2]   Yebes Observatory, Dirección General del Instituto Geográfico Nacional, 19141 Yebes, Spain
[3]   Indra Sistemas S.A., 28108 Madrid, Spain; mejara@indra.es (M.-E.J.-G.); egallo@minsait.com (E.G.-V.)
[4]   REPSOL S.A., 28045 Madrid, Spain; psanzs@repsol.com
*   Correspondence: dani@tsc.uc3m.es; Tel.: +34-91-624-8737

Abstract: The use of a low-cost split-ring resonator (SRR) passive sensor for the real-time permittivity characterization of hydrocarbon fluids is proposed in this paper. The characterization of the sensor is performed through both full-wave simulation and measurements. Thanks to the analysis of several crude samples, the possibility of discrimination between different types of crude and the estimation of several of their properties are demonstrated. Between them, the estimation of sulfur, aromatic hydrocarbons, and salt-water concentrations either in normal ambient conditions or in a high-pressure and high-temperature environment can be mentioned. Experiments were run both at normal ambient conditions and pressures up to 970 bar and temperatures up to 200 °C.

Keywords: effective permittivity; resonator; sensor; split-ring resonator (SRR); submersible sensor

Citation: Rivera-Lavado, A.; García-Lampérez, A.; Jara-Galán, M.-E.; Gallo-Valverde, E.; Sanz, P.; Segovia-Vargas, D. Low-Cost Electromagnetic Split-Ring Resonator Sensor System for the Petroleum Industry. *Sensors* **2022**, *22*, 3345. https://doi.org/10.3390/s22093345

Academic Editors: Longjun Dong, Yanlin Zhao and Wenxue Chen

Received: 14 March 2022
Accepted: 25 April 2022
Published: 27 April 2022

Publisher's Note: MDPI stays neutral with regard to jurisdictional claims in published maps and institutional affiliations.

1. Introduction

Intelligent well technology allows an efficient operation for both the oil and gas industry. The use of downhole sensors has become popular since it allows for the continuous and real-time monitoring of relevant reservoir factors, such as flow and pressure control, sand and water monitoring [1,2], and leakage detection [3], among others. Gathered data can be processed, analyzed, and used for closed-loop control, well management, and decision making in the extraction, transportation, and processing activities. Moreover, measuring the refractive index (or the permittivity) of the extracted fluid allows for the estimation of other relevant thermodynamic and physical properties, such as critical constants, and average molecular weight, density, viscosity, thermal conductivity, and boiling point [4].

The development of suitable downhole sensors is challenging since a well's depth can be between 1000 m and 4000 m [2]. These sensors must stand in a high-pressure (above 900 bar) and high-temperature (above 150 °C) environment. This makes it hard to obtain durable and reliable active sensors. Passive devices are more robust but suffer from long-distance communication issues, such as high losses and thermal noise.

Until now, available sensors mostly rely on optical [5–8], acoustic [9–12], and radio frequency (RF) [13,14] technologies. RF sensors are specially attractive for downhole applications, since they are easy to manufacture, cost-effective, and robust. Since they can be easily integrated inside well pipes, they are especially suitable for real-time crude properties estimation.

This paper proposes the use of RF split-ring resonators (SRR) as downhole passive sensors for real-time crude monitoring through permittivity estimation. SRR [15] and complementary SRR (CSRR) [16] can be used for measuring a wide variety of magnitudes, such as alignment [17], displacement [18], rotation [19], speed [20], blood glucose [21], and thickness [22,23]. Both SRR [23,24] and complementary SRR [25–27] sensors can be used for solids and liquids permittivity characterization. A cost-effective RF sensor interrogator is also introduced in this contribution. It can be placed in the well pad and connected to the downhole sensor trough a low-loss coaxial cable. This device can work as a stand-alone system that can be controlled via standard commands for programmable instruments (SCPI)-compatible commands through a TCP/IP network or can be integrated into a multi-sensor system. In our particular setup, a control system that implements a support vector machine (SVM) classifier combines the measurement of different sensors. Since all the data is normalized before the processing, calibrating the permittivity estimation in the interrogator is not required.

The rest of this manuscript is organized as follows. In Section 2, we introduce our proposed SRR sensor design for downhole crude permittivity estimation. The ability to discriminate between different crude samples and the determination of the sulfur, aromatic hydrocarbons, and salt-water concentrations is demonstrated in Section 3. In Section 4, we introduce an autonomous remotely controlled SRR sensor interrogator that allows real-time crude monitoring. The evaluation of the whole system in a high-pressure and high-temperature setup is shown in Section 5. The interrogator performance is compared in terms of stability and standard deviation with the ones obtained when using a conventional and expensive vector network analyzer (VNA).

2. Submersible Split-Ring Resonator-Based Sensor

An SRR resonator excited by an open-ended microstrip line is used for determining the electrical properties of the surrounding fluid. The electric parameter to be measured is the reflection coefficient $\Gamma = S_{11}$. Figure 1 shows the sensor model.

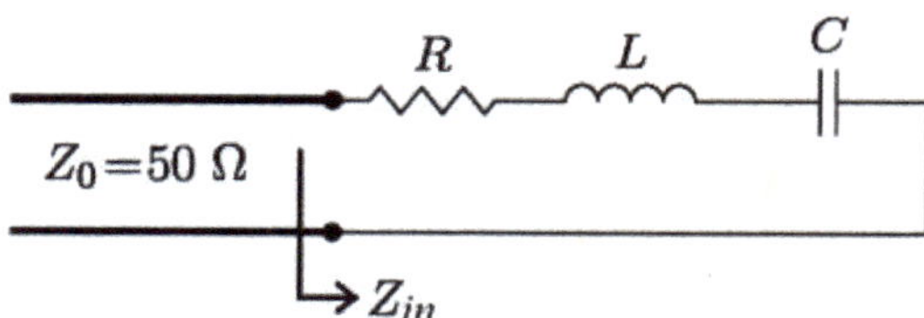

Figure 1. Model of a serial RLC resonator excited by a transmission line.

The complex SRR resonator input impedance $Z_{in}(\omega)$ has a frequency-independent real part, R, and an imaginary one, $X(\omega)$, which take both positive and negative values at different frequencies. Around the resonant frequency f_0, the input impedance can be modeled with the following equation

$$Z_{in}(\omega) = R + j\left(\omega L - \frac{1}{\omega C}\right) = R + jX(\omega) \tag{1}$$

The quality factor Q, the coupling factor s, and the resonant frequency f_0 are given as

$$Q_0 = \frac{\omega_0 L}{R} \tag{2}$$

$$s = \frac{Z_0}{R} \tag{3}$$

$$f_0 = \frac{\omega_0}{2\pi} = \frac{1}{2\pi\sqrt{LC}} \propto \frac{1}{\sqrt{\varepsilon_{\text{eff}}}} \tag{4}$$

f_0 is determined by the structure dimensions and the effective permittivity ε_{eff}, which depends on the substrate permittivity and the surrounding fluid permittivity ε_{LUT}. Any increase in the liquid-under-test (LUT) permittivity ε_{LUT} leads to a reduction in f_0, as will be shown later in our sensor full-wave simulations. The working frequency, f_0, has been set in the UHF frequency band in order to reach larger propagation distances and show higher resolution in the variations of fluid surrounding the sensor. However, the use of frequencies in the low microwave band makes the sensor have bigger sizes. In order to reduce the dimensions of the sensor's high permittivity substrates, Arlon AR1000 with a relative permittivity $\varepsilon_r = 10$ can be used. For avoiding a low sensitivity due to this relatively high permittivity, a thickness of 1.27 mm is chosen. Substrate losses are relatively low ($\tan(\delta) = 0.003$) at the working frequencies. Furthermore, according to our tests, this substrate can stand the expected level of pressure and temperature.

Several designs have been tested. In all of them, two SRRs are placed close to an open ending in a symmetric configuration around the microstrip line. Figure 2 shows a picture of the A-type (Figure 2a) and the B-type (Figure 2b,c sensor prototype. Both designs have a resonant frequency in the range of the 500 MHz band ($f_{0,air} = 474.5$ MHz for the A-Type and $f_{0,air} = 470.22$ MHz for the B-Type). Due to the manufacturing tolerances, each unit may have a different $f_{0,air}$. Such bias can be easily compensated by considering the frequency difference $\Delta f = f_0 - f_{0,air}$ instead of absolute frequency values for samples characterization. The A-type sensor has a dimension of 95.25×32.22 mm^2, while the B-type is a more compact version.

(**a**)

(**b**)

(**c**)

Figure 2. Manufactured A-type (**a**) SRR resonator-based sensor. B-type front (**b**) and back (**c**) view.

Figure 3 sketches the B-type sensor. Each resonator has a length L_{SRR} of 23 mm and a width W_{SRR} of 7.2 mm. Both line widths W_L and gaps G are 0.8 mm. The distance between the resonators and the microstrip line, D, is 0.93 mm. The microstrip line is 1.22 mm wide, which corresponds to a $Z_0 = 50\ \Omega$ characteristic impedance. The resulting substrate dimensions, $L \times W$, are 35×30 mm^2.

Due to the substrate porosity and roughness, the fluid to be measured may contaminate the sensor, which would make it unfit for future measurements. Furthermore, the presence of conductive compounds in the sample may short-circuit the resonators. Because of this, all manufactured sensors are protected with a 25 μm-thick Kapton layer (Figure 2b,c). This protective layer has a relative permittivity $\varepsilon_r = 3.4$ and a loss tangent $\tan(\delta) = 0.0018$, which must be taken into account in the resonator design.

The B-type sensor can be easily integrated inside a pipe for the real-time monitoring of a fluid flow. Figure 4 sketches this scenario. A Teflon cover must be placed above the sensor for ensuring a continuous axial flow. This cover was also considered in the full-wave simulations.

All the full-wave simulations were performed with Ansys HFSS v19 ©. Figure 5 shows the calculated $|S_{11}|$ when considering fluids of a relative permittivity from one to three. As expected, the resonant frequency f_0 shifts to lower frequencies when increasing the fluid permittivity ε_{LUT}. As expected, the Teflon cover creates a slight shift of f_0 that must be taken into account.

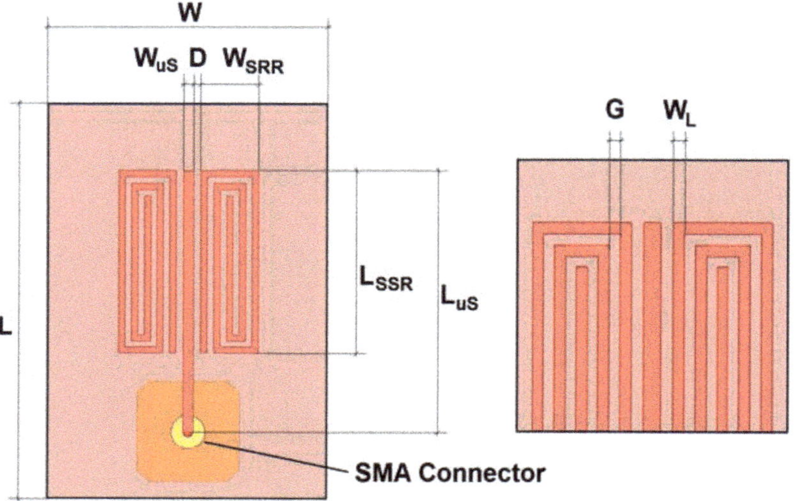

Figure 3. Sketch of the B-type SRR-based sensor. The inset shows a detail of the microstrip open termination.

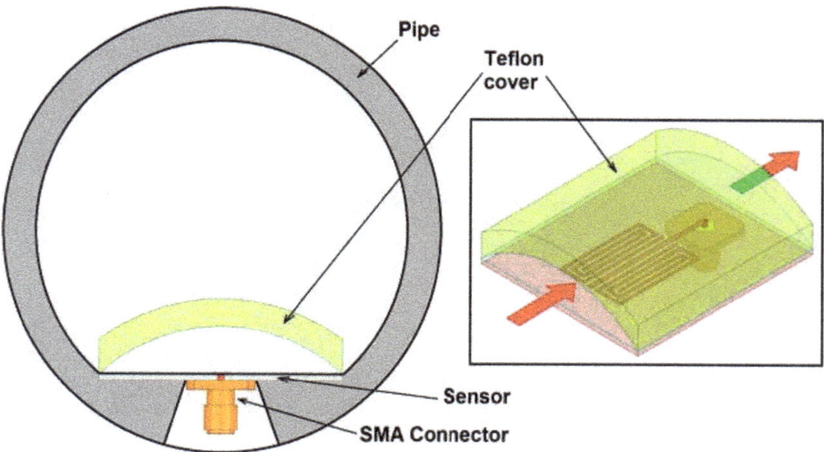

Figure 4. Integration of the sensor inside of a pipe (gray). A Teflon cover is placed above the sensor for ensuring an axial flow near the SRR resonator. Only part of the fluid (blue) contributes to the measurement. For obtaining a relevant sample, a homogeneous flow must be achieved. The inset shows a 3D view of the cover and the sensor.

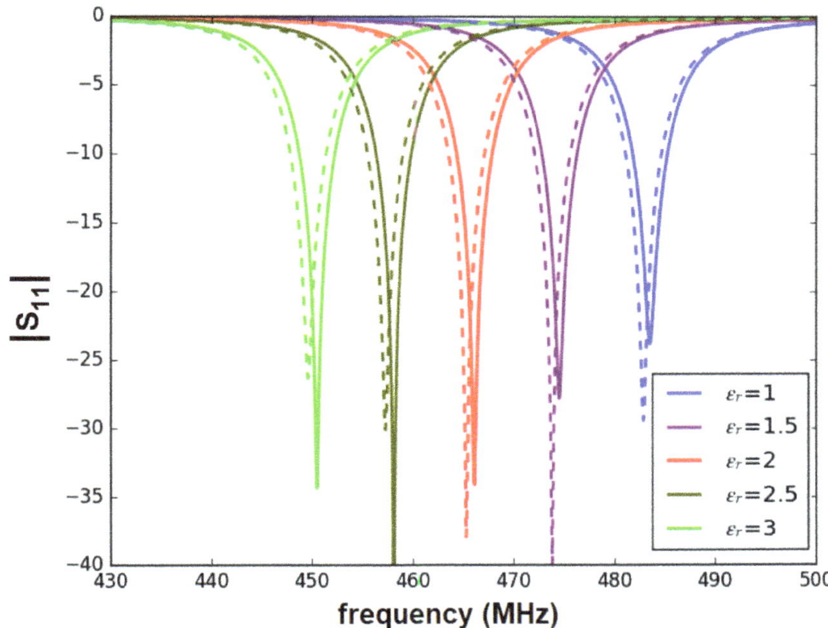

Figure 5. Simulated $|S_{11}|$ for the uncovered (solid) and covered (dashed) sensor. Fluids of permittivities between 1 and 3 are considered.

Only a one-dimension measurement, related to the complex magnitude $\Gamma = S_{11}$, is required for determining the resonant frequency. In Section 4, a cost-effective 1D interrogator for the SRR sensor is introduced. It is an interesting alternative to a 2D S-parameter determination using a one-port VNA, especially for multiple-sensors systems.

Next, we demonstrate that the resonant frequency measurement of SRR sensors allows the determination of different sample parameters.

3. Crude Properties Estimation

All measurements shown in this section have been undertaken with a Keysight N9914A Fieldfox VNA. The module of the S_{11} parameter of the resonator was obtained in order to monitor its resonant frequency f_0. The mixtures were prepared by taking into account the mass fraction of each of its components. Using a magnetic mixer ensures sample homogeneity. All measurements discussed in this section were taken under normal ambient temperature and pressure.

3.1. Discrimination between Different Crude Samples

During the experiments, five different crude samples, labeled as A, B, C, D, and E were used. Each of them were obtained from different wells around the world. Due to the differences in their composition, the permittivity is different and, therefore, also the measured resonant frequency (Figure 6a). As expected, when mixing crudes, newer permittivities are obtained. All possible combinations (binary, ternary, quaternary, and quinary) were also characterized. All mixtures have an equal mass fraction for each of their compounds. Figure 6a is a box plot that summarizes 100 measurements per sample of the immersed sensor f_0. Since $f_{0,air}$ (Figure 6b) keeps the same before (AIR 1) and after (AIR 2) measuring the crudes, it is obvious that no damage or contamination has been produced.

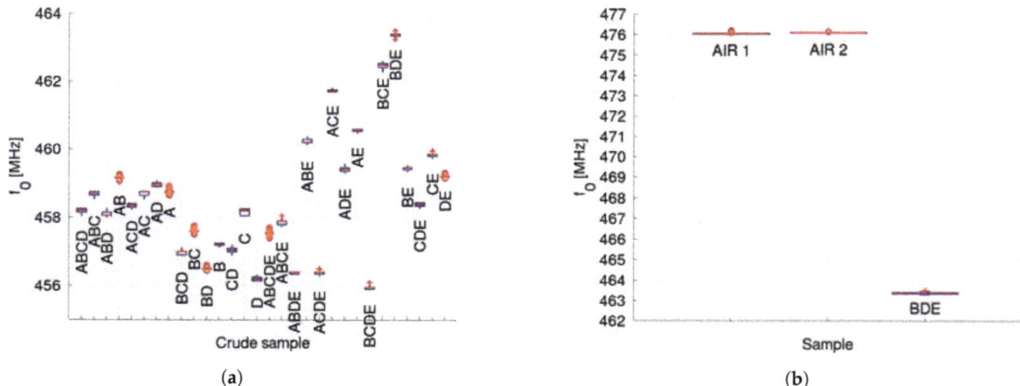

Figure 6. Resonant frequencies f_0 for different mixtures (**a**). Air was measured before and after the crudes (**b**). Since the same resonant frequency was obtained, the SRR sensor survived all the measurements between without any damage or substrate contamination. BDE sample is also shown.

3.2. Sulfur Concentration Estimation

Next, a group of samples with different sulfur concentrations were prepared from the same crude. Figure 7 shows the resonant frequency over the mass fraction. The best sensitivity is achieved for mass fractions of sulfur between 0.05% and 0.36%. Actual limits of measurable concentrations are related to the achieved frequency resolution. For concentrations below 0.05%, the effect of sulfur in the permittivity of the sample is negligible.

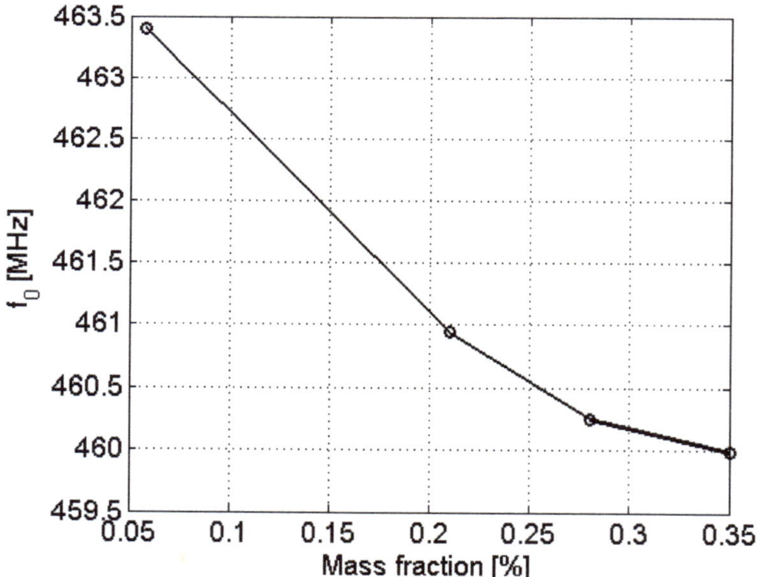

Figure 7. Evolution of the resonant frequency f_0 with the mass fraction of sulfur.

3.3. Aromatic Hydrocarbons Concentration Estimation

The same crude sample was used for evaluating the ability to detect aromatic hydrocarbons concentrations with SRR sensors. Because of this, a low concentration leads to the same resonant frequency ($f_0 = 463.4$ MHz), which corresponds to the specific crude sample. The highest sensitivity is achieved for mass fractions between 20% and 36% (Figure 8).

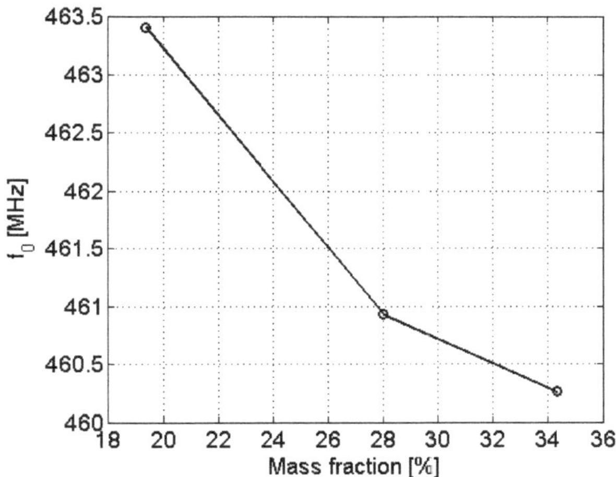

Figure 8. Evolution of the resonant frequency f_0 with the mass fraction of aromatic hydrocarbons.

3.4. Salt-Water Concentration

Finally, we prove that the sensor can estimate the amount of salt-water inside the crude fluid. Two different crude samples (CRUDE-1 and CRUDE-2, with resonant frequencies of $f_{0,C} = 459.2$ MHz) were used for generating this series of mixtures. A sample of Mediterranean water (resonant frequency of 282.6 MHz) was used as the salt-water. For each crude, four different mixtures of 0%, 5%, 20%, and 35% salt-water mass fractions were prepared. Figure 9 shows the frequency deviation in MHz from the 0% salt-water concentration. As can be seen, the frequency deviation is dependent on the salt-water concentration and independent of the CRUDE sample used in the mixture preparation.

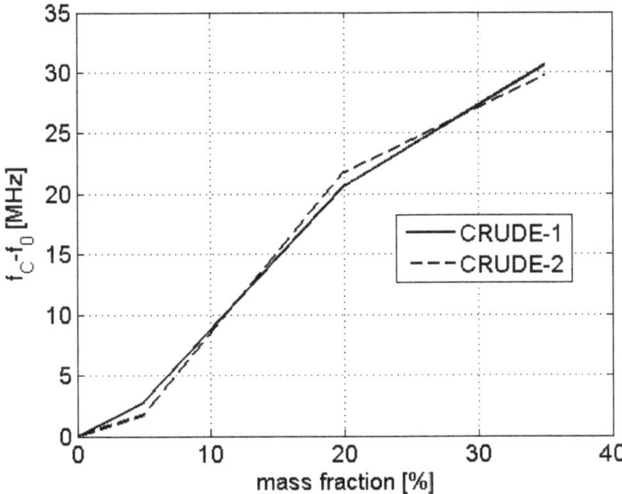

Figure 9. Evolution of the resonant frequency difference $f_{0,C} - f_0$ with the mass fraction of salt-water for mixtures created from two different samples: CRUDE-1 (solid) and CRUDE-2 (dashed).

Since many crude properties can be obtained from the SRR-resonant frequency, we have developed a cost-affordable sensor interrogator as an alternative to the S_{11} measurement using VNAs. Our proposed system can be remotely operated and does not require trained personnel, since it is fully autonomous after being installed.

4. SRR Sensor Interrogator

For a real-time permittivity measurement, the following topology is proposed (Figure 10). A continuous-wave voltage-controlled source injects power P_{TX} into the sensor through the circulator. The reflected power P_{RX} in the SRR is sent back to the detector through the circulator. An Atmega2560 microcontroller sets the VCO frequency via the transmitter digital to the analog converter (DAC). The receiver output voltage V_{RX} is then digitized by using the microcontroller ADC. All the measurement parameters (frequency, integration time, and scanning bandwidth, etc.) are configured via HTTP commands and sent through a standard TCP/IP Ethernet connection. A low-cost Arduino MEGA2560 and a W5100-based Ethernet shield are used for implementing the control electronics.

A Mini-Circuits ZX95-625+ is used as the VCO. A DPVCC45A circulator working between 410 MHz and 500 MHz is used for connecting the sensor to both the transmitters and receivers. The power detector is a Mini-Circuits ZX47-60LN+.

Next, both the transmitter and the receiver design is highlighted.

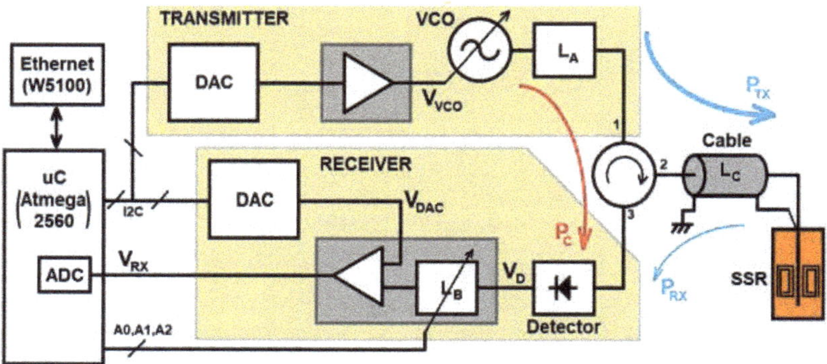

Figure 10. Sketch of the system.

4.1. Transmitter

The VCO is commanded by an I^{2}C 12-bit MCP4725 DAC. The DAC outputs voltage varies from 0 to 5 V. To maximize the frequency resolution, the V_{OSC} voltage range is tailored according to its response curve for the frequency range between 421.13 MHz and 506.5 MHz, which corresponds to a voltage level from 5.8 V to 9 V. This is achieved via offset and amplitude compensation with a TL081 set in a non-inverting op-amp configuration. Therefore, $2^{12} = 4096$ frequency steps imply a maximum resolution of $\approx$20.85 kHz.

The VCO generates a power level of 6 dBm for frequencies from 400 MHz to 520 MHz. Due to the measured circulator isolation between ports one and three ($max(|S_{31}|) = -24.6$ dB), an attenuator of $L_A = 16.3$ dB is placed at the VCO output. The VCO power coupled to the detector P_C is then reduced to -34.3 dBm. The maximum power level in the detector is 6 dBm $-$ 16.3 dB = -10.3 dBm.

4.2. Receiver

The power detector response curve is shown in Figure 11 [28]. As can be seen, the output voltage V_D is inversely proportional to the input power in dBm. The red line shows the power level P_C. Depending on the phase between the VCO-coupled power and the received signal, the power level in the detector can increase or decrease (blue). Both the cable and the sensor affect the receiver signal amplitude and phase. In this design, the cable impact is assumed to be constant and can be compensated by proper signal conditioning after the detector.

The receiver offset and amplitude compensation is performed with a variable attenuator and a TL081 non-inverting op-amp configuration. The schematic of this topology is shown in Figure 12. An MCP4725 DAC synthesizes the voltage V_{DAC} for the offset compensation. A resistive divider attenuates both V_{DAC} and the signal-detected V_D. A set of resistors, R2, R3, R4, R5, and R6, has been added to conform a variable resistor controlled by the three ports A0, A1, and A2. Each port can be configured as a low-state output or a high, which corresponds to low- and high-impedance status. This allows four different attenuation L_B values. Since the amplifier has a constant gain $G = 151$, a lower attenuation can be used for compensating higher cable losses.

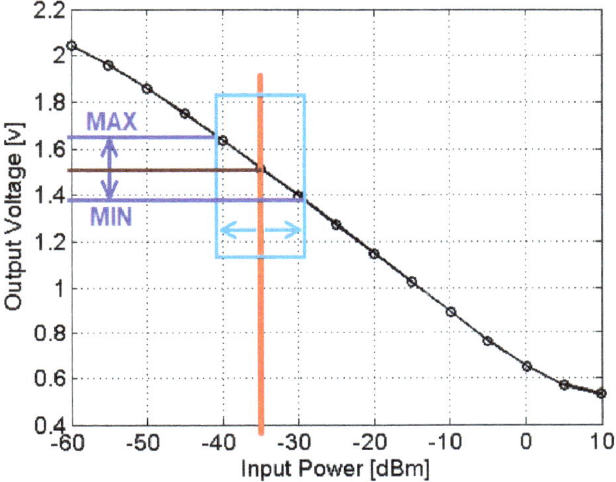

Figure 11. Response curve of the Mini-Circuits ZX95-625+ power detector [28].

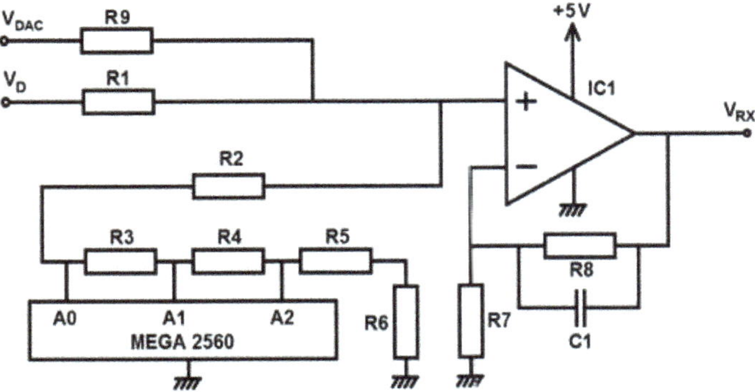

Figure 12. Simplified schematic of the receiver signal conditioner.

The capacitor C_1 limits the measurement speed and filters the high-frequency noise ($f_C = 48.22$ Hz). For flattening the microcontroller consumption, the Atmega2560 ADC is kept measuring continuously, which drastically reduces the measurement noise. All the samples taken during the measurement time t_m are averaged. t_m can be set from 300 ms to 30 s. Higher values can further improve the cable loss compensation. When using a 25 dB attenuator for simulating cable losses (total loss of 50 dB) and $t_m = 400$ ms, the sensor resonance can be still detected. With proper cable selection (i.e., Times Microwave HP1200 or Commscope AVA7-50, with losses of 3.4 dB/km and 1.53 dB/km, respectively) distances between the interrogator and the sensor above 1 km are feasible.

Figure 13 shows the prototype. It is integrated into a 19-inch universal case. The radio frequency elements are integrated into a dedicated aluminum block as an independent sub-assembly. This metal block is designed to have a large heat capacity. It allows thermal stability for both the VCO and the detector without active temperature control through either thermoelectric cooler or fans. The system stability is demonstrated in Section 5. The front panel has a single SMA female port on which the sensor is connected.

All the required operations are performed remotely via SCPI-compatible HTTP commands. It can work as a stand-alone device or as a part of a multi-sensor system. Both the SRR-resonant frequency and the full measurement vector can be obtained through HTTP commands. When measuring the SRR-resonant frequency in a stand-alone mode, only one-dimension data are extracted from the crude sample, so it is not possible to monitor the change of more than one property at the same time.

Figure 13. Manufactured SRR sensor interrogator.

5. High Pressure and High Temperature Measurements

The system was tested in a laboratory environment. The sensor is integrated into a bottle where the high pressures (up to 970 bar) and the high temperatures (up to 200 °C) are generated when filling with the liquid under test (LUT). The LUT constantly recirculates to avoid stratification. The measurement setup is sketched in Figure 14. Only the SRR sensor and a coaxial segment work in a high-pressure and high-temperature environment. The interrogator, the personal computer (PC), and the TCP/IP network run at a normal ambient pressure and temperature.

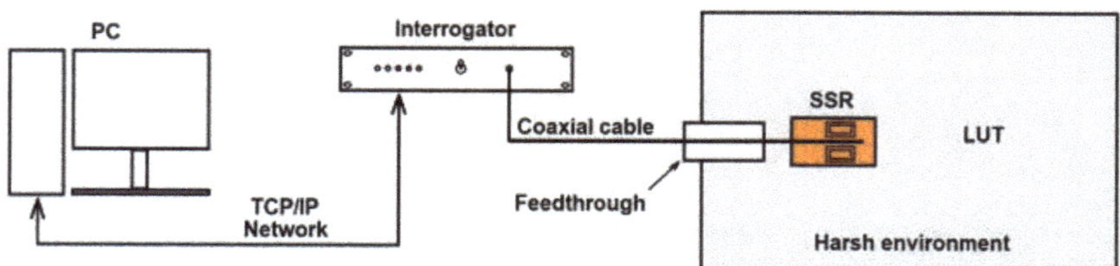

Figure 14. Sketch of the high-pressure and high-temperature measurement setup.

Figure 15 shows the experimental setup. The SRR sensor is fitted inside the high-pressure bottle (Figure 15a). The feed-through (Figure 15b) allows the radio frequency connection between the high-pressure and the low-pressure coaxial cable sections. It can also accommodate several optic fibers for multi-sensor characterization systems. The bottle is fixed to a holder that controls the experiment temperature (Figure 15c).

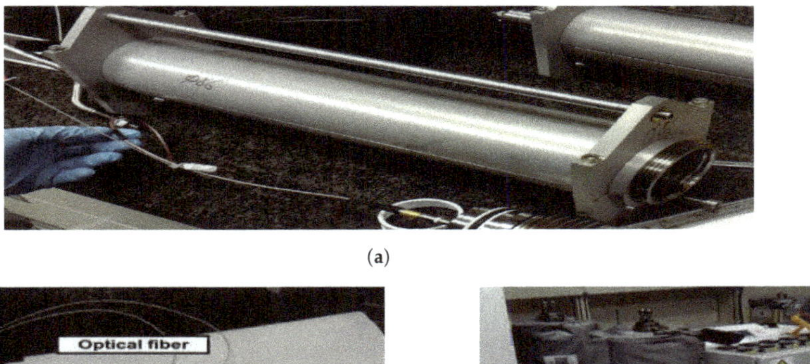

(**a**)

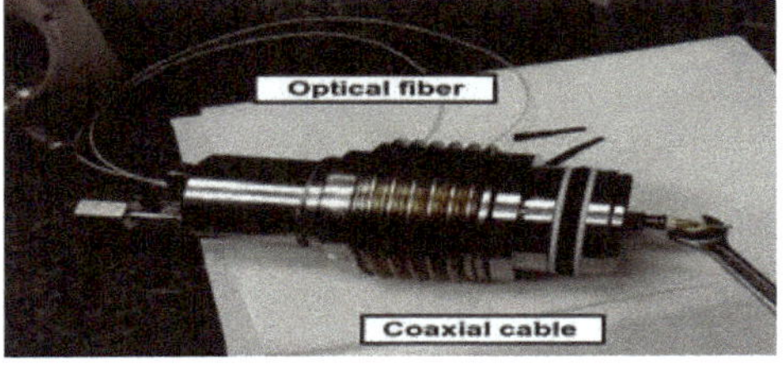

(**b**)   (**c**)

Figure 15. High-pressure and high-temperature measurement setup: high-pressure bottle (**a**), feed-through (**b**), and bottle holder and heater (**c**).

5.1. System Stability

Several crude samples and mixtures were characterized in different measurement rounds. The interrogator was kept on and measuring during the days that the experimental work was undertook, which allowed us to test the stability of the system after some time. Figure 16 shows the evolution of the two main peaks of V_{RX} when measuring heptane continuously during 11 h 30 min. The frequency resolution was 358.1 kHz so the frequency determination was inside the $\overline{f_{0,H}} \pm 895.2$ kHz interval, $\overline{f_{0,H}}$ being the mean of all resonant frequencies.

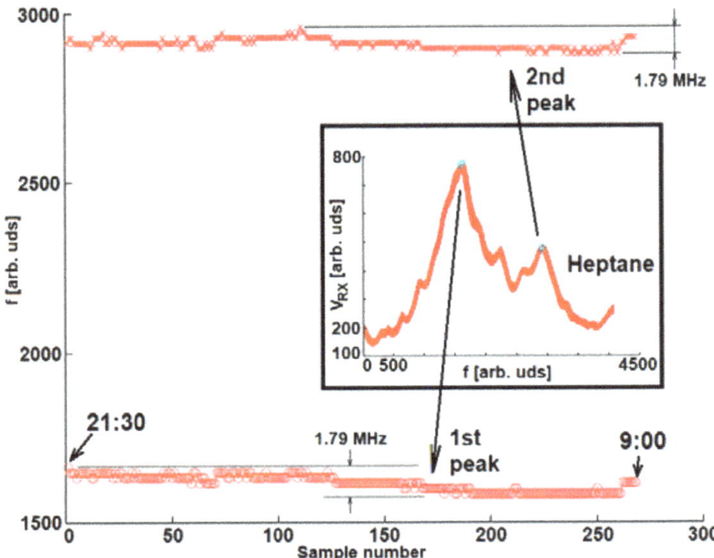

Figure 16. Frequency of the first and second V_{RX} peaks along time. No drifts can be appreciated when no parameters are changed in the high-pressure bottle. The inset shows all V_{RX} sweeps overlapped.

5.2. VNA and Interrogator Comparison

Six crude samples were measured with the same SRR sensor driven by using both the Keysight N9914B VNA and the developed interrogator in order to qualitatively compare their discrimination capabilities. They are plotted in Figure 17 in the frequency range from 445 MHz to 475 MHz. Figure 17a shows the module of the S_{11} parameter measured with the VNA in dB. Figure 17b shows the measured voltage V_{RX} (see Figures 10 and 12).

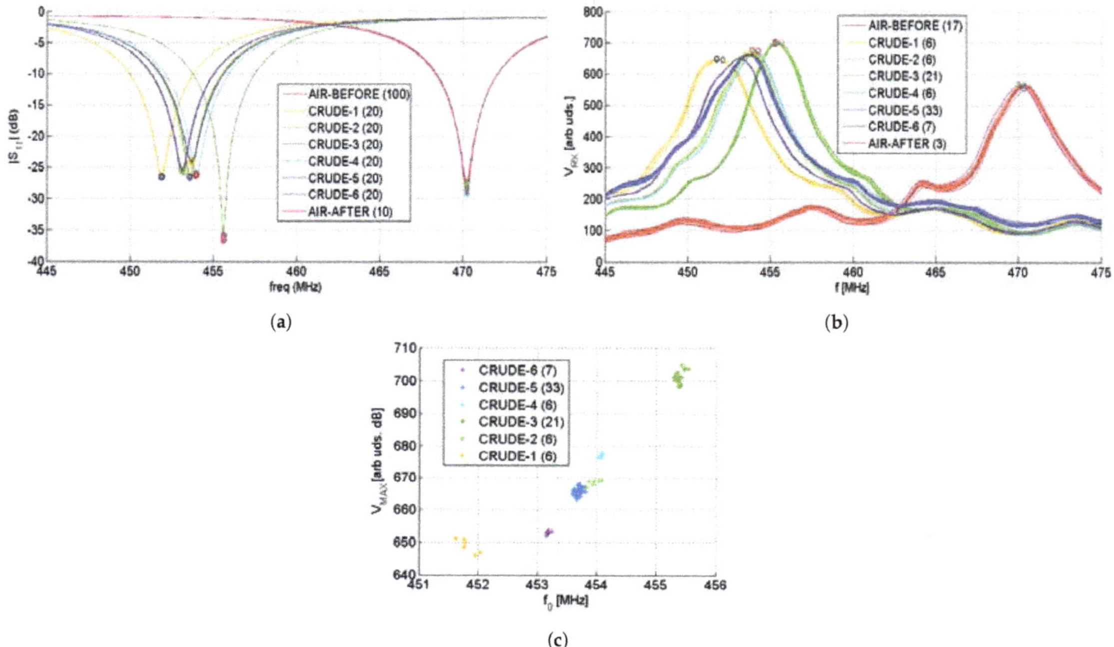

Figure 17. Measurements of six crude samples using the same SRR sensor, the VNA (**a**), and the Interrogator (**b**). The inset shows the number of measurements performed for each sample. The 2D plot of V_{MAX} and f_0 of each measurement (**c**). The air measurements are not shown.

The resonant frequency can be obtained by finding the minimum in the module of the S_{11} curve or the peak V_{MAX} in the V_{RX} curve. The resonant frequencies f_0 and the achieved standard deviations σ are summarized in Table 1. The VNA achieves a standard deviation that is one order of magnitude below the one achieved by our solution. For this experiment, the interrogator was configured with a measurement time of $t_m = 400$ ms. Smaller σ are achievable when increasing t_m. For the achieved value of σ, it is not obvious that a crude differentiation can be performed when considering only f_0. In our application, it was possible to do it thanks to the SVM that takes into account all the measured points. If our interrogator is used as a stand-alone system, a rudimentary classifier can be implemented in the microcontroller. Besides f_0, other parameters, such as the peak amplitude V_{MAX} and the peak width, can be considered. As an example, Figure 17c shows f_0 and V_{MAX} for each measurement. Although crude samples two, four, and five have a close f_0, the classification is still possible in the 2D plane.

As it can be appreciated, there is an agreement between the VNA and the interrogator measurements. Air was measured before and after all crude characterization. Since there is an agreement between both air measurements, we can conclude that the sensor is not contaminated or degraded and both the VNA and the implemented interrogator remain frequency-stable during the whole experiment, as expected after the results shown in Figure 16. Both systems were able to discriminate between the different samples.

Table 1. List of the measured resonant frequencies and standard deviation $\sigma = |f_0 - \overline{f_0}|^2 / N$ for the same sensor driven by the VNA and the implemented interrogator for different crude samples.

Sample	VNA $\overline{f_0}$ [MHz]	Interrogator $\overline{f_0}$ [MHz]	VNA σ	Interrogator σ
Air (before)	470.22	470.35	0.025	0.000
Air (after)	470.22	470.11	0.024	0.207
CRUDE-1	451.88	451.85	0.038	0.185
CRUDE-2	453.60	453.88	0.016	0.000
CRUDE-3	455.61	455.33	0.022	0.078
CRUDE-4	453.97	454.06	0.038	0.196
CRUDE-5	453.71	453.63	0.026	0.167
CRUDE-6	453.14	453.16	0.037	0.000

6. Conclusions

In this document, an SRR sensor-based system is proposed for the real-time monitoring of crude properties for the petroleum industry. The sensor can be integrated into the well's pipes and can work at high-pressure and high-temperature conditions.

Experimental work has been carried out for demonstrating that the system can discriminate between different crude samples. It is also able to determine the sulfur, the aromatic hydrocarbons, and the salt-water concentrations. It is obtained by detecting permittivity changes by measuring the corresponding SRR-resonant frequency.

A cost-affordable sensor interrogator has been developed. It is an autonomous system that can be remotely operated via HTTP commands sent through a TCP/IP network.

The whole system has been validated at high-pressure and high-temperature working conditions. The system stability has been tested through several weeks of continuous measurements. Nevertheless, further work is required in order to test our solution in a more realistic environment, since the effects of high pressure and high temperature on coaxial connectors could degrade the performance of the whole system. Furthermore, our solution should be compared with other state-of-the-art alternatives in on-site tests.

Author Contributions: A.R.-L., investigation, simulation, and writing; A.G.-L., mathematical modeling and simulation; M.-E.J.-G., review and editing; E.G.-V., review and editing; P.S., review and editing; D.S.-V., investigation, supervision, project administration, and responsible for the funding. All authors have read and agreed to the published version of the manuscript.

Funding: This work has been financially supported by the REMO project.

Institutional Review Board Statement: Not applicable.

Informed Consent Statement: Not applicable.

Data Availability Statement: Not applicable.

Acknowledgments: Section 3 measurements were performed at the Carlos III University of Madrid laboratories, in Leganés, Madrid (Spain). Section 4 experiments were performed at the Repsol Technology Lab, in Móstoles, Madrid (Spain). The authors would like to thank José-Miguel Salazar-Gutierrez for their support during the experimental work carried out at the Repsol Technology Lab.

Conflicts of Interest: The authors declare no conflict of interest.

References

1. Tubel, P.; Hopmann, M. Intelligent Completion for Oil and Gas Production Control in Subsea Multi-lateral Well Applications. In *SPE Annual Technical Conference and Exhibition*; OnePetro: Richardson, TX, USA, 1996.
2. Huiyun, M.; Chenggang, Y.; Liangliang, D.; Yukun, F.; Chungang, S.; Hanwen, S.; Xiaohua, Z. Review of intelligent well technolog. In *Petroleum*; Elsevier: Amsterdam, The Netherlands, 2019.
3. Adegboye, M.A.; Fung, W.-K.; Karnik, A. Recent Advances in Pipeline Monitoring and Oil Leakage Detection Technologies: Principles and Approaches. *Sensors* **2019**, *19*, 2548. [CrossRef] [PubMed]
4. Riazi, M.R.; Roomi, Y.A. Use of the Refractive Index in the Estimation of Thermophysical Properties of Hydrocarbons and Petroleum Mixtures. *Ind. Eng. Chem. Res.* **2001**, *40*, 1975–1984. [CrossRef]

5. Kersey, A. Optical Fiber Sensors for Permanent Downwell Monitoring Applications in the Oil and Gas Industry. *IEICE Trans. Electron.* **2001**, *84*, 400–404.
6. Jones, C.M.; Dai, B.; Price, J.; Li, J.; Pearl, M.; Soltmann, B.; Myrick, M.L. A New Multivariate Optical Computing Microelement and Miniature Sensor for Spectroscopic Chemical Sensing in Harsh Environments: Design, Fabrication, and Testing. *Sensors* **2019**, *19*, 701. [CrossRef] [PubMed]
7. Peng, G.; He, J.; Yang, S.; Zhou, W. Application of the fiber-optic distributed temperature sensing for monitoring the liquid level of producing oil wells. *Measurement* **2014**, *58*, 130–137. [CrossRef]
8. Feo, G.; Sharma, J.; Kortukov, D.; Williams, W.; Ogunsanwo, T. Distributed Fiber Optic Sensing for Real-Time Monitoring of Gas in Riser during Offshore Drilling. *Sensors* **2020**, *20*, 267. [CrossRef]
9. Shannon, K.; Li, X.; Wang, Z.; Cheeke, J.D.N. Mode conversion and the path of acoustic energy in a partially water-filled aluminum tube. *Ultrasonics* **1999**, *37*, 303–307. [CrossRef]
10. Rodríguez-Olivares, N.A.; Cruz-Cruz, J.V.; Gómez-Hernández, A.; Hernández-Alvarado, R.; Nava-Balanzar, L.; Salgado-Jiménez, T.; Soto-Cajiga, J.A. Improvement of Ultrasonic Pulse Generator for Automatic Pipeline Inspection. *Sensors* **2018**, *18*, 2950. [CrossRef]
11. Dong, L.; Hu, Q.; Tong, X.; Liu, Y. Velocity-Free MS/AE Source Location Method for Three-Dimensional Hole-Containing Structures. *Engineering* **2020**, *6*, 827–834. [CrossRef]
12. Dong, L.; Tong, X.; Ma, J. Quantitative Investigation of Tomographic Effects in Abnormal Regions of Complex Structures. *Engineering* **2021**, *7*, 1011–1022. [CrossRef]
13. Vutukury, J.N.; Donnell, K.M.; Hilgedick, S. Microwave sensing of sand production from petroleum wells. In Proceedings of the 2015 IEEE International Instrumentation and Measurement Technology Conference (I2MTC) Proceedings, Pisa, Italy, 11–14 May 2015; pp. 1210–1214.
14. Tayyab, M.; Sharawi M. S.; Al-Sarkhi, A. A Radio Frequency Sensor Array for Dielectric Constant Estimation of Multiphase Oil Flow in Pipelines. *IEEE Sens. J.* **2017**, *17*, 5900–5907. [CrossRef]
15. Pendry, J.B.; Holden, A.J.; Robbins, D.J.; Stewart, W.J. Magnetism from conductors and enhanced nonlinear phenomena. *IEEE Trans. Microw. Theory Tech.* **1999**, *47*, 2075–2084. [CrossRef]
16. Falcone, F.; Lopetegi, T.; Baena, J.D.; Marqués, R.; Martín, F.; Sorolla, M. Effective negative-epsilon stop-band microstrip lines based on complementary split ring resonators. *IEEE Microw. Wireless Compon. Lett.* **2004**, *14*, 280–282. [CrossRef]
17. Naqui, J.; Durán-Sindreu, M.; Martín, F. Alignment and position sensors based on split ring resonators. *Sensors* **2012**, *12*, 11790–11797. [CrossRef]
18. Horestani, A.K.; Fumeaux, C.; Al-Sarawi, S.F.; Abbott, D. Displacement sensor based on diamond-shaped tapered split ring resonator. *IEEE Sens. J.* **2013**, *13*, 1153–1160. [CrossRef]
19. Horestani, A.K.; Abbott, D.; Fumeaux, C. Rotation sensor based on horn-shaped split ring resonator. *IEEE Sens. J.* **2013**, *13*, 3014–3015. [CrossRef]
20. Naqui, J.; Coromina, J.; Karami-Horestani, A.; Fumeaux, C.; Martín, F. Angular displacement and velocity sensors based on coplanar waveguides (CPWs) loaded with S-shaped split ring resonators (S-SRR). *Sensors* **2015**, *15*, 9628–9650. [CrossRef]
21. Choi, H.; Nylon, J.; Luzio, S.; Beutler J.; Porch, A. Design of continuous non-invasive blood glucose monitoring sensor based on a microwave split ring resonator. In Proceedings of the 2014 IEEE MTT-S International Microwave Workshop Series on RF and Wireless Technologies for Biomedical and Healthcare Applications (IMWS-Bio2014), London, UK, 18–20 December 2018; pp. 1–3.
22. Lee C.-S.; Yang, C.-L. Thickness and permittivity measurement in multi-layered dielectric structures using complementary Split-Ring Resonators. *IEEE Sens. J.* **2014**, *14*, 695–700. [CrossRef]
23. Galindo-Romera, G.; Herraiz-Martínez, F.J.; Gil, M.; Martínez-Martínez, J.J.; Segovia-Vargas, D. Submersible printed Split-Ring Resonator-based sensor for thin-film detection and permittivity characterization. *IEEE Sens. J.* **2016**, *16*, 3587–3596. [CrossRef]
24. Vélez, P.; Su, L.; Grenier, K.; Mata-Contreras, J.; Dubuc, D.; Martín, F. Microwave Microfluidic Sensor Based on a Microstrip Splitter/Combiner Configuration and Split Ring Resonators (SRRs) for Dielectric Characterization of Liquids. *IEEE Sens. J.* **2017**, *17*, 6589–6598. [CrossRef]
25. Lee, C.-S.; Yang, C.-L.; Complementary Split-Ring Resonators for measuring dielectric constants and loss tangents. *IEEE Microw. Wireless Compon. Lett.* **2014**, *24*, 563–565. [CrossRef]
26. Kulkarni, S.; Joshi, M.S. Design and analysis of shielded vertically stacked ring resonator as complex permittivity sensor for petroleum oils. *IEEE Trans. Microw. Theory Techn.* **2015**, *63*, 2411–2417. [CrossRef]
27. Lijuan, S.; Mata-Contreras, J.; Vélez, P.; Fernández-Prieto, A.; Martín, F. Analytical Method to Estimate the Complex Permittivity of Oil Samples. *Sensors* **2018**, *18*, 984.
28. Minicircuits ZX95-625+ Voltage Controlled Oscillator Datasheet. Available online: Https://www.minicircuits.com/pdfs/ZX95-625+.pdf (accessed on 11 March 2022).

sustainability

Article

Experimental Study on the Mechanical Characteristics of Saturated Granite under Conventional Triaxial Loading and Unloading Tests

Zelin Liu and Wei Yi *

School of Civil Engineering, Central South University, Changsha 410075, China; zelinl@csu.edu.cn
* Correspondence: yi.wei@csu.edu.cn

Abstract: It is essential to study the mechanical properties of saturated rock under different loading and unloading paths for strength calculation, safety assessment and disaster prevention; however, current literature rarely mentions conventional triaxial loading and unloading conditions. To analyze the mechanical properties, strain energy evolution characteristics and failure mode, a series of conventional triaxial unloading tests (with axial loading rate v_a of 0.06–6 mm/min and circumferential unloading rate v_u of 0.1–10 MPa/s) and conventional triaxial compression tests were carried out on saturated granite. The test results showed that the damage sources of specimens in the conventional triaxial unloading test were mainly related to circumferential deformation, while in the conventional triaxial compression test, it was related to the axial deformation. Under the same v_a, the confining pressure and axial stress at the failure point decreased with the increase of v_u, and the stress coordinate of the failure point was located outside the conventional triaxial compression envelope of $\sigma_1-\sigma_3$. As v_u increases, except for the variation of circumferential strain energy ΔU_c decreasing slowly, the trend of strain energy changes must be determined together with v_a. As v_a increases, the relationship between the magnitude of each energy changes from $\Delta U_a > \Delta U > \Delta U_d > \Delta U_e > \Delta U_c$ to $\Delta U_d > \Delta U_a > \Delta U > \Delta U_e > \Delta U_c$, while the change of dissipated energy is dominated by v_u and v_a together to become dominated by v_a. In addition, with the increase of v_u and v_a, the damage pattern of the specimen also changes from shear damage in a single shear plane to mixed damage with tensile strain failure and shear plane during which the dilation angle of the specimen increases in total except for v_u = 10 MPa/s, v_a = 0.6 mm/min and 6 mm/min.

Keywords: saturated granite; loading and unloading tests; failure mode; energy evolution; dilatancy angle

Citation: Liu, Z.; Yi, W. Experimental Study on the Mechanical Characteristics of Saturated Granite under Conventional Triaxial Loading and Unloading Tests. *Sustainability* **2022**, *14*, 5445. https://doi.org/10.3390/su14095445

Academic Editor: Giovanna Pappalardo

Received: 2 March 2022
Accepted: 29 April 2022
Published: 30 April 2022

Publisher's Note: MDPI stays neutral with regard to jurisdictional claims in published maps and institutional affiliations.

1. Introduction

With the expansion of granite mining to deeper mines and the construction of large hydropower stations with high slopes, the stability of granite excavation under high stress has become an essential issue in engineering [1–4]. The final failure state of granite depends not only on the stress state of the rock, but also on the stress path and loading rate as well as geothermal, groundwater and other environmental factors [5–8]. There are apparent differences in the mechanical properties of granite under the loading and unloading paths, which is one of the crucial reasons for the lack of uniformity in the laws obtained from the current unloading test studies [9,10]. Since the underground deposits are excavated and unloaded, studying the rock damage under the unloading path may seem more meaningful than relying on the loading test [11].

The excavation of underground rocks is divided into blasting or mechanical crushing [12–14]. Accordingly, the rocks are damaged under different unloading rates. Many scholars have studied the mechanical properties, failure modes, and energy dissipation characteristics of rocks under different unloading rates, and the specific objects and stress paths are shown in Table 1. This paper mainly discusses the conventional triaxial rather than the true triaxial [15]. In [16], the author first proposed the virtual uniaxial compressive strength considering the unloading problem. He found that the Hoek–Brown material parameters of the rock specimens before the peak load after unloading changed more than the loading test. The virtual uniaxial compressive strength increased, and the Hoek–Brown material constant m decreased. Before the peak load, a larger unloading rate responds to a smaller axial strain and cohesion c and a larger lateral strain and a slight increase in the angle of internal friction φ. In contrast, the deformation modulus increases gradually after the peak load and decreases rapidly with the increasing unloading rate [13,14]. The stress adjustment hysteresis mainly affected the unloading rate, and additional unloading stress affects the rock strength, c and φ values. Higher unloading rates may lead to more dramatic damage. The Hoek–Brown criterion can predict the damage-confining pressure when the rock is unloaded laterally at lower unloading rates but may overestimate the stress value at higher unloading rates [17]. In addition, some authors [18–21] have also further analyzed the damage evolution law of rocks under unloading conditions after high temperatures. Furthermore, it is worth mentioning that some researchers [22–25] have employed theoretical or numerical methods to evaluate the characteristics of deeply buried rock masses.

During the triaxial unloading test, the rock's two failure modes are the tensile–shear failure and the shear failure, and the volume strain changes from compression to expansion [9]. In addition, the higher the initial confining pressure, the more severe the failure of the specimen at the same unloading stress path, and the degree of rupture of the specimen becomes more complicated as the unloading rate increases [26]. The failure evolution characteristics vary for different types of rock [27–29]. For example, shale [27] exhibits obvious elastic–plastic characteristics under conventional triaxial compression tests, while it shows apparent elastic–brittle features under triaxial unloading tests, and the brittle failure characteristics increase with the increase of unloading rate and initial confining pressure. Columnar jointed rock masses [30] exhibit strong volume expansion during unloading, which becomes more severe and the damage pattern becomes more complex with an increasing unloading rate, while unloading relaxation can be observed by decreasing the unloading rate. The later the inflection point of negative volume strain growth occurs when the unloading level is closer to the peak load, the damage caused by tensile cracking in the specimen is more severe than the damage caused by compression shear during unloading. In general, the corresponding expansion rate decreases with increasing confining compression at the same axial stress and increases with increasing axial stress at the same confining compression [10]. The expansion boundary of the unloading test starts from the unloading point, which is different from the expansion rate of the uniaxial and triaxial compression tests. Rapid unloading promotes the growth of cracks, and larger inelastic strains appear at lower unloading rates [31,32], which will increase the permeability of the rock [33–36]. In addition, the response of the ratio of height to diameter [37] and the acoustic emission behaviour [38] are also investigated, and the damage process can be observed by CT technique [39] and nuclear magnetic resonance technique [40,41].

From the view of energy conversion, the energy changes from the three principal stress directions in the test acted together to damage the rock. Wang [42] conducted triaxial unloading tests and post-test CT scan analysis on fine-grained marble and found that the specimens' total energy, elastic energy and dissipated energy almost all increased with the increase of deformation, and the elastic energy and dissipated energy decreased slightly before increasing again. After the unloading point, the dissipated energy increased sharply, and the elastic energy increment rate lowered; the crack pattern and energy dissipation and release in CT images depend on the unloading rate and time. The strain energy absorbed in the axial direction of the rock is mainly transformed into circumferential dilation to consume strain energy [43]. The degree of dilation is: reduce axial stress and confining pressure > maintain axial stress and reduce confining pressure > increase axial stress and reduce confining pressure. In contrast, it is transformed into less dissipated energy, and the dissipated energy increases significantly only near the time of destruction. The dissipation energy is significantly influenced by the unloading path and the initial confining pressure.

The initial confining pressure has a considerably more significant influence on the axial strain energy, the circumferential expansion strain energy and the elastic strain energy than the unloading path; all of them increase approximately linearly with the increase of the initial confining pressure. This also indicates that the dissipated strain energy determines the damage during the unloading test. In contrast, the damage during the triaxial compression test is mainly determined by the released elastic strain energy [44–46]. In addition, the magnitude of the initial confining pressure and the unloading rate have significant effects on the strain energy conversion, rock burst and limited storage energy [11], while at the same initial confining pressure and unloading rate, the variation of the axial stress has little effect on the ultimate storage energy of the rock. The higher the unloading rate, the smaller the ultimate storage energy.

As one of the most common rocks in underground rock engineering, granite is usually saturated in deep water-bearing environments. However, the existing literature mainly focuses on studying specimens in their natural state. This study conducted conventional triaxial tests with different loading and unloading paths on the saturated granite. The unloading rate varies between 0.1–10 MPa/s to study the mechanical properties and the evolution of strain energy of the surrounding rock under blasting excavation and static excavation in underground tunnels. It is of great significance for understanding the mechanism of unloading rockburst occurrence in high-stress rock masses and even the permeability characteristics of the engineering surrounding rock masses.

Table 1. Summary of the conventional triaxial unloading tests.

Rock Type	Stress Path	σ_3 (MPa)	Unloading Rate v_u (MPa/s)	Reference
Sandstone	1. Increase σ_1, unload σ_3	4–19	0.02–0.14	[47]
		4–10	0.05	[48]
		5–30	0.005	[49]
		15–30	0.05, 0.1, 0.2	[50]
		15–45	2	[51]
	2. Keep σ_1, unload σ_3	15–30	0.0003–0.1667	[52]
	1. Increase σ_1, unload σ_3 2. Keep σ_1, unload σ_3	10–30	0.1, 0.5	[53]
Marble	3. Unload σ_1, σ_3	10–30	0.0008	[33]
	1. Increase σ_1, unload σ_3	40	0.05	[37]
	2. Keep σ_1, unload σ_3	20	0.01–0.2	[40,42]
		10–60	0.01–1	[54]
	3. Unload σ_1, σ_3	20–60	0.05	[55]
		20–40	0.1–10	[11]
	1. Increase σ_1, unload σ_3 3. Unload σ_1, σ_3	20–40	0.26-1.28	[26]
Granite	1. Increase σ_1, unload σ_3	5–30	0.0017–0.0333	[56]
		30–60	0.05	[57]
	2. Keep σ_1, unload σ_3	10	0.005–0.0115	[19]
		10–30	0.1	[17,58]
	1. Increase σ_1, unload σ_3 3. Unload σ_1, σ_3	10–60	0.2	[45]
	1. Increase σ_1, unload σ_3 2. Keep σ_1, unload σ_3 3. Unload σ_1, σ_3	10–30	-	[43]
Mudstone	1. Increase σ_1, unload σ_3	6-15	0.05, 0.1, 0.2	[50]
		10–50	0.005–0.5	[59]
Limestones	2. Keep σ_1, unload σ_3	15	0.0217	[39]
Shale	1. Increase σ_1, unload σ_3	20–60	0.4–1.0	[27]
	2. Keep σ_1, unload σ_3	20–60	0.4	[27]

Table 1. *Cont.*

Rock Type	Stress Path	σ_3 (MPa)	Unloading Rate v_u (MPa/s)	Reference
Basalt	1. Increase σ_1, unload σ_3	3–12	0.008	[60]
	3. Unload σ_1, σ_3	3–12	0.008	[60]
		30	0.001–0.005	[34]
Coal	1. Increase σ_1, unload σ_3	4–8	0.012–0.024	[35]
		4–10	0.02–0.14	[47]
Dacite	2. Keep σ_1, unload σ_3	5–15	0.0083	[44]
Rock salt	2. Keep σ_1, unload σ_3	15–25	0.005	[10]
		23	0.001–0.5	[31]
Mudstone	2. Keep σ_1, unload σ_3	8–20	0.05	[61]
	3. Unload σ_1, σ_3	8–20	0.05	[61]
Rock-Like	2. Keep σ_1, unload σ_3	4–8	0.0083–0.0333	[30]

2. Materials and Methods

2.1. Specimen Preparation

As shown in Figure 1, all the granite specimens were taken from a quarry in Mianning district, Sichuan Province, China, about 400 km from the capital of Sichuan Province. The dimensions of the specimens were φ50 mm $\times$ h100 mm, and all specimens were drilled and cut in the same direction from a single rock mass to ensure minimum variability between different specimens. In addition, the specimens were carefully smoothed to meet the requirements of the International Society of Rock Mechanics (ISRM) [62], and the specimens were soaked in tap water for 48 h to reach a saturated state. Meanwhile, all operations were performed at room temperature. The XRD test result indicated that mineral compositions of the specimens in nature are quartz (49%), potassium feldspar (35%), black mica (8%), white mica (5%), apatite (2%) and others (1%). The SEM test demonstrated the specimens had a dense microstructure. The physical parameters were particle size ranging from 0.05–4 mm, average dry density of 2630 kg/m^3, the average saturated density of 2635 kg/m^3 and average p-wave velocity of 4027 m/s. The basic mechanical parameters were: uniaxial compressive strength was 104.93 MPa, Brazilian tensile strength was 4.53 MPa, cohesion was 21.83 MPa, and internal friction angle was 49.82°.

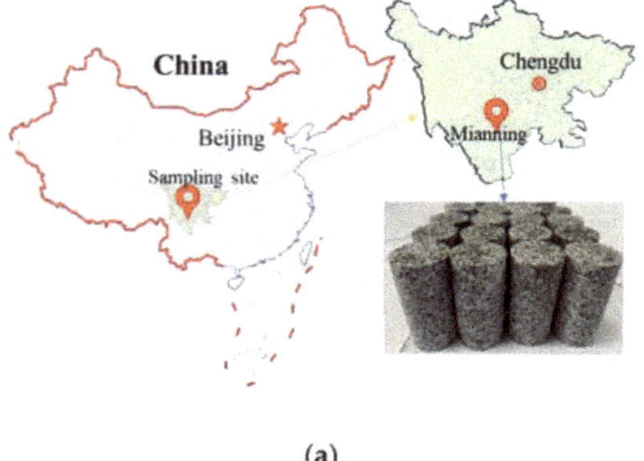

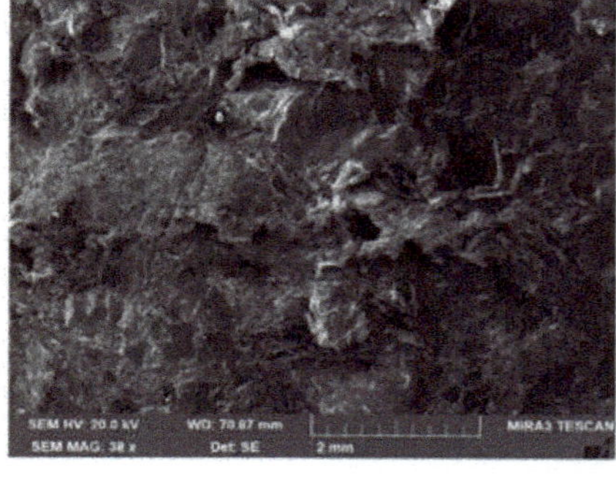

(a)

(b)

Figure 1. Geographical location and microstructure of the specimens: (**a**) specimens; (**b**) SEM image.

2.2. Test Scheme

As shown in Figure 2, the test was performed by the MTS815 testing machine, and the test path was generally divided into three paths, Path-I, Path-II and Path-III. As shown in Table 2, for Group-I, i.e., Path-I, conventional triaxial compression test, we first loaded σ_1

and σ_3 at 0.5 MPa/s to the corresponding initial hydrostatic pressure O′ with the help of the confining pressure system in Figure 2a. Four confining pressures of 0, 10, 30 and 50 MPa were selected to obtain the failure envelope under the conventional triaxial compression test. The specimen was then loaded in the axial direction at a 0.12 mm/min rate along the O′A path until the specimen was destroyed. For Group-II, the conventional triaxial unloading test, which includes Path-II and Path-III, the initial confining pressure was selected as 50 MPa, which corresponds to the ground stress at a burial depth of about 2000 m, and the difference from Group-I was that the confining pressure started unloading when the axial load reached 80% (O″) of the triaxial compression strength, the unloading rates were selected as 0.1, 1, 5 and 10 MPa/s, respectively. Meanwhile, the axial loading of different specimens was carried out at a rate of 0.06, 0.6, and 6 mm/min, until the specimen failed.

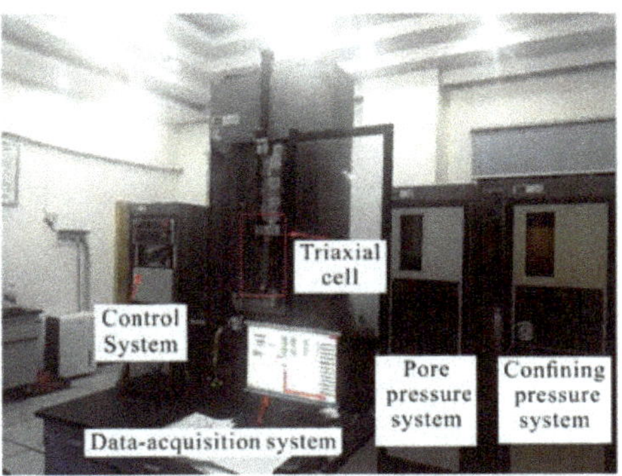

(a)

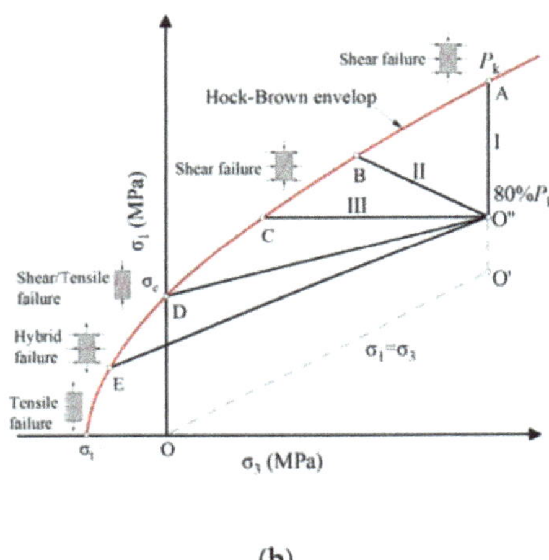

(b)

Figure 2. Test machine and scheme: (**a**) MTS815 test machine; (**b**) unloading paths.

Table 2. Conventional triaxial loading and unloading test scheme.

Test Scheme	Specimen Number	Initial Confining Pressure (MPa)	Axial Loading Rate	Unloading Rate of σ_3
Group-I	TC-0	0		-
	TC-10	10	0.12 mm/min	-
	TC-30	30		-
	TC-50	50		-
Group-II	G-0.06-X *	50	0.06 mm/min	0.1 MPa/s,
	G-0.6-X	50	0.6 mm/min	1 MPa/s,
	G-6-X	50	6 mm/min	5 MPa/s, 10 MPa/s

* X is the value of the axial loading rate.

3. Results and Discussion

3.1. Mechanical Characteristics

3.1.1. Stress–Strain Curve

Figure 3 shows the relationship between the axial strain ε_a, the circumferential strain ε_c, the body strain ε_v and the principal stress σ_1 for the saturated granite in the triaxial unloading test. In the unloading stage, the change rate of strain values increased significantly with the increase of the unloading rate, even though there was not much correlation with the final strain value, while the change of axial stress values was closely related to the axial loading rate v_a and the unloading rate v_u. When $v_a = 0.06$ mm/min, the principal stress σ_1 decreased gradually with the ε_a increase in the unloading stage. In particular, the unloading rate obviously affected the strain value; the strain value of the specimen changed rapidly when $v_u > 0.1$ MPa/s. At $v_a = 0.6$ mm/min, σ_1 gradually increased with ε_a except for $v_u = 0.1$ MPa/s and then gradually decreased with the increase of v_u at the rest of v_u; at $v_a = 6$ mm/min, σ_1 gradually increased with the increase of ε_a, and the magnitude of increase increased with the increase of v_u. The increased range decreased with the increase of v_u, until $v_u = 10$ MPa/s. When it was converted to σ_1, σ_1 increased first and then decreased with the increase of the ε_a. In addition, the ε_c-σ_1 and ε_v-σ_1 curves have the same trend as the ε_v-σ_1 curve.

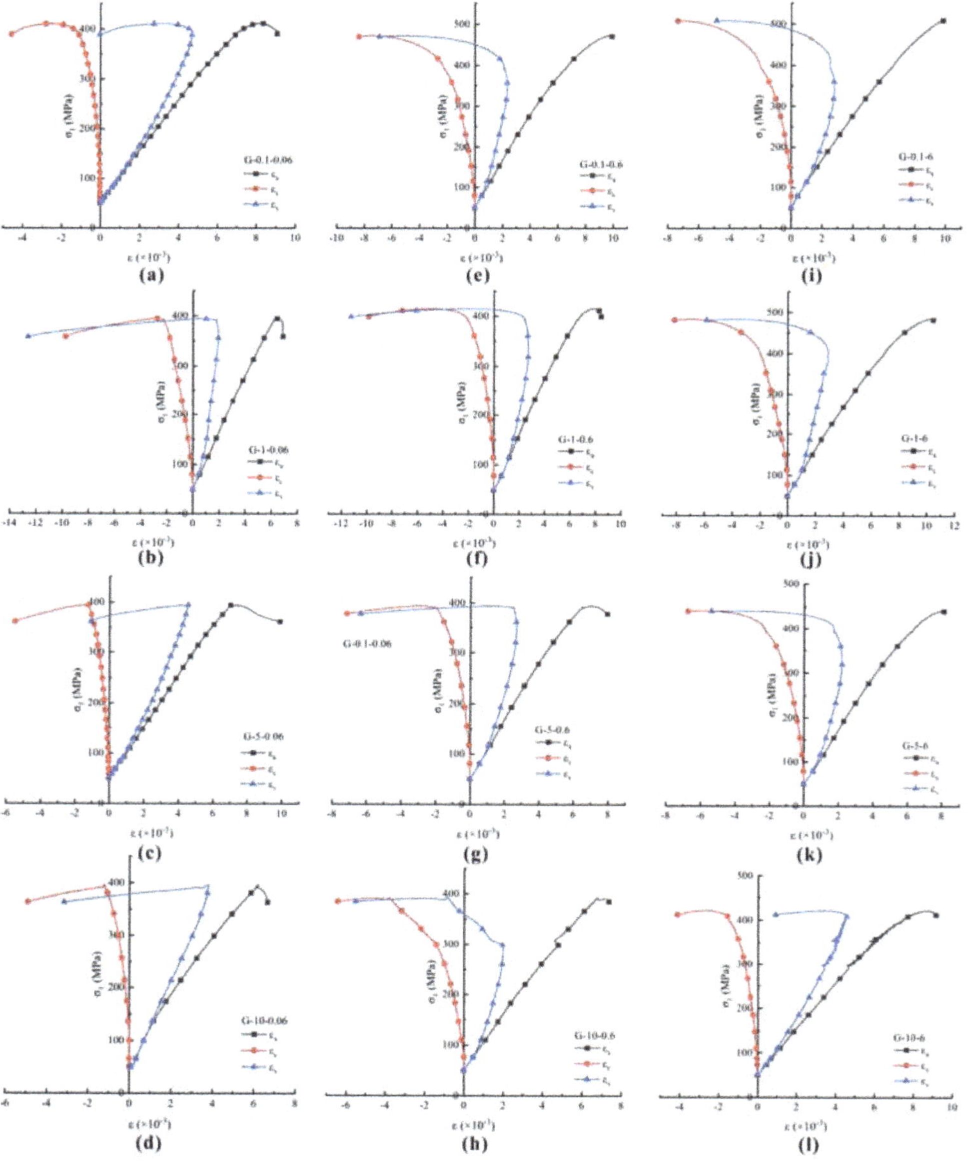

Figure 3. Stress−strain curves of saturated granite under a conventical triaxial unloading test with different v_u: (**a**−**d**) $v_a = 0.06$ mm/min; (**e**−**h**) $v_a = 0.6$ mm/min; (**i**−**l**) $v_a = 6$ mm/min.

3.1.2. Strength Characteristics

Figure 4 illustrates the variation of $\sigma_1-\sigma_3$ of all specimens before damage in the unloading test, in which the red scribed line is the conventional triaxial compression strength fitting line of saturated granite specimens under 0, 10, 30 and 50 MPa confining pressure. Combined with Table 3, it can be seen that in the triaxial unloading test, all the failure points (σ_1, σ_3) fall to the left side of the fitting line of the triaxial loading test except for G-0.6-1. This indicates that the bearing capacity of the specimen was improved to some extent under the unloading test, and this phenomenon is more obvious in Path-III. The main reason is that during the unloading test, the circumferential of the rock specimen has a significant expansion. When the stress change rate of the unloading rate is greater than that of the axial loading, for example, G-0.06-10, the confining pressure decreases quickly. The axial pressure decreases at a rate lower than the amount of the confining pressure; therefore, a smaller σ_3 and a relatively larger σ_1 are obtained. When the change rate of σ_3 is less than σ_1, for example, G-6-0.1, the specimen is loaded at a relatively larger rate in the axial direction. This disguisedly increases the specimen's strength. Only when the rate of stress change due to the unloading rate is similar to that of axial loading, i.e., G-0.6-1, is the axial loading rate of the specimen lower than the rate of conventional triaxial compression. Therefore, the strength is less than the conventional triaxial compression strength fitting line value.

In addition, when the ratio of axial loading rate to unloading rate $a \leq 0.001$ mm/MPa, the test paths correspond to Path-III, i.e., keep σ_1 reducing σ_3. When the ratio of the axial loading rate to the unloading rate $a \leq 0.02$ mm/MPa, the test paths are corresponding to Path-II and are closer to Path-III, when $a > 0.1$ mm/MPa, the test paths are corresponding to Path-II and are closer to Path-I. This phenomenon is more and more significant with the increase of a.

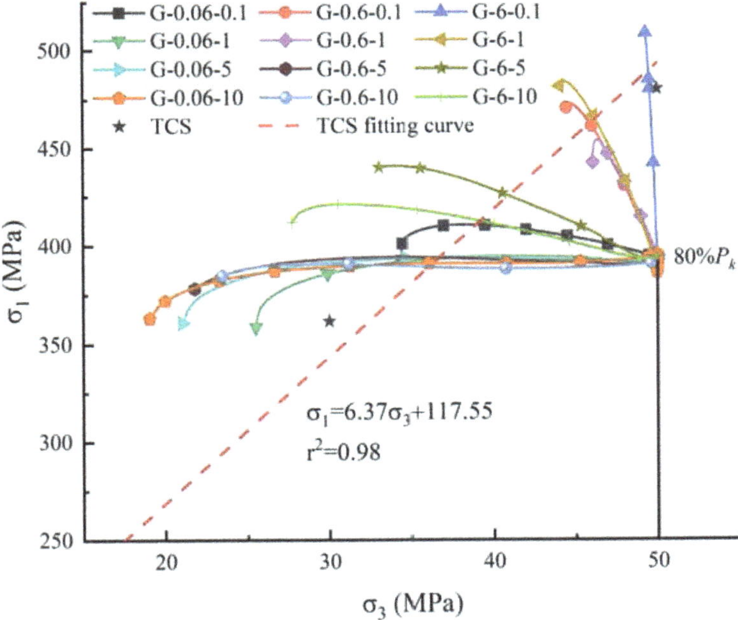

Figure 4. $\sigma_1-\sigma_3$ curves of saturated granite during an unloading test.

Table 3. Saturated granite's physical and mechanical parameters under conventional triaxial loading and unloading tests.

Test Scheme	Specimen No.	Dry Density (kg/m^3)	Saturated Density (kg/m^3)	σ_3^0 (MPa)	σ_3^f (MPa)	σ_1^f (MPa)	E (GPa)	ν
Group I	TC-0	2637.49	2644.55	0	0	104.93	43.04	0.10
	TC-10	2631.32	2637.24	10	10	202.99	42.67	0.10
	TC-30	2635.72	2641.16	30	30	361.84	53.96	0.11
	TC-50	2635.67	2639.43	50	50	480.32	50.57	0.12
Group II	G-0.06-0.1	2635.13	2638.36	50	34.42	390.08	51.4	0.10
	G-0.06-1	2637.42	2641.45	50	25.53	358.82	58.04	0.28
	G-0.06-5	2635.50	2641.38	50	21.06	361.00	51.71	0.11
	G-0.06-10	2638.97	2643.01	50	18.77	354.14	62.64	0.18
	G-0.6-0.1	2632.97	2637.00	50	44.48	470.63	58.39	0.17
	G-0.6-1	2651.75	2655.78	50	28.83	398.89	59.08	0.23
	G-0.6-5	2638.89	2642.93	50	21.81	378.21	59.03	0.21
	G-0.6-10	2636.85	2640.88	50	23.47	384.67	56.05	0.20
	G-6-0.1	2632.28	2636.32	50	49.31	508.61	60.45	0.24
	G-6-1	2602.96	2606.98	50	43.97	465.05	56.32	0.18
	G-6-5	2620.01	2624.07	50	32.07	400.16	62.90	0.18
	G-6-10	2592.90	2596.95	50	27.74	412.31	53.09	0.26

3.1.3. Strain Characteristics

Table 4 lists the strain axial strain ε_a^0, circumferential strain ε_c^0 and body strain ε_v^0 of the specimen at the unloading point and the strain axial strain ε_a^f, circumferential strain ε_c^f and body strain ε_v^f at the failure point, and the difference in body strain $\Delta\varepsilon_v^f$ between the failure and unloading points. The difference in circumferential strain $\Delta\varepsilon_c$ and body strain $\Delta\varepsilon_v$ between the failure point and unloading points are shown in Figure 5. It was found that the axial strain of the specimen in conventional triaxial compression was the leading cause of specimen failure. Specifically, the axial strain difference $\Delta\varepsilon_a = \varepsilon_a^f - \varepsilon_a^0$ of the specimen gradually increased with the increase of confining pressure, while the circumferential strain difference $\Delta\varepsilon_c = \varepsilon_c^f - \varepsilon_c^0$ did not change much. The body strain difference $\Delta\varepsilon_v = \varepsilon_v^f - \varepsilon_v^0$ gradually increases with increasing circumferential pressure, except for uniaxial compression. Where $\Delta\varepsilon_a/\Delta\varepsilon_c = 15.94\%$, the specimen mainly occurs in circumferential tensile strain failure. Under confining pressure, $\Delta\varepsilon_a/\Delta\varepsilon_c$ gradually increases with increasing circumferential pressure from 21.12% to 50.24%. At this time, the confining pressure has a good restraint effect on the circumferential deformation. The circumferential deformation is the leading cause of specimens' damage in the triaxial unloading test. At the axial loading rate of 0.06 mm/min, the maximum value of $\Delta\varepsilon_a/\Delta\varepsilon_c$ is only 33.37% (G-0.06-5), and the minimum value of $\Delta\varepsilon_a/\Delta\varepsilon_c$ is 4.23%, i.e., G-0.06-1, which is the minimum value in all the tests. At the axial loading rate of 6 mm/min, the axial loading rate is five times the loading rate of the conventional triaxial compression test, which causes the axial deformation of the specimen to be significantly enhanced, with the maximum value of $\Delta\varepsilon_a/\Delta\varepsilon_c$ even reaching 35.69% at G-6-10.

The stress state achieved by unloading the confining compression is equivalent to superimposing a circumferential tensile stress on the original stress state, resulting in a significant circumferential expansion of the specimen. In plasticity theory, the dilatancy angle ψ is usually used to characterize the inelastic volume change. As suggested by Vermeer [63], the dilatancy angle ψ can be expressed as Equation (1):

$$\psi = \arcsin\left(\frac{\Delta\varepsilon_v^p}{\Delta\varepsilon_v^p - 2\Delta\varepsilon_a^p}\right) \tag{1}$$

where $\Delta\varepsilon_a^p$ and $\Delta\varepsilon_v^p$ are the axial and volumetric plastic strain increments, respectively, and can be calculated by Equations (2)–(4).

$$\varepsilon_a^e = \frac{\sigma_1 - 2v\sigma_3}{E} \tag{2}$$

$$\varepsilon_v^e = \frac{(1 - 2v)\sigma_1 - 2(1 - v)\sigma_3}{E} \tag{3}$$

$$\varepsilon_a^p = \varepsilon_a - \varepsilon_a^e, \varepsilon_v^p = \varepsilon_v - \varepsilon_v^e \tag{4}$$

Table 4. Strains at the unloading point and the failure point of saturated granite specimens.

Specimen No.	σ_3^0 (MPa)	Unloading Point			Failure Point			$\Delta\varepsilon_v^f$ ($\times 10^{-3}$)
		ε_a^0 ($\times 10^{-3}$)	ε_c^0 ($\times 10^{-3}$)	ε_v^0 ($\times 10^{-3}$)	ε_a^f ($\times 10^{-3}$)	ε_c^f ($\times 10^{-3}$)	ε_v^f ($\times 10^{-3}$)	
TC-0	0	2.19	−0.14	1.91	2.92	−2.43	−1.94	−3.85
TC-10	10	3.55	−0.49	2.57	4.53	−2.81	−1.09	−3.66
TC-30	30	5.17	−0.97	3.23	7.38	−3.70	−0.02	−3.25
TC-50	50	8.16	−1.84	4.48	10.24	−3.91	2.42	−2.06
G-0.06-0.1	50	7.14	−1.22	4.70	9.06	−4.55	−0.05	−4.75
G-0.06-1		6.26	−2.30	1.67	6.89	−9.75	−12.60	−14.27
G-0.06-5		7.07	−1.28	4.51	9.83	−5.49	−1.09	−5.60
G-0.06-10		6.12	−1.24	3.64	6.67	−5.66	−4.65	−8.29
G-0.6-0.1	50	6.55	−2.24	2.07	9.90	−8.39	−6.88	−8.95
G-0.6-1		6.56	−2.07	2.42	8.43	−9.84	−11.25	−13.67
G-0.6-5		6.53	−2.04	2.46	7.94	−7.16	−6.37	−8.83
G-0.6-10		6.78	−3.86	−0.94	7.34	−6.44	−5.53	−4.59
G-6-0.1	50	6.46	−1.99	2.49	9.85	−7.35	−4.84	−7.33
G-6-1		6.74	−1.95	2.85	10.59	−10.05	−9.50	−12.35
G-6-5		6.19	−2.25	1.68	7.20	−11.92	−16.63	−18.31
G-6-10		7.23	−1.43	4.38	9.15	−4.12	0.91	−3.47

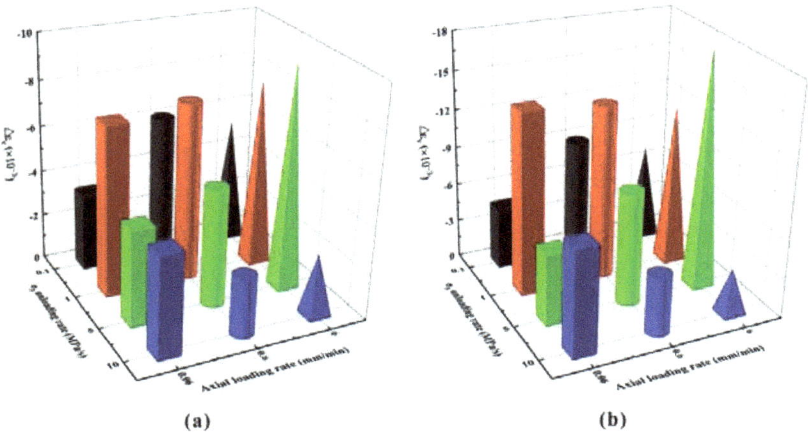

Figure 5. Strain difference between failure point and unloading point of saturated granite during a conventional triaxial unloading test: (**a**) circumferential strain difference; (**b**) volume strain difference.

The evolution of the dilatancy angle ψ with the normalized plastic shear strain increment $\Delta\gamma^p / \Delta\gamma^p_{max}$ from the beginning of unloading to the ultimate bearing strength under different unloading rates is given in Figure 6. For the influence of the loading rate on the dilatancy angle at the axial loading rate $v_a = 0.06$ mm/min, the change in the dilatancy angle ψ of the specimens was not significantly related to the unloading rate at $\Delta\gamma^p / \Delta\gamma^p_{max} < 0.2$, while at $\Delta\gamma^p / \Delta\gamma^p_{max} \geq 0.2$, the dilatancy angle ψ increased slowly overall with the increase of $\Delta\gamma^p / \Delta\gamma^p_{max}$. The size of the dilatancy angle was closely related to the initial dilatancy angle and less related to the unloading rate. Finally, the dilatancy angle exhibits a small decrease as the plastic shear strain $\Delta\gamma^p / \Delta\gamma^p_{max}$ approaches 1. At $v_a = 0.6$ mm/min, the ψ of the specimen slowly decreases with the increase of plastic shear strain when the unloading rate v_u is 10 MPa/s and increases first before $\Delta\gamma^P/\Delta\gamma^P{}_{max} = 0.45$, then decreases when v_u is 0.1 MPa/s. It slowly increases with the increase of plastic shear strain at the unloading rate of 1 MPa/s and 5 MPa/s. From the late stage of damage, the ψ of the specimen will gradually increase with the increase of the v_u except for G-6-10. When the axial loading rate $v_a = 6$ mm/min, the ψ of the specimen increases slowly with the increase of the plastic shear strain, except for $v_u = 10$ MPa/s. At this time, a higher unloading rate will correspond to a larger ψ of the specimen. At $v_u = 10$ MPa/s, the ψ of the specimen decreases rapidly with the increase of the plastic shear strain. It is noteworthy that at an unloading rate of 1 MPa/s, the ψ of the specimen decreases with an increasing axial loading rate, while at an unloading rate of 5 MPa/s, the ψ of the specimen increases with an increasing axial loading rate. This also indicates that slight plastic damage in the unloading process can cause a high expansion process, which is the reason for the expansion deformation in the annulus caused by the unloading stress path.

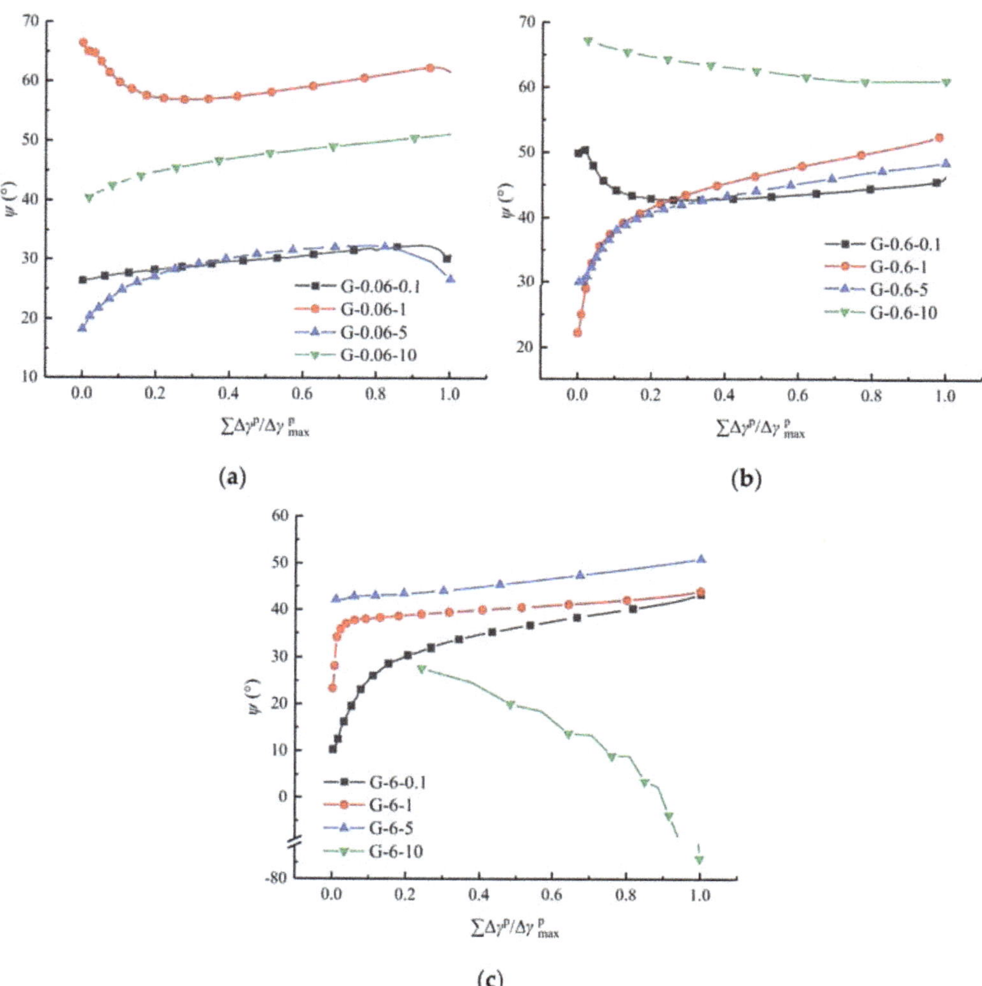

Figure 6. The evolution of the dilatancy angle ψ under triaxial unloading tests: (**a**) $v_a = 0.06$ mm/min; (**b**) $v_a = 0.6$ mm/min; (**c**) $v_a = 6$ mm/min.

3.2. Energy Evolution Characteristics

The total energy U, axial strain energy U_a, circumferential strain energy U_c, elastic energy U_e and dissipation energy U_d of the specimen during the tests can be calculated from the following Equations (5)–(8) [43], and the calculated results are presented in Table 5.

$$U = U_a + U_c = U_e + U_d \tag{5}$$

$$U_a = \int_0^{\varepsilon_a^t} \sigma_1 d\varepsilon_a = \sum_{i=1}^{n} \frac{1}{2}\left(\sigma_1^i + \sigma_1^{i+1}\right)\left(\varepsilon_a^{i+1} - \varepsilon_a^i\right) \tag{6}$$

$$U_c = 2\int_0^{\varepsilon_c^t} \sigma_3 d\varepsilon_c = \sum_{i=1}^{n}\left(\sigma_3^i + \sigma_3^{i+1}\right)\left(\varepsilon_c^{i+1} - \varepsilon_c^i\right) \tag{7}$$

$$U_e = \frac{1}{2E_u^t}\left[\sigma_1^2 + 2\sigma_3^2 - 2\mu_u^t\left(2\sigma_1\sigma_3 + \sigma_3^2\right)\right] \tag{8}$$

where n is the total number of segments of the stress–strain curve, i is the segmentation points, and the segment interval time is 0.1 s; E_u^t and μ_u^t are the unloaded elastic modulus and Poisson's ratio at time t, respectively.

Table 5. Energy under conventional triaxial loading and unloading test (unit: MJ/m^3).

Specimen No.	σ_3^0 (MPa)	Unloading Point					Failure Point				
		U_a^0	U_c^0	U^0	U_e^0	U_d^0	U_a^f	U_c^f	U^f	U_e^f	U_d^f
TC-0	0	-	-	-	-	-	0.16	-	-	0.13	0.03
TC-10	10	-	-	-	-	-	0.91	0.03	0.94	0.55	0.39
TC-30	30	-	-	-	-	-	2.55	0.14	2.70	1.74	0.95
TC-50	50	-	-	-	-	-	4.88	0.24	5.11	3.35	1.77
G-0.06-0.1	50	1.66	−0.06	1.60	1.32	0.28	2.44	−0.19	2.25	1.47	0.79
G-0.06-1		1.43	−0.11	1.32	1.02	0.29	1.68	−0.34	1.34	0.73	0.60
G-0.06-5		1.63	−0.06	1.56	1.30	0.26	2.69	−0.17	2.51	1.56	0.96
G-0.06-10		1.46	−0.06	1.39	1.11	0.28	1.67	−0.16	1.50	0.97	0.53
G-0.6-0.1	50	1.53	−0.11	1.42	1.08	0.34	3.02	−0.40	2.62	1.59	1.53
G-0.6-1		1.51	−0.10	1.40	1.11	0.30	2.27	−0.37	1.90	1.11	1.51
G-0.6-5		1.53	−0.10	1.43	1.10	0.32	2.08	−0.24	1.84	1.19	1.53
G-0.6-10		1.57	−0.19	1.37	0.96	0.41	1.79	−0.26	1.52	1.11	1.57
G-6-0.1	50	1.51	−0.10	1.41	−0.34	1.74	3.07	−0.36	2.71	−1.34	1.51
G-6-1		1.57	−0.10	1.47	1.16	0.31	3.31	−0.46	2.85	1.58	1.57
G-6-5		1.44	−0.11	1.33	1.01	0.32	1.90	−0.45	1.45	0.67	1.44
G-6-10		1.69	−0.07	1.62	1.32	0.30	2.49	−0.16	2.33	1.67	1.69

3.2.1. Evolution of Strain Energy in the Triaxial Compression Test

From the energy statistics in Table 5, it can be seen that the confining pressure σ_3 performs positive work in the conventional triaxial compression test to stop the destruction of the rock specimen, while in the conventional triaxial unloading test, the σ_3 does negative work to help the rock destruction. Meanwhile, $U_a > U_c$, and $U_e > U_d$. In the conventional triaxial compression test, the relationship between energy and confining pressure is shown in Equation (9), where U_c and U_d increase approximately linearly with the increase of the confining pressure, while U, U_a and U_e increase quadratically with the increase of the confining pressure.

$$\begin{cases} U = 0.0007\sigma_3^2 + 0.0646\sigma_3 + 0.1831 & r^2 = 0.9999 \\ U_a = 0.0007\sigma_3^2 + 0.0595\sigma_3 + 0.1904 & r^2 = 0.9996 \\ U_c = 0.0049\sigma_3 + 0.0085 & r^2 = 0.9946 \\ U_e = 0.0005\sigma_3^2 + 0.0374\sigma_3 + 0.1274 & r^2 = 0.9999 \\ U_d = 0.0341\sigma_3 + 0.0185 & r^2 = 0.9933 \end{cases} \tag{9}$$

3.2.2. Evolution of Strain Energy in a Triaxial Compression Unloading Confining Pressure Test

The variation of each energy in the conventional triaxial unloading confining pressure test is shown in Figure 7. It can be found that the relationship of each energy throughout the test is $U_a > U > U_d > U_e > U_c$, while the change of energy after the unloading point becomes more complicated.To explore the change characteristics, we define $\Delta U_\alpha = U_\alpha{}^f - U_\alpha{}^0$, which represents different energy types, and thus obtain the relationship between different unloading rates and the amount of energy change under the same axial loading rate as shown in Figure 8. It can be found that when the axial loading rate is 0.06 mm/min and 0.6 mm/min, the relationship between various energy changes is roughly $\Delta U_a > \Delta U > \Delta U_d > \Delta U_e > \Delta U_c$. When the axial loading is 0.6 mm/min, the relationship between various energy changes is $\Delta U_d > \Delta U_a > \Delta U > \Delta U_e > \Delta U_c$. The reason for this phenomenon is that when the axial loading rate is 0.06 mm/min and 0.6 mm/min, except for the unloading rate of 0.1 MPa/s and the axial loading rate of 0.6 mm/min, the σ_1 of the specimens is slowly reduced from the unloading point under the rest of the loading rate. The confining pressure has been doing negative work from the unloading point, so $\Delta U_a > \Delta U > \Delta U_d > \Delta U_e > \Delta U_c$, due to the reduction of the confining pressure. The bearing capacity of the specimen is gradually reduced, and part of the elastic energy U_e has been transformed into dissipated energy; U_{d1} was released. In addition, the dissipated energy U_{d2} released during the unloading process increases ΔU_d to a point greater than the change of elastic energy ΔUe. Furthermore, the circumferential strain energy U_c that has been doing negative work enables $U_a < U_d$, and finally, the relationship between each energy change amount is obtained by Equation (5). It is worth noting that at the axial loading rate of 0.6 mm/min with the unloading rate of 5 MPa/s and 10 MPa/s, the relationship of each energy change is consistent with that of the axial loading rate of 6 mm/min, i.e., $\Delta U_d > \Delta U_a > \Delta U > \Delta U_e > \Delta U_c$.

When the axial loading rate v_a is 6 mm/min, despite the unloading rate of the confining pressure gradually increasing, the growth of σ_1 slowly decreases, and σ_1 gradually increases overall, the axial stress increases rapidly, the confining pressure does negative work. As a result, ΔU_c becomes the smallest, and the amount of change gradually decreases with the increase of the unloading rate. The axial stress increases rapidly, the restriction of the circumferential direction becomes smaller and smaller. The dissipation energy increases sharply, and ΔU_d is the largest. From Figure 8c, it can be seen that the amount of change of ΔU_d is not much on the whole, i.e., it is not much correlated with the unloading rate, which also means that with the increase of the axial loading rate, the dissipation energy is determined from the original by the unloading rate and the axial loading rate to the axial loading rate alone. ΔU_a is slightly lower than ΔU_d, while the elastic energy change ΔU_e has a slight increase. The ΔU_a and ΔU decrease gradually with an increasing unloading rate, while ΔU_c decreases until the unloading rate is 1 MPa/s and increases slowly after that.

With the increase of v_u, the variation of circumferential energy ΔU_c tends to decrease slowly, independent of the axial loading rate. While the variation of the rest of the energy at $v_a = 0.06$ mm/min and $v_a = 0.6$ mm/min shows inverted V-shaped and V-shaped changes, respectively. Moreover, when $v_a = 6$ mm/min, except for the variation of total energy ΔU, axial energy ΔU_a decreases with the increase of the unloading rate, and the variation of the rest of the energy has increased in different degrees.

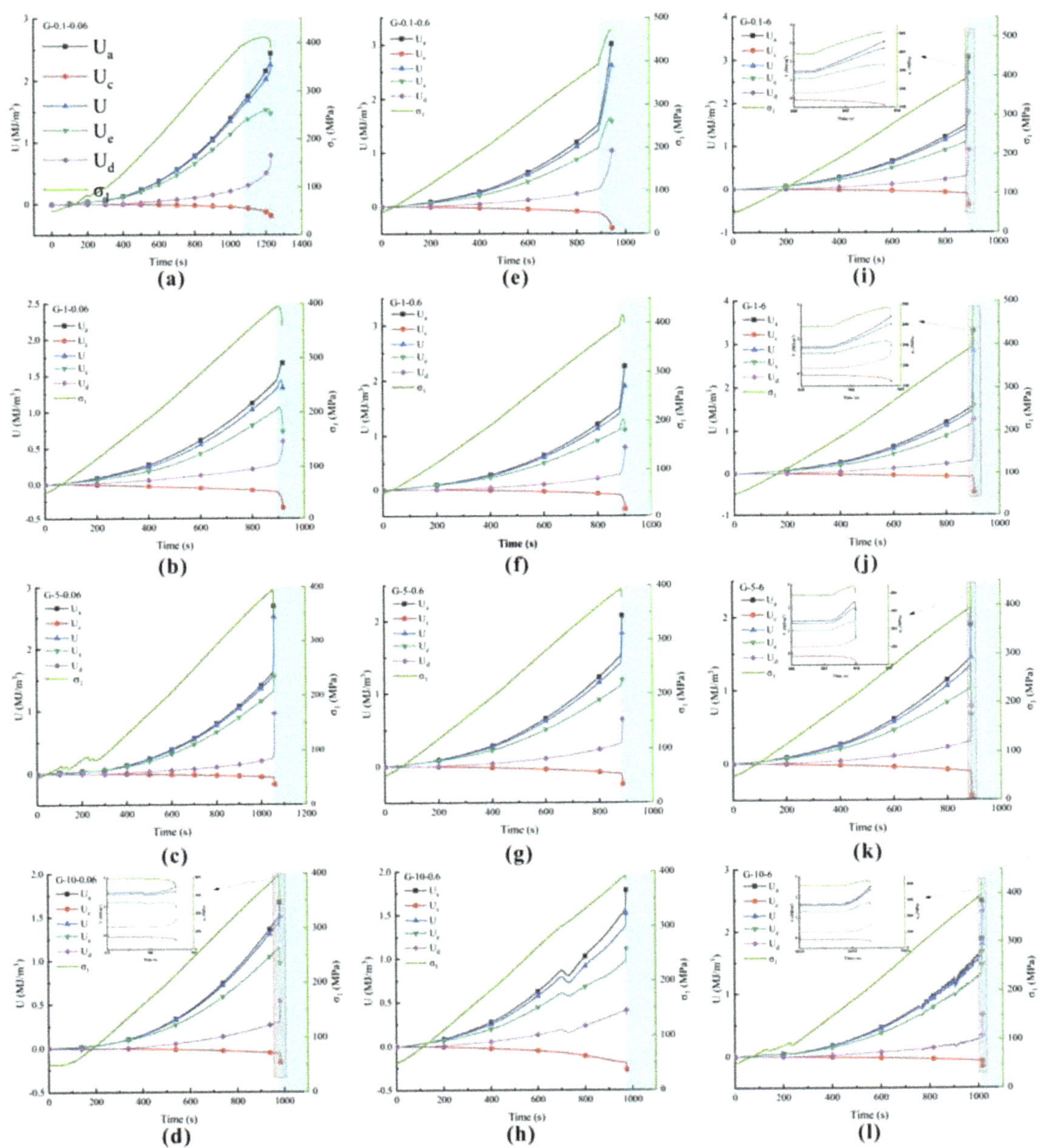

Figure 7. Energy variation under a triaxial compression test with different unloading rates: (**a–d**) axial loading rate 0.06 mm/min; (**e–h**) axial loading rate 0.6 mm/min; (**i–l**) axial loading rate 6 mm/min.

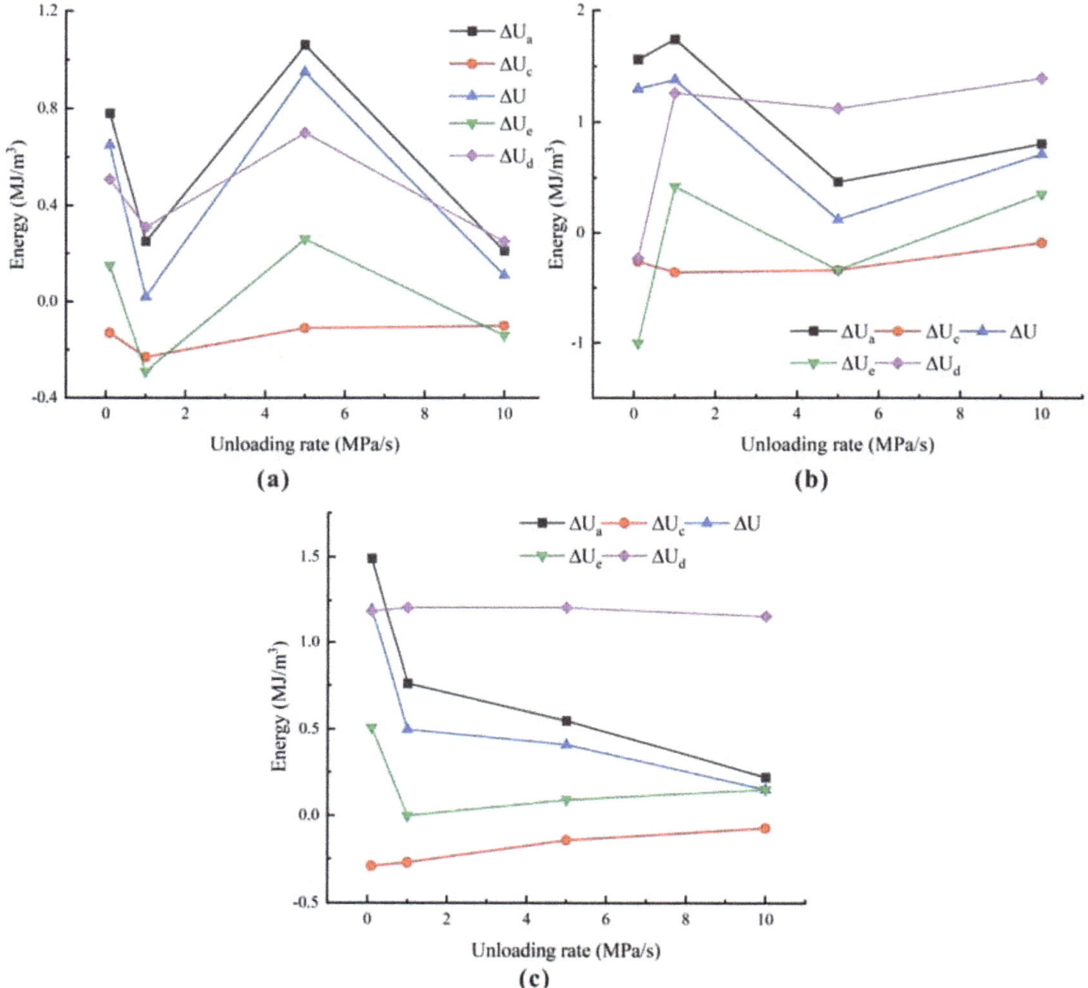

Figure 8. Energy variation at different loading rates and different unloading rates: (**a**) axial loading rate 0.06 mm/min; (**b**) axial loading rate 0.6 mm/min; (**c**) axial loading rate 6 mm/min.

3.3. Rock Failure Characteristics and Destruction Mechanism

The failure patterns of the saturated granite specimens under conventional triaxial unloading tests are shown in Figure 9. As you can see, at the axial loading rate of 0.06 mm/min, the failure surface of the specimen is a single shear failure plane, which is similar to the failure pattern of the specimen under conventional triaxial compression. At the axial loading rate of 0.6 mm/min, the failure surface of the specimen is complex, with a form of multiple shear surface compound. However, when the axial loading rate is 6 mm/min, the specimens show conjugate shear failure accompanied by a few tension strain failure, indicating that before the axial loading rate is 0.6 mm/min, the failure form of the specimen is mainly controlled by the axial stress, while after the axial loading rate of 0.6 mm/min, the failure mode of the specimen is controlled by both the axial stress and the circumferential stress.

Figure 9. Failure patterns of saturated granite specimens under a conventional triaxial unloading test with different unloading rates.

Figure 10 shows the damage evolution process of the granite specimen. The damage to the specimen is attributed to the combined effect of internal microcracks, axial stress and circumferential stress. Due to many microcracks inside the granite specimen, the microcracks have started to crack or tend to crack during the initial triaxial compression stage. As the triaxial unloading test begins, σ_1/σ_3 gradually increases with the unloading of the circumferential stress and the change of the axial stress. Meanwhile, the specimen is gradually compressed in the axial direction. At the same time, the reduction of the lateral constraint of the specimen provides help for the lateral deformation of the specimen. The specimen tended to be compressed axially and expanded laterally. The internal microcracks are cracked and expanded continuously, leading to penetration between step cracks and crack penetration along the σ_1 direction, which eventually leads to the specimen's shear failure and tensile strain failure.

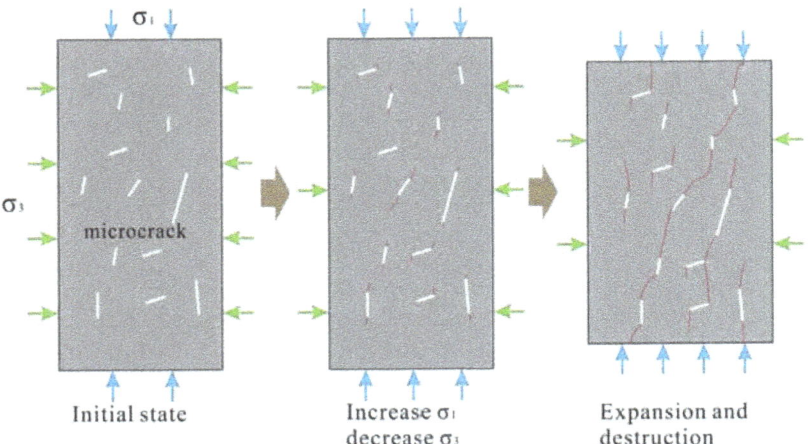

Figure 10. Damage evolution process.

Although a series of unloading tests with different axial loading rates and circumferential unloading rates have been successfully carried out on saturated granite, these tests are all based on conventional triaxial test equipment, which still cannot achieve the three unequal principal stresses in the underground space. Moreover, considering that the explosion is completed instantaneously during blasting excavation, the unloading rate is much larger than the unloading rate that can be achieved in the laboratory. Nevertheless, the damage to the surrounding rock results from the combined effect of axial and lateral stresses after unloading. Our study also provides some references for understanding the geological hazards such as rockburst in deeply buried tunnels. Subsequent research should consider the geological discontinuity and diversity, including single or multiple cracks or different materials included in one sample. Moreover, it is also meaningful to consider more environmental factors such as thermal shock and to study the microscopic characteristics and acoustic emission [64,65] characteristics of the fracture surface of the specimen using SEM and other equipment.

4. Conclusions

(1) In the conventional triaxial unloading test, the circumferential deformation is the leading cause of failure, and its strain is greater than that of conventional triaxial loading. Under the same axial loading rate, the faster the unloading rate and axial loading rate, the smaller the σ_3^f and σ_1^f at the time of failure, and the failure point is outside the envelope of conventional triaxial compression tests. While under the same unloading rate, a faster axial loading corresponds to a larger σ_3^f and σ_1^f.

(2) The variation of circumferential energy ΔU_c decreases with the unloading rate, and ΔU_d, ΔU_a, ΔU and ΔU_e showed inverted V-shaped and V-shaped $v_a = 0.06$ and 0.6 mm/min, respectively. When $v_a = 6$ mm/min, except for a slight increase in ΔU_a, the rest of the variation of energy decreases with the increase of the unloading rate. The variation of dissipation energy changes from being determined by the unloading rate and the axial loading rate together to being determined by the axial loading rate with the axial loading rate increasing.

(3) The failure modes of the specimens were mainly controlled by the axial stress and showed a single shear crack before $v_a = 0.6$ mm/min, while it is influenced by both axial and circumferential stress after the axial loading rate of not less than 6 mm/min, and the failure surface is a conjugate shear failure and tensile strain failure.

(4) In the triaxial unloading test, the specimens were in the high expansion process most of the time, and the dilatancy angle of the specimens showed an overall upward trend except for $v_u = 10$ MPa/s, $v_a = 0.6$ mm/min and 6 mm/min.

Author Contributions: Conceptualization, methodology, investigation, Z.L. and W.Y.; data curation, visualization, Z.L.; writing—original draft preparation, Z.L.; writing—review and editing, W.Y.; supervision, project administration, funding acquisition, W.Y. All authors have read and agreed to the published version of the manuscript.

Funding: The authors are grateful for the financial supports from the Excellent Postdoctoral Innovative Talents Project of Hunan Province, China (No. 2020RC2001).

Institutional Review Board Statement: Not applicable.

Informed Consent Statement: Not applicable.

Data Availability Statement: Data available on request due to privacy restrictions.

Acknowledgments: The authors would like to acknowledge the staff who processed the specimens and helped with testing, and many thanks to the three unknown reviewers for their careful review.

Conflicts of Interest: The authors declare no conflict of interest.

References

1. Li, X.; Gong, F.; Tao, M.; Dong, L.; Du, K.; Ma, C.; Zhou, Z.; Yin, T. Failure mechanism and coupled static-dynamic loading theory in deep hard rock mining: A review. *J. Rock Mech. Geotech. Eng.* **2017**, *9*, 767–782. [CrossRef]
2. Liu, Z.; Ma, C.; Wei, X.A.; Xie, W. Experimental Study of Rock Subjected to Triaxial Extension. *Rock Mech. Rock Eng.* **2022**, *55*, 1069–1077. [CrossRef]
3. Yi, W.; Rao, Q.-H.; Sun, D.-L.; Shen, Q.-Q.; Zhang, J. A new integral equation method for calculating interacting stress-intensity factors of multiple holed-cracked anisotropic rock under both far-field and arbitrary surface stresses. *Int. J. Rock Mech. Min. Sci.* **2021**, *148*, 104926. [CrossRef]

4. Fereidooni, D. Determination of the Geotechnical Characteristics of Hornfelsic Rocks with a Particular Emphasis on the Correlation Between Physical and Mechanical Properties. *Rock Mech. Rock Eng.* **2016**, *49*, 2595–2608. [CrossRef]

5. Gautam, L.; Jain, J.K.; Jain, A.; Kalla, P. Valorization of bone-china ceramic powder waste along with granite waste in self-compacting concrete. *Constr. Build. Mater.* **2022**, *315*, 125730. [CrossRef]

6. Gautam, P.K.; Dwivedi, R.; Kumar, A.; Kumar, A.; Verma, A.K.; Singh, K.H.; Singh, T.N. Damage Characteristics of Jalore Granitic Rocks After Thermal Cycling Effect for Nuclear Waste Repository. *Rock Mech. Rock Eng.* **2021**, *54*, 235–254. [CrossRef]

7. Gautam, P.K.; Verma, A.K.; Sharma, P.; Singh, T.N. Evolution of Thermal Damage Threshold of Jalore Granite. *Rock Mech. Rock Eng.* **2018**, *51*, 2949–2956. [CrossRef]

8. Gautam, P.K.; Jha, M.K.; Verma, A.K.; Singh, T.N. Evolution of absorption energy per unit thickness of damaged sandstone. *J. Therm. Anal. Calorim.* **2019**, *136*, 2305–2318. [CrossRef]

9. Xie, H.Q.; He, C. Study of the unloading characteristics of a rock mass using the triaxial test and damage mechanics. *Int. J. Rock Mech. Min. Sci.* **2004**, *41*, 366. [CrossRef]

10. Chen, J.; Jiang, D.Y.; Ren, S.; Yang, C.H. Comparison of the characteristics of rock salt exposed to loading and unloading of confining pressures. *Acta Geotech.* **2016**, *11*, 221–230. [CrossRef]

11. Huang, D.; Li, Y. Conversion of strain energy in Triaxial Unloading Tests on Marble. *Int. J. Rock Mech. Min. Sci.* **2014**, *66*, 160–168. [CrossRef]

12. Ma, C.; Xie, W.; Liu, Z.; Li, Q.; Xu, J.; Tan, G. A New Technology for Smooth Blasting without Detonating Cord for Rock Tunnel Excavation. *Appl. Sci.* **2020**, *10*, 6764. [CrossRef]

13. Sato, T.; Kikuchi, T.; Sugihara, K. In-situ experiments on an excavation disturbed zone induced by mechanical excavation in Neogene sedimentary rock at Tone mine, central Japan. *Eng. Geol.* **2000**, *56*, 97–108. [CrossRef]

14. Wei, X.A.; Li, Q.; Ma, C.; Dong, L.; Zheng, J.; Huang, X. Experimental investigations of direct measurement of borehole wall pressure under decoupling charge. *Tunn. Undergr. Space Technol.* **2022**, *120*, 104280. [CrossRef]

15. Xiao, F.; Jiang, D.; Wu, F.; Zou, Q.; Chen, J.; Chen, B.; Sun, Z. Effects of prior cyclic loading damage on failure characteristics of sandstone under true-triaxial unloading conditions. *Int. J. Rock Mech. Min. Sci.* **2020**, *132*, 104379. [CrossRef]

16. He, J.; Wang, Z.; Bai, J.; Zhang, L. Study on the Hoek-Brown parameters under loading and unloading triaxial laboratory tests. In Proceedings of the 1st International Conference on Civil Engineering, Architecture and Building Materials (CEABM 2011), Haikou, China, 18–20 June 2011; pp. 2074–2078.

17. Duan, K.; Ji, Y.L.; Wu, W.; Kwok, C.Y. Unloading-induced failure of brittle rock and implications for excavation induced strain burst. *Tunn. Undergr. Space Technol.* **2019**, *84*, 495–506. [CrossRef]

18. Ding, Q.-L.; Ju, F.; Mao, X.-B.; Ma, D.; Yu, B.-Y.; Song, S.-B. Experimental Investigation of the Mechanical Behavior in Unloading Conditions of Sandstone After High-Temperature Treatment. *Rock Mech. Rock Eng.* **2016**, *49*, 2641–2653. [CrossRef]

19. Kang, P.; Jing, Z.; Zou, Q.L.; Xiao, S. Deformation characteristics of granites at different unloading rates after high-temperature treatment. *Environ. Earth Sci.* **2020**, *79*, 343. [CrossRef]

20. Yi, W.; Rao, Q.-H.; Li, Z.; Sun, D.-L.; Shen, Q.-Q. Thermo-hydro-mechanical-chemical (THMC) coupling fracture criterion of brittle rock. *Trans. Nonferrous Met. Soc. China* **2021**, *31*, 2823–2835. [CrossRef]

21. Ma, Y.; Rao, Q.; Huang, D.; Yi, W.; He, Y. A new theoretical model of local air-leakage seepage field for the compressed air energy storage lined cavern. *J. Energy Storage* **2022**, *49*, 104160. [CrossRef]

22. Azarafza, M.; Ghazifard, A.; Akgun, H.; Asghari-Kaljahi, E. Geotechnical characteristics and empirical geo-engineering relations of the South Pars Zone marls, Iran. *Geomech. Eng.* **2019**, *19*, 393–405. [CrossRef]

23. Faradonbeh, R.S.; Taheri, A.; Ribeiro e Sousa, L.; Karakus, M. Rockburst assessment in deep geotechnical conditions using true-triaxial tests and data-driven approaches. *Int. J. Rock Mech. Min. Sci.* **2020**, *128*, 104279. [CrossRef]

24. Fern, E.J.; Di Murro, V.; Soga, K.; Li, Z.L.; Scibile, L.; Osborne, J.A. Geotechnical characterisation of a weak sedimentary rock mass at CERN, Geneva. *Tunn. Undergr. Space Technol.* **2018**, *77*, 249–260. [CrossRef]

25. Yi, W.; Rao, Q.-H.; Luo, S.; Shen, Q.-Q.; Li, Z. A new integral equation method for calculating interacting stress intensity factor of multiple crack-hole problem. *Theor. Appl. Fract. Mech.* **2020**, *107*, 102535. [CrossRef]

26. Huang, R.-Q.; Huang, D. Experimental research on affection laws of unloading rates on mechanical properties of Jinping marble under high geostress. *Chin. J. Rock Mech. Eng.* **2010**, *29*, 21–33.

27. Guo, Y.; Wang, L.; Chang, X. Study on the damage characteristics of gas-bearing shale under different unloading stress paths. *PLoS ONE* **2019**, *14*, e0224654. [CrossRef]

28. Liu, Z.; Ma, C.; Wei, X.A.; Xie, W. Experimental study on mechanical properties and failure modes of pre-existing cracks in sandstone during uniaxial tension/compression testing. *Eng. Fract. Mech.* **2021**, *255*, 107966. [CrossRef]

29. Liu, Z.; Ma, C.; Wei, X.-A. Electron scanning characteristics of rock materials under different loading methods: A review. *Geomech. Geophys. Geo* **2022**, *8*, 80. [CrossRef]

30. Xu, J.; Li, H.; Meng, Q.; Xu, W.; He, M.; Yang, J. A Study on Triaxial Unloading Test of Columnar-Jointed-Rock-Mass-Like Material with AW Velocity Analysis. *Adv. Civ. Eng.* **2020**, *2020*, 6693544. [CrossRef]

31. Han, Y.; Ma, H.; Yang, C.; Li, H.; Yang, J. The mechanical behavior of rock salt under different confining pressure unloading rates during compressed air energy storage (CAES). *J. Pet. Sci. Eng.* **2021**, *196*, 107676. [CrossRef]

32. Cong, Y.; Wang, Z.Q.; Zheng, Y.R.; Zhang, L.M. Effect of Unloading Stress Levels on Macro- and Microfracture Mechanisms in Brittle Rocks. *Int. J. Geomech.* **2020**, *20*, 04020066. [CrossRef]

33. Chen, X.; Tang, C.A.; Yu, J.; Zhou, J.F.; Cai, Y.Y. Experimental investigation on deformation characteristics and permeability evolution of rock under confining pressure unloading conditions. *J. Cent. South Univ.* **2018**, *25*, 1987–2001. [CrossRef]

34. Xue, Y.; Ranjith, P.G.; Gao, F.; Zhang, D.; Cheng, H.; Chong, Z.; Hou, P. Mechanical behaviour and permeability evolution of gas-containing coal from unloading confining pressure tests. *J. Nat. Gas Sci. Eng.* **2017**, *40*, 336–346. [CrossRef]

35. Wang, K.; Zheng, J.-Y.; Zhu, K.-S. Analysis of deformation failure characteristics and energy of anthracite under two kinds of stress paths. *Rock Soil Mech.* **2015**, *36*, 259–266. [CrossRef]

36. Yi, W.; Rao, Q.; Ma, W.; Sun, D.; Shen, Q. A new analytical-numerical method for calculating interacting stresses of a multi-hole problem under both remote and arbitrary surface stresses. *Appl. Math. Mech. Engl. Ed.* **2020**, *41*, 1539–1560. [CrossRef]

37. Chen, J.; Zhou, H.; Zeng, Z.Q.; Lu, J.J. Macro- and Microstructural Characteristics of the Tension-Shear and Compression-Shear Fracture of Granite. *Rock Mech. Rock Eng.* **2020**, *53*, 201–209. [CrossRef]

38. Yang, Y.J.; Zhou, Y.; Ma, D.P.; Ji, H.Y.; Zhang, Y.D. Acoustic Emission Characteristics of Coal under Different Triaxial Unloading Conditions. *Acta Geodyn. Geomater.* **2020**, *17*, 51–60. [CrossRef]

39. Zhou, X.P.; Zhang, Y.X.; Ha, Q.L. Real-time computerized tomography (CT) experiments on limestone damage evolution during unloading. *Theor. Appl. Fract. Mech.* **2008**, *50*, 49–56. [CrossRef]

40. Zhou, K.P.; Liu, T.Y.; Hu, Z.X. Exploration of damage evolution in marble due to lateral unloading using nuclear magnetic resonance. *Eng. Geol.* **2018**, *244*, 75–85. [CrossRef]

41. Wu, H.; Zhu, H.-H.; Zhang, C.-C.; Zhou, G.-Y.; Zhu, B.; Zhang, W.; Azarafza, M. Strain integration-based soil shear displacement measurement using high-resolution strain sensing technology. *Measurement* **2020**, *166*, 108210. [CrossRef]

42. Wang, Y.; Zhao, Q.H.; Xiao, Y.G.; Hou, Z.Q. Influence of time-lagged unloading paths on fracture behaviors of marble using energy analysis and post-test CT visualization. *Environ. Earth Sci.* **2020**, *79*, 217. [CrossRef]

43. Zhao, G.Y.; Dai, B.; Dong, L.J.; Yang, C. Energy conversion of rocks in process of unloading confining pressure under different unloading paths. *Trans. Nonferrous Met. Soc. China* **2015**, *25*, 1626–1632. [CrossRef]

44. Wang, S.; Wang, H.; Xu, W.; Qian, W. Investigation on mechanical behaviour of dacite under loading and unloading conditions. *Geotech. Lett.* **2019**, *9*, 130–135. [CrossRef]

45. Li, D.Y.; Sun, Z.; Xie, T.; Li, X.B.; Ranjith, P.G. Energy evolution characteristics of hard rock during triaxial failure with different loading and unloading paths. *Eng. Geol.* **2017**, *228*, 270–281. [CrossRef]

46. Yi, W.; Rao, Q.; Zhu, W.; Shen, Q.; Li, Z.; Ma, W. Interacting Stress Intensity Factors of Multiple Elliptical-Holes and Cracks Under Far-Field and Arbitrary Surface Stresses. *Adv. Appl. Math. Mech.* **2022**, *14*, 125–154. [CrossRef]

47. Ma, T.R.; Ma, D.P.; Yang, Y.J. Fractal Characteristics of Coal and Sandstone Failure under Different Unloading Confining Pressure Tests. *Adv. Mater. Sci. Eng.* **2020**, *2020*, 2185492. [CrossRef]

48. Yang, R.; Ma, D.P.; Yang, Y.J. Experimental Investigation of Energy Evolution in Sandstone Failure during Triaxial Unloading Confining Pressure Tests. *Adv. Civ. Eng.* **2019**, *2019*, 7419752. [CrossRef]

49. Wang, X.X.; Ma, W.J.; Huang, J.W.; Liao, Z.Y. Mechanical properties of sandstone under loading and unloading Conditions. In Proceedings of the 3rd International Conference on Materials Science and Engineering (ICMSE 2014), Jiujiang, China, 24–26 January 2014; pp. 441–446.

50. Chen, X.J.; Li, L.Y.; Wang, L.; Qi, L.L. The current situation and prevention and control countermeasures for typical dynamic disasters in kilometer-deep mines in China. *Saf. Sci.* **2019**, *115*, 229–236. [CrossRef]

51. Liu, X.-R.; Liu, J.; Li, D.-L.; He, C.-M.; Wang, Z.-J.; Xie, Y.-K. Experimental research on the effect of different initial unloading levels on mechanical properties of deep-buried sandstone. *Rock Soil Mech.* **2017**, *38*, 3081–3088. [CrossRef]

52. Dong, X.H.; Yang, G.S.; Tian, J.F.; Rong, T.L.; Jia, H.L.; Liu, H. Characteristics of deformation properties of frozen sandstone under lateral unloading condition. *Rock Soil Mech.* **2018**, *39*, 2518–2526. [CrossRef]

53. Xue, J.H.; Du, X.H.; Ma, Q.; Zhan, K.L. Experimental study on law of limit storage energy of rock under different confining pressures. *Arab. J. Geosci* **2021**, *14*, 62. [CrossRef]

54. Qiu, S.; Feng, X.; Zhang, C.Q.; Zhou, H.; Sun, F. Experimental research on mechanical properties of deep-buried marble under different unloading rates of confining pressures. *Chin. J. Rock Mech. Eng.* **2010**, *29*, 1807–1817.

55. Li, X.; Zhao, H.; Wang, B.; Xiao, T. Mechanical properties of deep-buried marble material under loading and unloading tests. *J. Wuhan Univ. Technol. Mater. Sci. Ed.* **2013**, *28*, 514–520. [CrossRef]

56. Wang, C.S.; Zhou, H.W.; He, S.S.; Wang, Z.H.; Liu, J.F. Effect of unloading rates on strength of Beishan granite. *Rock Soil Mech.* **2017**, *38*, 151–157. [CrossRef]

57. Wang, C.; Liu, J.; Chen, L.; Liu, J.; Wang, L. Mechanical behaviour and damage evolution of Beishan granite considering the transient and time-dependent effects of excavation unloading. *Eur. J. Environ. Civ. Eng.* **2020**, 1–17. [CrossRef]

58. Peng, C.; Wang, J.; Liu, H.; Li, G.; Zhao, W. Investigating the Macro-Micromechanical Properties and Failure Law of Granite under Loading and Unloading Conditions. *Adv. Civ. Eng.* **2021**, *2021*, 9983427. [CrossRef]

59. Huang, X.; Liu, Q.S.; Liu, B.; Liu, X.W.; Pan, Y.C.; Liu, J.P. Experimental Study on the Dilatancy and Fracturing Behavior of Soft Rock Under Unloading Conditions. *Int. J. Civ. Eng.* **2017**, *15*, 921–948. [CrossRef]

60. Xu, X.; Huang, R.; Li, H.; Huang, Q. Determination of Poisson's Ratio of Rock Material by Changing Axial Stress and Unloading Lateral Stress Test. *Rock Mech. Rock Eng.* **2015**, *48*, 853–857. [CrossRef]

61. Ren, Q.Y.; Zhang, H.M.; Liu, J.S. Rheological properties of mudstone under two unloading paths in experiments. *Rock Soil Mech.* **2019**, *40*, 127–134. [CrossRef]

62. Ulusay, R. *The ISRM Suggested Methods for Rock Characterization, Testing and Monitoring: 2007–2014*; Springer International Publishing: Berlin, Germany, 2014.
63. Vermeer, P.A.; De Borst, R. Non-associated plasticity for soils, concrete and rock. *HERON* **1984**, *29*, 3–64.
64. Liu, Z.; Ma, C.; Wei, X.A.; Xie, W. Experimental Study on the Mechanical Characteristics of Single-Fissure Sandstone Under Triaxial Extension. *Rock Mech. Rock Eng.* **2022**, *2020*, 9374352. [CrossRef]
65. Srinivasan, V.; Gupta, T.; Ansari, T.A.; Singh, T.N. An experimental study on rock damage and its influence in rock stress memory in a metamorphic rock. *Bull. Eng. Geol. Environ.* **2020**, *79*, 4335–4348. [CrossRef]

Article

Study on the Distribution Law of Coal Seam Gas and Hydrogen Sulfide Affected by Abandoned Oil Wells

Xiaoqi Wang [1,2], Heng Ma [1,2,*], Xiaohan Qi [1,2], Ke Gao [1,2] and Shengnan Li [1,2]

1   College of Safety Science and Engineering, Liaoning Technical University, Huludao 125105, China; 472010101@stu.lntu.edu.cn (X.W.); qixiaohan@lntu edu.cn (X.Q.); gaoke@lntu.edu.cn (K.G.); 472010092@stu.lntu.edu.cn (S.L.)
2   Key Laboratory of Mine Thermo-Motive Disaster and Prevention, Ministry of Education, Huludao 125105, China
*   Correspondence: maheng@163.com; Tel.: +86-138-4189-2792

Abstract: This paper is devoted to solving the problem of how to comprehensively control coal seam gas and hydrogen sulfide in the mining face, distributed from the coal seam in abandoned oil wells in coal mining resource areas. The abandoned oil wells of Ma tan 30 and Ma tan 31 in the No. I0104$_1$05 working face of the Shuang Ma Coal Mine were taken as examples. Through parameter testing, gas composition analysis, field investigation at the source distribution, and the influence range of gas and hydrogen sulfide in coal seam in the affected range of the abandoned oil wells were studied. The results show that the coal-bearing strata in Shuang Ma coal field belong to the coal–oil coexistence strata, and the emission of H_2S gas in the local area of the working face is mainly affected by closed and abandoned oil wells. Within the influence range of the abandoned oil wells, along the direction of the working face, the concentration of CH_4 and H_2S gas in the borehole increases as you move closer to the coal center, and the two sides of the oil well show a decreasing trend. In the affected area of the abandoned oil well, the distribution of the desorption gas content in coal seam along the center distance of the oil well presents a decreasing trend in power function, particularly the closer the working face is to the center of the oil well. The higher the concentration of CH_4 and H_2S, the lower the concentration when the working face moves further away from the oil well. The influence radius of CH_4 and H_2S gas on the coal seam in the affected area of Ma tan 31 abandoned oil well is over 300 m. The results provide a theoretical basis for further understanding the law of gas and hydrogen sulfide enrichment in the mining face and the design of treatment measures within the influence range of abandoned oil wells.

Keywords: abandoned oil well; gas treatment; hydrogen sulfide; coal-bearing strata; coexistence of kerosene

Citation: Wang, X.; Ma, H.; Qi, X.; Gao, K.; Li, S. Study on the Distribution Law of Coal Seam Gas and Hydrogen Sulfide Affected by Abandoned Oil Wells. *Energies* **2022**, *15*, 3373. https://doi.org/10.3390/en15093373

Academic Editors: Longjun Dong, Adam Smoliński and Amparo López Jiménez

Received: 22 March 2022
Accepted: 29 April 2022
Published: 5 May 2022

1. Introduction

As more and more producing oilfields in the world enter the mature stage, abandoned oil wells have become a problem that the world needs to deal with urgently. Oil drilling activities that have lasted for more than a century have left great hidden dangers. Millions of abandoned oil and gas wells scattered in many countries have become the most serious environmental challenge in the world [1]. Taking the United States, Canada, and China as examples, the United States does not have specific statistics on abandoned oil and gas wells, but, according to the EPA (Environmental Protection Agency, Washington, DC, USA), there are about 3 million in California, Arizona, New Mexico, and Texas alone. Abandoned oil and gas wells pose a serious threat to local water sources, soil, and wildlife. An unknown number of aging, unplugged, and abandoned oil and gas wells are rampantly releasing methane and other greenhouse gases across the continental United States, bringing toxins to the surface, polluting groundwater and the surrounding ecosystem. Furthermore, these oil and gas wells are spread all over North America [2]. EPA estimates that the annual emissions of methane leaking from abandoned oil wells in the United States are about 263,000 tons, equivalent to the carbon emissions of more than five coal-fired power plants. According to Columbia University in the United States, methane emissions from abandoned oil and gas wells in the United States are as high as 280,000 tons per year, equivalent to the carbon dioxide emissions of 2.1 million cars. More than half of Canada's oil and gas wells have been out of service for more than a decade, Canada's energy regulator said, with a more than 50 per cent increase in idle wells in the province between 2015 and 2021, with 6014 more expected to be idled in 2022. The surrounding environment of abandoned oil wells is full of hidden dangers, which increases the difficulty and cost of treatment of abandoned oil wells in various countries [3].

China's abandoned oil wells are increasing year by year. According to incomplete statistics, the number of abandoned oil wells increased by 25,119 from 2005 to 2010, with an annual growth rate of 19.7%; by 2021, the total number of abandoned oil wells will reach 150,000. In northwest China, there are numerous mining areas with superimposed resources and a large amount of coal, oil, natural gas and other resources existing in the area [4].

Fossil fuel companies around the world emit a large amount of toxic and harmful gases into the atmosphere every year, destroying the natural and ecological climate environment on which human beings depend [5]. The production of sulfur dioxide, carbon monoxide, soot, radioactive dust, nitrogen oxides, carbon dioxide and other substances during the combustion of fossils will directly harm human health, cause human cancer and cause radiation damage to organisms. Fossil fuels contain sulfur, which will produce toxic sulfur dioxide when burned, causing acid rain and damaging the ecological balance. Burning fossil fuels also produces carbon particles, which pollute the air and cause respiratory diseases. Burning fossil fuels also releases carbon dioxide, which eventually creates a greenhouse gas effect that causes sea level rise and threatens human habitation. Global warming will induce certain diseases, threatening the safety of people, especially the elderly. Frequent extreme climates, such as droughts, floods, storms and heat waves, threaten the living environment of animals and plants, increase the mortality rate, disability rate and infectious disease rate, and lead to climate change in some areas, thereby affecting the production and living environment of people in the area. At the same time, it also increases social psychological pressure [6].

Due to the early development of oil, natural gas and other resources, there are many abandoned oil wells in coal mining areas, and there is a large amount of water, gas, oil, etc. in the oil wells, which brings major safety hazards to the mining of mines. Among them, gas and hydrogen sulfide are the main toxic and harmful gases in the process of oil and gas well field exploitation, especially hydrogen sulfide, which, as a highly toxic gas, is seriously harmful [7]. The Shen fu coalfield in the Ordos basin is the largest coalfield proven in China and the largest integrated gas field proven on land. The mining area is rich in coal, oil, natural gas, and rock salt resources [8]. The coexistence of coal resources and oil exists in the Ma jia tan mining area in the western margin of the Ordos Basin [9]. There are also symbiosis phenomena of coal resources and oil in coal mining areas such as Huangling in Shanxi and Yao jie in Gansu [10]. Due to the early development of oil and natural gas resources, there are many abandoned oil and gas wells in overlapping resource areas. These abandoned oil and gas wells have brought great difficulties to the mining design of coal mines. It not only causes a waste of resources but also has significant safety hazards for the mining of coal seams adjacent to oil and gas wells [11].

Once the excavation roadway exposes the oil and gas wells, there will be significant safety hazards such as water, gas, and oil pouring into the working face, which seriously restricts coal mines' safe and efficient production [12]. The Shuangma Wellfield overlaps with the "Hu jianshan oil and gas exploration in Shaanxi, Ningxia, and Mongolia Ordos Basins" block of Petro-China. Shuangma coal mine is located in the overlapping area of coal and oil exploitation. There are 170 abandoned oil wells in the minefield [13].

In the past ten years, ten abandoned oil wells have affected the mining face. Many harmful gases such as gas and hydrogen sulfide have gradually escaped into the coal seams in the oil wells [14,15]. The concentration of gas and hydrogen sulfide in the affected area of abandoned oil wells suddenly increases, bringing great hidden dangers to the safe mining of the mining face of the mine [16,17]. The sudden emission of hydrogen sulfide in the affected area of abandoned oil wells causes not only personal injury but also has explosive and corrosive coal mine equipment and monitoring and testing facilities, and other hazards [18]. Hydrogen sulfide is a standard toxic and harmful gas in coal mines [19,20]. In response to the severe harm of hydrogen sulfide in coal mines, domestic and foreign research on prevention and control technologies for coal mine hydrogen sulfide ventilation and dilution, coal extraction, spraying of alkaline solution, and coal pre-injection of alkaline absorption liquid have been carried out [21,22]. At the same time, excellent results have been achieved. However, only a few studies on the law of hydrogen sulfide enrichment in coal mines, especially the occurrence of toxic and harmful gases such as gas and hydrogen sulfide within the influence range of abandoned oil wells, have been carried out. Only some scholars have carried out preliminary research on oil well plugging and coal seam hydrogen sulfide treatment [23–25]. In this regard, the author researches the source and distribution of gas and hydrogen sulfide under the influence of oil wells in Shuangma coal mine and the scope of power to provide a theoretical basis for the design of gas and hydrogen sulfide treatment measures within the influence scope of abandoned oil wells. Moreover, the research theory of this paper provides technical support for the coordinated exploitation of coal and oil, reduces the potential safety hazards in the process of coal mining near oil and gas wells, and ensures the safe, environmentally friendly and efficient development and production of coal, oil, natural gas and other mineral resources.

2. Analysis of Gas Sources

2.1. Gas Source Statistics during Face Excavation

According to on-site statistics, since the Shuangma coal mine was produced, there have been no abandoned oil wells in or near the I0104$_1$02 working face and I0104$_1$03 working face that have been stopped. There has never been H_2S in the working face during the tunneling and mining period. The first occurrence of H_2S in the mining face is the return air lane of the I0104$_1$04 working face, the return air lane of the I0104$_1$05 working face, the transportation lane I0104$_1$05 working face, and the return air lane of the I0104$_1$06 working face. According to the mining layout of the mine, this area is mainly the abandoned oil well of Ma tan 31, with a strike distance of 460 m. The relationship between the driving face and the H_2S gushing position is shown in Figure 1.

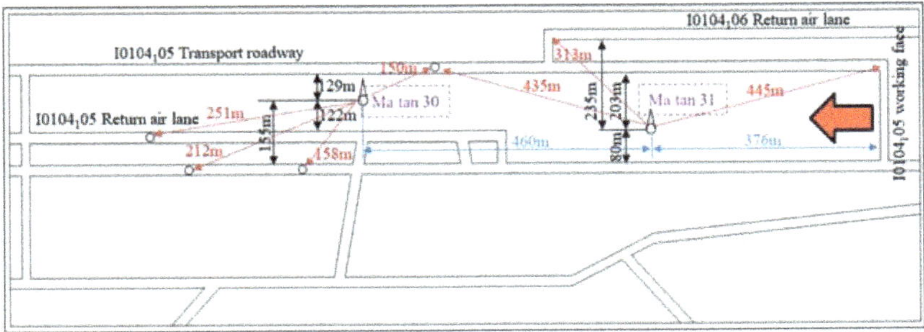

Figure 1. Emission position of H_2S on driving face.

According to the H_2S gushing position relationship shown in Figure 1, the H_2S gushing situation during the excavation of the working face is counted, as shown in Table 1. According to the statistical results shown in Table 1, H_2S will gush out only when the areas near the abandoned oil wells of Ma tan 30 and Ma tan 31 are operated, and the closer the distance to the abandoned oil well, the greater the concentration of H_2S gushing from the tunneling face [26].

Table 1. Statistics of H_2S emission at different positions of the tunneling face.

Location and Location	Gushing Gas Volume Fraction/10^{-6}	Actual Site Situation
I0104$_1$04 return air tunnel excavation, 404–508 m away from no. 5 connecting lane, 212 m away from ma tan 30 oil well	–	Local roof fissure oil seepage
I0104$_1$04 return air tunnel excavation, 158 m away from ma tan 30 oil well	3.5~15.0	The smell of rotten eggs in the lane
I0104$_1$05 return air tunnel excavation, 331 m away from no. 5 connecting lane, 251 m away from ma tan 30 oil well	1.5~50.0	Smell of rotten eggs
I0104$_1$05 transport lane excavation, 627 m away from no. 2 ventilation measures lane, 150,435 m away from ma tan 30 and 31 oil wells	60~80	The volume fraction of H_2S gushing out at 445 m is 45×10^{-6}
I0104$_1$06 excavation of return air lane, 296 and 313 m away from ma tan 30 and ma tan 31 oil wells	60~80	H_2S gushes out during tunneling

2.2. Analysis of Gas Composition in Detection Boreholes

Taking the abandoned oil well of Ma tan 31 as an example, the center of the oil well is 80 m (vertical distance) from the return airway of the I0104$_1$05 working face, 203 m (vertical distance) from the transportation lane, and the distance to the open-cut is 376 m, passing through 4-1 coal seam, 4-2 coal seam and 4-3 coal seam from top to bottom. When section II of the return air lane of the I0104$_1$05 working face was excavated to a position of 1050 m away from the No.5 connecting lane, two detection boreholes were constructed on the working face (the K_1 borehole and K_2 borehole were, respectively, 85 and 60 m away from the center of the abandoned oil well of Ma tan 31), The drilling layout is shown in Figure 2. Gas samples were then collected in the holes for gas detection and analysis. At the same time, a detection borehole was constructed in the return airway of the I0104$_1$05 working face (the borehole is 41 m south of the abandoned oil well in Ma tan 31), and gas samples were collected in the borehole for detection and analysis to obtain the gas component. The measurement results are shown in Table 2.

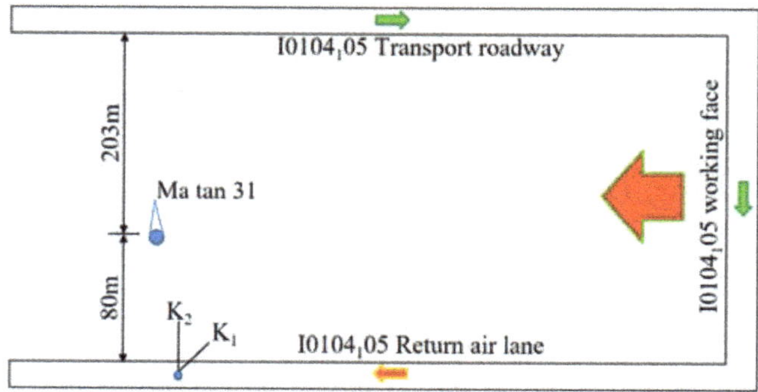

Figure 2. Schematic diagram of detection drilling layout.

Table 2. Detecting results of borehole gas detection in return airway of I0104$_1$05 working face.

Detection Location	Inclination/(°)	Length/m	Aperture/mm	μ (H$_2$S)/10^{-6}
K$_1$ drilling	7	40	94	38,033.52
K$_2$ drilling	6	35	94	25,599.14
The borehole is located 41 m south of the abandoned oil well of ma tan 31	-	-	-	10,200.00

According to the gas composition analysis results in Table 2, the volume fraction of H$_2$S in borehole K$_1$ is $38{,}033.52 \times 10^{-6}$, and the volume fraction of H$_2$S in borehole K$_2$ is $25{,}599.14 \times 10^{-6}$.

As shown in Table 3, according to the gas analysis results of the borehole 41 m to the south from the abandoned oil well of Ma tan 31, the main harmful gas components in the borehole are CH$_4$, H$_2$S, CO, and C$_2$ alkanes (the H$_2$S gas component has a volume fraction of $10{,}200 \times 10^{-6}$), which is roughly the same as the main gas components (CH$_4$, H$_2$S, CO, C$_2$H$_6$ et al.) commonly found in oil wells.

Table 3. Analysis results of borehole gas 41 m to the south from the abandoned oil well of ma tan 31.

Serial Number	Analysis Project	Analysis Result/%	Serial Number	Analysis Project	Analysis Result/%
1	H$_2$	<0.01	7	COS	<0.005
2	CO$_2$	0.90	8	CH$_4$	75.51
3	O$_2$	1.98	9	C$_2$ alkanes	0.46
4	N$_2$	21.13	10	C$_2$ alkanes	0.01
5	CO	<0.01	11	C$_3$ olefin	<0.01
6	H$_2$S	1.02	12	C$_4$ alkanes	<0.01

Based on further analysis of the literature [27,28], it is concluded that the leading harmful gases in the mining activities affected by abandoned oil wells in the Shuangma coal mine are H$_2$S and CH$_4$, which further shows that H$_2$S at the working face mainly comes from abandoned oil wells and is enriched in a specific range of the affected area of abandoned oil wells.

3. Analysis of the Positional Relationship between Oil Wells and Coal Seams

3.1. Wellbore Structure of Well Field Oil Well

According to relevant research data, the oil wells in the Shuangma mining area include exploration wells and production wells, all of which are vertical wells. The depth of oil wells (oil-bearing layers) is generally 700~900 m [29], and the depth of some oil wells is more than 2000 m, and the casing pressure is about 24~27 MPa. Most oil wells were constructed in the 1970s and 1980s, or even the 1950s and 1960s [30,31]. Most of the surveyed oil boreholes only have coordinate positions and cannot describe the sealing, casing setting, and borehole depth [32]. Figure 3 shows the actual situation of the abandoned Ma tan 31 oil well. It can be determined that the depth of the oil well is relatively small. Abandoned oil wells in the Shuangma coal mine field generally have only one layer of production casing (the depth of the oil well is small), and the surface of the wellhead is covered with surface sealing cement. There is the cement for fixing the casing between the bottom of the case and the cracks in the rock, and there may or may not be plugging cement inside the container. In addition, some of the abandoned oil wells are directly drilled bare, and the construction depth is different.

Due to the old age of the abandoned oil wells, most of the casings in the abandoned oil wells have been damaged [33,34]. After the open hole and casing are damaged, many harmful gases such as hydrogen sulfide and methane that may be stored in the oil well gradually invade the coal seam. Migration channels can be divided into three types, as shown in Figure 4.

Figure 3. Records of underground exposure of Ma tan 31 oil well.

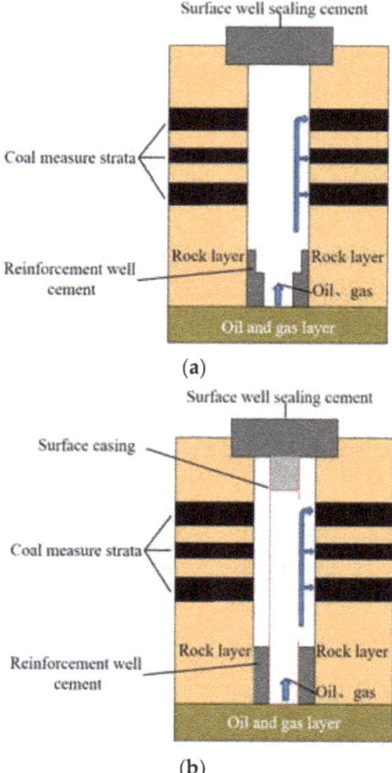

Figure 4. *Cont.*

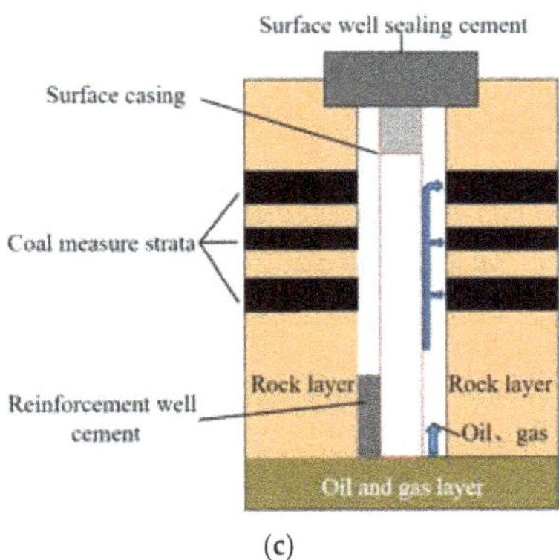

Figure 4. Schematic diagram of the wellbore structure of an abandoned oil well. (**a**) No casing (open hole); (**b**) Broken casing; (**c**) Reinforced wells are not toghtly sealed by ceme.

It can be seen from Figure 4 that after the oil well is abandoned, the well may be filled with harmful gases under pressure. In no casing, harmful gases directly contact the coal-measure formation and gradually penetrate the coal seam, forming an escape ring. In the case of the casing, if the casing is damaged, harmful gas will penetrate the coal seam through the damaged point. If the casing is intact, the harmful gas in the abandoned oil well will not contact the coal seam. If the gap between the casing and the wall of the oil well is not tightly sealed, the oil and gas in the oil-bearing layer will gradually migrate to the coal-measure formation through the pores.

In addition, if the oil well has a shallow depth and is abandoned before being constructed to the oil-bearing layer, oil and gas in the oil-bearing layer will not enter the coal seam. Therefore, whether the mining face in the area adjacent to the abandoned oil well is affected by harmful gases such as hydrogen sulfide in the abandoned oil well must be determined according to the completion status, damage type, and layer relationship of the abandoned oil well.

3.2. Position Relationship between Oil Wells and Coal-Measure Strata

According to the relevant geological data of Ma tan 30 and Ma tan 31 abandoned oil wells, the crossing conditions and thickness of the abandoned oil wells of Ma tan 30 and Ma tan 31 are shown in Table 4. The coal-bearing strata in the Shuangma minefield are the Middle Jurassic Yan a Formation, with nearly 30 coal-bearing layers.

Table 4. Abandoned oil wells pass through the formation and statistics of thickness.

| Oil Well | Layer Thickness/m | | Layer Thickness/m | | | Triassic | Cumulative Depth/m |
| | | | Jurassic | | | | |
	Fourth Series	Paleogene	Stability Group	Zhiluo Formation	Yan'an Formation	Yan'an Formation	
Ma tan 30	10	81.8	-	134.7	344.7	666	1237.2
Ma tan 30	15	64.0	63.1	67.9	290.0	666	1166.0

According to geological survey data, the buried depth of coal seams in all areas near the Ma tan 30 and Ma tan 31 oil wells within the I0104$_1$05 working face is 185.71–513.66 m [35]. Among them, the average thickness of the 4-1 coal seam mined at the I0104$_1$05 working face is 3.8 m, the buried depth is 226.86 m, and the distance between the underlying 4-2 and 4-3 coal seams is 10 and 35 m, respectively. The average inclination angle of the coal seam is 7°, and the lithology of the roof and floor of the coal seam is mainly siltstone and sandy mudstone [36]. Figure 5 shows the positional relationship between the Ma tan 31 oil well and the coal-measure strata.

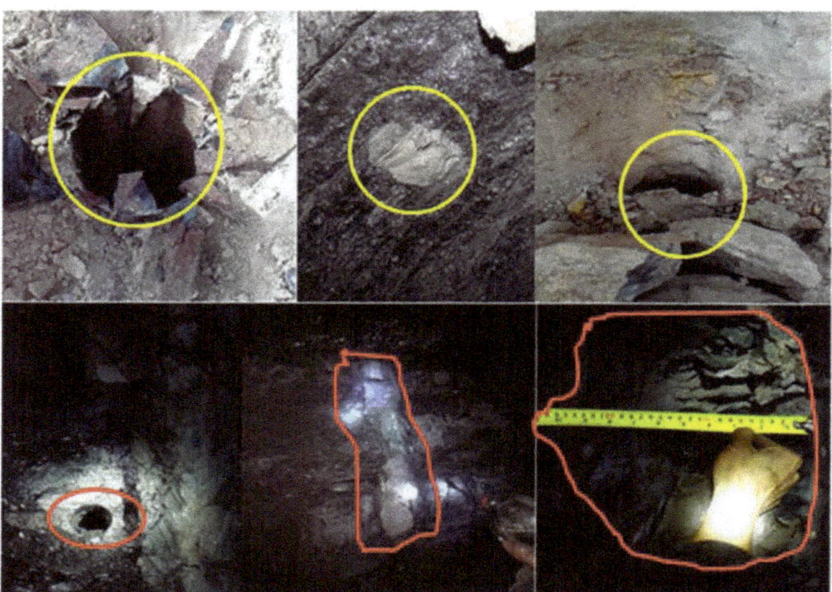

Figure 5. The positional relationship between the Ma tan 31 oil well and the coal-measure strata.

Therefore, based on the analysis of the relationship between oil wells and coal seams and field demonstrations, there are oil-bearing layers in some areas of the coal-bearing strata, but they have little impact on mining. The oil layer of the Shuangma well field is a Triassic stratum, which is located below the bottom layer of the mine (18-2 coal seam, buried depth of about 520 m). Its impact on the coal seam in the abandoned oil well area mainly depends on the oil well's form of destruction [37].

3.3. Oil Well Damage Form and Impact

Abandoned oil wells are mainly production wells, water injection wells, and other oil and gas wells [38]. Some oil wells are open hole wells, and other oil wells have casings. Due to the long period of completion of wells, most of the casings in the abandoned oil wells are damaged due to rust, corrosion, etc. A significant amount of H_2S, methane, and other gases in the petroleum reservoir gradually escape into the coal seam. According to the classification of oil well risk levels, the main factors that affect the H_2S, methane, and other gas enrichment in coal seams near abandoned oil wells are the distance between the oil-bearing stratum and the coal-bearing stratum, the presence or absence of casing, the damage of casing, and the location of the breaking point [39,40]. Generally speaking, when the oil well casing is undamaged, or the damage point is below 100 m in the coal-bearing strata, H_2S, methane, and other gases in the abandoned oil well will not easily escape into the coal seam.

When the casing is damaged and is within 100 m below the coal-bearing strata, and when the casing is damaged and located in the coal-bearing strata, it will have varying degrees of influence on the mining of the working face. According to the conditions of the exposed oil wells and gas analysis results and the analysis of the sealing conditions of the Ma tan 30 and Ma tan 31 oil well, the casing of the Ma tan 30 oil well is intact. The Ma tan 31 Oil well were wholly damaged by corrosion in the coal-measure formation, so the Ma tan 31 oil wells have a more significant impact on the H_2S and methane enrichment of the coal seams.

Based on the above analysis, it can be seen that because the oil-bearing strata and coal-bearing strata in the Shuangma coal mine are the same set of strata, the coal seam roof contains a certain amount of petroleum. In terms of content, the roof may contain oil-poor layers, but it has little impact on the mining face. H_2S gushing in local areas of the mining face is mainly affected by poor sealing, damaged casing and within 100 m of coal-bearing strata, and abandoned oil wells with broken casing and located in the coal-bearing strata.

4. Distribution Laws Regarding Gases

4.1. The Distribution Law of Gas and Hydrogen Sulfide Gas Concentration in Boreholes

In view of the current technology and equipment limitations of coal seam hydrogen sulfide content testing [41,42]. To obtain the distribution law of coal seam gas and hydrogen sulfide in the area affected by abandoned oil wells, by measuring the gas and hydrogen sulfide gas concentration in coal seam boreholes in the area affected by abandoned oil wells, the distribution law of gas and hydrogen sulfide can be indirectly reflected. With the Ma tan 31 abandoned oil and the center, 26 boreholes were constructed within 100~150 m of the I0104₁05 return airway along the strike. Twenty-one drilling holes were constructed within 100 m before and after the I0104₁05 transportation lane along the strike. After the drilling was completed, the holes were immediately plugged, and after the gas escaped and balanced in the holes, on-site sampling tests and ground gas sample analysis were carried out.

The layout diagram of the gas concentration test borehole is shown in Figure 6.

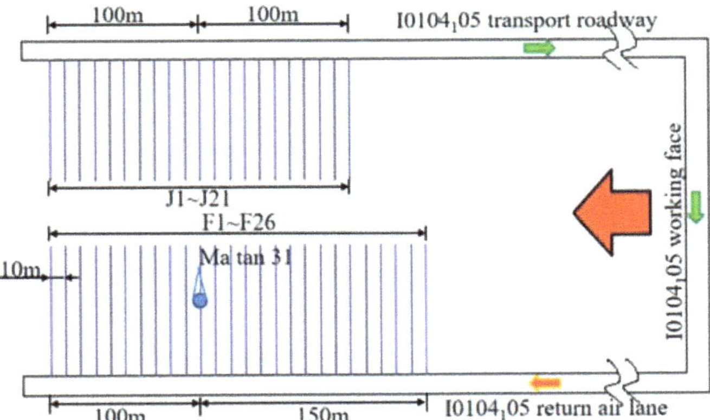

Figure 6. H_2S concentration test drilling layout of Ma tan 31 oil well.

The oil-bearing strata and coal-bearing strata in the Shuangma coal minefield are the same set of strata. The roof of the coal seam contains a certain amount of oil. The top may include oil-poor strata in terms of content, but it has little impact on the mining face. H_2S gushing in local areas of the mining face is mainly affected by poor sealing, damaged casing and within 100 m of coal-bearing strata, and abandoned oil wells with broken casing located in coal-bearing strata. According to the test and analysis results, taking the center of the Ma tan 31 oil wells as the benchmark, the gas concentration distribution law of H_2S and gas in each borehole of the return airway and transportation lane of the I0104₁05 working face is shown in Figures 7 and 8.

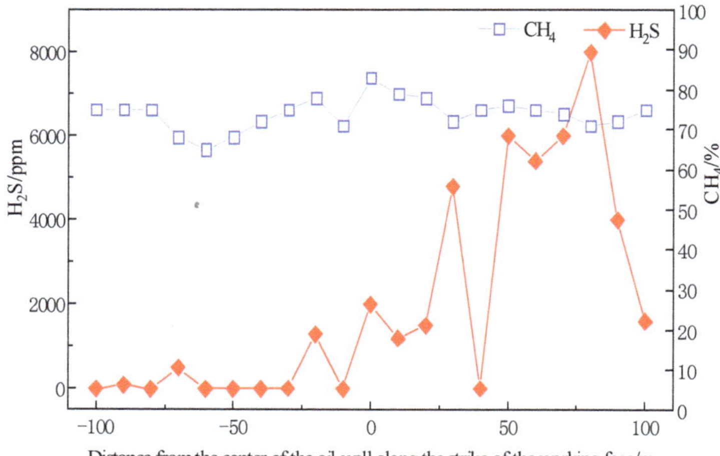

Figure 7. Distribution law of gas concentration in each borehole at the side of the transportation road in I0104₁05 working face.

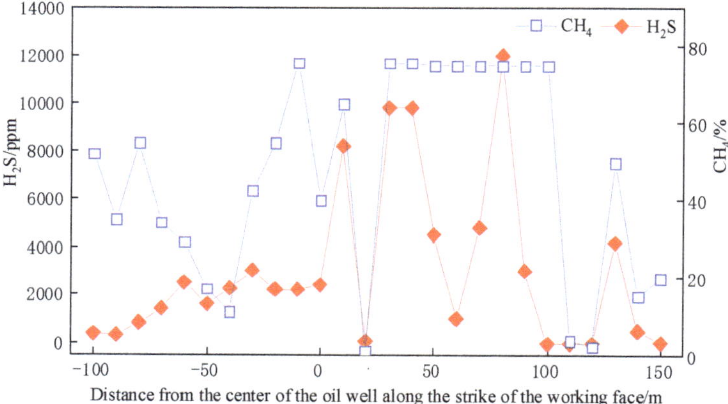

Figure 8. Gas concentration distribution law in each borehole on the side of return airway in I0104₁05 working face.

According to the analysis results of the borehole gas composition measured in area 100–150 m before and after the strike of the abandoned oil well of Ma tan 31, it can be concluded that the I0104₁05 working face transportation lane side, along the working face, the closer to the center of the oil well, the higher the hydrogen sulfide gas concentration, but the hydrogen sulfide concentration distribution in the range of −80~80 m shows a gradually increasing trend. On the side of I0104₁05 transportation lane and return air lane, along the strike direction, within the range of 100~150 m, the area with high hydrogen sulfide concentration is located within 80 m of the cut hole from the center of the oil well to the working face.

Since the coal seam contains gas, the distribution of gas concentration in the borehole on the side of the transportation lane relatively far away from the oil well is relatively stable with little change. The distribution on the side of the return airway where the oil well is relatively close in the horizontal distance shows fluctuations. Still, near the center of the oil well, the gas concentration is relatively high.

Analysis of the Source of Hydrogen Sulfide Gushing

According to the data analysis measured before and after the absorbent injection and ten days later, it was concluded that the hydrogen sulfide absorption effect after the absorbent injection is almost 100%. When the injection of absorbent into the borehole was stopped, it was found that there were signs of rapid rebound of hydrogen sulfide in boreholes F6, F12, F14, and F18, which fully proved that the source of hydrogen sulfide had a supply channel, as shown in Figure 9.

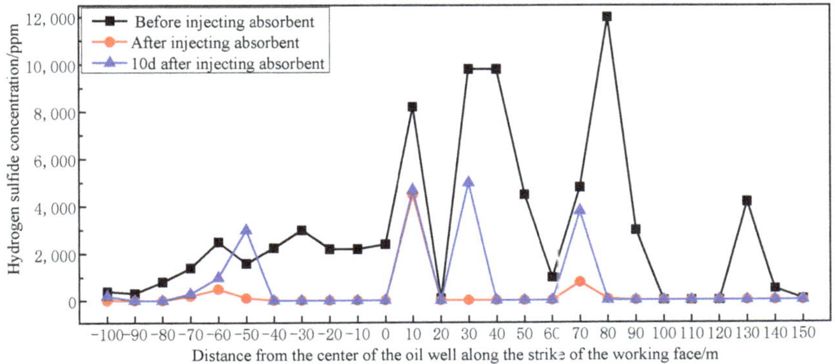

Figure 9. The trend diagram of hydrogen sulfide concentration before and after the absorbent injection in the abandoned oil well of Ma tan 31.

4.2. Determination and Distribution Law of Coal Seam Gas Content

Coal is a porous dual-medium with a lot of pores and cracks [43]. Due to coal pores and crevices, gas generally occurs in coal in two forms, including free and adsorbed states, and is in a state of dynamic equilibrium and constant exchange [44,45].

Under the action of gas pressure, gas, hydrogen sulfide, and other gases existing in abandoned oil wells flow through reverse seepage for an extended period and enter the coal seam from the oil well. Therefore, the gas content of the coal seam in the affected area of the abandoned oil well is generally more significant than that outside the affected area.

Measurement Results of Coal Seam Gas Content

To grasp the distribution law of gas content in the affected area of abandoned oil wells, by drilling test boreholes in the affected area of abandoned oil wells and outside the affected area of abandoned oil wells, the original gas content determination and distribution law of the coal seams in the affected area were analyzed.

Due to the small gas content of the coal seam itself, to avoid the difference in the distribution of coal seam gas content in different mining areas of the working face and the influence of measurement errors, when sampling on-site, it is important to avoid the geological structure area and adopt the method of direct and rapid gas content test as far as possible to make the measurement data accurate and reliable. Five boreholes were constructed in the return airway of the I0104$_1$05 working face, and the original gas content of the coal seam in the 120 m area of the abandoned oil well was measured. Each borehole had a depth of 60 m, and coal samples were taken at 20 m, 40 m, and 60 m for testing.

As shown in Figure 10, according to the measurement results, the maximum desorb gas content of coal seam 4-1 in the affected area of the abandoned oil well is 1.28 m^3/t.

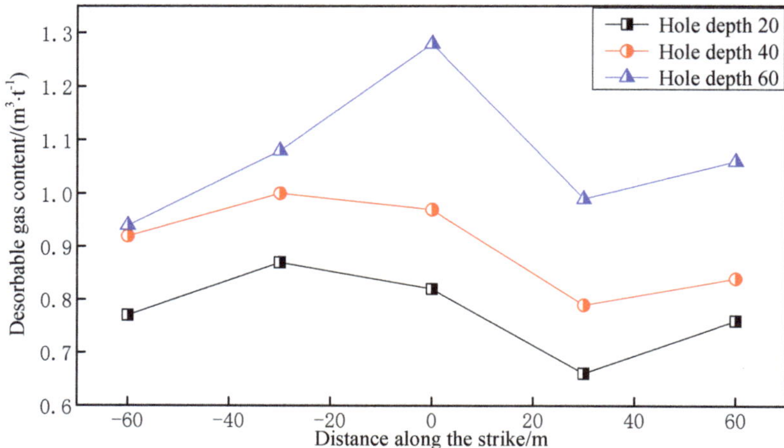

Figure 10. Variation law of gas content near Ma tan 31 abandoned oil well.

Taking the Ma tan 31 abandoned oil well as the research object, the influence of the abandoned oil well on the distribution of coal seam gas content is studied. By sampling the gas concentration test borehole shown in Figure 6 in the affected area of the abandoned oil well (within the radius of 160 m), the desorb gas content of the coal seam was tested, and the distribution of desorbing gas within the influence range of the abandoned oil well was obtained.

As shown in Figure 11, the closer to the abandoned oil well of Ma tan 31, the higher the desorb gas content of the coal seam. The distribution of coal seam gas content in the affected area of abandoned oil wells along the distance from the center of the oil well shows a decreasing power function trend. Through data fitting, the distribution model of coal seam desorb gas content can be obtained, as shown by Equation (1) [46]:

$$y = -17.439x^{0.0149} + 19.4076, R^2 = 0.7999 \tag{1}$$

where y is the desorb gas content of the 4-1 coal seam, m^3/t, x is the radial distance from the center of the abandoned oil well, m.

To verify the influence of abandoned oil wells on coal seam gas content distribution, three boreholes at 100, 150, and 200 m south of the 05B4 drilling field in the return airway of the I0104₁06 working face (the distance between the borehole and the abandoned oil well is more than 300 m) were constructed with a drilling depth of 80 m, and coal is used at a depth of 40, 60, and 80 m, respectively. Direct testing of the coal seam gas content of the sample shows that the maximum desorb gas content of coal seam 4-1 outside the affected area of abandoned oil wells is 0.72 m^3/t (average value 0.64 m^3/t).

Therefore, the desorb gas content of the coal seam in the affected area of the abandoned oil well is generally more significant than the desorb gas content of the coal seam outside the affected area of the abandoned oil well. However, due to the relatively small gas content of coal seam 4-1 in Shuangma coal mine, after fully considering the differences in coal seam gas content distribution in different mining areas and reducing errors, abandoned oil wells have a particular impact on coal seam gas content, but the degree of effects is relatively small.

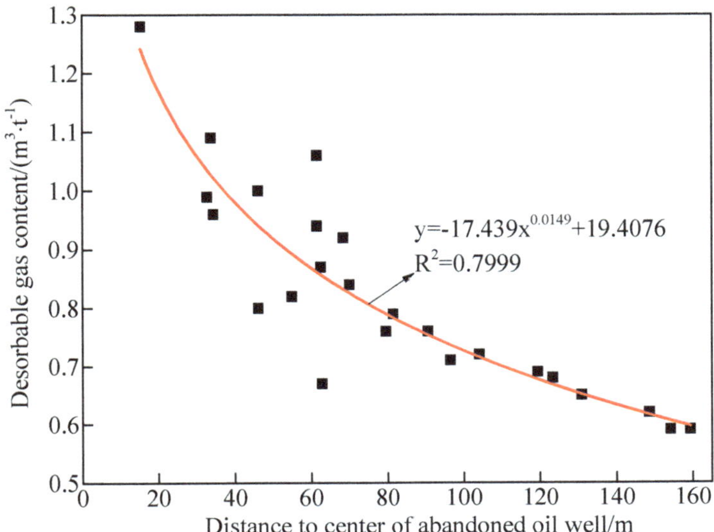

Figure 11. The distribution law of coal seam descrb gas content along the radial direction.

5. The Law of Harmful Gas Gushing

5.1. Investigation on Distribution Law of Gas and Hydrogen Sulfide

The coal body is a porous dual-medium that is rich in pores and cracks. Therefore, the gas occurs in the coal body in two forms: free and adsorbed [47,48]. They are mainly adsorbed and in a dynamic equilibrium state. Regarding the occurrence of hydrogen sulfide in coal seams, relevant scholars have carried out coal adsorption experiments on CO_2, CH_4, and N_2 [49,50]. The results show that the adsorption capacity of adsorbed substances increases with the increase in gas boiling point. Since the lowest boiling point of H_2S gas is -60.33 °C, which is higher than CO_2, CH_4, N_2, and other gases, coal has a more substantial adsorption capacity for H_2S than the above gas [51].

At the same time, because the polarization rate of hydrogen sulfide 3.64×10^{-30} m^3 is greater than that of methane gas 2.60×10^{-30} m^3, the adsorption capacity of hydrogen sulfide gas is more vital than that of methane gas coal [52]. According to the actual measurement results of the attenuation law of hydrogen sulfide emission in the area affected by abandoned oil wells during the tunneling and mining of the I0104₁05 working face of Shuang Ma coal mine, it is concluded that after the coal seam is mined, the escape of hydrogen sulfide will be accelerated. Still, it will be emitted when coal cutting is stopped. The hydrogen sulfide decreases rapidly and quickly decays until no hydrogen sulfide is detected.

Therefore, it is further proved that the hydrogen sulfide in the coal body mainly exists in an adsorbed state [53,54]. It is not easy to release the hydrogen sulfide in the adsorbed state in the coal body without the effect of considerable energy. In addition, studies have shown that some hydrogen sulfide gas exists in a free state in the coal body. Because hydrogen sulfide gas is easily soluble in water, hydrogen sulfide also occurs in a water-soluble state to form sulfuric acid or sulfurous acid. The attenuation law of hydrogen sulfide gushing after mining is shown in Figure 12.

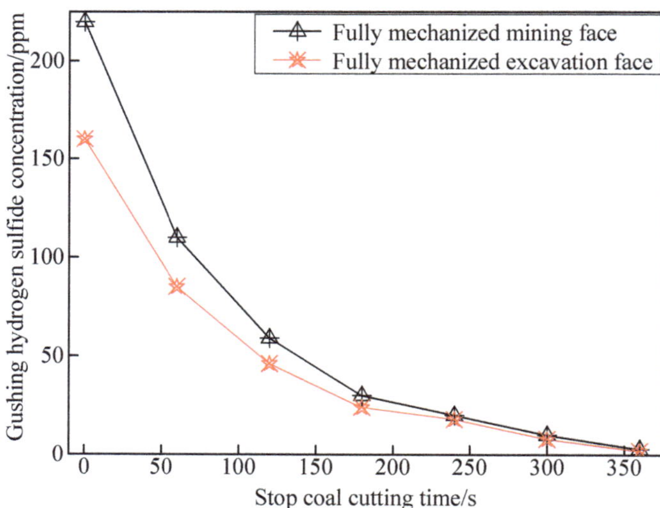

Figure 12. The attenuation law of hydrogen sulfide gushing during the mining process of the working face.

5.2. Investigation on the Influence Scope of Gas and H2S Mining and Gushing

Taking the abandoned oil well Ma tan 31 as the research object, the gas, and hydrogen sulfide emission distribution law during the mining of the working face through the area affected by the abandoned oil well was studied. The abandoned oil well is 375 m away from the I0104$_1$05 working face. The maximum hydrogen sulfide concentration measured by boreholes in the affected area of the abandoned oil well reaches 0.012, and the maximum desorb gas content is 1.28 m^3/t. At the initial mining stage, the air distribution volume is about 1260 m^3/min, and after the return air gas concentration increases, the air distribution volume is adjusted to 1800–2300 m^3/min.

As shown in Figure 13, according to the collected field data, before the working face is advanced to the abandoned oil well, the gas concentration rises. As the working face advances away from the abandoned oil well, the gas concentration decreases. Affected by abandoned oil wells, there is abnormal gas emission in the upper corners of the working surface. It reaches the maximum value before approaching the center of the oil well.

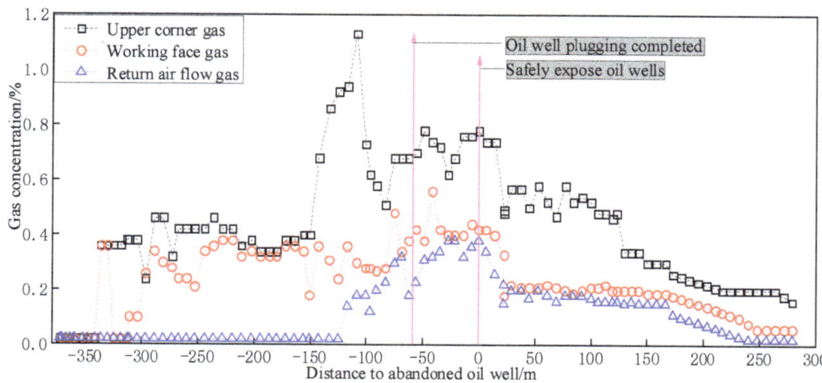

Figure 13. Gas concentration change curve during the period of the impact of the I0104$_1$05 working face passing through the Ma tan 31 abandoned oil well.

In the areas before and after passing through the center of the oil well, the gas concentration in the working face and the return airway has an increasing trend. The abandoned oil well of Ma tan 31 is an open hole with no casing, and ground plugging has not been carried out in advance. The gas in the oil and gas formation penetrates the coal-measure formation through the cracks and is adsorbed in the coal–rock layer. The gas in the coal–rock layer is affected by mining, with a large amount of escaping ito the mining space.

As shown in Figure 14, from the beginning of recovery to the 150 m of the oil well, the working face tends to have a relatively large hydrogen sulfide gushing concentration. The average gushing concentration is between 40×10^{-6} and 350×10^{-6} ppm. When the working face starts to be mined, the coal seam is affected by mining, and a large amount of hydrogen sulfide gushes out. After that, the concentration of hydrogen sulfide gushes out to maintain a high concentration value and fluctuates. After the oil well is blocked until the working face is far away from the oil well, the sulfide concentration of hydrogen gushing gradually decreases.

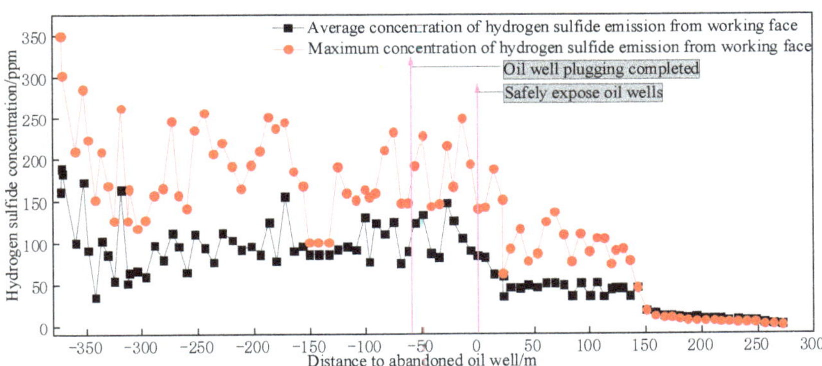

Figure 14. The concentration change curve of hydrogen sulfide emission during the I0104$_1$05 working face passing through the Ma tan 31 abandoned oil well.

As shown in Figure 15, the gas and H_2S gushing trend of the I0104$_1$05 working face when passing through the Ma tan 31 abandoned oil well can be obtained. Combining the location of H_2S during the excavation and the distribution of gas and H_2S within the affected area of abandoned oil wells, it can be concluded that:

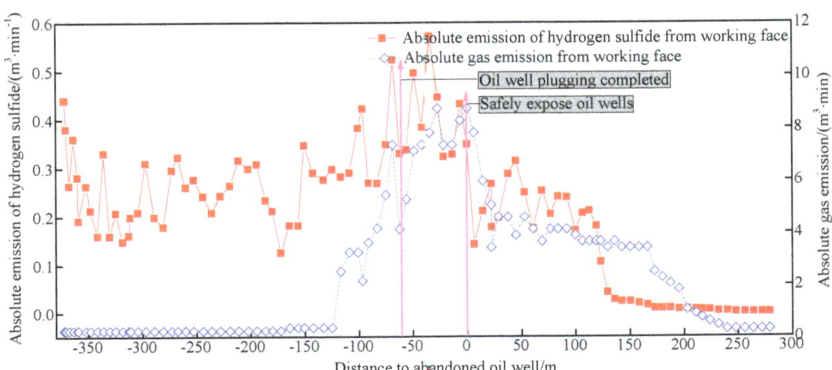

Figure 15. The trend diagram of absolute gas and hydrogen sulfide emission changes in the I0104$_1$05 working face passing through the Ma tan 31 abandoned oil well.

(1) Before the working face passes the abandoned oil well, the concentration of gas and H_2S rises. As the working face is far away from the abandoned oil well, the concentration of gas and H_2S shows a downward trend.

Since the abandoned oil well of Ma tan 31 is an open hole without casing, and the ground has not been plugged in advance, the gas and H_2S of the oil and gas formation invade into the coal-measure formation through the fissures and are adsorbed in the coal seam. When the coal seam is gradually moving forward, a large amount of toxic and harmful gas slowly escapes to the mining space.

(2) When the abandoned oil well of Ma tan 31 is about 376 m away from the I0104₁05 working face, according to the changing trend of H_2S concentration, H_2S appeared when the I0104₁05 working face started to advance, which is completely consistent with the gas source statistics during the tunneling of the aforementioned working face. When the I0104₁05 working face passes through the Ma tan 31 abandoned well, the impact range of the Ma tan 31 abandoned oil well on strike is more than 600 m.

(3) In the I0104₁05 working face, the gas and H_2S concentration exceeded the limit in varying degrees during tunneling and coal mining. At the same time, gas was exceeded in the upper corner after advancing 80 m on the working face, and the highest volume fraction reached 1.13%. Based on the changes in the location and concentration of gas and hydrogen sulfide during excavation and coal mining, as well as long-term tracking and determination of gas and hydrogen sulfide, it is proposed that the affected area of the Ma tan 31 abandoned oil well has a "300 m radius of escape circle", that is the abandoned oil well. As the center of the circle, the diffusion radius of H_2S and other harmful gases in the oil well is above 300 m.

6. Conclusions

(1) According to the structure of the oil well and the positional relationship with the coal seam, the source of hydrogen sulfide in the affected area of the abandoned oil well in the coal mining face of Shuangma Coal Mine is mainly affected by the abandoned oil well, and the casing of the oil well is damaged and within 100 m below the coal-bearing formation, and when the casing is damaged and located in the coal-bearing formation, it will have a greater impact on the working face.

(2) The hydrogen sulfide in the coal in the affected area of the abandoned oil well mainly exists in the adsorbed state. The petroleum reservoir contains H_2S, CH_4 and other gases with a certain pressure (concentration), which gradually escape to the coal-measure strata after a long period of time and are adsorbed in the coal-measure formation. In the coal seam, a large amount of desorption is released under the influence of mining. The distribution of coal seam desorbs gas content along the distance from the center of the oil well, showing a decreasing trend in the power function.

(3) The study shows that the gas and H_2S in the affected area of the Ma tan 31 abandoned oil well in Shuangma coal mine have an impact radius of at least 300 m on the I0104₁05 working face. The hypothesis of gas "escape circle" distribution in the affected area of abandoned oil wells are proposed. However, due to the influence of coal physical properties, hydrogen sulfide adsorption characteristics, geological structure, and other factors, the distribution of gas and H_2S in the affected area of abandoned oil wells and the scope of influence may be different in the next step. The above factors will be comprehensively considered in further in-depth research.

(4) The concentration distribution and variation trend of coal seam gas and hydrogen sulfide gushing in the area affected by abandoned oil wells are generally consistent with the distribution and changing trend of coal seam gas content distribution. The hydrogen sulfide gushing has a greater impact on the working face. Due to the particularity of the occurrence of hydrogen sulfide in the coal seam, the problem of gushing control cannot be solved only by ventilation and dilution. Active and passive control methods such as injection of absorbing liquid into the coal seam before mining and spray absorption during mining can be adopted.

Author Contributions: Conceptualization, H.M.; writing—original draft preparation, H.M. and X.W.; writing—review and editing, H.M., X.W., X.Q., K.G. and S.L.; supervision, H.M., X.W. and X.Q. All authors have read and agreed to the published version of the manuscript.

Funding: Project (52104194) supported by the National Science Foundation for Distinguished Young Scholars of China and project (52074148) supported by Natural Science Foundation of China.

Institutional Review Board Statement: Not applicable.

Informed Consent Statement: Not applicable.

Data Availability Statement: The datasets generated during and analyzed during the current study are available from the corresponding author on reasonable request.

Conflicts of Interest: The authors declare no conflict of interest.

References

1. Raimi, D.; Krupnick, A.J.; Shah, J.S. Decommissioning orphaned and abandoned oil and gas wells: New estimates and cost drivers. *Environ. Sci. Techn.* **2021**, *55*, 10224–10230. [CrossRef] [PubMed]
2. Romanzo, N.N. Locating Undocumented Abandoned Oil and Gas Wells. Master's Thesis, State University of New York at Binghamton, New York, NY, USA, 18 March 2020.
3. Lebel, E.D.; Lu, H.S.; Vielstädte, L. Methane emissions from abandoned oil and gas wells in California. *Environ. Sci. Techn.* **2020**, *54*, 14617–14626. [CrossRef] [PubMed]
4. Riddick, S.N.; Mauzerall, D.L.; Celia, M.A. Variability observed over time in methane emissions from abandoned oil and gas wells. *Int. J. Greenh. Gas Control* **2020**, *100*, 103116. [CrossRef]
5. Zanocco, C.; Boudet, H.; Nilson, R. Place, proximity, and perceived harm: Extreme weather events and views about climate change. *Clim. Chang.* **2018**, *149*, 349–365. [CrossRef]
6. Luca, C.; Antonio, Z. Constraints of fossil fuels depletion on global warming projections. *Energy Policy.* **2011**, *39*, 5026–5034.
7. Zhang, W.H.; Yan, Q.Y.; Yuan, J.H.; He, G.; Teng, T.L.; Zhang, M.J.; Zeng, Y. A realistic pathway for coal-fired power in China from 2020 to 2030. *J. Clean. Prod.* **2020**, *275*, 122859. [CrossRef]
8. Zhang, G.; Suo, Y.L.; Xu, G. Distribution Characteristics of H_2S in Coal Seams around Abandoned Oil Wells and Division of Its Mining Influence Range in Shuangma Coal Mine. *Sci. Technol. Eng. J. Sci. Technol. Eng.* **2021**, *21*, 5759–5766. Available online: https://kns.cnki.net/kcms/detail/detail.aspx?FileName=KXJS202114017&DbName=CJFQ2021 (accessed on 6 January 2021). (In Chinese)
9. Zhang, P.; Lan, H.Q.; Yu, M. Reliability evaluation for ventilation system of gas tunnel based on Bayesian network. *Tunn. Undergr. Space Technol.* **2021**, *112*, 103882. [CrossRef]
10. Wang, G.F.; Xu, Y.X.; Ren, H.W. Intelligent and ecological coal mining as well as clean utilization technology in China: Review and prospects. *Int. J. Min. Sci. Technol.* **2019**, *29*, 161–169. [CrossRef]
11. Wang, K.; Jiang, S.G.; Zhang, W.Q.; Wu, Z.Y.; Shao, H.; Kou, L.W. Destruction mechanism of gas explosion to ventilation facilities and automatic recovery technology. *Int. J. Min. Sci. Technol.* **2012**, *22*, 417–422. [CrossRef]
12. Huang, D.; Liu, J.; Deng, L.J. A hybrid-encoding adaptive evolutionary strategy algorithm for windage alteration fault diagnosis. *Process Saf. Environ. Prot.* **2020**, *136*, 242–252. [CrossRef]
13. Gao, K.; Deng, L.J.; Liu, J.; Wen, L.X.; Wong, D.; Liu, Z.Y. Study on Mine Ventilation Resistance Coefficient Inversion Based on Genetic Algorithm. *Arch. Min. Sci.* **2018**, *63*, 813–826.
14. Hu, J.H.; Zhao, Y.; Zhou, T.; Ma, S.W.; Wang, X.L.; Zhao, L. Multi-factor influence of cross-sectional airflow distribution in roadway with rough roof. *J. Cent. South Univ.* **2021**, *28*, 2067–2078. [CrossRef]
15. Mayala, L.P.; Veiga, M.M.; Khorzoughi, M.B. Assessment of mine ventilation systems and air pollution impacts on artisanal tanzanite miners at Merelani, Tanzania. *J. Clean. Prod.* **2016**, *116*, 118–124. [CrossRef]
16. Liang, P.Y.; Han, Y.; Zhang, Y.Y.; Wen, Y.T.; Gao, Q.F.; Meng, J. Novel non-destructive testing method using a two-electrode planar capacitive sensor based on measured normalized capacitance values. *Measurement* **2021**, *167*, 108455. [CrossRef]
17. Rodriguez-Vega, M.; Canudas-de-Wit, C.; Fourati, H. Location of turning ratio and flow sensors for flow reconstruction in large traffic networks. *Transp. Res. Part B Methodol.* **2019**, *121*, 21–40. [CrossRef]
18. Hu, X.; Han, Y.M.; Yu, B.; Geng, Z.Q.; Fan, J.Z. Novel leakage detection and water loss management of urban water supply network using multiscale neural networks. *J. Clean. Prod.* **2021**, *278*, 123611. [CrossRef]
19. Lau, P.-W.; Cheung, B.-Y.; Lai, W.-L.; Sham, J.-C. Characterizing pipe leakage with a combination of GPR wave velocity algorithms. *Tunn. Undergr. Space Technol.* **2021**, *109*, 103740.
20. Liang, L.P.; Xu, K.J.; Wang, X.F.; Zhang, Z.; Yang, S.L.; Zhang, R. Statistical modeling and signal reconstruction processing method of EMF for slurry flow measurement. *Measurement* **2014**, *54*, 1–13. [CrossRef]
21. Lu, H.F.; Iseley, T.; Behbahani, S.; Fu, L.D. Leakage detection techniques for oil and gas pipelines: State-of-the-art. *Tunn. Undergr. Space Technol.* **2020**, *98*, 103249. [CrossRef]

22. Aida-zade, K.R.; Ashrafova, E.R. Localization of the points of leakage in an oil main pipeline under nonstationary conditions. *J. Eng. Phys. Thermophys.* **2012**, *85*, 1148–1156. [CrossRef]
23. Zhang, Z.W.; Hou, L.F.; Yuan, M.Q.; Fu, M.; Qian, X.M.; Duanmu, W.; Li, Y.Z. Optimization monitoring distribution method for gas pipeline leakage detection in underground spaces. *Tunn. Undergr. Space Technol.* **2020**, *104*, 103545. [CrossRef]
24. Santos, R.B.; de Sousa, E.O.; da Silva, F.V.; da Cruz, S.L.; Fileti, A.M.F. Detection and on-line prediction of leak magnitude in a gas pipeline using an acoustic method and neural network data processing. *Braz. J. Chem. Eng.* **2014**, *31*, 145–153. [CrossRef]
25. Li, J.; Li, Y.L.; Huang, X.J.; Ren, J.H.; Feng, H.; Zhang, Y.; Yang, X.X. High-sensitivity gas leak detection sensor based on a compact microphone array. *Measurement* **2021**, *174*, 109017. [CrossRef]
26. Singh, K.R.; Dutta, R.; Kalamdhad, A.S.; Kumar, B. An investigation on water quality variability and identification of ideal monitoring locations by using entropy based disorder indices. *Sci. Total Environ.* **2019**, *647*, 1444–1455. [CrossRef]
27. Huang, D.; Liu, Y.; Liu, Y.H.; Song, Y.; Hong, C.S.; Li, X.Y. Identification of sources with abnormal radon exhalation rates based on radon concentrations in underground environments. *Sci. Total Environ.* **2022**, *807*, 150800. [CrossRef]
28. Castillo, E.; Jiménez, P.; Menendez, J.M.; Conejo, A.J. The Observability Problem in Traffic Models: Algebraic and Topological Methods. *IEEE Trans. Intell. Transp. Syst.* **2008**, *9*, 275–287. [CrossRef]
29. Castillo, E.; Calviño, A.; Lo, H.K.; Menéndez, J.M.; Grande, Z. Non-planar hole-generated networks and link flow observability based on link counters. *Transp. Res. Part B Methodol.* **2014**, *68*, 239–261. [CrossRef]
30. Rinaudo, P.; Paya-Zaforteza, I.; Calderón, P.A. Improving tunnel resilience against fires: A new methodology based on temperature monitoring. *Tunn. Undergr. Space Technol.* **2016**, *52*, 71–84. [CrossRef]
31. Balaji, S.; Anitha, M.; Rekha, D.; Arivudainambi, D. Energy efficient target coverage for a wireless sensor network. *Measurement* **2020**, *165*, 108167. [CrossRef]
32. Li, Y.T.; Bao, T.F.; Chen, H.; Zhang, K.; Shu, X.S.; Chen, Z.X.; Hu, Y.H. A large-scale sensor missing data imputation framework for dams using deep learning and transfer learning strategy. *Measurement* **2021**, *178*, 109377. [CrossRef]
33. Ng, M. Synergistic sensor location for link flow inference without path enumeration: A node-based approach. *Transp. Res. Part B Methodol.* **2012**, *46*, 781–788. [CrossRef]
34. He, S.X. A graphical approach to identify sensor locations for link flow inference. *Transp. Res. Part B Methodol.* **2013**, *51*, 65–76. [CrossRef]
35. Muduli, L.; Jana, P.K.; Mishra, D.P. A novel wireless sensor network deployment scheme for environmental monitoring in longwall coal mines. *Process Saf. Environ. Prot.* **2017**, *109*, 564–576. [CrossRef]
36. Wang, K.; Jiang, S.G.; Wu, Z.Y.; Shao, H.; Zhang, W.Q.; Pei, X.D.; Cui, C.B. Intelligent safety adjustment of branch airflow volume during ventilation-on-demand changes in coal mines. *Process Saf. Environ. Prot.* **2017**, *111*, 491–506. [CrossRef]
37. Song, Y.W.; Yang, S.Q.; Hu, X.C.; Song, W.X.; Sang, N.W.; Cai, J.W.; Xu, Q. Prediction of gas and coal spontaneous combustion coexisting disaster through the chaotic characteristic analysis of gas indexes in goaf gas extraction. *Process Saf. Environ. Prot.* **2019**, *129*, 8–16. [CrossRef]
38. Lyu, P.Y.; Chen, N.; Mao, S.J.; Li, M. LSTM based encoder-decoder for short-term predictions of gas concentration using multi-sensor fusion. *Process Saf. Environ. Prot.* **2020**, *137*, 93–105. [CrossRef]
39. Foorginezhad, S.; Mohseni-Dargah, M.; Firoozirad, K.; Aryai, V.; Razmjou, A.; Abbassi, R.; Garaniya, V.; Beheshti, A.; Asadnia, M. Recent Advances in Sensing and Assessment of Corrosion in Sewage Pipelines. *Process Saf. Environ. Prot.* **2021**, *147*, 192–213. [CrossRef]
40. Hu, Y.N.; Koroleva, O.I.; Krstić, M. Nonlinear control of mine ventilation networks. *Syst. Control Lett.* **2003**, *49*, 239–254. [CrossRef]
41. Khan, K.S.; Tariq, M. Accurate Monitoring and Fault Detection in Wind Measuring Devices through Wireless Sensor Networks. *Sensors* **2014**, *14*, 22140–22158. [CrossRef]
42. Sun, J.P.; Tang, L.; Li, C.S.; Zhu, N.; Zhang, B. Application of air-volume Proportion rule in optimal placement of gas sensor in mine. *J. China Coal Soc.* **2008**, *33*, 1126–1130. (In Chinese)
43. Zhao, D.; Liu, J.; Pan, J.T.; Li, Z.X. Application study of air velocity fault source diagnosis technology for ventilation system in Daming Mine. *Chin. J. Saf. Environ.* **2012**, *12*, 204–207. (In Chinese)
44. Dong, X.L.; Chen, S.; Zhao, D.; Pan, J.T. Study on Application of Minimum Tree Principle in Layout of Wind Speed Sensor in Mine. *Chin. World Sci-Tech R D* **2015**, *37*, 680–683. (In Chinese)
45. Liang, S.H.; He, J.; Zheng, H.; Sun, R.H. Research on the HPACA Algorithm to Solve Alternative Covering Location Model for Methane Sensors. *Procedia Comput. Sci.* **2018**, *139*, 464–472. [CrossRef]
46. Zhao, D.; Zhang, H.; Pan, J.T. Solving Optimization of A Mine Gas Sensor Layout Based on A Hybrid GA-DBPSO Algorithm. *IEEE Sens. J.* **2019**, *19*, 6400–6409. [CrossRef]
47. Wu, C.Q.; Wang, L. On Efficient Deployment of Wireless Sensors for Coverage and Connectivity in Constrained 3D Space. *Sensors* **2017**, *17*, 2304. [CrossRef]
48. Semin, M.A.; Levin, L.Y. Stability of air flows in mine ventilation networks. *Process Saf. Environ. Prot.* **2019**, *124*, 167–171. [CrossRef]
49. Liu, J.; Jiang, Q.H.; Liu, L.; Wang, D.; Huang, D.; Deng, L.J.; Zhou, Q.C. Resistance variant fault diagnosis of mine ventilation system and position optimization of wind speed sensor. *J. China Coal Soc.* **2021**, *46*, 1907–1914. (In Chinese)
50. Jia, J.Z.; Liu, J.; Geng, X.W. Mathematical model of mine ventilation simulation system. *J. Liaoning Tech. Univ.* **2003**, *22* (Suppl. S1), 88–90. (In Chinese)

51. Dong, L.; Hu, Q.; Tong, X. Velocity-free MS/AE source location method for three-dimensional hole containing structures. *Engineering* **2020**, *6*, 827–834. [CrossRef]
52. Dong, L.; Tong, X.; Ma, J. Quantitative investigation of tomographic effects in abnormal regions of complex structures. *Engineering* **2021**, *7*, 1011–1022. [CrossRef]
53. Dong, L.; Tong, X.; Hu, Q.; Tao, Q. Empty region identification method and experimental verification for the two-dimensional complex structure. *Int. J. Rock Mech. Min. Sci.* **2021**, *147*, 104885. [CrossRef]
54. Dong, L.J.; Chen, Y.C.; Sun, D.Y.; Zhang, Y.H. Implications for rock instability precursors and principal stress direction from rock acoustic experiments. *Int. J. Min. Sci. Technol.* **2021**, *31*, 789–798. [CrossRef]

Article

A Triple-Helix Intervention Approach to Direct the Marble Industry towards Sustainable Business in Mexico

Teodoro Alarcón-Ruíz [1,2], María Evelinda Santiago-Jiménez [2], Loecelia Guadalupe Ruvalcaba-Sánchez [3], Fabiola Sánchez-Galván [4], Luis Enrique García-Santamaría [1] and Gregorio Fernández-Lambert [1,*]

[1] Division of Postgraduate Studies and Research, Tecnológico Nacional de México/ITS de Misantla, Km. 1.8 Carretera a Loma del Cojolite, Misantla, Veracruz 93821, Mexico; 202t0642@itsm.edu.mx (T.A.-R.); legarcias@itsm.edu.mx (L.E.G.-S.)

[2] Division of Postgraduate Studies and Research, Avenida Tecnológico Número 420 Col. Maravillas, Puebla 72220, Mexico; evelinda.santiago@puebla.tecnm.mx

[3] CentroGeo, Contoy 137, Lomas de Padierna, Tlalpan, Mexico City 14240, Mexico; lruvalcaba@centrogeo.edu.mx

[4] Postgraduate and Research Department, Tecnológico Nacional de México/ITS de Tantoyuca, Desviación Lindero Tametate S/N, Colonia La Morita, Tantoyuca, Veracruz 92100, Mexico; fsgalvan01@gmail.com

* Correspondence: gfernandezl@itsm.edu.mx; Tel.: +52-235-103-1175

Abstract: The marble industry in Mexico, similarly to the international market, is going through some problems which are characterized by low productivity performance, inconsistency in management and administrative organization, high raw material waste, and negative social and environmental impact. The methodology used in this paper is based on a systematic review of the strategies and solutions used to address these problems reported between 2014 and 2021, including the results of the application of in situ surveys to three marble companies in the Mixteca Poblana region. These surveys are collected in this article alongside industry experience to propose, in a structured way, a three-pronged management approach with the aim of directing the marble industry towards a sustainable industry model. The structure of this approach, based on forms of capital and sustainability dimensions, engages governments, companies, schools and society to guide this industrial sector to become a sustainable business, integrating knowledge and experience of the marble industry processes. We recommend adding performance metrics to this approach to assess the value chain of the marble industry.

Keywords: sustainability; marble; capital forms; production

Citation: Alarcón-Ruíz, T.; Santiago-Jiménez, M.E.; Ruvalcaba-Sánchez, L.G.; Sánchez-Galván, F.; García-Santamaría, L.E.; Fernández-Lambert, G. A Triple-Helix Intervention Approach to Direct the Marble Industry towards Sustainable Business in Mexico. *Sustainability* **2022**, *14*, 5576. https://doi.org/10.3390/su14095576

Academic Editor: Piera Centobelli

Received: 20 March 2022
Accepted: 15 April 2022
Published: 6 May 2022

Publisher's Note: MDPI stays neutral with regard to jurisdictional claims in published maps and institutional affiliations.

1. Introduction

Internationally speaking, the exploitation, extraction, and transformation of marble is a model of profitable economic development which has seen exponential demand since the 1980s. However, this industry has reported low productivity levels, high raw material waste levels, and a negative environmental impact generated by its dimension stone extraction and transformation processes [1,2]. In the last decade [3], an assessment of the environmental, social, and cultural factors from a perspective of sustainability within this industry has been provided. The results show the formation of pollutants being discharged into bodies of water, the emission of dust from the cutting processes, the presence of occupational hazards, the creation of environmental noise, and damage to health in Palestine. Similar circumstances have been reported in the marble industry in Turkey [4], where remedial actions have been proposed in order to reduce the damage to workers' health and the environment, and strengthen economic aspects in the marble industry, resulting in a 7% improvement in the productive efficiency of the marble process. In this context [2], to assess the problems of air, soil, and water pollution resulting from stone and marble activities in China, a methodology was proposed that lays the foundations to create a clean production scheme.

In Mexico, the problems in the industry are characterized by low productive performance, a lack of management and organization structures [5], high amounts of raw material waste [6], and negative social and environmental impacts in the two representative marble exploitation areas in Mexico: the Comarca Lagunera and the Mixteca Poblana [7,8]. In this industrial context, actions towards sustainability must be part of the corporate strategy of the industrial sector, with the objectives of raising the quality of life, reducing and eradicating poverty, and minimizing the environmental impact created by this industry [9]. The publication *Governance of Mineral Resources in the 21st Century: Guiding the Extractive Industry towards a Sustainable Development at the 2020 International Resources Panel* has offered suggestions to improve the economic state of the mining industry, alongside encouraging companies to comply with social and environmental standards in order to generate trust among the population and avoid social conflicts. In this context, the 2030 Agenda for Sustainable Development promotes action plans to eradicate poverty, protect the environment, and reduce problems related to climate change. Within the 17 objectives set out in the 2030 Agenda, objective 9—"Industry, Innovation, and Infrastructures"—refers to generating and applying actions to ensure that industry is more productive while significantly reducing carbon dioxide emissions and the amount of waste generated in its processes.

This research is based on the method of the Business Council (BS) for Sustainable Development (2002), and the framework is a diagnostic study of the context and the challenges of the Mexican marble industry in these dimensions: social, political, economic, cultural, environmental, research, and development. Thus, this article contributes a comprehensive management approach of three helices for the integration of forms of capital reported in the Mining, Minerals, and Sustainable Development (MMSD) project of 2006, and the Multidimensional Model proposed by [10], to synergize government, business, school, and society strengths, with the aim of transforming the marble industry sector in the following principles of sustainability. The integration of the strengths of the marble industry in the international arena under a comprehensive management approach of three helices—government–business–school—allows the experiences and knowledge related to the marble industry processes reported internationally to be aligned with the local experience where this comprehensive three-helix management (THM) approach is replicated. With this purpose in mind, the objectives of this article are the following:

- Synthesize the internationally written research articles about the marble industry from 2014 to 2021 and accumulate knowledge to help address problems in this mining sector;
- Suggest the government–business–school approach as a useful intervention guide for researchers to address problems in the marble industry value chain.

In the following sections, we summarize the experience reported by international writings related to problem-solving strategies in environmental, economic, social, cultural, political, and social dimensions, and integrate them into the comprehensive three-helix intervention, or triple-helix intervention (THI), approach for the sustainable development of the marble industry.

2. Materials and Methods

This exploratory and descriptive research integrates the socio-economic and environmental dynamics of the marble industry, reported in the scientific writings from 2014 to 2021, with the objective of relating the dimensional elements of a sustainable company reported by [10] and described in Figure 1, with the forms of capital of the Mining, Minerals and Sustainable Development (MMSD) project reported in the United Nations through ECLAC, i.e., natural capital, manufactured capital, human capital, social capital, and financial capital. As shown in Figure 2, this research uses secondary information sources organized according to [11] which provides a model for the revision of the articles and written research defined by stages of evaluation. This model is relevant to describe the technological, productive and environmental context of the Mexican marble industry, supported with international articles. In this research, the primary information was collected in situ from three marble companies in Tepexi de Rodríguez-Puebla in Mexico by applying five hedonic-level Likert scale surveys validating aspects related to the processes of marble extraction and processing, and which have been reported by international and national research articles.

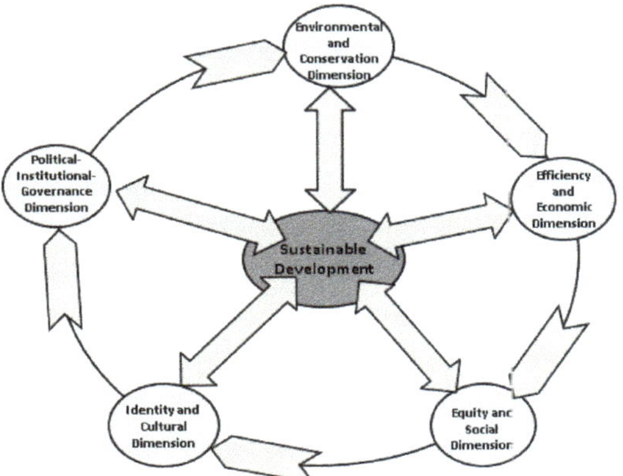

Figure 1. Multidimensional approach (MA) for sustainable development.

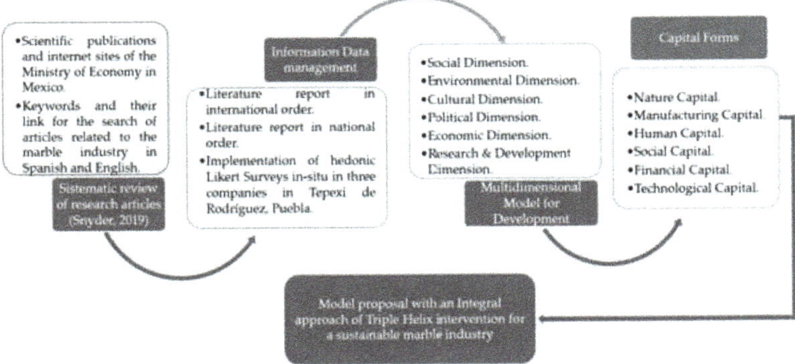

Figure 2. Model proposal with an integral approach and its influence field in the forms of capital for the marble industry in Mexico.

Findings identified in the literature framework and from the in situ research are used in this study for a THI approach to manage the processes of the marble industry towards becoming a sustainable enterprise in relation to economic, social, and environmental progress. The scientific literature has reported the evolution of models with a quadruple- or quintuple-helix approach, in which knowledge of society is highlighted to create synergies between economy, society, and democracy [12]; however, this industry in Mexico is in the process of development. Therefore, it is considered that the mining sector, and specifically the marble industry, in the process of this development, must be oriented so that products' exploitation, production, and transformation are sustainable in order "to" preserve those resources that are essential for human subsistence.

3. Results

3.1. Socioeconomic and Environmental Contexts of the National Marble Industry and Its International Contrast

One of the industrial activities with a great negative impact to ecosystems is the mining industry and, specifically, the extraction and transformation of stone materials, such as marble. Worldwide, the countries with the largest deposits are Italy, Spain, Portugal, Turkey, and Greece. From these countries, the three main producers are Turkey, with 39%, followed by Italy with 16%, and Greece with 7% [13]. In this industry, Palestine, Turkey, and Pakistan have identified problems inlcuding the unsustainable management of waste resulting from the extraction, transformation, transportation and distribution of marble [14], as well as the demand for large amounts of water for the cutting and polishing processes of the marble cuts, with an estimated annual average of 38,700,000 m^3 [15]. According to [16], the high demand for this resource negatively impacts the communities in which this industry operates due to the discharge of contaminated water characterized by a high content of 95% calcium carbonate, with the remaining 5% comprising chemicals such as calcium sulfide, chrome, zinc, and iron used in the marble cutting and finishing processes [16,17]. This section is divided by subheadings. It should provide a concise and precise description of the experimental results and their interpretation, as well as the experimental conclusions that can be drawn.

On the other hand, the low professionalization of workers in this industry is also a concern, as are the poor conditions in the ventilation and lighting of the facilities, compromising the medium- and long-term health of the workers in this industry [4]. In this context, the absence of good practices in the management and exploitation of quarries, the absence of formally established environmental controls, companies with changing structures and organizations, negative externalities to public health, and push production systems are all factors that affect the integration of the supply chain in the marble industry, and can be divided into economic, social, and environmental dimensions. In this regard [1], it is highlighted that the lack of strategic association with suppliers and customers, the lack of communication and integration between areas, as well as the absence of marketing plans and inventory management, consequently impact the disarticulation of processes which, in turn, greatly affect the profitability of the company, and increase negative externalities. Recently [3], it was reported that, in Palestine, the marble and stone industry has negative effects on public health, the environment, water resources, and ecosystems, as well as causing noise pollution and radical changes to the environment landscape where this industry is located.

In Mexico, the marble industry has been growing in economic activity since the 1960s, with two areas dedicated to exploitation and extraction of marble: Comarca Lagunera, with 15 municipalities [18], and the state of Puebla with 5 municipalities [19]. The statistical yearbook of 2019 Mexican Mining reported a production of 1,964,041.14 tons, with an estimated value of 12 billion dollars [20], which shows that this type of industry meets a demand that strengthens the economy of the country. However, the negative externalities caused by its exploitation, transformation, storage, and distribution must be considered [21]. A disarticulation between economic agents and, consequently, a push type of production is reported, which generates the excessive waste of natural and human resources, negatively impacting economic, social, cultural, and environmental factors. Meanwhile, [8] it has been emphasized that the marble industry has collaterally created large amounts of waste from its extraction to finishing processes, without having a responsible final disposal for them, with the waste being thrown into open-air landfills.

The Ministry of Economy of Mexico in 2020 reported that, in the extraction and exploitation stage, there have been performance problems in terms of the efficient use of resources, low levels of safety, and the effective use of the minerals, as well as low economic profitability, due to the fact that artisanal and improvised techniques, as well as the use of machinery in poor technological conditions, continue to be used. Hence, ref. [6] it has also been identified that small companies in this industrial sector are technologically backwards.

The recommendations identified through the research performed on this sector place special emphasis on the fact that the marble industry requires an in-depth analysis and an action plan to remedy its negative externalities and to lead this sector towards being a sustainable industry. Figure 3 shows that the marble industry in Mexico had a rise in 2014 and a significant drop in production in the following years.

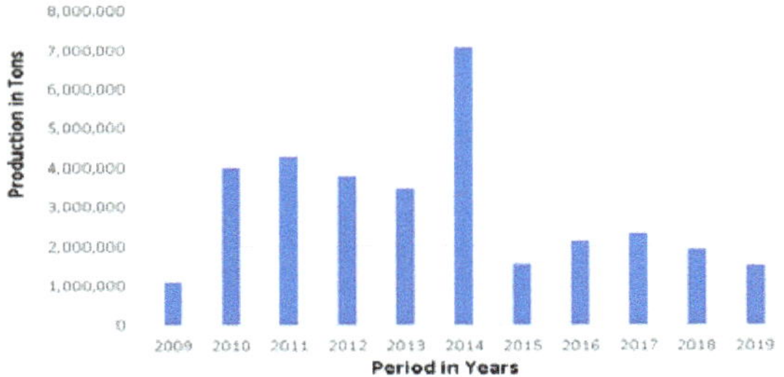

Figure 3. Marble production volume in Mexico from 2009 to 2019.

Various authors agree that the problems faced by the Mexican marble industry can be summarized by social, political, cultural, and environmental dimensions. The Ministry of Economy of Mexico have emphasized the importance of giving this industry an impulse from the start point of exploitation of the stone so that problems hindering its progress can be addressed. To achieve this, experts in the sector, as well as Mexican mining organizations and the Mexican Ministry of Economy, highlight the importance of learning from the experiences of other countries such as Pakistan, Turkey, Iran, and Italy which, despite encountering the same problems as Mexico, have shown some signs of progress in the productive and environmental dynamics of the marble industry.

Table 1 illustrates the challenges the marble industry faces in Mexico that require the attention of the scientific community and stakeholders in this industrial sector. The common problems in the two marble industrial zones in Mexico are summarized by negative externalities to public health and damages to the green mantle, soil, and air; inefficient processes from excessive exploitation of the rocks in the quarry; and waste in the extraction of the dimension rocks and in the lamination process.

Table 1. Problems and challenges of the marble industry in Mexico.

Dimension	Problem	Challenge
Social	The marble industry in Mexico is characterized by the participation of the community and agricultural actors who have this marble industry as their primary source of income. This industry's operation has negative effects on public health caused by dust, environmental noise, and occupational hazards [22]. Technical advice is not available for this industry; there are no financing programs with the scope of the productive investment required by this sector [5,23].	• Manage and promote effective links with research centers that provide advice to guide and exploit the marble industry in a sustainable manner; • Within a three-helix framework (THF), manage advisory programs for the exploitation of marble reserves; • Identify those public policies that promote and bring together marble entrepreneurs and the *ejido* communities to address the social problems arising from the activities of the marble industry to find suggestions and solutions; • Implement occupational hygiene and safety management systems that guarantee the physical health of the workers who provide their services in the quarries, and marble processing companies.
Political and Economic	The productive and technological situation faces problems characterized by inefficient processes, poor organization and administration, added to technological obsolescence and conventional methods still being applied. The previous mentioned aspects impact the low use of the dimensional marble stone, and more quality defects also start showing [8]. On the other hand, as a non-concessionable activity, the extraction of marble is exempt from the payment of mining rights at a federal level. Hence, there has been an increase in excessive exploitation without considering a geological analysis of the mine and land where these deposits are located.	• Establish public policies that regulate the exploitation of marble deposits; • Identify programs that promote the marble industry in its different stages of the value chain; • Encourage school–company relationships for the development of competitive advantages in the marble industrial sector.

Table 1. *Cont.*

Dimension	Problem	Challenge
Cultural	Various externalities of the marble industry that transcend in the community´s cultural changes, for example, working in marble factories and leaving lands to cultivate [24]. For this reason, populations living near the areas of marble exploitation and extraction demand that their rights, customs and traditions are not altered by the marble mining activities. There are also reports of water scarcity in the municipalities where this industry is located [17].	• Develop strategic alliances between interested parties and organizations that participate directly and indirectly in the marble industry (*ejidatarios*, business owners, entities at local, state, and national government levels); as well as private organizations, higher education universities and research centers to create collaboration networks for mutual government–society–business benefit.
Environmental	The marble industry in Mexico has an impact on green mantles [7] and the irregular consumption of water resources [17], as well as affecting local landscaping [5]. These problems are attributed to the extraction, cutting and transformation processes of dimension stones.	• Promote practices and processes, under a socially responsible culture, to regulate environmental policies so that the marble industry can be led towards sustainable development; • Identify three-helix association mechanisms aimed to manage and promote circular economy models based on marble industry waste; • Identify linking mechanisms between public and private organizations to establish strategies for the treatment of discharges to soil, air, and water that come from polishing, smoothing, and special processes; • Develop collaborative strategies between public and private organizations and the marble industry to carry out environmental impact evaluations before and during industry operations, as well as in quarry operations.

3.2. International Marble Industry: Problems and Solutions

Table 2 shows that one of the problems of great concern in the marble industry in the international context is waste management; secondly, the low quality in the finishing process and the productivity related to the efficient use of raw and in-process materials are also concerning. On the other hand, as shown in Figure 4, in this industrial sector, job security is a large problem, yet it receives little attention in the industry: most of the proposals for improvement are frequently focused on reducing waste generated by this industry. Likewise, the least managed issues are those oriented to environmental safety, and the treatment of industrial effluents from this industry. Those problems raise the possibility of waste management and the treatment of its effluents from a circular economy approach. The proposed solutions show that they are focused, with greater importance, on addressing the problems related to occupational health and safety, as well as negative externalities caused by dust and environmental noise [4]. The waste generated by the marble industry—essentially pieces of laminated marble and sludge from marble cutting—have caused obvious environmental damage that range from changing the landscape [6] to impact on flora and fauna, in addition to the pollution caused by the emission of marble dust into the air, causing respiratory diseases [25]. Given the fact that these events have been reported in international articles [2], the authors recommend carrying out environmental performance evaluations and looking for alternative economies that are useful and marketable [4].

The reality faced by the marble mining sector in the international order, and its contrast in the national order, show similar scenarios. In Mexico and other parts of the world, the low efficient use of marble in its extraction and transformation processes [6], the damage to the health of workers and of local people in the community in this industry [25], and the high amounts of waste and sludge created [3] are among the most recurrent problems. To that effect, given the scenario of declining production shown by the industry in Mexico (Figure 3), several authors recommend improving the processes of marble extraction, transformation, and distribution.

Table 2. Problems and solutions of the marble industry in some countries.

Author	Country	Problems					Solutions						
		Supply Flow	Water	Safety	Health	Waste Management and Handling	Quality and Productivity	Logistics and Productive Strategies	Recycling and Water Treatment	Safety and Hygiene Programs	Professionalization of Collaborators	Circular Economy Strategies for Waste	Environmental Performance Evaluation
[14]	Palestine	X	X	X	X		X	X	X		X		X
[26]	Cuba	X				X		X				X	
[27]	Turkey		X		X				X				
[16]	Pakistan		X		X				X				X
[4]	Turkey	X				X	X			X	X	X	
[8]	Spain						X	X					
[28]	Peru					X						X	
[29]	Spain					X						X	
[1]	Ethiopia	X					X	X			X		
[30]	Spain					X						X	X
[31]	Egypt					X							
[2]	China					X							X
[32]	Pakistan		X						X			X	
[33]	Guatemala						X	X					
[34]	Spain					X						X	
[35]	India					X						X	
[36]	Egypt					X						X	
[3]	Palestine			X	X	X	X			X	X	X	
[37]	Turkey						X	X					

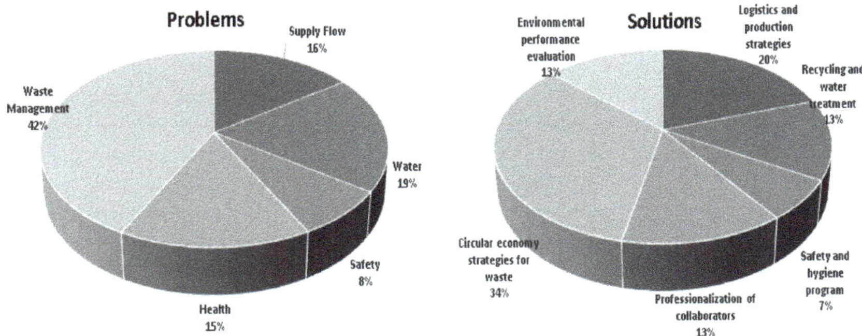

Figure 4. Proportion of the problems and solutions of the marble industry internationally.

4. Discussion

Conceptual Approach of Triple-Helix Intervention for the Marble Industry in Mexico

The dynamics of emerging economies motivate industries to innovate to increase competitiveness and promote the economic development of the region where they are embedded [38]. Currently, academics and researchers observe the convergence of the activities of citizens, companies, the government, universities, and society with great attention and interest. This idea describes the conceptual approach of the quadruple- and quintuple-helix models. However, in the current context of the marble industry, especially in emerging countries such as Mexico, social intervention behaves only as a regulatory body of industrial practice. To that effect, we propose initiating the transition towards the sustainability of this industry through a conceptual model of THI: government–company–school. In this regard, the findings reported by international research on the problems of the marble industry and the fieldwork of three marble companies located in Tepexi de Rodríguez in Puebla, Mexico, coincide with the importance of the intervention of the government, companies, and universities to manage an efficient and friendly marble industry in a socially responsible environment. It is evident that the problems and challenges of the marble industry in Mexico are not unrelated to those presented by industries in other countries. Some experiences in the different dimensions of sustainability have been internationally covered in different articles. These experiences have made it possible to address specific problems when public policies, research, and responsible societies are presented.

The Ministry of Economy and the Ministry of Mining recognize that Mexico has an important diversity of marble reserves that have not yet been quantified, and that its quality allows it to compete internationally. Therefore, it is important for direct and indirect agents to participate in this industry, and to tackle the global environmental regulatory challenges to transition towards a sustainable industry. Countries in the European community are already on this path towards a sustainable marble industry.

In the 1990s, the United Nations—through the division of natural resources and infrastructure—carried out a reform of the mining laws in Latin-American and Caribbean countries. These agreements essentially focused on the economic aspects and, later, relevance was given to the environmental dimension [9]. However, the social dimension was not addressed, causing the mining industry to reflect an imbalance affecting sustainable development. The report *Governance of Mineral Resources in the 21st century: Gearing the Extractive Industry towards Sustainable Development at the 2020 International Resources Panel* identified suggestions to improve the economic performance of the mining industry to ensure compliance with social and environmental standards. Moreover, this report observed a series of guidelines so that this sector could establish a governance structure that addresses safety and resource efficiency. These approaches motivated replacing the concept of "Social License" with "Sustainable Development license" and "Operate". The first built trust among the population and avoided social conflicts, whereas the second works under an integrating approach, with the objective for companies to achieve positive environmental, social, and economic results with fairer agreements, taking part in positive actions for the environment and stakeholders so that local customs and traditions can be preserved.

In this train of thought, it is necessary to create "alternatives for life and the general welfare without compromising the ability of future generations to meet their own needs" [39], which is the case of the mining industry and, especially, of the marble industry. In this respect, the report of the project MMSD (2006) describes that, to satisfy the needs of current contexts without affecting future generations, things must be sorted in an effective way [40]; in other words, the available capitals must be produced and exploited in a reasonable manner, considering the preservation of those fundamental for human life. The MMSD project (2006) defines five forms of capital as a valuable investment for the sustainable operation of the mining sector: natural, manufactural, human, social, and financial.

Natural capital provides a sustained income from ecosystem benefits, such as biological diversity, mineral resources, clean air, and water. Manufacturing capital transforms natural capital to create consumption value, and is defined by machines, facilities, and infrastructure. Social capital is represented by groups and institutions that make collaborations between people and groups possible. Financial capital is the representation and result of the natural, manufacturing, human, and social capitals [40].

Figure 5 concentrates this set of elements in terms of dimensions and forms of capital with which [10] positive impacts will contribute to improve production practices along with environmental quality, resulting in a positive impact on the community by addressing common interests from a multidimensional systematic perspective with the collaboration of specialists and social actors. This intervention approach seeks to integrate synergies based on partnerships between stakeholders so that the marble industry can transition to sustainability. The author of [9] notes that collaboration agreements and the creation of communication channels are two necessary aspects in a world with an empathetic and responsible society. This work scenario is strengthened by [23], which recommends creating committees as managers to monitor the operation of this industry. With this purpose in mind, the design and permanent assessment of the performance of the different operations from the supply of dimension stone and throughout the chain of economic actors of this industry is necessary.

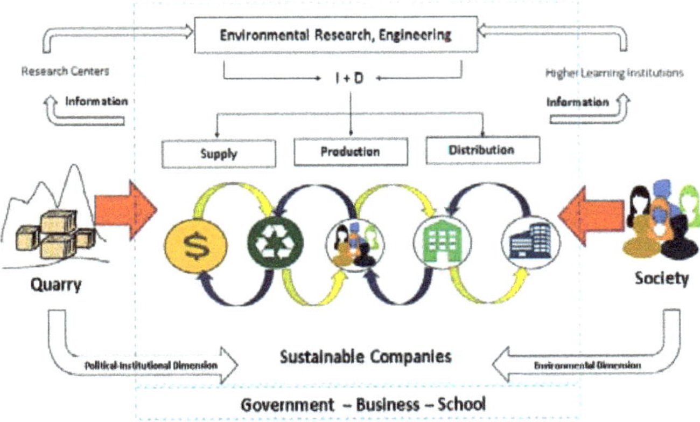

Figure 5. Comprehensive THI approach for sustainable development in the marble industry.

In essence, the sum of the efforts of all intervening actors, directly and indirectly, is fundamental for the development of this national industry. Hence, many research centers and universities have created connections or partnerships in regard to research and development (R&D).

This school–company relationship can be seen, for example, in India [35], where they have created a recycling alternative for approximately 12 million tons of marble waste under the concept of a sustainable circular economy approach. In Turkey, the successful transformation of marble and the costs involved were documented by the computational program of the mathematical model of linear programming of mixed integers, which solved the problem of marble cutting waste and the planning of marble cutting [37].

In Ethiopia [1], a research project was carried out to determine the factors that affect the performance of the supply chain of the marble industry, supported by a review of information from research documents, the application of surveys, interviews, and descriptive statistical analysis. The results provided recommendations with emphasis on creating a strategic partnership between customers and suppliers through an electronic communication network for the management of operations in the extraction and transformation of marble. In Peru, similarly to the case in India, a study that reused marble waste for the construction industry, reported by [28], demonstrated the usefulness of these residues to manufacture concrete with a resistance of 279.18 kg/cm^2 through mechanical resistance and comprehension tests, showing greater resistance compared to the formula of conventional concrete.

In regard to health, safety, and the environment in Turkey, problems related to human factors (noise and lighting) and to the environment (reuse of waste, water, sludge) were solved by using the Inventive Problem-Solving Theory (TRIZ), which resulted in an increase in productivity of more than 7.5%. The author of [5] documented the characterization of the marble production chain in the state of Puebla so areas of opportunity could be identified regarding raw material, transformation processes, and organizational knowledge. From this research, a series of recommendations, such as those reported in Ethiopia by [1], could be emphasized, leading to the efficient integration of the production chain, the creation of new management and organizational practices, and the diversification of marble products towards new national and international selling markets.

These findings highlight the importance of government–business–school conections to undertake improvements throughout the value chain in the marble industry. For this purpose, Figure 5 shows a comprehensive THI approach for the development of the sustainable marble company. This approach, based on a government–company–school relationship as an intervention manager, seeks to bring research centers and universities closer together to carry out research and the innovation of these processes throughout the supply value chain, including supply, production, and distribution. The improvement interventions in the marble industry supply chain are managed as capital reported by the MMSD project (2006), and the dimensions are oriented towards those of a sustainable company, as reported by [10]. In this THI approach, society has an important role in which permanent communication is maintained to assess the impact of the benefits and externalities caused by the marble industry.

Given the dynamics of emerging economies, some studies have argued for the need to apply actions aimed at innovation, entrepreneurship, and economic development through the synergistic work of companies, governments, and universities in a triple-helix management approach [41]. However, given the development observed by academic and research bodies, the need in these economies to incorporate the social sector as a manager oriented towards business innovation stands out [42].

Based on the comprehensive proposal of the triple-helix intervention to guide the marble industry in Mexico towards sustainable development, it is important to establish a descriptive model of the stages where all components, dimensions, and forms of capital converge to create processes that are efficient and friendly to the social and environmental surrounding contexts (Figure 6).

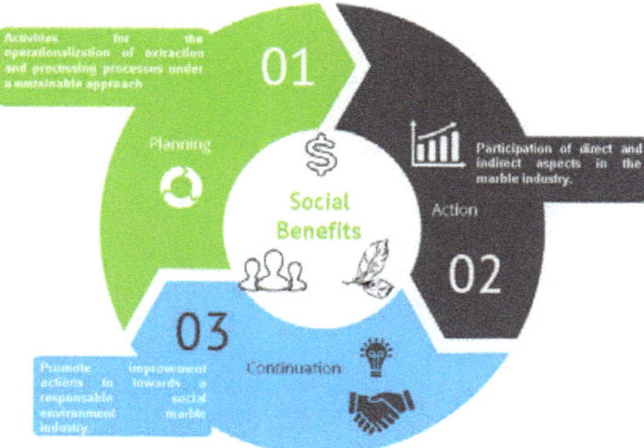

Figure 6. Descriptive functional model of the THI aproach in the marble industry.

Figure 6 shows the three stages of the descriptive functional model. The planning phase, from a systemic approach, allows the comprehension of the context of the productive and environmental dynamics in the marble industry, leading it towards sustainability. In the next section, the action stage, the direct and indirect variables are fundamental so that the transformation, logistics, and service processes can generate social benefits with activities that elicit more efficient and productive dynamics, considering social and ecological implications. Based on the forgoing, the third phase of "continuation" emphasizes the universal measures that help to quantitatively assess the performance of the marble industry in regard to the labor, production, and environmental aspects, through a permanent evaluation in which indicators will be identified, managed, and referenced to ensure continuous improvement.

5. Conclusions

The exploitation, extraction, and transformation of marble have structured a model of profitable economic development which has observed a significant economic exponential increase since the 1980s. However, this industry in the international market has reported low levels of productivity, high levels of raw material waste, and a negative environmental impact generated by its extraction and transformation processes. In this context, the marble industry in Mexico presents similar problems with its two representative zones: the Comarca Lagunera and the Mixteca Poblana. To solve these problems, the literature reports various strategies. In this article, an approach is presented to intervene in the value chain of the marble industry based on social dimensions and forms of capital, with the aim of integrating the knowledge of public organisations and the experience of stakeholders in this industry to lead it in a sustainable manner. This integrating approach suggests the improvement of the productive processes, including the extraction, transformation, and the distribution of marble. To this end, the integration of the social sector to stimulate synergy between operations should follow the triple-helix model proposed here as a manager of innovation, and leave behind actors with only social claim.

Finally, although the identification of the performance metrics of the value chain have not yet been identified, this approach strives to intervene so that the value chain can be complemented.

Author Contributions: Conceptualization, T.A.-R.; methodology, F.S.-G., L.E.G.-S., and G.F.-L.; supervision, L.E.G.-S.; validation, F.S.-G.; writing—original draft preparation, G.F.-L.; writing—review and editing, T.A.-R., M.E.S.-J., L.G.R.-S., and G.F.-L. All authors have read and agreed to the published version of the manuscript.

Funding: This research received no external funding.

Institutional Review Board Statement: This study did not involve biological human experiments.

Informed Consent Statement: Informed consent was obtained from all subjects involved in the study.

Data Availability Statement: Not applicable.

Acknowledgments: The authors of this study thank the companies "Intemármol S.A de C.V", "Mármoles RocaDura S.A de C.V", and "Mármoles Tres Hermanos S.A de C.V", located in Tepexi de Rodríguez Puebla, Mexico, for the facilities in accessing its facilities and answering the measuring instruments at the end.

Conflicts of Interest: The authors declare no conflict of interest.

References

1. Teklu, M. Factors Affecting Supply Chain Integration in Marble Industry (The Case of Ethiopian Marble Processing Enterprise) A Research Paper Submitted to Addis Ababa University School of Commerce in Partial Fulfillment of the Requirements for the Degree of Maste. Master's Thesis, Addis Ababa University, Addis Ababa, Ethiopia, 2018.
2. Bai, S.; Hua, Q.; Cheng, L.J.; Wang, Q.Y.; Elwert, T. Improve sustainability of stone mining region in developing countries based on cleaner production evaluation: Methodology and a case study in Laizhou region of China. *J. Clean. Prod.* **2019**, *207*, 929–950. [CrossRef]
3. Salem, H.S. Evaluation of the Stone and Marble Industry in Palestine: Environmental, geological, health, socioeconomic, cultural, and legal perspectives, in view of sustainable development. *Environ. Sci. Pollut. Res.* **2021**, *28*, 1–23. [CrossRef] [PubMed]
4. Ayber, S.; Ulutas, B.H. Assessing Sustainable Manufacturing Related Problems for Marble Facilities: An Application. *Procedia Manuf.* **2017**, *8*, 129–135. [CrossRef]
5. Aguilar, C. Caracterización de la cadena productiva del mármol-travertino en el Estado de Puebla, México. *Int. Red Congr. Compet.* **2020**, *12*, 990–1011.
6. Garcia-Galicia, J.A.; Morales-Olán, G.; Cadena-Tecayehuatl, M.H.; Moreno-Zarate, P. Experimental and Theoretical Modelling of Waste Produced By the Marble Industry of Tepexi De Rodríguez. *Eur. J. Sustain. Dev.* **2017**, *6*, 405–412. [CrossRef]

7. Olán, G.M.; Zarate, P.M.; Bautista, C.d.R.; Galindo, M.d.L.; Marín, A.P. Proyectos sustentables desarrollados para la región Mixteca Baja del estado de Puebla, México. *Ing. Solidar* **2017**, *13*, 9–26. [CrossRef]

8. Carretero-Gómez, A.; de Pablo-Valenciano, J.; Velasco-Muñoz, J.F. Recursos endógenos mineros y desarrollo territorial. El caso de la comarca del Mármol (Almería, España). *Univ. De Andal.* **2018**, *111*, 51–75.

9. Polo, C. *Los ejes Centrales para el Desarrollo de una Minería Sostenible;* CEPAL-SERIE Recursos Naturales e Infraestructura: Santiago, Chile, 2006; ISBN 9213228953.

10. Márquez, L. *Propuesta y Aplicación de un Sistema de Indicadores para Determinar el Índice de Desarrollo Sostenible Global (IDSG) de un Destino Turístico. caso: Patanemo, Venezuela;* Universidad Camilo Cienfuegos: Matanzas, Cuba, 2006; Volume 17.

11. Snyder, H. La revisión de la literatura como metodología de investigación: Descripción general y directrices. *Rev. Investig. Empres.* **2019**, *104*, 333–339. [CrossRef]

12. Carayannis, E.G.; Barth, T.D.; Campbell, D.F. The Quintuple Helix innovation model: Global warming as a challenge and driver for innovation. *J. Innov. Entrep.* **2012**, *1*, 2. [CrossRef]

13. de Minería, S.G.M.S. Anuario Estadístico de la Minería Mexicana, Edición 2020. 2020. Available online: http://www.sgm.gob.mx/productos/pdf/Anuario_2019_Edicion_2020.pdf (accessed on 8 September 2020).

14. Hanieh, A.A.; AbdElall, S.; Hasan, A. Sustainable development of stone and marble sector in Palestine. *J. Clean. Prod.* **2014**, *84*, 581–588. [CrossRef]

15. CONAGUA. Estadísticas del Agua en México. 2018. Available online: http://sina.conagua.gob.mx/publicaciones/EAM_2018.pdf (accessed on 10 September 2020).

16. Mulk, S.; Korai, A.L.; Azizullah, A.; Shahi, L.; Khattak, M.N.K. Marble industry effluents cause an increased bioaccumulation of heavy metals in Mahaseer (Tor putitora) in Barandu River, district Buner, Pakistan. *Environ. Sci. Pollut. Res.* **2017**, *24*, 23039–23056. [CrossRef] [PubMed]

17. Miranda-Trejo, M.; Ocampo-Fletes, I.; Escobedo-Castillo, J.F.; Hernández-Rodríguez, M.D.L. La distribución del agua potable en Tepexi de Rodríguez, Puebla. *Agric. Soc. Desarro.* **2015**, *12*, 261. [CrossRef]

18. Taydelopez. Servicio Geológico Mexicano Anuario Estadístico de la Minería Mexicana, 2018. 2019. Available online: www.gob.mx/sgm (accessed on 8 September 2020).

19. de Puebla, S.d.G.G. Tepexi de Rodríguez. 2019. Available online: https://planeader.puebla.gob.mx/pdf/ProgramasRegionales2020/0_ProRegionales18TepexideRodríguez.pdf (accessed on 2 February 2020).

20. de Economía, S.M. Perfil de Mercado del Mármol. 2020. Available online: https://www.gob.mx/cms/uploads/attachment/file/624818/17Perfil_M_rmol_2020__T_.pdf (accessed on 16 June 2021).

21. Sandoval, C.R.; Moncayo, L.; del Rosario, M. Diagnóstico de la industria del mármol en la Comarca Lagunera. *RICEA Rev. Iberoam. Contad. Econ. Adm.* **2017**, *5*, 56.

22. Betancourt Chávez, J.R.; Lizárraga Mendiola, L.G.; Narayanasamy, R.; Olguín Coca, F.J.; Sáenz López, A. Revisión sobre el uso de residuos de mármol, para elaborar materiales para la construcción. *Rev. Arquit. E Ing.* **2015**, *9*, 1–12.

23. de Economía, M.S. Estudio de la Cadena Productiva del Mármol. 2021. Available online: https://www.gob.mx/cms/uploads/attachment/file/624823/Cadena_Productiva_M_rmol_2020.pdf (accessed on 16 June 2021).

24. Alarcón, T. *Generación de Estrategia para la Alfabetización Socioecológica de los Alumnos de la Carrera de Ingeniería Industrial del Instituto Tecnológico Superior de Tepexi de Rodríguez Puebla;* Universidad de Málaga: Málaga, Spain, 2012.

25. Hita López, F. El Polvo y la Sílice Cristalina en la Industria Extractiva de la Piedra Natural. 2013. pp. 1–149. Available online: https://www.ugt-fica.org/images/proyectosl/sl/indirectas/2013/piedranatural/ELCONTROLDELPOLVOYLASÍLICECRISTALINAENLAINDUSTRIAEXTRACTIVADELAPIEDRANATURALmanual.pdf (accessed on 8 September 2020).

26. García, C.A. *Aplicación de un Procedimiento Para el Diseño de la Cadena de Suministros de los Residuos en La empresa 'Mármoles Centro' del Municipio de Fomento, Provincia Sancti Spiritus;* Universidad Central "Marta Abreu" de Las Villas: Santa Clara, Cuba, 2015.

27. Ozcelik, M. Environmental pollution and its effect on water sources from marble quarries in western Turkey. *Environ. Earth Sci.* **2016**, *75*, 796. [CrossRef]

28. Cruz, O.A.H. *Comparación de la Resistencia Mecánica a la Compresión del Concreto Elaborado con Residuos de Mármol;* Universidad de Huánuco: Huánuco, Peru, 2017.

29. Zornoza, R.; Faz, Á.; Martínez-Martínez, S.; Acosta, J.; Gómez-López, M.; Muñoz, M.; Sánchez-Medrano, R.; Murcia, F.J.; Cortés, F.J.F.; Martínez, E.L.; et al. Rehabilitación de una presa de residuos mineros mediante la aplicación de lodo de mármol y purín de cerdo para el desarrollo de una fitoestabilización asistida. *Bol. Geol. Min.* **2017**, *128*, 421–435. [CrossRef]

30. Abad, C.R. *El Sector del Mármol en el Medio Vinalopó: Análisis de las últimas Décadas y sus Impactos Económicos, Sociales, Territoriales y Paisajísticos;* Universidad de Alicante: Alicante, Spain, 2018.

31. El-Sayed, H.A.; Farag, A.B.; Kandeel, A.M.; Younes, A.A.; Yousef, M.M. Characteristics of the marble processing powder waste at Shaq El-Thoaban industrial area, Egypt, and its suitability for cement manufacture. *HBRC J.* **2018**, *14*, 171–179. [CrossRef]

32. Noreen, U.; Ahmed, Z.; Khalid, A.; Di Serafino, A.; Habiba, U.; Ali, F.; Hussain, M. Water pollution and occupational health hazards caused by the marble industries in district Mardan, Pakistan. *Environ. Technol. Innov.* **2019**, *16*, 100470. [CrossRef]

33. Figueroa, R.F.A. *Desarrollo de un Plan de Mejoramiento Para la Reducción de Fallas en el Proceso de Fabricación de Materia Derivada del Mármol y Granito en la Industria Ornamental;* Universidad de San Carlo Guatemala: Guatemala City, Guatemala, 2019.

34. Cobo-Ceacero, C.J.; Cotes-Palomino, M.T.; Martínez-García, C.; Moreno-Maroto, J.M.; Uceda-Rodríguez, M. Use of marble sludge waste in the manufacture of eco-friendly materials: Applying the principles of the Circular Economy. *Environ. Sci. Pollut. Res.* **2019**, *26*, 35399–35410. [CrossRef]
35. Pappu, A.; Chaturvedi, R.; Tyagi, P. Sustainable approach towards utilizing Makrana marble waste for making water resistant green composite materials. *SN Appl. Sci.* **2020**, *2*, 347. [CrossRef]
36. Awad, A.H.; Abdel-Ghany, A.W.; El-Wahab, A.A.A.; El-Gamasy, R.; Abdellatif, M.H. The influence of adding marble and granite dust on the mechanical and physical properties of PP composites. *J. Therm. Anal. Calorim.* **2020**, *140*, 2615–2623. [CrossRef]
37. Baykasoğlu, A.; Özbel, B.K. Modeling and solving a real-world cutting stock problem in the marble industry via mathematical programming and stochastic diffusion search approaches. *Comput. Oper. Res.* **2021**, *128*, 105173. [CrossRef]
38. Bares, L. PORTINNOVA: Un estudio de caso de un modelo de la cuadruple helice. *Rev. Contrib. A Las Cienc. Soc.* **2018**, 10. Available online: https://www.eumed.net/rev/cccss/2018/03/modelo-cuadruple-helice.html (accessed on 4 February 2020).
39. Brundtland. Informe de la Comisión Mundial sobre el Medio Ambiente y el Desarrollo Nota del Secretario General. 1987. Available online: http://www.ecominga.uqam.ca/PDF/BIBLIOGRAPHIE/GUIDE_LECTURE_1/CMMAD-Informe-Comision-Brundtland-sobre-Medio-Ambiente-Desarrollo.pdf (accessed on 4 February 2020).
40. Bacchetta Víctor, L. Business and Council for Sustainable Development, "MMSD, Proyecto Minería, Minerales y Desarrollo Sustentable en América del Sur", International Institute for Environment and Development (IIED), World Business Council for Sustainable Development (WBCSD) 2002, Reino Unido. 2002. Available online: https://www2.congreso.gob.pe/sicr/cendocbib/con4_uibd.nsf/7832DF547B40C2FF05257EF2006E308A/$FILE/Miner%C3%ADa_Minerales_y_Desarrollo_Sustentable.pdf (accessed on 4 February 2020).
41. Leydesdorff, L.; Etzkowitz, H. The Triple Helix as a model for innovation studies. *Sci. Public Policy* **1998**, *25*, 195–203.
42. Afonso, O.; Monteiro, S.; Thompson, M. A growth model for the quadruple helix. *J. Bus. Econ. Manag.* **2012**, *13*, 849–865. [CrossRef]

MDPI
St. Alban-Anlage 66
4052 Basel
Switzerland
Tel. +41 61 683 77 34
Fax +41 61 302 89 18
www.mdpi.com

MDPI Books Editorial Office
E-mail: books@mdpi.com
www.mdpi.com/books